M000312174

AAPS Advances in the Pharmaceutical Sciences Series

The AAPS Advances in the Pharmaceutical Sciences Series, published in partnership with the American Association of Pharmaceutical Scientists, is designed to deliver well written volumes authored by opinion leaders and authorities from around the globe, addressing innovations in drug research and development, and best practice for scientists and industry professionals in the pharma and biotech industries. For more details and to see a list of titles in the Series please visit http://www.springer.com/series/8825

Series Editors
Daan J.A. Crommelin
Robert A. Lipper

For further volumes:
http://www.springer.com/series/8825

Lawrence X. Yu • Bing V. Li
Editors

FDA Bioequivalence Standards

Editors
Lawrence X. Yu
Office of Pharmaceutical Science
Center for Drug Evaluation and Research
U.S. Food and Drug Administration
Silver Spring, MD, USA

Bing V. Li
Office of Generic Drugs
Center for Drug Evaluation and Research
U.S. Food and Drug Administration
Silver Spring, MD, USA

ISSN 2210-7371   ISSN 2210-738X (electronic)
ISBN 978-1-4939-1251-3   ISBN 978-1-4939-1252-0 (eBook)
DOI 10.1007/978-1-4939-1252-0
Springer New York Heidelberg Dordrecht London

Library of Congress Control Number: 2014945359

Springer is part of Springer Science+Business Media (www.springer.com)

Foreword

No area of drug development has been more active in recent years than that of generic drugs. More than four out of every five prescriptions dispensed in the USA are generic versions of drug products. In the past decade alone, generic drugs have generated more than a trillion dollars in savings to our nation's health care system. Central to the generic drug success story have been scientifically sound bioequivalence standards, developed within the FDA's Center for Drug Evaluation and Research (CDER), and explored here in *FDA Bioequivalence Standards*.

The primary economic tenet of generic drug development is that the majority of the costly animal and human studies required to ensure the safety and efficacy of innovator drugs need not be repeated for generic copies. Bioequivalence, among other FDA standards, ensures that generic drugs are safe, effective, and equivalent to the innovator drugs by providing the essential link between the data generated during innovator drug development and the generic copy.

The influence of our bioequivalence standards extends beyond safeguarding the safety and efficacy of generic drug products. New drug development also benefits from application of bioequivalence testing, which provides developmental avenues to formulation refinement, manufacturing scale-up, and other post-approval issues. Through applied bioequivalence concepts, scientists have addressed to varying extents many of the major problems that have recently confronted pharmaceutical regulation, including threats to public health stemming from drug shortages and product adulteration.

Bioequivalence standards must continue to evolve as the landscape of drug development incorporates new levels of complexity. I am proud to see the wealth of outstanding work that has emerged from CDER in support of this goal. *FDA Bioequivalence Standards* provide the specifics of bioequivalence studies so that developers are familiar with the thinking of CDER experts who have confronted bioequivalence issues in numerous diverse and challenging contexts. I, like the authors and editors of this book, hope that readers will appreciate the current contribution of bioequivalence standards to drug development and anticipate the scientific hurdles that lie ahead as we confront more complex products and diverse routes of administration.

Silver Spring, MD, USA   Janet Woodcock

Preface

The initial seed for publishing a book on FDA's bioequivalence standards was implanted at the "2008 American Association of Pharmaceutical Scientists Annual Meeting and Exposition" in Atlanta, Georgia. It was in November—the climate was pleasantly cool and gentle, but inside the convention center the mood was hot and lively because of the discourse among a group of pharmaceutical scientists from around the world regarding FDA's bioequivalence guidance. While appraising FDA's bioequivalence guidance for specific drug products, many of the attendees raised inquiries about the bioequivalence of highly variable drugs, a topic that has been in controversy for decades. The discussions revealed a need—in particular, a need for some sort of literature to make available to the public that would systemically and transparently expound on FDA's rationale on bioequivalence. After stepping out of the room where the discussion was taking place, a thought sparkled in the editors' minds: what about a book?

The desire to publish a book on FDA's bioequivalence standards continued to grow in 2009 and 2010, a vigorous period when FDA implemented the partial area under the plasma concentration time curve (AUC) approach for drugs with complex pharmacokinetic profiles and initiated discussions on bioequivalence for narrow therapeutic index drugs. Meanwhile, debates on bioequivalence approaches for locally acting gastrointestinal drugs indicated that the public bore tremendous misunderstanding of the FDA's bioequivalence approaches. Although papers and books touching on these topics were published over that period of time, the information was delivered sporadically and in an unsystemic manner.

With the recent development of bioequivalence approaches for locally acting gastrointestinal drugs, liposomes, and inhalation products, as well as the issuance of FDA guidance on bioanalytical method validation, the editors of this book felt it was time—in fact, even essential to publish a book that summarized the origin, current development, and future trends of FDA's bioequivalence standards. To date, no book had been published that systemically communicated FDA's bioequivalence approaches to the public.

FDA Bioequivalence Standards features a comprehensive selection: 16 chapters of the most current regulatory sciences in the bioequivalence area. These chapters are scrupulously selected to construct broad yet thorough coverage of the relevant topics in the field of bioequivalence. Chapter 1 discusses the origin of bioequivalence and reviews recent developments. Chapters 2 and 3 describe fundamentals of bioequivalence and detail statistical considerations. Chapter 4 explains the science of food effect on bioequivalence studies and elaborates on the study details. Chapter 5 discusses conditions for waivers of bioequivalence study, the Biopharmaceutics Classification Systems (BCS), and the Biopharmaceutics Drug Disposition Classification System (BDDCS). These five chapters are the foundation of bioequivalence. We recommend that beginning learners of this subject matter refer to these five chapters to garner the fundamentals of bioequivalence.

Chapters 6–8 introduce FDA approaches for highly variable drugs, the partial AUC concept, and narrow therapeutic index drugs. Chapters 9 and 10 focus on bioequivalence approaches with pharmacodynamics and clinical endpoints. Chapters 11–14 discuss the individual product classes that are considered more complex because the conventional pharmacokinetic approach alone is not sufficient to establish their bioequivalence. Because of their complexity, new approaches are developed to establish bioequivalence. The products discussed in these chapters are liposome, locally gastrointestinal drug products, topical products, and nasal and inhalation products. Chapter 15 is devoted to modeling and simulation, an area that has recently received considerable attentions as a tool in the demonstration of bioequivalence. Finally, Chapter 16 discusses the current best practices in bioanalytical method validation, introduces recent developments in bioanalysis, and highlights the challenges in bioanalysis.

These chapters are written, at least with our hopes and emphasis, in such way that beginning learners of bioequivalence can pick up a chapter, read through a subject of interest, and understand its overall contour and generate an outline of profile. Meanwhile, readers with years of experience in the bioequivalence area, when encountered with a puzzle, will be able to consult this book to help them find their answer. As such, we strived to ensure that the breadth and depth were appropriately measured.

FDA scientists who themselves develop regulatory policies and conduct regulatory assessment of bioequivalence studies contributed all of the chapters in this volume. Thus, fundamental sciences, as well as practical case studies, are highlighted in these chapters. The original contributions were then reviewed by renowned scientists who are respected experts in their fields to ensure the quality of the contributions. Herein, we would like to thank our chapter reviewers for their valuable time and effort. It was an intellectually gratifying experience to collaborate with them on this book.

We believe that the publication of this book will bring the most state-of-the-art regulatory science in bioequivalence and provide invaluable information to worldwide scientists who work in the pharmaceutical industry, regulatory agencies, and academia. Meanwhile, we affirm it will also serve as a valuable education resource for undergraduate and graduate students.

Silver Spring, MD, USA   Lawrence X. Yu
                                                                                                                          Bing V. Li

Contents

Contributors

April C. Braddy Center for Drug Evaluation and Research, U.S. Food and Drug Administration, Silver Spring, MD, USA

Mei-Ling Chen Center for Drug Evaluation and Research, U.S. Food and Drug Administration, Silver Spring, MD, USA

Dale P. Conner Center for Drug Evaluation and Research, U.S. Food and Drug Administration, Silver Spring, MD, USA

Barbara M. Davit Biopharmaceutics, Clinical Research, Merck, Sharp and Dohme Corp., Whitehouse Station, NJ, USA

Wayne I. DeHaven Center for Drug Evaluation and Research, U.S. Food and Drug Administration, Silver Spring, MD, USA

Stella C. Grosser Center for Drug Evaluation and Research, U.S. Food and Drug Administration, Silver Spring, MD, USA

Wenlei Jiang Center for Drug Evaluation and Research, U.S. Food and Drug Administration, Silver Spring, MD, USA

Xiaojian Jiang Center for Drug Evaluation and Research, U.S. Food and Drug Administration, Silver Spring, MD, USA

Sau L. Lee Center for Drug Evaluation and Research, U.S. Food and Drug Administration, Silver Spring, MD, USA

Bing V. Li Center for Drug Evaluation and Research, U.S. Food and Drug Administration, Silver Spring, MD, USA

Robert Lionberger Center for Drug Evaluation and Research, U.S. Food and Drug Administration, Silver Spring, MD, USA

Fairouz T. Makhlouf Center for Drug Evaluation and Research, U.S. Food and Drug Administration, Silver Spring, MD, USA

Mehul Mehta Center for Drug Evaluation and Research, U.S. Food and Drug Administration, Silver Spring, MD, USA

Devvrat T. Patel Office of Generic Drugs, Center for Drug Evaluation and Research, US Food and Drug Administration, Rockville, MD, USA

John R. Peters Center for Drug Evaluation and Research, U.S. Food and Drug Administration, Silver Spring, MD, USA

Bhawana Saluja Center for Drug Evaluation and Research, U.S. Food and Drug Administration, Silver Spring, MD, USA

Donald J. Schuirmann Center for Drug Evaluation and Research, U.S. Food and Drug Administration, Silver Spring, MD, USA

Ethan Stier Center for Drug Evaluation and Research, U.S. Food and Drug Administration, Silver Spring, MD, USA

Sriram Subramaniam Center for Drug Evaluation and Research, U.S. Food and Drug Administration, Silver Spring, MD, USA

Duxin Sun College of Pharmacy, University of Michigan, Ann Arbor, MI, USA

Ramana S. Uppoor Center for Drug Evaluation and Research, U.S. Food and Drug Administration, Silver Spring, MD, USA

Jayabharathi Vaidyanathan Center for Drug Evaluation and Research, U.S. Food and Drug Administration, Silver Spring, MD, USA

Yongsheng Yang Center for Drug Evaluation and Research, U.S. Food and Drug Administration, Silver Spring, MD, USA

Alex Yu College of Pharmacy, University of Michigan, Ann Arbor, MI, USA

Lawrence X. Yu Center for Drug Evaluation and Research, U.S. Food and Drug Administration, Silver Spring, MD, USA

Xinyuan Zhang Office of Generic Drugs, Center for Drug Evaluation and Research, Food and Drug Administration, Silver Spring, MD, USA

Nan Zheng Center for Drug Evaluation and Research, U.S. Food and Drug Administration, Silver Spring, MD, USA

Hao Zhu Center for Drug Evaluation and Research, U.S. Food and Drug Administration, Silver Spring, MD, USA

Peng Zou Center for Drug Evaluation and Research, U.S. Food and Drug Administration, Silver Spring, MD, USA

About the Editors

Lawrence X. Yu is the acting director of the Office of Pharmaceutical Science, Center for Drug Evaluation and Research (CDER), FDA, in Maryland, USA, where he oversees new, generic, and biotechnology product quality review functions as well as the FDA CDER quality labs. Dr. Yu is an adjunct professor at the University of Michigan, a fellow of the American Association of Pharmaceutical Scientists (AAPS), and an associate editor of *The AAPS Journal*. Dr. Yu received an M.S. in Chemical Engineering from Zhejiang University in Hangzhou, China; an M.S. in Pharmaceutics from the University of Cincinnati in Cincinnati, Ohio, USA; and a Ph.D. in Pharmaceutics from the University of Michigan in Ann Arbor, Michigan, USA. He is also the author/coauthor of more than 100 papers, abstracts, and book chapters and coeditor of *Biopharmaceutics Applications in Drug Development*.

Bing V. Li is a acting deputy director in the Division of Bioequivalence I, Office of Generic Drugs, Center for Drug Evaluation and Research (CDER), FDA, in Maryland, USA. Her current responsibility is to review drug products submitted in Abbreviated New Drug Applications (ANDAs) to determine the adequacy of the data from bioequivalence studies based on study design, analytical methodology, and statistical analysis. Dr. Li received her Ph.D. in Pharmaceutical Sciences from the University of Wisconsin in Madison, Wisconsin, USA. She has chaired numerous FDA working groups, including the bioequivalence "For-Cause" Inspection, bioequivalence for nasal product review template, and population bioequivalence of inhalation products. Dr. Li is also the author/coauthor of 40 papers, abstracts, and book chapters and winner of the Thomas Alva Edison Patent Award.

Chapter 1
Bioequivalence History

Alex Yu, Duxin Sun, Bing V. Li, and Lawrence X. Yu

1.1 Introduction

Bioequivalence (BE) is defined as the absence of a significant difference in the rate and extent to which the active ingredient or active moiety in pharmaceutical equivalents or pharmaceutical alternatives becomes available at the site of drug action when administered at the same molar dose under similar conditions in an appropriately designed study. Drug products are considered pharmaceutical equivalents if they contain the same active ingredient(s), are of the same dosage form, route of administration, are identical in strength or concentration, and meet the same or compendial or other applicable standards (i.e., strength, quality, purity, and identity). Drug products are considered pharmaceutical alternatives if they contain the same therapeutic moiety, but are different salt, esters, or complexes of that moiety, or are different dosage forms or strengths (21 CFR 320).

Bioequivalence studies are a major component in evaluating therapeutic equivalence by verifying that the active ingredient of the test drug product will be absorbed into the body to the same extent and at the same rate as the corresponding reference drug product. The significance of this study is that when two pharmaceutically equivalent products are shown to be bioequivalent, the two products are judged to be therapeutically equivalent. Therapeutically equivalent products are expected to have the same safety and efficacy profiles, when administered under the conditions listed in the product labeling. For generic drugs, bioequivalence studies confirm the clinical equivalence between the generic and reference products. For new drugs, these studies verify the clinical equivalence between different

A. Yu • D. Sun
College of Pharmacy, University of Michigan, Ann Arbor, MI 48109, USA
e-mail: alexmyu@med.umich.edu; duxins@med.umich.edu

B.V. Li • L.X. Yu (✉)
Center for Drug Evaluation and Research, U.S. Food and Drug Administration,
10903 New Hampshire Avenue, Silver Spring, MD 20993, USA
e-mail: bing.li@fda.hhs.gov; Lawrence.Yu@fda.hhs.gov

L.X. Yu and B.V. Li (eds.), *FDA Bioequivalence Standards*, AAPS Advances
in the Pharmaceutical Sciences Series 13, DOI 10.1007/978-1-4939-1252-0_1,
© The United States Government 2014

formulations and sometimes between different strengths. As such, bioequivalence is an integral part of development and regulations for both generic and new drugs.

This chapter discusses the evolution of bioequivalence by dividing the history of bioequivalence into three time periods: the 1970–1980s, the 1990s, and the 2000s. The 1970s and 1980s were when bioequivalence was first established with an important role in drug development and regulations. The 1990s marked an intense discussion of the individual bioequivalence concept as well as the development of the Biopharmaceutics Classification System (BCS) and its subsequent applications to regulatory guidances. This era also featured the development of the predictive compartmental absorption and transit (CAT) model. The turn of the millennia (2000s) saw the development of Biopharmaceutics Drug Disposition Classification System (BDDCS), evolution of BE standards for highly variable drugs, implementation of partial area under the curve (pAUC), creation of novel approaches for narrow therapeutic index (NTI) drugs, and the development of a number of BE approaches for locally acting drugs.

1.2   Bioequivalence Evolution in 1970s and 1980s

1.2.1   Bioequivalence Problems and Recognition

The early 1970s observed the start of serious investigation when patients that took digoxin had variable or poor responses to the medication. Lindenbaum et al. (1971) conducted a crossover study where 0.5 mg of digoxin was orally administered to four normal volunteers. There were significant variations observed in peak serum levels from the same drug in different products made by various manufacturers. One product exhibited sevenfold higher peak serum levels than the other manufacturer's formulation. Even within the same manufacturer, there was significant between-lot variation. Wagner et al. (1973), under the contract with the FDA (Skelly 1976), confirmed Lindenbaum's findings of lack of equivalence in plasma levels of digoxin tablets made by different manufacturers.

A likely reason for the variation was a formulation defect where there was an insufficient or excessive amount of active ingredient in the dosage form. This was confirmed by the FDA through a systematic testing program initiated in April 1970 (Vitti et al. 1971). When digoxin tablet lots from Lindenbaum's study were assayed, it was found that the tablets from B2 were out of potency specification (between 72 and 158.2 % of declared potency) whereas products A and B1 were within potency requirements (Vitti et al. 1971).

Other possible reasons for variation include particle size, disintegration time, dissolution rate, and the effects of various excipients. Wagner et al. (1973) found equivalence lacking in digoxin plasma levels even with tablets that met the acceptance criteria for both potency and disintegration. Similar observations were also made for other products such as tetracycline (Barnett et al. 1974; Barr et al. 1972), chloramphenicol (Glazko et al. 1968), phenylbutazone (Chiou 1972; Van Petten et al. 1971), and oxytetracycline (Barber et al. 1974). These drug products exhibited large variations in drug plasma levels exposing patients to potentially deadly hazards.

Recognizing the existence of bioequivalence problems in marketed products, the FDA Office of Technology Assessment (OTA) organized a drug bioequivalence study panel of ten senior clinicians and scientists in 1974. The panel examined the relationships between chemical and therapeutic equivalence of drug products on the market. They also assessed the capabilities of technology available at that time to determine whether drug products with the same physical and chemical composition produced comparable therapeutic effects (OTA 1974b). Among the panel's eleven conclusions and recommendations, five are critical to the establishment of bioequivalence regulations (OTA 1974a):

1. Current standards and regulatory practices do not assure bioequivalence.
2. Variations in bioavailability are recognized as responsible for a few therapeutic failures. It is probable that other therapeutic failures (or toxicity) of a similar origin have escaped recognition.
3. Most of the analytical methodology and experimental procedures for the conduct of bioavailability studies in man are available. Additional work may be required to develop means of applying them to certain drugs and to special situations of drug use.
4. It is neither feasible nor desirable that studies or bioavailability be conducted for all drugs or drug products. Certain classes of drugs for which evidence of bioequivalence is critical should be identified. Selection of these classes should be based on clinical importance, ratio of therapeutic to toxic concentration in blood, and certain pharmaceutical characteristics.
5. Additional research aimed at improving the assessment and prediction of bioequivalence is needed. This research should include efforts to develop in vitro tests or animal models that will be valid predictors of bioavailability in man.

1.2.2 FDA 1977 Bioequivalence Regulation

Based on the recommendations provided by the drug bioequivalence study panel, the FDA issued regulations that set forth procedures to establish bioequivalence requirements. Effective February 7, 1977, these regulations define the terms of drug product, pharmaceutical equivalent, pharmaceutical alternative, bioequivalent drug product, and bioequivalence requirement (Federal Register 1977).

A bioequivalence requirement may be one or more of the following (Federal Register 1977):

- An in vivo test in humans
- An in vivo test in animals other than humans that has been correlated with human in vivo data

- An in vivo test in animal other than humans that has not been correlated with human in vivo data
- An in vitro bioequivalence standard, i.e., an in vitro test that has been correlated with human in vivo bioavailability data
- A currently available in vitro test (usually a dissolution rate test) that has not been correlated with human in vivo bioavailability data
- In vivo testing in humans shall ordinarily be required if there is well-documented evidence that pharmaceutical equivalents or pharmaceutical alternatives intended to be used interchangeably for the same therapeutic effect meet one of the following conditions:

  - They do not give comparable therapeutic effect
  - They are not bioequivalent drug product
  - They exhibit a narrow therapeutic ratio, e.g., there is less than a twofold difference in LD50 and ED50 values, or there is less than twofold difference in minimum toxic concentration and minimum effective concentration in the blood, and safe and effective use of the product requires careful dosage titration and patient monitoring

These regulations also required that all bioequivalence in vivo or in vitro testing records of any marketed batch of drug products must be maintained until 2 years after the batch expiration date and remain available to be submitted to the FDA on request.

1.2.3 Drug Price Competition and Patent Term Restoration Act

The FDA 1977 bioequivalence regulations played an important role in the establishment of the 1984 Drug Price Competition and Patent Term Restoration Act, informally known as the "Hatch-Waxman Act." This act assumes that bioequivalence is an effective surrogate for safety and efficacy. It established the modern system of generic drugs where drug products must be therapeutically equivalent by meeting the following general criteria (FDA Orange Book 2013):

1. Products are approved as safe and effective.
2. Products are pharmaceutical equivalents in that they (a) contain identical amounts of the same active drug ingredient in the same dosage form and route of administration and (b) meet compendial or other applicable standards of strength, quality, purity, and identity.
3. Products are bioequivalent in that (a) they do not present a known or potential bioequivalence problem, and they meet an acceptable in vitro standard, or (b) if they do present such a known or potential problem, they are shown to meet an appropriate bioequivalence standard.
4. Products are adequately labeled.
5. Products are manufactured in compliance with Current Good Manufacturing Practice regulations.

Upon meeting these requirements, generic products are expected to have the same clinical effect and safety profile when administered to patients under the conditions specified in the labeling (FDA 2013a).

Continual refinement of in vivo and in vitro science has led the FDA to revise methods to demonstrate bioequivalence. As of publication date, current methods used to meet the statutory bioequivalence requirement include (FDA 2003a):

1. Pharmacokinetic (PK) studies
2. Pharmacodynamic (PD) studies
3. Comparative clinical trials
4. In vitro studies

The selection of the type of bioequivalence studies to be conducted is based on the drug's site of action and the study design's ability to compare drug delivery.

1.2.4 Bioequivalence Decision Rules

There is extensive literature discussing the criteria for establishing bioequivalence. The FDA Orange Book mentions a common notion that "based on the opinions of FDA medical experts, a difference of greater than 20 % for each of the above tests (area under the curve (AUC) and C_{max}) was determined to be significant, and therefore, undesirable for all drug products (FDA Orange Book 2013)." As such, the bioequivalence limits have generally been taken within 20 % of the standard (Hauck and Anderson 1984).

During the early development of bioequivalence, Skelly (2010) suggested the determination of AUC measurements by physically plotting serum concentration versus time on specially weighted paper, cutting out the respective plots, and weighing each plot separately for comparison. This method, known as the Canadian rule of ±20 %, requires that the mean AUC of the generic drug be within 20 % of the mean AUC of the approved product.

After the 1971 conference on bioavailability of drugs at the National Academy of Science (Brodie and Heller 1971), the FDA started using the power approach. This approach involved determining the AUC through integration instead of physical weights and required both AUC and C_{max} to be within ±20 % of the innovator product at an estimated power of 80 %.

However, the power approach is limited in that it only considers differences in the calculated averages of AUC and C_{max}. With this approach, two approved products can have equal AUC and C_{max} mean values but differ in variability, which may be problematic for some drugs such as NTI drugs. Consideration of variability was deemed necessary for these drugs at the time. As a result, the FDA developed an additional 75/75 rule, under which bioequivalence would be met if:

(a) There was no more than 20 % difference in mean AUC and C_{max} between the test and reference products.

(b) The relative bioavailability of the test product to the reference product exceeded 75 % in at least 75 % of the subjects studied.

The use of 75/75 rule would be responsible for ensuring that there is not a lack of efficacy in the event that there is variable plasma concentration (Patterson and James 2005; Cabana 1983). However, the opposite also applies in that it is possible that the 75/75 rule does not prevent side effects that result of potentially high concentrations. Haynes (1981) also demonstrated that the rule had undesirable performance characteristics and lacked statistical underpinning. As such, the 75/75 rule was later abandoned.

In 1983, Hauck and Anderson (1984) proposed the use of a bioequivalence analysis that incorporated two null hypotheses (H_0) t-tests as shown below:

$$H_0 : \mu_T - \mu_R \leq \theta_1 \quad \text{or} \quad \mu_T - \mu_R \geq \theta_2$$
$$H_1 : \theta_1 < \mu_T - \mu_R < \theta_2$$

For these equations, μ_T is the logarithmic mean for the test (i.e., generic drug), μ_R is the logarithmic mean for the reference (reference product), θ_1 is the lower limit (log 80 %), and θ_2 is the upper limit (log 125 %). By combining the two statistical one-sided tests, the null hypothesis (H_0) states that the means *are not* equivalent and the alternative hypothesis (H_1) states that the means *are* equivalent.

However, it is possible that the Hauck–Anderson t-test could conclude that two products are bioequivalent when they are not. Schuirmann (1987) proposed a solution called the "two one-sided tests procedure" that splits the alternative hypothesis into two parts:

$$H_{01} : \mu_T - \mu_R \leq \theta_1 \quad H_{02} : \mu_T - \mu_R \geq \theta_2$$
$$H_{11} : \mu_T - \mu_R > \theta_1 \quad H_{12} : \mu_T - \mu_R < \theta_2$$

This test eliminates the possibility of an infinitely large rejection region when certain criteria are met (typically when the observed means between the test and reference are similar). This two one-sided test procedure has been used to establish bioequivalence to this day.

To evaluate the performance of the two one-sided tests, Davit et al. (2009) collected a total over 2,000 single-dose bioequivalence studies of orally administered generic drug products approved by the Food and Drug Administration (FDA) from 1996 to 2007 for a period of 12 years. For each study, the measurements evaluated were drug plasma peak concentration (C_{max}) and drug concentration in plasma over time (AUC). The average difference in C_{max} and AUC between generic and innovator products was 4.35 % and 3.56 %, respectively. In addition, in nearly 98 % of the bioequivalence studies conducted during this period, the generic product AUC differed from that of the innovator product by less than 10 %. The resulting conclusion is that while the statistical test analyzes BE confidence from the limit of 80–125 %, the actual difference between test and reference drug is usually much smaller as noted by Fig. 1.1.

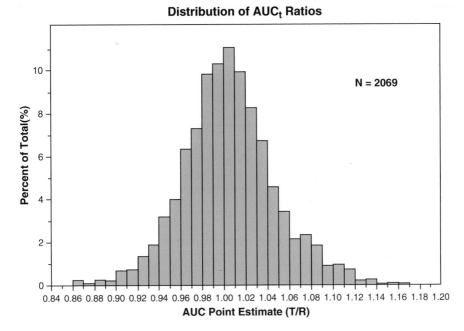

Fig. 1.1   AUC distributions over 12 years of FDA BE data (Yu 2013)

1.3   Bioequivalence Evolution in 1990–2000

1.3.1   Individual Bioequivalence

One of the potential weaknesses of the two one-sided tests procedure lies in the fact that it cannot address the question of whether the bioequivalence outcome is sufficient to guarantee that an individual patient could be expected to respond similarly to two different products.

This is because the two one-sided tests procedure only assesses the difference between the test and reference means (average bioequivalence) while individual bioequivalence assesses the difference of the mean and variability. Anderson and Hauck proposed an individual bioequivalence test to provide reasonable assurance that an individual patient could be switched from a therapeutically successful product to another (Anderson and Hauck 1990; Hauck and Anderson 1994).

In the 1990s, the FDA published guidance documents on the proposed criterion and statistical methodology for an individual bioequivalence approach (Chen and Lesko 2001; FDA 1999b). These guidances would allow comparison of intra-subject variances, scaling of bioequivalence criterion to the reference variability, and detection of possible subject-by-formulation interactions. The new criterion would also promote inclusion of heterogeneous population of volunteers in bioequivalence studies. Based on these considerations, the FDA had intended for use of individual bioequivalence to replace average bioequivalence (Chen et al. 2000; Hauck et al. 2000).

Despite the advantages and benefits, there were challenges for using the individual bioequivalence approach. Most questions were focused on the following three general areas (Chen and Lesko 2001):

1. Justification and need for an individual bioequivalence criterion
2. Financial and human resource burden of conducting replicate study designs
3. Appropriateness of the statistical methodology

To address these questions, there were many AAPS public workshops and conferences (AAPS 1997, 1998, 1999) as well as the FDA Advisory Committee for Pharmaceutical Science meetings (FDA 1996, 1997, 1998, 1999a, 2000a). The FDA Individual Bioequivalence Expert Panel chaired by Leslie Benet reported at the 1999 FDA Advisory Committee for Pharmaceutical Science meeting that (Benet 1999):

• Individual bioequivalence is a promising, clinically relevant method that should theoretically provide further confidence to clinicians and patients that generic drug products are indeed equivalent in an individual patient.
• Even today, considering the studies summarized and analyzed by the FDA, the data is inadequate to validate the theoretical approach and provide confidence to the scientific community that the methodology required and the expense entailed are justified.
• At this time, individual bioequivalence still remains a theoretical solution to solve a theoretical clinical problem. We have no evidence that we have a clinical problem, either a safety or an efficacy issue, and we have no evidence that if we have the problem that individual bioequivalence will solve the problem.

As a result, the average bioequivalence approach remains the key method for evaluation of bioequivalence today.

1.3.2   Biopharmaceutics Classification System

Amidon et al. (1995) developed a BCS for correlating in vitro drug product dissolution and in vivo bioavailability. This classification was derived from the physical properties of solubility and permeability on drug absorption. According to the devised BCS system, drug substances are classified into four classes, as shown in Table 1.1.

The BCS solubility classification is derived from an in vitro experiment that tests the highest strength of a drug product. If the highest strength drug of a specified dosage form is soluble in 250 mL or less of aqueous media over the pH range of 1.0–7.5, the drug is considered highly soluble. The 250 mL volume estimate is

Table 1.1 Biopharmaceutics classification system (Amidon et al. 1995)

Biopharmaceutics class	Solubility	Permeability
I	High	High
II	Low	High
III	High	Low
IV	Low	Low

derived from typical bioequivalence study protocols that prescribe administration of drug product with a glass of water (~8 oz) to fasted human volunteers.

The BCS permeability classification is based directly on the extent of intestinal absorption of a drug substance in humans or indirectly based on measurements of the mass transfer rate across the human intestinal membrane. Animal or in vitro models capable of predicting the extent of intestinal absorptions in humans may also be used as alternatives, e.g., in situ rat perfusion models and in vitro epithelial cell culture models (FDA 2000b). A drug substance is considered highly permeable when the extent of intestinal absorption is determined to be 90 % or higher.

1.3.3 Biowaiver Based on BCS

In 2000, the FDA issued a guidance describing the waiver of in vivo bioavailability and bioequivalence studies for immediate-release (IR) solid oral dosage forms based on the BCS. This guidance allows applicants to request biowaivers for highly soluble and highly permeable drug substances (Class I) in immediate-release solid oral dosage forms provided the following conditions are met (FDA 2000b):

(a) The drug must be stable in the gastrointestinal tract
(b) Excipients used in the IR solid oral dosage forms have no significant effect on the rate and extent of oral drug absorption
(c) The drug must not have an NTI
(d) The product is designed not to be absorbed in the oral cavity
(e) The drug dissolves rapidly in vitro

An IR drug product is considered to have a rapid dissolution when not less than 85 % of the labeled amount of the drug substance dissolves within 30 min using USP Apparatus I at 100 rpm or USP Apparatus II at 50 rpm in a volume of 900 mL or less of each of the following media (FDA 2000b):

(a) Acidic media, such as 0.1 N HCl or USP simulated gastric fluid without enzymes (SGF)
(b) A pH 4.5 buffer
(c) A pH 6.8 buffer or USP simulated intestinal fluid without enzymes (SIF)

If the drug product does not meet these requirements, it is not considered to be a rapidly dissolving product.

Based on these BCS scientific principles, the cause of two pharmaceutically equivalent solid oral products exhibiting in vivo differences in the rate and extent of drug absorption may be due to in vivo differences in drug dissolution. If the in vivo dissolution of an IR oral dosage form is rapid relative to gastric emptying, then the rate and extent of drug absorption is likely to be independent of drug dissolution. In terms of in vivo behavior, a highly soluble and rapidly dissolving drug product is similar to an oral solution. Demonstration of in vivo bioequivalence may not be necessary as long as the inactive ingredients used in the dosage form do not significantly affect absorption of the active ingredient. For BCS Class I (both high solubility and high permeability) drug products, demonstration of rapid in vitro dissolution using required test conditions is sufficient for assurance of similarly rapid in vivo dissolution. This avoids unnecessary costs and risks involved in conducting clinical trials to demonstrate bioequivalence.

1.3.4 CAT Model

Although it was well known that small intestine transit time plays an important role in absorption, there was little development in this area before 1990s. In 1996, Yu et al. developed a CAT model constructed from the understandings of small intestinal transit flow and its characterization (Yu et al. 1996a, b; Yu and Amidon 1998a). This model is able to predict both the rate and extent of absorption (Yu et al. 1996a; Yu and Amidon 1998b).

When compared to the dispersion and single compartment model, it was found that the CAT model was superior to the single-compartment model and less complex than the dispersion model. The single compartment model characterizes the drug as being distributed into the body as a single volume while the dispersion model characterizes the drug distribution through convection and dispersion.

To extend the original CAT model's capabilities in determining the rate, extent, and approximate gastrointestinal location of drug liberation (for controlled release formulations), an advanced compartmental absorption and transit (ACAT) model was developed later (Agoram et al. 2001). The ACAT model is essentially the same as the integrated absorption model which estimates fraction of dose absorbed and provides a framework to determine when the absorption is limited by permeability, dissolution, and solubility (Yu 1999).

The subsequent development of computer software transformed the ACAT models into commercially available software for research and evaluation. Continued development has led to more accurate prediction models of in vitro–in vivo correlations for oral absorption in comparison to previous models (Grbic et al. 2011). Combined with biorelevant solubility, the modern computer programs are also able to predict the magnitude of food effects and oral pharmacokinetics of different drugs in both fasted and fed conditions. In addition to its use for predicting oral drug absorption in the GI tract, whole body Physiologically Based Pharmacokinetic Modeling (PBPK) and combined

Table 1.2 Biopharmaceutics drug disposition classification system (Wu and Benet 2005)

Biopharmaceutics class	Solubility	Permeability	Predominant elimination	Transporter effects
I	High	High	Metabolism	Transport effect minimal
II	Low	High	Metabolism	Efflux transport effects predominate[a]
III	High	Low	Renal/Biliary elimination drug unchanged	Absorptive transporter effects predominate
IV	Low	Low	Renal/Biliary elimination drug unchanged	Absorptive and efflux transporter effects could be important

[a]Both absorptive and efflux transporter effects can occur in liver (Thompson 2011)

Pharmacokinetic/Pharmacodynamic models have been constructed for predicting whole body PK/PD consequences in humans (Huang et al. 2009). Recently, the computer models have been also used to conduct virtual bioequivalence simulations (Zhang et al. 2011).

1.4 Bioequivalence Evolution from 2000s to Present

1.4.1 Biopharmaceutics Drug Disposition Classification System

In 2005, Wu and Benet (2005) proposed the BDDCS. This work expanded on the BCS' foundations of solubility and permeability by incorporating transporter effects and elimination mechanisms. In particular, the BDDCS was developed to predict drug disposition and drug–drug interactions in both the intestine and liver. According to the BDDCS, a drug can be classified into one of the four classes as shown in Table 1.2.

An example of how inclusion of metabolic and transporter analysis can allow for predictions of drug delivery behavior when high-fat meals are taken into account is given by Fleisher et al. (1999). Similar thought processes can also be used to predict scenarios such as in vivo drug–drug interaction (i.e., competing transporters; Benet 2013). This approach leads to a significant distinction between BDDCS and BCS as the former focuses on metabolism and the latter on absorption.

1.4.2 Bioequivalence Approach for Highly Variable Drugs

Highly variable drugs are defined as those for which within-subject variability (%CV) of bioequivalence (BE) measures is 30 % or greater (Haidar et al. 2008). The sources of within-subject variability include:

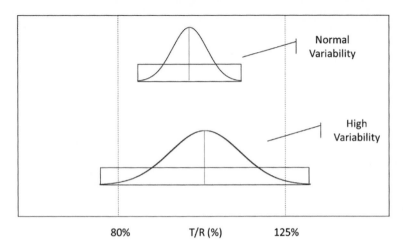

Fig. 1.2 Effect of variability on BE studies. With the same number of subjects, high variability will lead to the wide confident intervals, making the study more difficult to pass the BE limit of 80–125 %

- Physiological factors affecting bioavailability such as regional pH in the gastrointestinal tract, bile and pancreatic secretions, luminal and mucosal enzymes, gastrointestinal motility, gastric emptying, small intestinal transit time, and colonic residence time.
- Inherent properties of the drug such as distribution, first-pass metabolism, systemic metabolism, and elimination.
- Physicochemical properties of drug substance such as solubility.
- Formulation factors such as drug release.
- Other factors such as food intake.

Because of the nature of the average bioequivalence approach, bioequivalence studies for highly variable drugs may need to enroll a large number of subjects even when the generic and reference products have very little difference in mean bioavailability. This is a consequence of high within-subject variability as shown in Fig. 1.2. It is even possible that a highly variable reference product will fail to demonstrate bioequivalence when compared with itself in a bioequivalence study using the average bioequivalence approach and usual sample size (Midha et al. 2005).

The belief is that highly intra-subject variable drugs generally have a wide therapeutic window where products have been demonstrated to be both safe and effective despite the high variability (Benet 2006). With this in mind, applying the same average bioequivalence criteria to highly variable drugs/products may unnecessarily expose large number of healthy subjects to the drug (Benet 2006). To minimize unnecessary human testing, various approaches with alternate study designs, statistical methods, and other considerations have been proposed and investigated to demonstrate bioequivalence of highly variable drugs. These approaches include bioequivalence studies with multiple doses at steady state, a limited sampling method, individual bioequivalence, direct expansion of bioequivalence limits to prefixed values, and widening of bioequivalence limits by scaling approaches (Zhang et al. 2013). Because each method has its advantages and disadvantages, there is no universally accepted approach to demonstrating bioequivalence for highly variable drugs.

The FDA has chosen to evaluate the following approaches for demonstration of bioequivalence for highly variable drugs and products: direct expansion of BE limits, expansion of BE limits based on fixed sample size, widening of BE limits based on reference variability, and expansion of BE limits based on sample size and scaling (FDA 2004). Based on these evaluations, the FDA developed a reference-scaled average bioequivalence approach with a point-estimate constraint, where the bioequivalence acceptance limits are scaled to the variability of the reference product.

This approach adjusts the bioequivalence limits of highly variable drugs by scaling to the within-subject variability of the reference product in the study. The use of reference-scaling is based on the general concept that reference variability should be used as an index for setting the public standard of the bioequivalence limit. This effectively decreases the sample size needed to demonstrate bioequivalence of highly variable drugs.

The FDA's final approach includes the additional requirement of a point-estimate constraint that imposes a limit on the difference between the test and reference means. This eliminates the potential that a test product could enter the market with a large mean difference from the reference product. The use of the reference-scaling approach necessitates a study design that would allow for determination of reference variability, i.e., multiple administrations of the reference treatment to each subject. The FDA recommended partial replicate design as the most efficient way to obtain this information. The reference-scaled average bioequivalence approach has been used successfully at the FDA. To date, this new approach has supported many approvals of high variable generic drug products.

1.4.3   Bioequivalence for NTI Drugs

Although the use of 80–125 % bioequivalence limits has been historically proven to be a rigorous criterion after approval of thousands of generic drugs and post-marketing drug product changes, this criterion may not be conservative enough for NTI drugs as small changes in blood concentration of these drugs can potentially have serious therapeutic consequences and/or adverse drug reactions in patient use. Because of the risks that can arise from the NTI drugs, there have been debates, among health care professionals, pharmaceutical scientists, regulatory agencies, and consumer advocates, about how much assurance is needed for a generic NTI drug product to be considered bioequivalent to its reference product. In 2010 and 2011, the FDA held two advisory committee meetings to discuss the definition of NTI drugs and the BE approaches to establishing therapeutic equivalence of these drug products (FDA 2011d).

Historically, a variety of terms have been used to describe the drugs in which comparatively small differences in dose or concentration may lead to serious therapeutic failures and/or serious adverse drug reactions in patients. These may include NTI, narrow therapeutic range, narrow therapeutic ratio, narrow therapeutic window, and critical-dose drugs. The FDA advisory committee recommended the use of the term "narrow therapeutic index (NTI)" and defined NTI drugs as drugs where small differences in dose or blood concentration may lead to serious therapeutic failures and/or adverse drug reactions that are life-threatening or result in persistent or significant disability or incapacity (Yu 2011):

(a) There is little separation between therapeutic and toxic doses or associated blood/plasma concentrations.
(b) Subtherapeutic concentrations may lead to serious therapeutic failure and/or above-therapeutic concentrations may lead to serious adverse drug reactions in patients.
(c) Subject to therapeutic monitoring based on pharmacokinetic or pharmacodynamics measures.
(d) Possess low-to-moderate (i.e., no more than 30 %) within-subject variability.
(e) In clinical practice, doses are often adjusted in very small increments (less than 20 %).

Based on the input from the advisory committee, FDA conducted simulations to investigate the application of different BE approaches for NTI drugs, including the use of (1) direct tightening of average BE limits and (2) tightening BE limits based on the variability of the reference product (the reference-scaled average BE approach) (FDA 2011d). Variables evaluated in the simulations included within-subject variability, sample size, and point-estimate limit. The powers of a given study design were compared using the reference-scaled average BE approach versus the average BE approach. Simulation results indicated that an approach that tightens BE limits based on reference variability is the preferred approach for evaluating the BE of NTI drugs. A four-way, crossover, fully replicated study design is preferred because such a study design will permit variability comparison in addition to the mean comparison. Both comparisons have to be considered when declaring bioequivalent.

The baseline BE limits for NTI drugs is 90–111 %, which would be scaled based on the within-subject variability of the reference product. When the reference variability is ≤10 %, the BE limits will be narrower than 90–111 %. Conversely, when the reference variability is >10 %, the BE limits will be wider than 90–111 %, but are capped at 80–125 %. To ensure that the BE limits for NTI drugs are never wider than those for conventional drugs, it is critical that every study pass the scaled average BE and the unscaled average BE limits of 80–125 %. Because most NTI drugs have low within-subject variability, the BE limits for these drug products would almost always be tightened to less than 80–125 % accordingly.

The four-way, crossover, fully replicated study design will also permit the comparison of within-subject variability in the test and reference products to confirm that their variances do not differ significantly. FDA's simulation studies demonstrated that test and reference products with unacceptably large differences in within-subject variability may still pass the reference-scaled BE limits, suggesting that the reference-scaled average bioequivalence approach alone is not adequate to ensure the similarity of test and reference products for NTI drugs. FDA proposed an F-test to evaluate whether the within-subject variability of test and reference products are comparable by calculating the 90 % confidence interval of the ratio of the within-subject standard deviation of the test to reference product (FDA 2011d). To determine the appropriate upper limit of the confidence interval of the variability test, FDA evaluated the limit value of 2, 2.5, and 3 and concluded that the appropriate upper limit of the 90 % confidence interval should be ≤ 2.5.

1.4.4 Partial Area Under the Curve

Since the inception of bioequivalence, the peak exposure (C_{max}) and total exposure (AUC) have been used to measure the rate and extent of absorption. These two metrics along with the time to peak concentration (T_{max}) generally work well for immediate release and even for many modified release dosage forms. However, for some modified products that exhibit multiphasic pharmacokinetic behavior which is clinically important and meaningful, the traditional metrics of AUC and C_{max} may not be sufficient to ensure BE. In these cases, AUC and C_{max} may be equivalent for two products, but the rate or extent of exposure during a clinically relevant time interval may not be equivalent (Heald 2010). Consequently, an additional PK metric, such as a pAUC to assess partial exposure, may be necessary to demonstrate bioequivalence.

Chen proposed to use a pAUC approach for the evaluation of equivalence in the rate of absorption for immediate-release formulations (Chen 1992; Chen et al. 2011). For orally administered immediate-release drug products, bioequivalence can generally be demonstrated by measurements of peak and total exposure. An early exposure measure may be informative on the basis of appropriate clinical efficacy/safety trials and/or pharmacokinetic/pharmacodynamic studies that call for better control of drug absorption into the systemic circulation (e.g., to ensure rapid onset of an analgesic effect or to avoid an excessive hypotensive action of an antihypertensive). Although the FDA general BA/BE guidance recommended the use of partial AUC as an early exposure measure, it was rarely used (FDA 2003a).

In 2011 and 2012, the FDA implemented the use of pAUC for the determination of bioequivalence of zolpidem extended-release tablets and methylphenidate

Fig. 1.3 Absorption
process and action site of
locally acting and orally
administered
gastrointestinal drugs

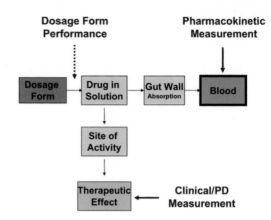

hydrochloride extended-release capsules and tablets (FDA 2011c, 2012c). Pharmacokinetic/pharmacodynamic relationship is the foundation for recommending use of pAUC for these products. Modeling and simulation studies were performed to aid in understanding the need for pAUC measures and also the proper pAUC truncation times (Lionberger et al. 2012; Stier et al. 2012; Fourie Zirkelbach et al. 2013).

The choice of truncation of the area under the curve is most appropriately based on PK/PD relationship or efficacy/safety data for the drug under examination. When PK/PD relationship is lacking, the selection of the truncation point for pAUCs is challenging. When pAUC is highly variable, the reference-scaling approach can be employed for bioequivalence evaluation.

1.4.5 Bioequivalence for Locally Acting Gastrointestinal Drugs

The function of locally acting gastrointestinal (GI) drug products is to deliver active ingredients directly to the site of action in the GI tract, which allows the intended therapeutic effect to occur without entering the systemic circulation as shown in Fig. 1.3.

While local delivery is excellent from a therapeutic effect standpoint, it presents challenges when attempting to evaluate bioequivalence using standard techniques. Some locally acting GI drugs such as mesalamine are permeable to the intestinal membrane and can enter the systemic circulation while others such as vancomycin hydrochloride are not as permeable and have very low systemic availability (Zhang et al. 2013). There is a strong possibility that systemic exposure may not be directly correlated to the local concentration of the drug in the GI tract. In order to confirm bioequivalence, a selection of BE methods are often used depending on considerations of various factors, such as mechanism of drug delivery, mechanism of drug

Table 1.3 Examples of locally acting GI drug products and respective BE methods (Lionberger 2004, 2008; Yu 2008)

Product category	Bioequivalence methods	Example drug/Drug product
Insoluble binding agents	In vitro disintegration and binding assay	Cholestyramine (FDA 2012a); Lanthanum carbonate (FDA 2011a); Calcium acetate (FDA 2009b); Sevelamer (FDA 2011b)
High solubility immediate release dosage forms	In vitro dissolution + studies to show that any difference in formulation does not affect the safety and efficacy of drug product	Vancomycin HCl oral capsules (FDA 2008a); Acarbose tablets (FDA 2009a)
Low solubility immediate release dosage forms	In vivo PK, in vivo PD, or clinical studies or combination of two methods	Rifaximin capsules (FDA 2012d); Lubiprostone capsules (FDA 2010b)
Modified release dosage forms	In vitro dissolution, in vivo PK or in vivo PD, or clinical studies, or combination of two methods	Mesalamine ER and DR products (FDA 2012b)

release, systemic absorption of the drug, drug physiochemical properties, and study feasibility.

It is currently recommended that bioequivalence methods for mesalamine include in vitro dissolution studies as well as in vivo BE studies with PK endpoints (Zhang et al. 2013). Because mesalamine is well absorbed from the GI tract, it is likely that the PK profiles obtained may reflect the local availability of the drug. In vitro dissolution in different solutions will confirm that the release profile is similar throughout the GI tract. On the other hand, vancomycin HCl is highly soluble and expected to be solubilized before reaching the site of action in the lower GI tract. As such, the FDA recommends that in vitro dissolution studies be conducted for the vancomycin HCl formulations that are quantitatively and qualitatively the same or an in vivo BE study be conducted with clinical endpoints if formulations are not quantitatively and qualitatively the same. The same quantitative and qualitative requirements ensure that there is no excipient interaction on the transport of vancomycin in vivo. Table 1.3 shows an example of BE methods for some locally acting gastrointestinal drug products.

It should be noted that if there is a safety concern related to systemic exposure or there are contributions of systemic exposure to efficacy, then the FDA Office of Generic Drugs (OGD) may recommend a PK study intended to demonstrate equivalent systemic exposure, in addition to any other study requested to demonstrate equivalent local delivery (FDA 2008b).

Detail discussions of locally acting GI drug bioequivalence will be presented in a subsequent chapter.

1.4.6   Bioequivalence for Nasal and Inhalation Products

1.4.6.1   Bioequivalence for Nasal Sprays for Local Action

Nasal spray products deliver drug to the nasal cavity by spraying a metered dose of the active ingredient that is dissolved or suspended in solutions or mixtures of excipients in nonpressurized or pressurized dispensers. Because of the delivery form and the target site of activity, the bioequivalence of locally acting nasal drug products is currently not believed to be evaluable via traditional bioequivalence methods used for systemically targeted drug products (i.e., blood plasma). For solution formulations of locally acting nasal drug products, the bioequivalence standard is based on the premise that in vitro studies would be more sensitive indicators of drug delivery to nasal sites of action than clinical studies (FDA 2003b) and there is no local in vivo drug dissolution step that might lead to differences in local bioavailability. The following in vitro tests can demonstrate equivalent product performance if there is formulation sameness and device comparability between test and reference products (Li et al. 2013):

1. Single actuation content through container life
2. Droplet size distribution (by laser diffraction)
3. Drug in small particles/droplet size distribution
4. Spray pattern
5. Plume geometry
6. Priming and repriming

In the case of formulation sameness, the inactive ingredient of the test and reference formulations must be qualitatively and quantitatively the same. Device comparability defines that the dimensions of all critical components that are involved in the dispensing of the formulation is comparable.

For suspension formulations, due to the presence of in vivo local dissolution of solid drug particles, the FDA's bioequivalence requirements are based on weight-of-evidence which includes the six in vitro tests as well as the following two in vivo studies (Li et al. 2013):

1. A clinical endpoint (PD) study to ensure equivalent delivery of drug substance to nasal sites of action.
2. A PK endpoint study to establish equivalence of systemic exposure and potential systemic toxicity of the drug.

The addition of in vivo bioequivalence testing for suspension formulations stems from the current inability of particle sizing technologies to adequately distinguish between the active ingredient and suspending agent. The result is a potential difference of the active ingredient's particle size distribution (PSD) between two formulations. Different particle size of drugs in different products could result in distinctive rate and extent of local in vivo dissolution, leading to different bioavailability/clinical results. Because of this concern, in vivo BE testing is needed for suspension formulations.

Fig. 1.4 The aggregate-weight-of-evidence approach for establishing bioequivalence of dry powder inhalers

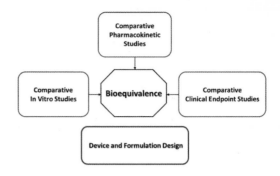

1.4.6.2 Bioequivalence for Locally Acting Orally Inhaled Drugs Products

Similar to locally acting nasal spray suspensions, locally acting orally inhaled drug products do not depend on systemic circulation for drug delivery and intended action. As such, bioequivalence for products such as dry powder inhalers (DPI) is established based on an aggregate weight of evidence approach that includes in vitro studies to demonstrate comparative in vitro performance, pharmacokinetic or pharmacodynamic studies to establish equivalence of systemic exposure, and pharmacodynamics or clinical endpoint studies to demonstrate equivalence in local action (Lee et al. 2009), as shown in Fig. 1.4.

Evaluation of formulation and device are considered to ensure bioequivalence. Because excipients can influence performance, such as the addition of magnesium stearate to drug-lactose mixture to improve particle deagglomeration, it is generally recommended that the qualitative and quantitative formulation aspects between test and reference products remain the same (within ±5 %). Pharmaceutical development data, involving in vitro testing of multiple drug-to-excipient ratios that encompass combinations below and above the ratios used in the test and reference products are needed to justify a test product formulation that is quantitatively different from the reference product. Likewise, although there are several types of DPI dosing systems (premetered single-dose units, drug reservoir (device metered), and premetered multiple dose units), it is recommended that the generic product device's mechanism of function remain the same as that of the reference product. Furthermore, the generic product device itself should maintain a similar shape to ensure equivalence and decrease patient confusion when a generic product is substituted (Chrystyn 2007; Molimard et al. 2003).

Because in vitro testing is less variable and more sensitive to differences in bioequivalence, the following tests can be conducted to detect differences in test and reference products (FDA 2013b):

1. Single inhalation content at different flow rates
2. PSD at different flow rates

Although similar to the locally acting gastrointestinal drug products in that systemic circulation occurs after delivery to the local site, the drug moieties detected from systemic circulation for locally acting orally inhaled products include drugs from potentially multiple sites including the lung, buccal, and GI tract areas. Therefore, a systemic BE study is recommended to ensure equivalent systemic exposure of generic and reference drugs (Adams et al. 2010).

An additional part of the bioequivalence approach for demonstrating equivalence for locally inhalation products is the pharmacodynamics or clinical endpoint study. An example of a typical measurement is the Forced Expiratory Volume in 1 s (FEV1) which is the maximal amount of air an individual can exhale in 1 s.

1.4.7   Bioequivalence for Liposomal Products

A liposome is an artificially prepared vesicle comprises a lipid bilayer shell and an inner core of aqueous compartment. The drug substance may be encapsulated in the lipid bilayer or inner core. Liposome drug products may be designed to release drug to a particular target tissue, or to act as a parenteral dosage form for sustained release in systemic circulation. Due to the engineered properties, these nanoparticle drug products have altered pharmacokinetic and pharmacodynamics profiles. The success of liposome use as a drug carrier has been reflected in a number of liposome-based products which are commercially available or currently undergoing clinical trials. The first liposome drug product Doxil, a PEGylated liposome formulation of doxorubicin HCl shown in Fig. 1.5, was approved by the FDA in 1995.

Liposomes such as Doxil can be biocompatible, biodegradable, and locally targeting. They can also avoid in vivo clearance by various mechanisms such as reticuloendothelial systems, renal clearance, and chemical or enzymatic inactivation (Scott 2008). Although the expected clinical behavior can be ideal, these liposomes are designed to exploit the enhanced permeability properties at the tissue site, and thus traditional bioequivalence methods such as pharmacokinetic measurements of systemic exposure alone may not be indicative of equivalent drug concentrations in the targeted tumor tissues. No direct correlations between plasma and target tissue concentrations have been established so far. As such, bioequivalence to Doxil can be demonstrated based on the following in vivo and in vitro tests recommended by FDA (FDA 2010a):

- Same drug product composition
- Same active loading process with an ammonium sulfate gradient
- Equivalent in vitro liposome characteristics including liposome composition, state of encapsulated drug, internal environment of liposome, liposome size distribution, number of lamellar, grafted PEG at the liposome surface, electrical surface potential or charge, and in vitro leakage
- Equivalent in vivo plasma pharmacokinetics of free and encapsulated drug

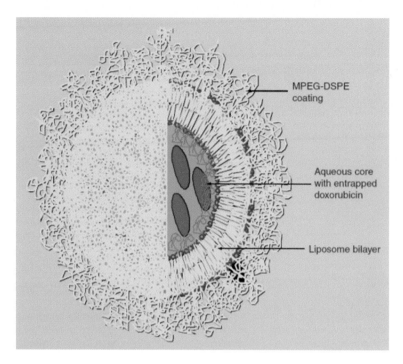

Fig. 1.5 Representation of a PEGylated liposomal doxorubicin (Jiang et al. 2011)

Requiring the same drug product composition, qualitatively and quantitatively, ensures that test and reference products use the same amounts of the same excipients. Requiring the same active loading manufacturing process with an ammonium sulfate gradient ensures equivalent contents within the liposome. Equivalent liposome size distribution and pharmacokinetics ensures equivalence in the mononuclear phagocyte system avoidance, long half-life, and liposome tumor distribution. Table 1.4 lists some of the proposed methods for evaluating in vitro leakage as well as the corresponding justification.

Once equivalent liposome distribution in target tissues is reached, equivalent in vivo pharmacokinetics and in vitro liposome characteristics will ensure equivalent drug delivery into cells. For example, the characterization of liposome surface chemistry can be used to assess liposome–cell interactions involved in liposome fusion or uptake mechanisms by tumor cells. In addition, equivalence in liposome internal environment, size distribution, state of encapsulated doxorubicin, and drug leakage can ensure equivalent drug leakage around tumor tissues or inside tumor cell endosomes or lysosomes. Plasma pharmacokinetics of free drug also accounts for drug release from liposomes.

Detail discussions of liposome product bioequivalence will be presented in a subsequent chapter.

Table 1.4 Proposed in vitro leakage evaluation conditions and their justification (Jiang et al. 2011)

In vitro drug leakage condition	Purpose	Rationale
At 37 °C in 50 % human plasma for 24 h	Evaluate liposome stability in blood circulation	Plasma mostly mimics blood conditions
At 37 °C with pH values 5.5, 6.5, and 7.5 for 24 h in buffer	Mimic drug release in normal tissues, around cancer cells, or inside cancer cells	Normal tissues: pH 7.3 Cancer tissues: pH 6.6 Insider cancer cells (endosomes and lysosomes): pH 5–6 (endosome and lysosomes of cancer cells may be involved in liposome uptake and induce drug release)
At a range of temperatures (43, 47, 52, and 57 °C) in pH 6.5 buffer for up to 12 h or until complete release	Evaluate the lipid bilayer integrity	The Tm of lipids is determined by lipid bilayer properties such as rigidity, stiffness, and chemical composition. Differences in release as a function of temperature (below or above Tm) will reflect small differences in lipid properties
At 37 °C under low-frequency (20 kHz) ultrasound for 2 h or until complete release	Evaluate the state of encapsulated drug in the liposome	Low-frequency ultrasound (20 kHz) disrupts the lipid bilayer via a transient introduction of pore-like defects and will render the release of doxorubicin controlled by the dissolution of the gel inside the liposome

Tm Phase-transition-temperature

1.5 Future Development

While the development of bioequivalence concept and standards, as well as its subsequent rise to become the regulatory requirement has made a monumental impact since the 1970s, there are still many unanswered questions in the field. For example, the current bioequivalence methods for many locally acting products use clinical endpoints for determination. The overall qualification and assessment of these endpoints may be lacking in sensitivity and need to be reevaluated. Similarly, questions have been raised regarding the sensitivity of in vitro testing for BE assessments, and how to develop better in vitro methods for implementation to improve BE standards. The benefits of in vitro testing, modeling, and simulation are enormous, but further investigations are needed to study the sensitivity, reliability, and correlation to clinical significance of these methods. For controlled release dosage forms such as monthly doses, the question is how the bioequivalence should be assessed for these dosage forms since it takes a long time to complete an in vivo study. These are only some of the questions that need to be answered to stimulate future improvements in bioequivalence methodology.

References

21 CFR 320. 21 CFR 320 Bioavailability and bioequivalence requirements. http://www.accessdata.fda.gov/scripts/cdrh/cfdocs/cfcfr/CFRSearch.cfm?fr=320.1. Accessed 9 July 2013

AAPS (1997) Science and regulations: individual and population bioequivalence—regulatory approaches and issues. America Association of Pharmaceutical Scientists (AAPS) Annual Meeting, Nov 2–6 1997, Boston

AAPS (1998) Scientific and regulatory issues in product quality: narrow therapeutic index drugs and individual bioequivalence. American Association of Pharmaceutical Scientists (AAPS) Workshop, Mar 16–18 1998, Arlington

AAPS (1999) Individual bioequivalence: realities and implementation. American Association of Pharmaceutical Scientists (AAPS) International Workshop, Aug 30–Sep 1 1999, Montreal

Adams WP, Ahrens RC, Chen ML, Christopher D, Chowdhury BA, Conner DP, Dalby R, Fitzgerald K, Hendeles L, Hickey AJ, Hochhaus G, Laube BL, Lucas P, Lee SL, Lyapustina S, Li B, O'Connor D, Parikh N, Parkins DA, Peri P, Pitcairn GR, Riebe M, Roy P, Shah T, Singh GJ, Sharp SS, Suman JD, Weda M, Woodcock J, Yu L (2010) Demonstrating bioequivalence of locally acting orally inhaled drug products (OIPs): workshop summary report. J Aerosol Med Pulm Drug Deliv 23:1–29

Agoram B, Woltosz WS, Bolger MB (2001) Predicting the impact of physiological and biochemical processes on oral drug bioavailability. Adv Drug Deliv Rev 50(Suppl 1):S41–S67

Amidon GL, Lennernas H, Shah VP, Crison JR (1995) A theoretical basis for a biopharmaceutic drug classification: the correlation of in vitro drug product dissolution and in vivo bioavailability. Pharm Res 12:413–420

Anderson S, Hauck WW (1990) Consideration of individual bioequivalence. J Pharmacokinet Biopharm 18:259–273

Barber HE, Calvey TN, Muir K, Hart A (1974) Biological availability and in vitro dissolution of oxytetracycline dihydrate tablets. Br J Clin Pharmacol 1:405–408

Barnett DB, Smith RN, Greenwood ND, Hetherington C (1974) Bioavailability of commercial tetracycline products. Br J Clin Pharmacol 1:319–323

Barr WH, Gerbracht LM, Letcher K, Plaut M, Strahl N (1972) Assessment of the biologic availability of tetracycline products in man. Clin Pharmacol Ther 13:97–108

Benet LZ (1999) Individual bioequivalence: have the opinions of the scientific community changed? http://www.fda.gov/ohrms/dockets/ac/01/slides/3804s2_09_benet.ppt. Accessed 8 Aug 2013

Benet LZ (2006) Therapeutic considerations of highly variable drugs. http://www.fda.gov/ohrms/dockets/ac/06/slides/2006-4241s2_2_files/frame.htm. Accessed 18 July 2013

Benet LZ (2013) The role of BCS (biopharmaceutics classification system) and BDDCS (biopharmaceutics drug disposition classification system) in drug development. J Pharm Sci 102:34–42

Brodie BB, Heller WM (1971) Bioavailability of drugs: proceedings. S. Karger, Basel

Cabana BE (1983) Assessment of 75/75 rule: FDA viewpoint. J Pharm Sci 72:98–100

Chen ML (1992) An alternative approach for assessment of rate of absorption in bioequivalence studies. Pharm Res 9:1380–1385

Chen ML, Davit B, Lionberger R, Wahba Z, Ahn HY, Yu LX (2011) Using partial area for evaluation of bioavailability and bioequivalence. Pharm Res 28:1939–1947

Chen ML, Lesko LJ (2001) Individual bioequivalence revisited. Clin Pharmacokinet 40:701–706

Chen ML, Patnaik R, Hauck WW, Schuirmann DJ, Hyslop T, Williams R (2000) An individual bioequivalence criterion: regulatory considerations. Stat Med 19:2821–2842

Chiou WL (1972) Determination of physiologic availability of commercial phenylbutazone preparations. J Clin Pharmacol New Drugs 12:296–300

Chrystyn H (2007) The Diskus: a review of its position among dry powder inhaler devices. Int J Clin Pract 61:1022–1036

Davit BM, Nwakama PE, Buehler GJ, Conner DP, Haidar SH, Patel DT, Yang Y, Yu LX, Woodcock J (2009) Comparing generic and innovator drugs: a review of 12 years of bioequivalence data from the United States Food and Drug Administration. Ann Pharmacother 43:1583–1597

FDA Orange Book Preface (2013) http://www.fda.gov/Drugs/DevelopmentApprovalProcess/ucm079068.htm. Accessed 7 Aug 2013

FDA (1996) FDA Advisory Committee for Pharmaceutical Science Meeting, August 1996, Gaithersburg

FDA (1997) FDA Advisory Committee for Pharmaceutical Science Meeting, May 1997, Gaithersburg

FDA (1998) FDA Advisory Committee for Pharmaceutical Science Meeting, October 1998, Gaithersburg

FDA (1999a) FDA Advisory Committee for Pharmaceutical Science Meeting, September 1999, Gaithersburg

FDA (1999b) Average, population, and individual approaches to establishing bioequivalence. http://www.fda.gov/OHRMS/DOCKETS/98fr/3657gd1.pdf. Accessed 8 Aug 2013

FDA (2000a) FDA Advisory Committee for Pharmaceutical Science Meeting, November 2000, Gaithersburg

FDA (2000b) Guidance for industry: waiver of in vivo bioavailability and bioequivalence studies for immediate-release solid oral dosage forms based on a biopharmaceutics classification system. http://www.fda.gov/downloads/Drugs/GuidanceComplianceRegulatoryInformation/Guidances/UCM070246.pdf. Accessed 12 July 2013

FDA (2003a) Guidance for industry bioavailability and bioequivalence studies for orally administered drug products general considerations. http://www.fda.gov/downloads/Drugs/.../Guidances/ucm070124.pdf. Accessed 26 Aug 2013

FDA (2003b) Guidance for industry: bioavailability and bioequivalence studies for nasal aerosols and nasal sprays for local action. http://www.fda.gov/downloads/Drugs/GuidanceComplianceRegulatoryInformation/Guidances/ucm070111.pdf. Accessed 31 July 2013

FDA (2004) Background information for advisory committee meeting on bioequivalence requirements for highly variable drugs and drug products. http://www.fda.gov/ohrms/dockets/ac/04/briefing/4034B1_07_Bioequivalence%20Requirments-Draft.doc. Accessed 21 Aug 2013

FDA (2008a) Draft guidance on vancomycin hydrochloride capsule. http://www.fda.gov/downloads/Drugs/GuidanceComplianceRegulatoryInformation/Guidances/UCM082278.pdf. Accessed 26 Aug 2013

FDA (2008b) Meeting of the advisory committee for pharmaceutical science and clinical pharmacology. FDA. http://www.fda.gov/ohrms/dockets/ac/08/briefing/2008-4370b1-01-FDA.pdf. Accessed 21 Aug 2013

FDA (2009a) Draft guidance on acarbose tablet. http://www.fda.gov/downloads/Drugs/GuidanceComplianceRegulatoryInformation/Guidances/UCM170242.pdf. Accessed 26 Aug 2013

FDA (2009b) Draft guidance on calcium acetate capsule. http://www.fda.gov/downloads/Drugs/GuidanceComplianceRegulatoryInformation/Guidances/UCM148185.pdf. Accessed 26 Aug 2013

FDA (2010a) Draft guidance on doxorubicin hydrochloride liposome injection. http://www.fda.gov/downloads/Drugs/GuidanceComplianceRegulatory%20Information/Guidances/UCM199635.pdf. Accessed 15 Aug 2013

FDA (2010b) Draft guidance on lubiprostone capsule. http://www.fda.gov/downloads/Drugs/GuidanceComplianceRegulatoryInformation/Guidances/UCM224220.pdf. Accessed 26 Aug 2013

FDA (2011a) Draft guidance on lanthanum carbonate tablet. http://www.fda.gov/downloads/Drugs/GuidanceComplianceRegulatoryInformation/Guidances/UCM270541.pdf. Accessed 26 Aug 2013

FDA (2011b) Draft guidance on sevelamer carbonate tablet. http://www.fda.gov/downloads/Drugs/GuidanceComplianceRegulatoryInformation/Guidances/ucm089620.pdf. Accessed 26 Aug 2013

FDA (2011c) Guidance on zolpidem tablet. http://www.fda.gov/downloads/Drugs/Guidance ComplianceRegulatoryInformation/Guidances/UCM175029.pdf. Accessed 26 Aug 2013

FDA (2011d) July 26th Topic 1: Bioequivalence (BE) and quality standards for narrow therapeutic index (NTI) drug products. http://www.fda.gov/downloads/AdvisoryCommittees/ CommitteesMeetingMaterials/Drugs/AdvisoryCommitteeforPharmaceuticalScienceandClinical Pharmacology/UCM263465.pdf. Accessed 13 Aug 2013

FDA (2012a) Draft guidance on cholestyramine power. http://www.fda.gov/downloads/Drugs/ GuidanceComplianceRegulatoryInformation/Guidances/UCM273910.pdf. Accessed 26 Aug 2013

FDA (2012b) Draft guidance on mesalamine tablet. http://www.fda.gov/downloads/Drugs/ GuidanceComplianceRegulatoryInformation/Guidances/UCM320002.pdf. Accessed 26 Aug 2013

FDA (2012c) Draft guidance on methylphenidate hydrochloride tablet. http://www.fda.gov/ downloads/Drugs/GuidanceComplianceRegulatoryInformation/Guidances/UCM320007.pdf. Accessed 26 Aug 2013

FDA (2012d) Draft guidance on rifaximin tablet. http://www.fda.gov/downloads/Drugs/ GuidanceComplianceRegulatoryInformation/Guidances/UCM291392.pdf. Accessed 26 Aug 2013

FDA (2013a) Approved drug products with therapeutic equivalence evaluations. http://www.fda. gov/downloads/Drugs/DevelopmentApprovalProcess/UCM071436.pdf. Accessed 31 July 2013

FDA (2013b) Draft guidance on fluticasone propionate: salmeterol xinafoate power/inhalation. http://www.fda.gov/Drugs/GuidanceComplianceRegulatoryInformation/Guidances/ucm081320. htm. Accessed 20 Oct 2013

Federal Register (1977) Bioavailability and bioequivalence requirements

Fleisher D, Li C, Zhou Y, Pao LH, Karim A (1999) Drug, meal and formulation interactions influencing drug absorption after oral administration. Clinical implications. Clin Pharmacokinet 36:233–254

Fourie Zirkelbach J, Jackson AJ, Wang Y, Schuirmann DJ (2013) Use of partial AUC (PAUC) to evaluate bioequivalence—a case study with complex absorption: methylphenidate. Pharm Res 30:191–202

Glazko AJ, Kinkel AW, Alegnani WC, Holmes EL (1968) An evaluation of the absorption characteristics of different chloramphenicol preparations in normal human subjects. Clin Pharmacol Ther 9:472–483

Grbic S, Parojcic J, Ibric S, Djuric Z (2011) In vitro-in vivo correlation for gliclazide immediate-release tablets based on mechanistic absorption simulation. AAPS PharmSciTech 12:165–171

Haidar SH, Davit B, Chen ML, Conner D, Lee L, Li QH, Lionberger R, Makhlouf F, Patel D, Schuirmann DJ, Yu LX (2008) Bioequivalence approaches for highly variable drugs and drug products. Pharm Res 25:237–241

Hauck WW, Anderson S (1984) A new statistical procedure for testing equivalence in two-group comparative bioavailability trials. J Pharmacokinet Biopharm 12:83–91

Hauck WW, Anderson S (1994) Measuring switchability and prescribability: when is average bioequivalence sufficient? J Pharmacokinet Biopharm 22:551–564

Hauck WW, Hyslop T, Chen ML, Patnaik R, Williams RL (2000) Subject-by-formulation interaction in bioequivalence: conceptual and statistical issues. FDA Population/Individual Bioequivalence Working Group. Food and Drug Administration. Pharm Res 17:375–380

Haynes JD (1981) Statistical simulation study of new proposed uniformity requirement for bioequivalency studies. J Pharm Sci 70:673–675

Heald D (2010) Conventional bioequivalence criteria may not ensure clinical equivalence and, therefore, interchangeability for products with complex pharmacokinetic profiles. FDA. http://www.fda. gov/downloads/AdvisoryCommittees/CommitteesMeetingMaterials/Drugs/AdvisoryCommitteefor PharmaceuticalScienceandClinicalPharmacology/UCM209322.pdf. Accessed 13 Aug 2013

Huang W, Lee SL, Yu LX (2009) Mechanistic approaches to predicting oral drug absorption. AAPS J 11:217–224

Jiang W, Lionberger R, Yu LX (2011) In vitro and in vivo characterizations of PEGylated liposomal doxorubicin. Bioanalysis 3:333–344

Lee SL, Adams WP, Li BV, Conner DP, Chowdhury BA, Yu LX (2009) In vitro considerations to support bioequivalence of locally acting drugs in dry powder inhalers for lung diseases. AAPS J 11:414–423

Li BV, Jin F, Lee SL, Bai T, Chowdhury B, Caramenico HT, Conner DP (2013) Bioequivalence for locally acting nasal spray and nasal aerosol products: standard development and generic approval. AAPS J 15:875–883

Lindenbaum J, Mellow MH, Blackstone MO, Butler VP Jr (1971) Variation in biologic availability of digoxin from four preparations. N Engl J Med 285:1344–1347

Lionberger R (2004) Bioequivalence of locally acting GI drugs. http://www.fda.gov/ohrms/dockets/ac/04/slides/2004-4078S2_11_Lionberger.ppt. Accessed 21 Aug 2013

Lionberger, R. 2008. Bioequivalence of Poorly Soluble Locally Acting GI Drugs [Online]. FDA. Available: http://www.fda.gov/ohrms/dockets/ac/08/slides/2008-4370s2-02-fda-lionberger.ppt. Accessed 21 Aug 2013

Lionberger RA, Raw AS, Kim SH, Zhang X, Yu LX (2012) Use of partial AUC to demonstrate bioequivalence of zolpidem tartrate extended release formulations. Pharm Res 29:1110–1120

Midha KK, Rawson MJ, Hubbard JW (2005) The bioequivalence of highly variable drugs and drug products. Int J Clin Pharmacol Ther 43:485–498

Molimard M, Raherison C, Lignot S, Depont F, Abouelfath A, Moore N (2003) Assessment of handling of inhaler devices in real life: an observational study in 3811 patients in primary care. J Aerosol Med 16:249–254

OTA (1974a) Drug bioequivalence. Recommendations from the Drug Bioequivalence Study Panel to the Office of Technology Assessment, Congress of the United States. J Pharmacokinet Biopharm 2:433–466

OTA (1974b) Drug bioequivalence: a report of the Office of Technology Assessment, Drug Bioequivalence Study Panel, Washington, The Office: for sale by the Superintendent of Documents, U.S. Government Printing Office

Patterson S, James B (2005) Bioequivalence and statistics in clinical pharmacology. CRC Press, Boca Raton

Schuirmann DJ (1987) A comparison of the two one-sided tests procedure and the power approach for assessing the equivalence of average bioavailability. J Pharmacokinet Biopharm 15:657–680

Scott RC (2008) Targeting immunoliposomes containing pro-angiogenic compounds to the infarcted rat heart. Philadelphia: Temple University

Skelly JP (1976) Bioavailability and bioequivalence. J Clin Pharmacol 16:539–545

Skelly JP (2010) A history of biopharmaceutics in the Food and Drug Administration 1968–1993. AAPS J 12:44–50

Stier EM, Davit BM, Chandaroy P, Chen ML, Fourie-Zirkelbach J, Jackson A, Kim S, Lionberger R, Mehta M, Uppoor RS, Wang Y, Yu L, Conner DP (2012) Use of partial area under the curve metrics to assess bioequivalence of methylphenidate multiphasic modified release formulations. AAPS J 14:925–926

Thompson T (2011) The clinical significance of drug transporters in drug disposition and drug interactions. In: Bonate PL, Howard DR (eds) Pharmacokinetics in drug development. Springer, New York

Van Petten GR, Feng H, Withey RJ, Lettau HF (1971) The physiologic availability of solid dosage forms of phenylbutazone. I. In vivo physiologic availability and pharmacologic considerations. J Clin Pharmacol New Drugs 11:177–186

Vitti TG, Banes D, Byers TE (1971) Bioavailability of digoxin. N Engl J Med 285:1433–1434

Wagner JG, Christensen M, Sakmar E, Blair D, Yates JD, Willis PW 3rd, Sedman AJ, Stoll RG (1973) Equivalence lack in digoxin plasma levels. JAMA 224:199–204

Wu CY, Benet LZ (2005) Predicting drug disposition via application of BCS: transport/absorption/elimination interplay and development of a biopharmaceutics drug disposition classification system. Pharm Res 22:11–23

Yu LX (1999) An integrated model for determining causes of poor oral drug absorption. Pharm Res 16:1883–1887

Yu LX (2008) Bioequivalence of locally acting gastrointestinal drugs: an overview. FDA. http://www.fda.gov/ohrms/dockets/ac/08/slides/2008-4370s2-01-FDA-YU.ppt. Accessed 21 Aug 2013

Yu LX (2011) Quality and bioequivalence standards for narrow therapeutic index drugs. FDA. http://www.fda.gov/downloads/Drugs/DevelopmentApprovalProcess/HowDrugsareDevelopedand Approved/ApprovalApplications/AbbreviatedNewDrugApplicationANDAGenerics/UCM292676. pdf. Accessed 13 Aug 2013

Yu LX (2013) Scientific and regulatory considerations for the new requirements for demonstrating bioequivalence of NTI drugs in US. AAPS

Yu LX, Amidon GL (1998a) Characterization of small intestinal transit time distribution in humans. Int J Pharm 171:157–163

Yu LX, Amidon GL (1998b) Saturable small intestinal drug absorption in humans: modeling and interpretation of cefatrizine data. Eur J Pharm Biopharm 45:199–203

Yu LX, Crison JR, Amidon GL (1996a) Compartmental transit and dispersion model analysis of small intestinal transit flow in humans. Int J Pharm 140:111–118

Yu LX, Lipka E, Crison JR, Amidon GL (1996b) Transport approaches to the biopharmaceutical design of oral drug delivery systems: prediction of intestinal absorption. Adv Drug Deliv Rev 19:359–376

Zhang X, Lionberger RA, Davit BM, Yu LX (2011) Utility of physiologically based absorption modeling in implementing Quality by Design in drug development. AAPS J 13:59–71

Zhang X, Zheng N, Lionberger RA, Yu LX (2013) Innovative approaches for demonstration of bioequivalence: the US FDA perspective. Ther Deliv 4:725–740

Chapter 2
Fundamentals of Bioequivalence

Mei-Ling Chen

2.1 Definition of Bioavailability and Bioequivalence

The US regulatory requirements for bioavailability (BA) and bioequivalence (BE) studies in drug applications originated from a report issued by the Congressional Office of Technology Assessment in 1974. Many recommendations in this report were adopted by the US Food and Drug Administration (FDA) and subsequently became the BA/BE regulations in 1977 (FDA 2013a). Statutory definitions for BA and BE are both expressed in terms of rate and extent of absorption, and thus they are interrelated to each other. Specifically, BA is defined in the regulations as "the rate and extent to which the active ingredient or active moiety is absorbed from a drug product and becomes available at the site of action" (FDA 2013a). Similarly, BE is defined as "the absence of a significant difference in the rate and extent to which the active ingredient or active moiety in pharmaceutical equivalents or pharmaceutical alternatives becomes available at the site of drug action when administered at the same molar dose under similar conditions in an appropriately designed study" (FDA 2013a). Both definitions describe the processes by which the drug substance is released from a dosage form followed by absorption and distribution to the site of action. As a result, similar approaches such as developing a systemic exposure profile by monitoring drug concentrations in plasma or serum over time have generally been applied to measure BA and demonstrate BE in drug applications.

The only difference between BA and BE definitions lies in the study goals, hence the study designs and statistical analysis of study outcome. BA studies can be employed to assess the pharmacokinetics and performance of a drug product related to the absorption, distribution, and elimination of the drug in vivo. In contrast, BE

M.-L. Chen (✉)
Center for Drug Evaluation and Research, U.S. Food and Drug Administration,
10903 New Hampshire Avenue, Silver Spring, MD 20993, USA
e-mail: meiling.chen@fda.hhs.gov

L.X. Yu and B.V. Li (eds.), *FDA Bioequivalence Standards*, AAPS Advances
in the Pharmaceutical Sciences Series 13, DOI 10.1007/978-1-4939-1252-0_2,
© The United States Government 2014

studies are primarily utilized for formulation comparisons, and thus data analysis focuses on the release of active ingredient (or moiety) from the drug product and subsequent absorption into the systemic circulation. Establishing BA is a benchmarking effort for drug products with a new molecular entity (NME), while demonstrating BE is a formal test that compares BA of various formulations with the same drug substance in the same dosage form, using specified criteria and acceptance limits for BE comparisons.

It is noteworthy that in the regulatory setting, BE can be established between drug products that are either pharmaceutical equivalents or pharmaceutical alternatives (Orange Book 2013). Drug products are considered as pharmaceutical equivalents when they are in identical dosage forms and contain identical amounts of the identical active drug ingredient. These products do not necessarily contain the same inactive ingredients (i.e., excipients) and they may differ in characteristics such as shape, scoring configuration, release mechanisms, packaging, expiration time, and within certain limits, labeling. In contrast, pharmaceutical alternatives contain identical therapeutic moiety (or its precursor) but not necessarily in the same amount or dosage form or as the same salt or ester. Based on the Drug Price Competition and Patent Term Restoration Act of 1984 (Hatch-Waxman Act), evidence of pharmaceutical equivalence and bioequivalence provides the assurance of therapeutic equivalence, hence interchangeability between a generic product and its innovator counterpart (Orange Book 2013).

2.2 Application of Bioavailability and Bioequivalence Studies

BA/BE information is deemed important in the drug development and for regulatory approval of pharmaceutical products (FDA 2003a). BA and/or BE studies are required in support of drug applications, including Investigational New Drug Applications (INDs), New Drug Applications (NDAs), Abbreviated New Drug Applications (ANDAs), and their amendments and supplements.

During the IND and NDA period, appropriately designed BA studies are necessary to assess performance of the drug product(s) used in clinical trials that provides evidence of safety and efficacy. As described earlier, BA studies can furnish pharmacokinetic information related to drug absorption, distribution, and elimination in vivo. BA studies can also be used to achieve many other objectives such as estimating fraction of dose absorbed from an orally administered drug product, providing information on dose proportionality and linearity in pharmacokinetics, and investigating the effect of various intrinsic/extrinsic factors on the pharmacokinetics of the drug under examination. For orally administered drug products with an NME, absolute BA is obtained by comparison to an intravenous dose, while relative BA can be accomplished by comparisons to an oral solution, oral suspension, or other formulation.

On the other hand, BE studies are often used as a bridging tool to support evidence for safety and efficacy between two drug products. During the IND and NDA period, BE studies can be utilized to provide links among formulations used in different phases of clinical trials, as well as to establish links between formulations used in stability studies and clinical trials. In addition, BE studies are critical to the approval of ANDAs. Manufacturers seeking approval to market a generic drug product must submit an ANDA, demonstrating that the drug product is both pharmaceutically equivalent and bioequivalent to the Reference Listed Drug (RLD, i.e., innovator product). Documentation of BE is also essential to ensure product quality throughout the shelf life of a drug product whenever changes occur in the manufacturing or formulation, which applies to both new and generic drug products. Depending on the level of changes, BE may be established through comparative in vivo or in vitro studies between products before and after change (FDA 2003a).

2.3   Approaches for Establishment of Bioequivalence

Based on the statutory definition of BE, several in vivo and in vitro methods can be employed for BE establishment. Nonetheless, the US FDA requires that drug applicants conduct BE testing using the most accurate, sensitive, and reproducible approach (FDA 2013b). Hence, in descending order of preference, the following methods have been recommended for BE documentation (FDA 2013b):

(a) Comparative pharmacokinetic studies
(b) Comparative pharmacodynamic studies
(c) Comparative clinical trials
(d) Comparative in vitro tests
(e) Any other approach deemed adequate by FDA

Experiences thus far have revealed that comparative pharmacokinetic studies are mostly used for BE demonstration of systemically absorbed drug products while pharmacodynamic studies and clinical trials are generally employed for locally acting drug products. Historically, in vitro tests alone are rarely utilized for the purpose of BE establishment. However, with the recent advances in modern science and technology, comparative in vitro studies have started to take on an added importance for BE demonstration of certain drug products (see Sect. 2.3.4).

2.3.1   Comparative Pharmacokinetic Studies

As indicated earlier, for systemically acting drug products, demonstration of BE between a test (T) and reference (R) product can be achieved by the conduct of comparative pharmacokinetic studies. These studies are generally performed with a limited number of healthy volunteers, e.g., 24–36 subjects (FDA 2003a). Most studies have a two-sequence, two-period, crossover design where each subject is randomly assigned to either sequence TR or RT with an adequate washout interval between the two treatment periods (FDA 2003a). Derived from the plasma or serum concentration–time profile, the rate of drug absorption is commonly expressed by maximum concentration (C_{max}) and time to maximum concentration (T_{max}) whereas the extent of absorption is expressed by the area-under-the-curve from time zero after drug administration to time infinity (AUC_∞) and/or to the last quantifiable drug concentration (AUC_t). AUC_t may be calculated using the simple trapezoidal rule (Gibaldi and Perier 1982) while AUC_∞ can be estimated by summing up AUC_t and $C_t/\lambda z$ where C_t is the last quantifiable concentration and λz is the terminal rate constant.

With the exception of T_{max} parameter, both AUCs and C_{max} are statistically analyzed using the two one-sided tests procedure to determine if the average values between the T and R products are comparable (Schuirmann 1987). These comparisons require the calculation of a 90 % confidence interval for the geometric mean ratios of the T and R products. BE is generally declared if the 90 % confidence interval is within the BE limit of 80.00–125.00 % (FDA 2003a). However, the BE limits for highly variable drugs and narrow therapeutic index drugs have been scaled to the intrasubject variability of the reference product in the study (Davit et al. 2012; FDA 2011c, 2012b). To obtain geometric means, the data of AUCs and C_{max} are log-transformed prior to conducting an analysis of variance (ANOVA), then back-transformed before calculating the T/R ratio (Davit et al. 2009). Currently, statistical comparison is not performed for T_{max} values due to the lack of an appropriate method for this discrete variable (Chen et al. 2001; Davit et al. 2009; Nightingale and Morrison 1987). However, if there is any notable difference in a BE study, consultation on the clinical relevance is sought with medical officers in the FDA.

Since systemic exposure of locally acting drug products may entail a risk of systemic adverse reactions, a comparative pharmacokinetic study is globally required for these products to ensure that systemic drug exposure for the T product is similar to the R product (Chen et al. 2011a). The BE limits of 80–125 % (based on 90 % confidence interval) can be applied to these studies.

2.3.1.1 Measures of Systemic Exposure

Despite the US regulations that dictate the reliance of rate and extent of drug absorption for BA/BE determination, there have been concerns regarding the use of C_{max} for assessment of absorption rate in BA/BE studies (Chen et al. 2001; FDA 2003a). For example, C_{max} is insensitive to changes in rate of input as generally expressed by a rate constant (ka). C_{max} is not a pure measure of absorption rate since it is confounded with the distribution (and perhaps elimination) of the drug. In addition, determination of C_{max} depends substantially on the sampling schedule and thus this parameter may not be accurate. In recent years, recognizing that systemic exposure is the key to the efficacy/safety of a drug and that there are multiple challenges inherent in identifying an appropriate pharmacokinetic measure to express both rate and exposure, the US FDA has recommended a change in focus from the measures of "absorption rate and extent" to measures of "systemic exposure" for BA and BE studies (FDA 2003a).

Systematic exposure measures can be used for drugs that achieve therapeutic effects after entry into the systemic circulation. In the FDA Guidance (2003a), these measures are defined relative to the total, peak, and early portions of the plasma/serum profile, which encompasses total exposure (AUC_∞ or AUC_t), peak exposure (C_{max}), and early exposure (partial AUC to the median T_{max} of the R product), respectively. In most cases, systemic exposure measures include AUC_∞ (or AUC_t) and C_{max}. Nonetheless, early exposure may be needed in some cases where a better control of the drug input rate is essential for achieving therapeutic effects or circumventing adverse reactions. Notably, these recommendations do not propose a statutory change, given that the conventional measures including C_{max} and AUC are still used for regulatory determination of BA/BE. More importantly, however, is the conceptual change and understanding that systemic exposure measures based on a concentration–time profile relate directly to efficacy and safety outcomes expressed by therapeutic effects or adverse reactions.

2.3.1.2 Measures of Partial Exposure

For immediate-release drug products, consideration of early exposure is needed when the control of drug input rate is critical to achieve a rapid onset of action such as analgesic effect, or avoid a toxic side effect such as hypotensive action from an antihypertensive (FDA 2003a). This notion is unequivocally applicable to modified-release drug products where an appropriate input rate of the drug is necessary to warrant the efficacy and safety profile in the patient (Chen et al. 2011b). In addition to the early exposure measure, the concept of "partial exposure" has recently been expanded to include "late exposure" and any segment of AUC with appropriate cutoff points for better PK/PD characterization and BA/BE assessment. This is exemplified by multiphasic, modified-release drug products that combine both immediate- and extended-release components in a formulation to achieve a quick onset of action as well as a sustained response from the drug afterwards (Chen et al. 2011b; Lionberger et al. 2012; Stier et al. 2012).

Methylphenidate HCl extended-release product is an example for the application of partial AUC measures in establishing BE between an innovator product and its generic versions. Currently, there are three distinct innovator products of extended-release methylphenidate on the market, including a tablet form (Concerta®) and two capsule forms (Ritalin LA® and Metadate CD®). Each product has its unique PK/PD relationship and thus the cutoff for partial AUC may be different from product to product. However, the general principles apply to all three products. For example, the drug labeling of Concerta® indicates that this is an extended-release formulation of methylphenidate with a bimodal release profile. Each Concerta[®] tablet comprises an immediate-release component and an extended-release component, thus providing an instant release followed by sustained release of methylphenidate. Therefore, it is a multiphasic modified-release formulation designed to release a bolus of the drug with a slower drug delivery later in the day. The clinical studies showed a statistically significant improvement in behavioral assessment scores throughout the day for Concerta[®] Tablet relative to placebo, following administration of a single morning dose.

In view of the fact that Concerta[®] Tablet is designed to achieve both rapid onset of action and sustained activity throughout the day, the US FDA has proposed two additional partial AUC metrics for BE demonstration (FDA 2011a). The first partial AUC metric provides assurance that a T and R product will be therapeutically equivalent over the early part of the daily dosing interval, corresponding to the onset of response. The second partial AUC metric ensures that the two products in comparison will be therapeutically equivalent over the later part of the daily dosing interval, corresponding to the duration of the sustained response.

The cutoff point for the first partial AUC metric has been determined using the estimate of T_{max} for the immediate-release component of Concerta[®] Tablet. Since the T_{max} values of this formulation is 2 ± 0.5 h in a fasting study and 3 ± 0.5 h in a fed study and it is believed that 95 % of observations would fall within two standard deviations of the mean, the cutoff of early partial AUC metric for BE determination was set to be 3 h and 4 h for the fasting and fed study, respectively. Based on the cutoff of the first partial AUC metric, the second partial AUC metric was then determined to be AUC_{3-t} and AUC_{4-t} for the respective fasting and fed BE study.

2.3.2 Comparative Pharmacodynamic Studies

The use of pharmacodynamic or clinical endpoints for BE demonstration is not recommended for a drug product when the drug is absorbed into the systemic circulation and pharmacokinetic approach can be used to assess systemic exposure for BE evaluation (FDA 2003a). However, in those instances where a pharmacokinetic approach is not possible, determination of BE may be achieved using suitably validated pharmacodynamic or clinical endpoints (FDA 2003a). This can occur to most locally acting drug products and some systemically acting drug products for which drug levels are too low to be measured in biological fluid or there is a safety concern for using the pharmacokinetic approach to assess BE. For locally acting drug products, another reason for not using pharmacokinetic approach to demonstrating BE lies in the fact that drug concentrations in the systemic circulation following administration of these products may not reflect the availability of the drug at the site of action although certain locally acting products are designed for systemic absorption (FDA 2003b). In addition, systemic absorption of some locally acting drug products may have an impact on the safety profile of the product.

2.3.2.1 Dose–Response Relationship

An essential component of BE studies based on a pharmacodynamic response is the documentation of a dose–response relationship (FDA 1995a; Holford and Sheiner 1981). Pharmacodynamic endpoints selected for BE studies are required to have the capacity of detecting potential differences between the test and reference products. This can be ascertained by a pilot study that demonstrates the existence of a clear dose–response relationship, which should be done before the conduct of pivotal BE studies (FDA 1995a). Depending on the drugs, the dose–response curve may be linear, nonlinear, steep, or shallow. A shallow dose–response curve may not allow for detection of potential formulation differences between products. Linearity may be obtained in some cases when the dose is expressed on logarithmic scale. For many drugs, however, the dose–response relationship based on a pharmacodynamic endpoint is nonlinear and can be fitted to a hyperbolic E_{max} model as follows (Holford and Sheiner 1981):

$$E = E_0 + \frac{E_{max} \times D}{ED_{50} + D},$$

where E is the estimated (fitted) value of pharmacodynamic response, E_0 is the baseline pharmacodynamic effect, E_{max} is the maximum pharmacodynamic effect, and ED_{50} is the dose where the pharmacodynamic effect is half-maximal.

Statistical analysis of BE studies using pharmacokinetic measures has been performed with the two one-sided tests procedure (Schuirmann 1987). This procedure, however, would not be appropriate for analysis of a pharmacodynamic endpoint if the dose–response relationship is nonlinear. To circumvent this problem, the US FDA has introduced a "dose-scale" approach where BE is determined based on the projected equivalent dose of the test product in lieu of the pharmacodynamic effect on the dose–response curve (Gillespie 1996; FDA 2010a, 2013c). Specifically, pharmacodynamic responses of the test and reference products determined in the BE study may be converted to estimates of delivered dose of the test and reference products by using the "dose-scale" method. The benefits of the "dose-scale" approach to BE assessment arise from the translation of nonlinear pharmacodynamic measurements to linear dose measurements.

2.3.2.2 Sensitivity of Pharmacodynamic Measures

The curvilinear dose–response relationship for pharmacodynamic measures may depend on a number of factors, including the mechanism of drug action and potency, pharmacodynamic measure, study population, and severity of the underlying disease. Therefore, conduct of pharmacodynamic studies warrants careful considerations of screening appropriate subjects for the BE study so that the likelihood of obtaining discernible response is enhanced (FDA 1995a, 2003b). The doses used in the BE study should be situated in the discriminative region of the dose–response curve, so lower doses are usually recommended for the study (FDA 1995a, 2003b). The basic pharmacodynamic study design for BE determination may include two doses of the reference product. Additional doses can be used to enhance precision in the estimated values. In the case of topical drug products, different doses are normally made by varying the duration of application when there is only one dose strength available for the product (FDA 1995a). For nasal/inhalation products, different doses may be given by single actuation from one or more products. However, multiple strengths are usually available for solid oral dosage forms. In general, a pilot study is first conducted using the reference product to determine the most sensitive dose for the pivotal BE study.

2.3.2.3 Examples of Pharmacodynamic Endpoints

The choice of pharmacodynamic endpoints for a drug product depends on the mechanism of drug action. For example, topical dermatologic corticosteroid products along with the comparators can be tested for BE using a vasoconstrictor assay to quantify the "topical bioavailability" between formulations (FDA 1995a). This pharmacodynamic approach is based on the property of corticosteroids to produce blanching or vasoconstriction in the microvasculature of the skin, which presumably relates to the amount of the drug entering the skin. The assay is sometimes referred to as the Stoughton–McKenzie test, vasoconstrictor assay, or skin blanching assay (Stoughton 1992). For most topical drug products, however, comparative clinical trials have been employed to determine BE due to the lack of appropriate pharmacodynamic measures.

Inhalation aerosols represent another example for which pharmacodynamic endpoints are used to evaluate BE. A case in point is short-acting beta-agonists (e.g., albuterol) that are indicated for prevention and treatment of bronchospasm in asthmatic patients. Based on the mechanism of action, pharmacodynamic effects of these drug products are measured in terms of bronchodilation or prevention of experimentally induced bronchoconstriction (FDA 2013c). The most commonly used measure of bronchodilation is an increase in forced expiratory volume within one second (FEV_1). In this case, bronchoprovocation with methacholine challenge has been employed to compare the protective effects of beta agonists through the estimation of provocative dose (PD_{20}) or concentration (PC_{20}) that produces a 20 % decrease in FEV_1 (FDA 2013c).

Many inhalation drug products combine a drug(s) and device in the dosage form. Because of the complexity of these dosage forms, establishment of BE by the US FDA has been based on an "aggregate weight of evidence" approach that utilizes (a) pharmacodynamic or clinical endpoint studies to demonstrate equivalence in local action, (b) pharmacokinetic studies to ensure minimal systemic exposure, and (c) a battery of in vitro studies to support equivalent performance of the device (FDA 2003b).

2.3.3 Comparative Clinical Trials

Clinical responses are often located near or at the plateau of the dose–response curve, thus insensitive to distinguish the therapeutic difference between a test and reference formulation (FDA 2003b). As a result, conduct of these studies for BE assessment requires a large number of patients to detect formulation differences. Demonstration of dose–response relationships is not required for clinical BE studies since they are intended only to confirm the lack of important clinical differences between products in comparison. Because of all the reasons mentioned above, BE studies using clinical endpoints will be considered only when both pharmacokinetic and pharmacodynamic approaches are impossible for BE determination.

Several FDA guidance documents for industry are available on the application of clinical approaches to document BE for topical drug products (FDA 2010b). Typically, a randomized, double-blind, placebo-controlled, parallel group study is required. However, placebo treatments are not needed for drugs treating infectious diseases. BE is established if the T product is equivalent to the R product and superior to the placebo treatment. In the case of nasal sprays for local action, the US FDA may waive the in vivo BE studies for solution-based products as BA/BE is self-evident for these products. However, such testing is required for suspension-based nasal sprays due to the lack of a suitable method for particle size determination in suspension formulations (FDA 2003b). Moreover, in vivo BE testing cannot be exempted for nasal solutions in metered dose devices because they are drug-device combination products (FDA 2013c). For establishment of equivalence in local delivery of suspension-based nasal sprays, the US FDA has recommended clinical trials in seasonal allergic rhinitis patients. The study design is a randomized, double-blind, placebo-controlled, parallel group of 14-day duration. The clinical endpoints for equivalence and efficacy analyses are patient self-rated mean total nasal symptom scores.

In general, for drug products that BE determination is made on the basis of pharmacodynamic or clinical endpoints, measurement of the active ingredients, or active moieties in an accessible biological fluid (i.e., pharmacokinetic approach) is necessary to ensure comparable systemic exposure (albeit minimal) between the T and R product (FDA 2003b). However, for some locally acting drug products, such pharmacokinetic studies may be limited by the labeled maximum dose, drug bioavailability, and sensitivity of the bioassay used. In such circumstances, pharmacodynamic or clinical studies could be used to document comparable systemic effects of these drug products.

2.3.4   Comparative In Vitro Studies

Traditionally, in vitro studies are seldom used alone for BE determination except with some special cases where (1) the drug of interest was approved before 1962 and was determined to be a nonbioproblem drug, or (2) scientific evidences have shown that in vitro test data are correlated with in vivo results (FDA 1997a). Over the decades, however, the evolution in pharmaceutical science and technology may have provided opportunities for relying more on in vitro tests to support BE demonstration. Indeed, this can be exemplified by the recent application of a Biopharmaceutics Classification System (BCS) that classifies drugs based on their biopharmaceutical attributes and predicts BA/BE of the drug products in an immediate-release dosage form. In this case, biowaiver can be granted for a BCS Class I (highly soluble and highly permeable) drug formulated in a rapidly dissolving, immediate-release drug product (FDA 2000). Apart from the enhanced role for in vitro dissolution/release testing, the FDA guidance on BCS has indicated certain in vitro approaches (such as in vitro epithelial cell culture methods) that can be used to determine the permeability class of individual drugs (FDA 2000).

2.3.4.1   In Vitro Dissolution/Release Testing

Dissolution/release testing is the most commonly used in vitro method for BE assessment. Although in vitro dissolution/release testing has seldom been used alone as a tool for BE demonstration, dissolution/release information along with the in vivo study data is routinely submitted by drug sponsors for BE documentation of orally administered drug products (FDA 2003a). Dissolution/release data have often been employed to substantiate BE when there is a minor change to formulation or manufacturing (FDA 1995b, c, 1997a, b, 2003a). In addition, in vitro dissolution/release data are utilized to support waiver of BA/BE studies for lower strengths of a drug product, provided that an acceptable in vivo study has been conducted for a higher strength and compositions of these strengths are proportionally similar (FDA 2003a). Together with the use of BCS, in vitro dissolution/release testing has played an increasingly important role in the regulatory determination as to whether the waiver of in vivo BE studies can be granted for an immediate-release drug product (FDA 2000).

In the regulatory arena, to serve as an indicator for BE, an in vitro dissolution/release test should be correlated with and predicative of in vivo BA (FDA 1995c, 2003a). In this setting, the in vitro dissolution/release methodology should be optimized to closely mimic the physiological environment in vivo. For a drug product, proper in vitro dissolution/release behavior in the presence of different formulations with defined in vivo absorption characteristics will be useful to facilitate the establishment of an in vitro–in vivo correlation (IVIVC)

(FDA 1995c). The in vitro dissolution/release method developed in such a manner may be utilized as a surrogate for BA/BE studies when a change occurs in manufacturing or formulation.

2.3.4.2 Other In Vitro Methods

To date, with the better understanding of pharmaceutical attributes, formulation characteristics, and mechanism of action, in vitro studies have taken on an added importance for BE evaluations. A case in point is cholestyramine resin that lowers cholesterol by sequestering bile acid in the gastrointestinal tract (FDA 2012a). For these products, the US FDA has recommended the use of both in vitro equilibrium and in vitro kinetic binding studies of bile acid salts for BE evaluation. The application of these in vitro assays takes advantage of the mechanism of action from resin to assess its binding behavior between the innovator and generic formulation of cholestyramine. Similarly, the Agency has recommended the use of in vitro dissolution, phosphate equilibrium binding, and phosphate kinetic binding studies for BE establishment of lanthanum carbonate chewable tablets (FDA 2011b). Lanthanum is a compound used as a phosphate binder to treat hyperphosphatemia in patients with kidney disease. Lanthanum works in the acid environment of the upper gastrointestinal tract by binding dietary phosphate to form an insoluble complex, which is then eliminated via feces. BE determination with a pharmacokinetic approach is inappropriate for lanthanum because it has an extremely low BA (less than 0.002 %) and the site of drug action lies in the gastrointestinal tract. Likewise, in vitro test methods have been widely used to support BE determination of other locally acting drug products. For example, several in vitro test methods are currently used to support BE assessment of nasal and inhalation products (FDA 2003b). For these products, the key parameters that can be assessed through in vitro tests may include (a) delivered or emitted dose, (b) aerodynamic particle size distribution, (c) spray pattern and plume geometry, and (d) impurities and/or microbial contaminants in formulations and devices during storage or use.

As indicated earlier, pharmaceutical equivalence plays an integral part of therapeutic equivalence between a generic and an innovator product (Orange Book 2013). For simple dosage forms or drug products, pharmaceutical equivalence can be made by a qualitative (Q1) and quantitative (Q2) comparison of composition between formulations. However, this approach may not be sufficient for complex dosage forms or drug products. Use of comparative in vitro test methods may furnish additional evidence to support pharmaceutical equivalence of these products. For instance, the US FDA has suggested the use of a higher level of comparison (Q3) that examines the arrangement of matter (or microstructure) in drug products to supplement the traditional approach for evaluating pharmaceutical equivalence of topical drug products (Lionberger 2005). In this case, the in vitro data for Q3 assessment may include comparisons of physicochemical characteristics as well as in vitro drug release pattern to show structural similarity between formulations.

2.4   Design and Conduct of BE Studies

Currently, the US FDA recommends use of (a) a two-period, two-sequence, two-treatment, single-dose, crossover study design, (b) a single-dose, parallel study design, or (c) a replicate study design for BE studies (FDA 2001, 2003a). Several factors may be considered when choosing appropriate designs for a BE study. For instance, the two-way crossover study design is generally conducted with healthy subjects for most drug products that release drug into the systemic circulation. In this design, each subject will receive each treatment (T or R product) in random order as follows:

	Period	
	1	**2**
	T	R
Sequence		
	R	T

For crossover designs, an adequate washout interval is required between the two periods so that drug level at the beginning of each period is almost zero or negligible. In contrast, for parallel designs, each treatment will be administered to a separate group of subjects with similar demographics and no washout period is needed. Parallel designs are often used for BE studies conducted in patients or for drugs with a long half-life where crossover studies are difficult or impossible to perform.

Replicated crossover designs allow for estimation of intrasubject variability of the T and/or R products using a partial (three-way) or full (four-way) replication of treatment as shown below.

	Period		
	1	**2**	**3**
	T	R	T
Sequence			
	R	T	R

	Period			
	1	**2**	**3**	**4**
	T	R	T	R
Sequence				
	R	T	R	T

For replicate designs, one or both treatments will be administered to the same subjects on two separate occasions. Replicate design has the advantage of using fewer subjects to achieve the same statistical power compared to the regular two-treatment, two-period crossover design. Replicate designs are particularly useful for highly variable drugs and narrow therapeutic index drugs in that the BE of these drugs can be assessed using a scaling approach based on the intrasubject variability of the R product determined from the study (FDA 2011c, 2012b).

2.4.1 Crossover Versus Parallel Design

Single-dose, crossover designs with a washout period between treatments may not be employed for BE studies conducted in patients due to ethical concerns. In such circumstances, parallel designs can be used. Additionally, the crossover design of BE studies may not be practical for drugs with a long half-life because of two reasons. First, adequate characterization of the half-life calls for blood sampling over a long period of time. Secondly, pharmacokinetic principles dictate a washout interval of more than 5 half-lives of the moieties to be measured, which may last for several weeks or months for some drugs. In cases where the conduct of a crossover study is problematic, single-dose parallel designs can be an alternative choice since the latter do not need a washout period between treatments (FDA 2003a) although more subjects are necessary to achieve the same statistical power with parallel designs compared to crossover designs.

Monte Carlo simulations with crossover design studies have demonstrated that using truncated area (such as $AUC_{0-72 h}$) had the power and accuracy equivalent to those obtained using AUC_{0-t} (sampling up to the last quantifiable concentration) for a long half-life drug with low intrasubject variability in distribution and clearance (Kharidia et al. 1999). Similarly, simulations using parallel design studies for drugs with a half-life of 30 h or more revealed that truncation time range between 60 and 96 h was most informative for BE determination, and that sampling beyond 120 h would not affect BE decision (El-tahtawy et al. 2012). It appears that these simulation results are in agreement with the general belief that completion of gastrointestinal transit of a solid, oral, immediate-release drug product, and absorption of its drug substance will occur within approximately 2–3 days after dosing, regardless of the length of half-life for the drug.

The US FDA has recommended that sample collection be truncated at 72 h for long half-life drugs (≥ 24 h) in oral solid dosage forms, using either a crossover or parallel study (FDA 2003a). However, for drugs demonstrating high intrasubject variability in distribution and/or clearance, AUC truncation cannot be used (FDA 2003a).

2.4.2   Single Dose Versus Multiple Doses

Several simulations have been conducted to investigate the sensitivity of single-dose versus multiple-dose studies in detecting formulation differences using a typical crossover design for BE evaluation. Most simulation results revealed that single-dose studies are more sensitive than multiple-dose studies to detect rate differences between a T and R product, which appears to be consistent with the results found in experimental data. In essence, drugs characterized by low accumulation indices showed virtually no change in the 90 % confidence intervals of AUC and C_{max} from single-dose to multiple-dose (El-Tahtawy et al. 1994). However, drugs with higher accumulation indices had smaller confidence interval at steady state, and thus the probability of failing a BE test is dramatically decreased upon multiple dosing (El-Tahtawy et al. 1994).

The US FDA has generally recommended single-dose pharmacokinetic studies for BE demonstration of both immediate- and modified-release products (FDA 2003a). However, steady-state studies may be needed for BE demonstration in some cases (FDA 2003a). As an example, safety considerations for healthy volunteers may suggest the use of patients who are already receiving the medication and it is possible to establish BE without disrupting the ongoing treatment of a patient using a steady-state study. This scenario can be illustrated by clozapine, a drug used to treat the symptoms of schizophrenia (FDA 2005). To demonstrate BE of clozapine tablets, applicants are requested to conduct a single-dose (100 mg), two-treatment, two-period crossover study at steady state. In this case, subjects recruited are patients receiving a stable daily dose of clozapine administered in equally divided doses at 12-h intervals. In addition, patients who are receiving multiples of 100 mg every 12 h can participate in the study of the 100 mg strength by continuing their established maintenance dose. The US FDA recommends that these studies not be conducted using healthy subjects because of safety concerns. According to the crossover randomization schedule, an equal number of patients would receive either the generic or reference formulation in the same dose as administered prior to the study every 12 h for 10 days. Patients would then be switched to the other product for a second period of 10 days. No washout period is necessary between the two treatment periods since it is a steady-state study. After the study is completed, patients could be continued on their current dose of clozapine using an approved clozapine product as prescribed by their clinicians. In all cases where a steady-state study is indicated, applicants are required to carry out appropriate dosage administration and sampling to document the attainment of steady state.

2.4.3   Healthy Subjects Versus Patients

A common practice in conducting pharmacokinetic studies for BE evaluation has been to recruit healthy subjects with 18 years of age or older, which reflects the common interest of having a homogeneous group of individuals to participate in the study and enhance the likelihood of demonstrating BE. However, recent experiences have revealed that in some instances, albeit rare, there is a lack of subject-to-subject similarity in the difference between the T and R product, the so-called *subject-by-formulation interaction* in statistical term (Hauck et al. 2000). Such interactions can arise when the products (or formulations) differ in a subgroup but not in the remaining subjects of the population.

An earlier report on subject-by-formulation interactions may be related to age (Carter et al. 1993). In this study, one of the generic products had AUC and C_{max} values 43 and 77 % higher in the elderly than in the young subjects, while the innovator and another generic product had similar values in the elderly and young. The cause of this interaction had been attributed to the age-related differences in pH, gastric emptying, and/or transit time in the gastrointestinal tract between the two populations. Another example of subject-by-formulation interactions was found from FDA data base with a drug (calcium-channel blocking agent) in two modified-release products (Chen 2005). The drug was a substrate of both CYP3A4 and P-gp. The mean ratio of the T over R product was significantly different between males and females from single-dose and multiple-dose studies, suggesting the presence of a sex-based, group-by-formulation interaction. The in vitro dissolution testing using varying pH media also revealed a pronounced difference in the dissolution behavior of the two products. Based on these data, the interaction was postulated to occur because of different pH-dependent in vivo release profiles between the two products, as well as sex differences in intestinal epithelial drug metabolism and/or transport. In a recent FDA contract study, an apparent subject-by-formulation interaction was also found for ranitidine solution in the presence of a large amount of sorbitol as opposed to sucrose (Chen et al. 2007). A relevant factor accounting for such an interaction may relate to the unique osmotic effect of sorbitol on gastrointestinal physiology observed in various subgroups of the general population (Jain et al. 1985, 1987).

The US FDA currently recommends that in vivo BE studies be conducted in individuals representative of the general population, taking into account age, sex, and race (FDA 2003a). The rationale for having healthy volunteers in most BE studies with pharmacokinetic measures relies on the use of crossover designs where each subject can serve as his/her own control, and thus the conclusion drawn from these study results with respect to BE determination is unbiased, regardless of the populations used. Only under certain circumstances will safety considerations preclude the use of healthy subjects. In such situations, applicants are generally advised to enroll targeted patients with stable disease process and treatments for the duration of the BE study. Depending on the drug characteristics, indications, safety and/or efficacy profiles, the studies may be conducted with crossover and/or parallel designs. Using everolimus as an example, 10 mg tablet of this drug may be dosed once daily for oncology use. Patients who are already receiving everolimus with such dosing regimen can continue on the same dose for both periods of the crossover or parallel study at steady state without disrupting the course of therapy in the patient (FDA 2012c).

2.4.4 Administered Dose

In the USA, when a drug product is in the same dosage form, but in a different strength and is proportionally similar in its active and inactive ingredients to the higher strength product on which BE testing has been conducted, an in vivo BE demonstration of one or more lower strengths can be waived based on appropriate dissolution data (FDA 2003a). Hence, the recommended dose used in a BE study is generally the dose corresponding to the highest marketed strength administered as a single unit (FDA 2003a). However, at times a lower strength may have to be administered due to toxicity concerns, as exemplified by clozapine (FDA 2005). The RLD product of clozapine tablets has five dose strengths (12.5, 25, 50, 100, and 200 mg) available on the market. Yet, the BE study of clozapine has been recommended to be performed on 100 mg (instead of 200 mg) strength because of safety considerations. The US FDA has allowed biowaivers for the rest of strengths (including 200 mg) of clozapine tablets, providing that (a) linear elimination kinetics has been established over the therapeutic dose range; (b) acceptable in vivo BE studies on the 100 mg strength; (c) proportional similarity of the formulations across all strengths; and (d) acceptable in vitro dissolution testing of all strengths. Similarly, if warranted for analytical reasons, multiple units of the highest strength can be administered, as long as the total single dose remains within the labeled dose range and the total dose is safe for administration to the study subjects.

For an in vivo BE study, the US FDA has recommended that the assayed drug content of the T product batch should not differ from the R product by more than $\pm 5~\%$. This is to ensure that comparable doses will be given in the BE study so that no dose correction is necessary for subsequent analysis of study data (FDA 2003a).

2.4.5 Sampling

In a typical BE study, the T and R product are generally administered with 8 oz (i.e., 240 mL) of water to each participating subject under fasting conditions, unless the study is to be conducted under fed conditions where a high-fat meal will be given (FDA 2002, 2003a). For fasting studies, subjects are usually fasted overnight before drug administration in the following day and standardized meals will be provided to subjects no less than 4 h after dosing.

For BE studies with pharmacokinetic measures, under normal circumstances, a series of blood samples (rather than urine or tissue samples) will be collected after dosing and parent drug (and major metabolites) concentrations in serum or plasma will be measured. However, depending on the drug kinetics, whole blood may be more appropriate for analysis of some drugs, e.g., tacrolimus (FDA 2012d). Tacrolimus is extensively bound to red blood cells with a mean blood to plasma ratio of about 15, while albumin and alpha 1-acid glycoprotein appear to primarily bind tacrolimus in plasma (Venkataramanan et al. 1995).

In a single-dose pharmacokinetic study, collection of blood samples should be scheduled at appropriate times in such a manner that the absorption, distribution, and elimination phases of the drug can be well described. This is generally achieved by collecting 12–18 samples (including a pre-dose sample) for each subject after each dose. More frequent sampling should be made around the anticipated peak time (T_{max}) so that C_{max} can be determined with accuracy. The sampling schedule should continue for at least three or more terminal elimination half-life of the drug to ensure complete characterization of the entire pharmacokinetic profile. The exact timing for sample collection depends on the kinetics of the drug and the input rate from the drug product. However, at least three to four samples should be obtained during the terminal log-linear phase to allow for an accurate estimate of terminal rate constant (λz) from linear regression so that AUC_{∞} can be calculated without difficulty.

2.4.6   Parent Drug Versus Metabolites

For most drugs, one or more primary metabolites are formed as a result of biotransformation. Primary metabolites often undergo further metabolic transformation to one or more secondary metabolites. The administered substance (parent drug) and/or its primary/secondary metabolites may produce either desired therapeutic effect or undesired adverse effect or both. If the administered substance is inactive (i.e., has neither therapeutic nor adverse effects), it is termed a pro-drug. After oral administration, biotransformation may occur pre-systemically when the gastrointestinal mucosa and/or liver contribute to the overall metabolism of the administered substance.

The debate over measuring the parent drug versus metabolite(s) is similar to the debate over whether blood level measures or clinical outcomes should be used in BE studies. From a regulatory perspective, reliance on measurement of the parent drug as a marker of rate and extent of release is preferred, even when the parent drug has no clinical activity or the metabolite has a significant therapeutic effect. The rationale for this approach is that the concentration–time profile of the parent drug is more sensitive to changes in formulation performance than the metabolite. The parent drug data mirror the absorption process of the active moiety in the formulation whereas the metabolite data are more reflective of the processes of metabolite formation, distribution, and elimination (FDA 2003a). In many cases, the formation of metabolite(s) is a sequence secondary to the absorption of parent drug, and thus metabolite(s) data are not useful for distinguishing small differences existing, if any, between formulations. From a clinical perspective, measurement of a metabolite may be desirable when the metabolite possesses most of the clinical activity. Nevertheless, consideration of parent drug versus metabolite for BE evaluation should be focused on the accuracy, sensitivity, and reproducibility of the approach used for assessment.

Indeed, the above notion of using parent drug (rather metabolites) data in BE assessment has been supported by the experimental data and extensive simulations conducted over the years (Chen and Jackson 1991, 1995; Jackson 2000; Jackson et al. 2004; Braddy and Jackson 2010). In most cases, it has been found that 90 % confidence intervals for AUC and/or C_{max} of the metabolite are smaller than those of the parent drug, regardless of the drug kinetics and level of error contained in the data. Exceptions arise only when a high degree of intrasubject variability exists in the first-pass metabolism compared to the absorption process of the drug (Chen and Jackson 1995). Under such conditions, the metabolite data is needed in addition to the parent drug data for BE assessment.

In general, it has been concluded that concentration–time profile of the parent drug, as compared to its metabolite(s), is more sensitive to changes in formulation performance, and thus pharmacokinetic data from parent drug should be used for BE assessment. However, metabolite data may be important and should be obtained if a primary metabolite(s) is formed substantially through pre-systemic metabolism (e.g., first-pass, gut wall, or gut lumen metabolism) and contributes significantly to the safety and efficacy of the drug product. This approach should be applied to all drug products, including pro-drugs. To determine BE, the US FDA currently only requires statistical analysis using a confidence interval approach for parent drug while metabolite data are used to provide supportive evidence of comparable therapeutic outcome.

2.4.7 Enantiomers Versus Racemates

In chemistry, stereoisomers have the same molecular formula with the same atoms, connected in the same sequence, but their atoms are positioned differently in space. Enantiomers are two stereoisomers that are related to each other by a reflection and thus they are mirror images of each other, but they are not superimposable. Analytically, one enantiomer will rotate the plane of polarized light to the right (dextrorotatory, d or +), while its antipode will rotate it to the left with the same magnitude (levorotatory, l or −). The prefixes R- and S- are assigned to the enantiomers on the basis of their absolute configuration. However, there are no relationships between the d/l versus R-/S- nomenclatures.

A drug molecule can be obtained either from natural sources or by chemical synthesis. Natural source drugs may have only one enantiomer whereas chemically synthesized drugs are generally racemates. Many drugs have been developed and marketed as a racemic (50:50) mixture of the R- and S-enantiomers. For example, nonsteroidal anti-inflammatory drugs (NSAIDS) are an important group of racemic drugs with the S-isomer generally associated with clinical efficacy (Evans 1992). The systemic exposure of many NSAID enantiomers such as ketoprofen and flurbiprofen are found comparable in terms of AUC and the S-/R-concentration ratio in plasma remains constant over time (Ariens 1984). However, it has been observed that for some other NSAIDs, such as fenoprofen and ibuprofen, the AUC of S-isomer may exceed that of the R-isomer (Rubin et al. 1985; Cox 1988; Evans et al. 1990). Due to the low solubility of ibuprofen at acidic pH, different formulations may show different in vivo dissolution rates that in turn, translate into different absorption rates. Substantial unidirectional inversion of the R-(−) to S-(+) enantiomer occurs systemically, which may be influenced by the absorption rate of ibuprofen (Jamali et al. 1988; Davies 1998). In a study comparing two formulations of racemic ibuprofen tablets, results from both chiral (enantiospecific) and achiral (non-enantiospecific) assays showed BE of the two products. However, compared to the achiral assay, the chiral assay detected a larger difference in the eutomer (Garcia-Arieta et al. 2005). In another study with two ibuprofen oral suspensions (2 %), achiral method showed BE of two products for both AUC and C_{max}. However, the chiral method showed differences in AUC and C_{max}, resulting in non-bioequivalence for the individual enantiomers (Torrado et al. 2010).

Measurement of racemates in plasma or serum using an achiral assay is generally sufficient for BE studies if identical BE outcome can be obtained with the use of racemate or enantiomer data. However, depending on the pharmacokinetic and pharmacodynamic characteristics of the drug under study, BE decision may vary with the use of racemate or enantiomers. As a result, the FDA Guidance (2003a) currently recommends analysis of individual enantiomers for a BE study when all of the following conditions have been met:

- The enantiomers exhibit different pharmacokinetic characteristics
- The enantiomers exhibit different pharmacodynamic characteristics
- Primary efficacy and safety activity reside with the minor enantiomer
- Nonlinear absorption is present for at least one of the enantiomers, as expressed by a change in the enantiomer concentration ratio with change in the input rate of the drug

2.4.8   Endogenous Compounds

Some drug substances are endogenous compounds either because they are naturally produced in the body or because they are present in the normal diet. If the endogenous compound is identical to the drug, BE determination may be difficult since the exogenous drug cannot be distinguished from the endogenous compound.

Baseline-corrected data is generally recommended for BE evaluation when the endogenous levels are fairly constant before and during the study. The baseline levels are often determined by averaging the data from multiple samples taken in the time period before administration of the study drug. In addition, baseline levels should be determined at each dosing interval if they are period specific. Provided below are two examples of endogenous compounds with one (estradiol) produced naturally and the other (potassium chloride) derived from diet intake.

Endogenous estrogens are largely responsible for the development and maintenance of the female reproductive system and secondary sexual characteristics. Although circulating estrogens exist in a dynamic equilibrium of metabolic interconversions, estradiol is the principal intracellular estrogen with substantially higher potency than its metabolites, estrone and estriol, at the receptor level. The primary source of estrogens in premenopausal women is ovarian follicles. However, after menopause, most endogenous estrogen is produced by conversion of androstenedione to estrone in peripheral tissues. Therefore, estrone and its sulfate-conjugated form are the most abundant circulating estrogens in postmenopausal women.

In the case of estradiol tablets, a single-dose, two-way, crossover design has been recommended for the BE study in healthy, physiologically or surgically postmenopausal women (FDA 2010d). This population is preferred because estradiol is often used to treat symptoms of menopause and the baseline levels in these subjects are fairly constant. The FDA Guidance on estradiol (2010d) has indicated that BE evaluation of estradiol tablets should be based on 90 % confidence interval of baseline-adjusted data of total estrone, with estradiol (unconjugated) and estrone (unconjugated) data as supportive evidence of comparable therapeutic outcome.

Potassium chloride represents an endogenous compound that comes from dietary intake. In this case, it is best to conduct the BE study by strictly controlling the intake before and during the study. The FDA Guidance on potassium chloride (2011d) recommends that subjects be placed on a standardized diet, with known amounts of potassium, sodium, calories, and fluid intake. Strict control and knowledge of the actual intakes of potassium, sodium, calories, and fluid are critical for study success. In addition, subjects should be placed in a climate-controlled environment, remaining in-house as much as possible. Physical activity should be restricted to avoid excessive sweating and thus potassium loss. Meals, snacks, and fluids should be given at standard times, and subjects are strongly encouraged to ingest the recommended amounts while refraining from unnecessary physical activity.

While baseline-correction can be done for pharmacokinetic data of those endogenous compounds that have constant baseline levels in the body, the issue of whether baseline adjustment is appropriate for BE determination may arise when (a) it is not possible to determine baseline concentrations with accuracy; or (b) a feedback mechanism prevails during the study. Presumably, if the interest is to know whether the exogenous compound administered results in the comparable systemic levels that are within the normal physiological range, baseline-uncorrected data may be sufficient for BA/BE assessment. However, if the contribution of baseline levels to the total levels in the blood/plasma is substantial for the compound, it may be problematic to use baseline-uncorrected data for BE determination.

2.5 Conclusions

BE studies have played an important role in the drug development as well as during the post-approval period for both pioneer and generic drugs. The main objectives of these studies may be twofold. First, they serve as bridging studies in the presence of formulation or manufacturing changes to provide supportive evidence for safety and efficacy of a drug product. Second, they can be utilized to assure product quality and performance throughout the life time of a drug product. In the USA, with the passage of the 1984 Hatch-Waxman Act, considerable interest and attention has been added to focus on the use of these studies for approval of generic drugs.

The statutory definition of BE, expressed in rate and extent of absorption of the active moiety or ingredient to the site of action, emphasizes the use of pharmacokinetic measures to indicate release of the drug substance from the drug product with absorption into the systemic circulation. This approach rests on an understanding that measurement of the active moiety or ingredient at the site(s) of action is generally not possible and that there is some relationship between the drug concentrations at the site of action relative to those in the systemic circulation. In cases where pharmacokinetic approach is impossible, BE studies can be conducted using pharmacodynamic measures, clinical endpoints, or in vitro tests, with due considerations.

Extraordinary progress has been made in pharmaceutical science and technology since the enactment of 1977 BA/BE regulations in the USA. The contemporary knowledge and methodologies may provide an opportunity to enhance the regulatory approaches for BE demonstration. An ideal paradigm of BE evaluation may take into account the therapeutic index, clinical importance, and pharmaceutical characteristics of the drug substance and drug product under examination. This can be illustrated by the recent changes in the BE approaches for highly variable drugs and narrow therapeutic index drugs. With modern science and technology, an enhanced reliance on in vitro methods for BE demonstration may be possible in the future. Further refinement of the BCS approach may expand the horizon of using in vitro studies for establishment of BE. Multiple in vitro methods may also be developed to substantiate BE demonstration of complex dosage forms or drug products.

References

Ariens EJ (1984) Stereochemistry as a basis for sophisticated nonsense in pharmacokinetics and clinical pharmacology. Eur J Clin Pharmacol 26:663–668

Braddy AC, Jackson AJ (2010) Role of metabolites for drugs that undergo nonlinear first-pass effect: impact on bioequivalency assessment using single-dose simulations. J Pharm Sci 9:515–523

Carter BL, Noyes MA, Demmler RW (1993) Differences in serum concentrations of and responses to generic verapamil in the elderly. Pharmacotherapy 13:359–368

Chen ML, Jackson AJ (1991) The role of metabolites in bioequivalency assessment—I. Linear pharmacokinetics without first-pass effect. Pharm Res 8:25–32

Chen ML, Jackson AJ (1995) The role of metabolites in bioequivalency assessment—II. Drugs with linear pharmacokinetics and first-pass effect. Pharm Res 12:700–708

Chen ML, Lesko L, Williams RL (2001) Measures of exposure versus measures of rate and extent of absorption. Clin Pharmacokinet 40:565–572

Chen ML (2005) Confounding factors for sex differences in pharmacokinetics and pharmacodynamics: Focusing on dosing regimen, dosage form and formulation. Clin Pharmacol Ther 78:322–329

Chen ML, Straughn AB, Sadrieh N, Meyer M, Faustino PJ, Ciavarella AB, Meibohm B, Yates CR, Hussain AS (2007) A modern view of excipient effects on bioequivalence: case study of sorbitol. Pharm Res 24:73–80

Chen ML, Shah VP, Crommelin DJ, Shargel L, Bashaw D, Bhatti M, Blume H, Dressman J, Ducharme M, Fackler P, Hyslop T, Lutter L, Morais J, Ormsby E, Thomas S, Tsang YC, Velagapudi R, Yu LX (2011a) Harmonization of regulatory approaches for evaluating therapeutic equivalence and interchangeability of multisource drug products: workshop summary report. AAPS J 13:556–564

Chen ML, Davit B, Lionberger R, Wahba J, Ahn HY, Yu LX (2011b) Using partial area for evaluation of bioavailability and bioequivalence. Pharm Res 28:1939–1947

Cox SR (1988) Effect of route administration on the chiral inversion of R-(−) ibuprofen [abstract]. Clin Pharmacol Ther 43:146

Davies NM (1998) Clinical pharmacokinetics of ibuprofen—the first 30 years. Clin Pharmacokinet 34:101–154

Davit BM, Nwakama PE, Buehler GJ, Conner DP, Haidar SH, Patel DT, Yang Y, Yu LX, Woodcock J (2009) Comparing generic and innovator drugs: a review of 12 years of bioequivalence data from the United States Food and Drug Administration. Ann Pharmacother 43:1583–1597

Davit BM, Chen ML, Conner DP, Haidar SH, Kim S, Lee CH, Lionberger RA, Makhlouf FT, Nwakama PE, Patel DT, Schuirmann DJ, Yu LX (2012) Implementation of a reference-scaled average bioequivalence approach for highly variable generic drug products by the US Food and Drug Administration. AAPS J 14:915–924

El-Tahtawy AA, Jackson AJ, Ludden TM (1994) Comparison of single and multiple dose pharmacokinetics using clinical bioequivalence data and Monte Carlo simulations. Pharm Res 11:1330–1336

El-Tahtawy A, Harrison F, Zirkelbach JF, Jackson AJ (2012) Bioequivalence of long half-life drugs—informative sampling determination—using truncated area in parallel-designed studies for slow sustained-release formulations. J Pharm Sci 101:4337–4346

Evans AM, Nation RL, Sansom LN et al (1990) The relationship between the pharmacokinetics of ibuprofen enantiomers and the dose of racemic ibuprofen in humans. Biopharm Drug Dispos 11:507–518

Evans AM (1992) Enantioselectivity pharmacodynamics and pharmacokinetics of chiral non-steroidal anti-inflammatory drugs. Eur J Clin Pharmacol 42:237–256

FDA (1995a) Guidance for industry: topical dermatologic corticosteroids: in vivo bioequivalence. http://www.fda.gov/downloads/Drugs/GuidanceComplianceRegulatoryInformation/Guidances/UCM070234.pdf. Accessed Sept 2013

FDA (1995b) Guidance for industry: SUPAC-IR immediate release solid oral dosage forms—scale-up and postapproval changes: chemistry, manufacturing, and controls, in vitro dissolution testing, and in vivo bioequivalence documentation. http://www.fda.gov/downloads/Drugs/GuidanceComplianceRegulatoryInformation/Guidances/UCM070636.pdf. Accessed Sept 2013

FDA (1995c) Guidance for industry: SUPAC-MR modified release solid oral dosage forms—scale-up and postapproval changes: chemistry, manufacturing, and controls, in vitro dissolution testing, and in vivo bioequivalence documentation. http://www.fda.gov/downloads/Drugs/GuidanceComplianceRegulatoryInformation/Guidances/UCM070640.pdf. Accessed Sept 2013

FDA (1997a) Guidance for industry: extended release oral dosage forms: development, evaluation, and application of in vitro/in vivo correlations. http://www.fda.gov/downloads/Drugs/GuidanceComplianceRegulatoryInformation/Guidances/UCM070239.pdf. Accessed Sept 2013

FDA (1997b) Guidance for Industry: dissolution testing of immediate release solid oral dosage forms. Available from http://www.fda.gov/downloads/Drugs/GuidanceComplianceRegulatory Information/Guidances/UCM070237.pdf. Accessed Sept 2013

FDA (2000) Guidance for industry: waiver of in vivo bioavailability and bioequivalence studies for immediate release solid oral dosage forms based on a biopharmaceutics classification system. http://www.fda.gov/downloads/Drugs/GuidanceComplianceRegulatoryInformation/Guidances/UCM070246.pdf. Accessed Sept 2013

FDA (2001) Guidance for industry: statistical approaches to establishing bioequivalence. http://www.fda.gov/downloads/Drugs/GuidanceComplianceRegulatoryInformation/Guidances/UCM070244.pdf. Accessed Sept 2013

FDA (2002) Guidance for industry: food-effect bioavailability and fed bioequivalence studies. http://www.fda.gov/downloads/Drugs/GuidanceComplianceRegulatoryInformation/Guidances/UCM070241.pdf. Accessed Sept 2013

FDA (2003a) Guidance for industry: bioavailability and bioequivalence studies for orally administered drug products—general considerations. http://www.fda.gov/downloads/Drugs/GuidanceComplianceRegulatoryInformation/Guidances/UCM070124.pdf. Accessed Sept 2013

FDA (2003b) Guidance for industry: bioavailability and bioequivalence studies for nasal aerosols and nasal sprays for local action. http://www.fda.gov/downloads/Drugs/GuidanceComplianceRegulatoryInformation/Guidances/UCM070111.pdf. Accessed Sept 2013

FDA (2005) Draft guidance on clozapine. http://www.fda.gov/downloads/Drugs/GuidanceComplianceRegulatoryInformation/Guidances/UCM249219.pdf. Accessed Sept 2013

FDA (2010a) Draft guidance on orlistat. http://www.fda.gov/downloads/Drugs/GuidanceComplianceRegulatoryInformation/Guidances/UCM201268.pdf. Accessed Sept 2013

FDA (2010b) Bioequivalence recommendations for specific products. http://www.fda.gov/Drugs/GuidanceComplianceRegulatoryInformation/Guidances/ucm075207.htm. Accessed Sept 2013

FDA (2010c) Draft guidance on amiodarone hydrochloride. http://www.fda.gov/downloads/Drugs/GuidanceComplianceRegulatoryInformation/Guidances/UCM238049.pdf. Accessed Sept 2013

FDA (2010d) Draft guidance on estradiol. http://www.fda.gov/downloads/Drugs/GuidanceComplianceRegulatoryInformation/Guidances/UCM238055.pdf. Accessed Oct 2013

FDA (2011a) Draft guidance on methylphenidate hydrochloride. http://www.fda.gov/downloads/Drugs/GuidanceComplianceRegulatoryInformation/Guidances/UCM320007.pdf. Accessed Sept 2013

FDA (2011b) Draft guidance on lanthanum carbonate. http://www.fda.gov/downloads/Drugs/GuidanceComplianceRegulatoryInformation/Guidances/UCM270541.pdf. Accessed Sept 2013

FDA (2011c) Draft guidance on progesterone. http://www.fda.gov/downloads/Drugs/GuidanceComplianceRegulatoryInformation/Guidances/UCM209294.pdf. Accessed Sept 2013

FDA (2011d) Draft guidance on potassium chloride. http://www.fda.gov/downloads/Drugs/GuidanceComplianceRegulatoryInformation/Guidances/UCM270390.pdf. Accessed Oct 2013

FDA (2012a) Draft guidance on cholestyramine. http://www.fda.gov/downloads/Drugs/
    GuidanceComplianceRegulatoryInformation/Guidances/UCM273910.pdf. Accessed Sept 2013
FDA (2012b) Draft guidance on warfarin sodium. http://www.fda.gov/downloads/Drugs/
    GuidanceComplianceRegulatoryInformation/Guidances/UCM201283.pdf. Accessed Sept 2013
FDA (2012c) Draft guidance on everolimus. http://www.fda.gov/downloads/Drugs/Guidance
    ComplianceRegulatoryInformation/Guidances/UCM249239.pdf. Accessed Sept 2013
FDA (2012d) Draft guidance on tacrolimus. http://www.fda.gov/downloads/Drugs/Guidance
    ComplianceRegulatoryInformation/Guidances/UCM181006.pdf. Accessed Sept 2013
FDA (2013a) Title 21 Code of Federal Regulations (CFR) 320.1, Office of the Federal Register,
    National Archives and Records Administration. http://www.gpo.gov/fdsys/pkg/CFR-2013-
    title21-vol5/pdf/CFR-2013-title21-vol5.pdf. Accessed Sept 2013
FDA (2013b) Title 21 Code of Federal Regulations (CFR) 320.24, Office of the Federal Register,
    National Archives and Records Administration. http://www.gpo.gov/fdsys/pkg/CFR-2013-
    title21-vol5/pdf/CFR-2013-title21-vol5.pdf. Accessed Sept 2013
FDA (2013c) Draft guidance on albuterol sulfate. http://www.fda.gov/downloads/Drugs/
    GuidanceComplianceRegulatoryInformation/Guidances/UCM346985.pdf. Accessed Sept 2013
Garcia-Arieta A, Abad-Santos F, Rodriquez-Martinez MA, Varas-Polo Y, Novalbos J, Laparidis
    N, Gallego-Sandin S, Orfanidis K, Torrado J (2005) An eutomer/distomer ratio near unity does
    not justify non-enantiospecific assay methods in bioequivalence studies. Chirality 17:470–475
Gibaldi M, Perier D (1982) Estimation of areas. In: Pharmacokinetics, 2nd edn. Marcel Dekker,
    New York
Gillespie WR (1996) Bioequivalence assessment based on pharmacodynamic response—
    bioequivalence on the dose scale: rationale, theory and methods. Presentation to a joint session
    of the Advisory Committee for Pharmaceutical Science and Pulmonary-Allergy Drugs Advi-
    sory Committee, Gaithersburg, October 16, 1996, Transcript, pp 40–52
Hauck WW, Hyslop T, Chen ML, Patnaik R, Williams RL (2000) Subject-by-formulation
    interaction in bioequivalence: conceptual and statistical issues. Pharm Res 17:375–380
Holford NHG, Sheiner LB (1981) Understanding of the dose-response relationship: clinical
    application of pharmacokinetic-pharmacodynamic models. Clin Pharmacokinet 6:429–453
Jackson AJ (2000) The role of metabolites in bioequivalency assessment—III. Highly variable
    drugs with linear kinetics and first-pass effect. Pharm Res 17:1432–1436
Jackson AJ, Robbie G, Marroum P (2004) Metabolites and bioequivalence: Past and present. Clin
    Pharmacokinet 43:655–672
Jain NK, Rosenberg DB, Ulahannan MJ, Glasser MJ, Pitchumoni CS (1985) Sorbitol intolerance
    in adults. Am J Gastroenterol 80:678–681
Jain NK, Patel VP, Pitchumoni CS (1987) Sorbitol intolerance in adults: prevalence and patho-
    genesis on two continents. J Clin Gastroenterol 9:317–319
Jamali F, Singh NN, Pasutto FM, Russell AS, Coutts RT (1988) Pharmacokinetics of ibuprofen
    enantiomers in man following oral administration of tablets with different absorption rates.
    Pharm Res 5:40–43
Kharidia J, Jackson AJ, Ouderkirk LA (1999) Use of truncated areas to measure extent of drug
    absorption in bioequivalence studies: effects of drug absorption rate and elimination rate
    variability on this metric. Pharm Res 16:130–134
Lionberger RA (2005) Topical bioequivalence update. Presentation to the Advisory Committee for
    Pharmaceutical Science Meeting, Rockville, May 4, 2005. http://www.fda.gov/ohrms/dockets/
    ac/05/transcripts/2005-4137T2.htm. Accessed Sept 2013
Lionberger RA, Raw AS, Kim SH, Zhang X, Yu LX (2012) Use of partial AUC to demonstrate
    bioequivalence of zolpidem tartrate extended release formulations. Pharm Res 29:1110–1120
Nightingale SL, Morrison JC (1987) Generic drugs and the prescribing physician. JAMA 258
    (9):1200–1204
Orange Book (2013) Approved drug products with therapeutic equivalence evaluations, 33rd edn.
    U.S. Department of Health and Human Services, Food and Drug Administration, Center for

Drug Evaluation and Research, Office of Generic Drugs, 2013. http://www.fda.gov/down loads/Drugs/DevelopmentApprovalProcess/UCM071436.pdf. Accessed Sept 2013

Rubin A, Knadler MP, Ho PPK et al (1985) Stereoselective inversion of R-fenoprofen to S-fenoprofen in humans. J Pharm Sci 74:82–85

Schuirmann DJ (1987) A comparison of the two one-sided tests procedure and the power approach for assessing the equivalence of average bioavailability. J Pharmacokinet Biopharm 15:657–680

Stier EM, Davit DM, Chandaroy P, Chen ML, Fourie-Zirkelbach J, Jackson A, Kim S, Lionberger R, Mehta M, Uppoor RS, Wang Y, Yu L, Conner DP (2012) Use of partial area under the curve metrics to assess bioequivalence of methylphenidate multiphasic modified release formulations. AAPS J 14:925–926

Stoughton RB (1992) Vasoconstrictor assay—specific applications. In: Maibach HI, Surber C (eds) Topical corticosteroids. Karger, Basel, pp 42–53

Torrado JJ, Blanco M, Farre M, Roset P, Garcia-Arieta A (2010) Rationale and conditions for the requirement of chiral bioanalytical methods in bioequivalence studies. Eur J Clin Pharmacol 66:599–604

Venkataramanan R, Swaminathan A, Prasad T, Jain A, Zuckerman S, Warty V, McMichael J, Lever J, Buckart G, Starzel T (1995) Clinical pharmacokinetics of tacrolimus. Clin Pharmacokinet 29:404–430

Chapter 3
Basic Statistical Considerations

Fairouz T. Makhlouf, Stella C. Grosser, and Donald J. Schuirmann

3.1 Introduction

According to the Code of Federal Regulations (CFR) 21, Part 320.1, two drug products are considered bioequivalent if they are pharmaceutical equivalents or pharmaceutical alternatives whose rate and extent of absorption do not show a significant difference when administered at the same molar dose of the active moiety under similar experimental conditions, either single dose or multiple dose. CFR 21, 320.1 also specifies that the statistical techniques used should be of sufficient sensitivity to detect difference in rate and extent of absorption that are not attributable to subject variability. This chapter discusses three statistical approaches for bioequivalence (BE) comparisons: average, population, and individual. Many of the principles described here also apply to the design and analysis of clinical endpoint studies; however, we do not discuss such studies specifically.

Defined as *relative bioavailability* (BA), BE involves comparison between a Test (T) and Reference (R) drug product, where T and R can vary, depending on the comparison to be performed (e.g., to-be-marketed dosage form versus clinical trial material, generic drug versus reference listed drug, and drug product changed after approval versus drug product before the change). Although BA and BE are closely related, BE comparisons normally rely on a criterion, a confidence interval for the criterion, and a predetermined BE limit. BE comparisons could also be used in certain pharmaceutical product line extensions, such as additional strengths, new dosage forms (e.g., changes from immediate release to extended release), and new

Disclaimer: The views presented in this chapter by the authors do not necessarily reflect those of the US FDA.

F.T. Makhlouf (✉) • S.C. Grosser • D.J. Schuirmann
Center for Drug Evaluation and Research, U.S. Food and Drug Administration,
10903 New Hampshire Avenue, Silver Spring, MD 20993, USA
e-mail: Fairouz.Makhlouf@fda.hhs.gov

routes of administration. In these settings, the approaches described in this chapter can be used to determine BE. The general approaches discussed in this chapter may also be useful when assessing therapeutic equivalence or performing equivalence comparisons in clinical pharmacology studies and other areas.

A standard in vivo BE study design is based on the administration of either single or multiple doses of the T and R products to healthy subjects on separate occasions, with random assignment to the two possible sequences of drug product administration.

Statistical analysis for pharmacokinetic measures, such as area under the blood-level versus time curve (AUC) and peak concentration (C_{max}), is based on the *two one-sided tests procedure* (Schuirmann 1989) to determine whether the average values for the pharmacokinetic measures determined after administration of the T and R products are comparable. This approach is termed *average bioequivalence* and involves the calculation of a 90 % confidence interval for the ratio of the averages (population geometric means) of the measures for the T and R products. To establish BE, the calculated confidence interval should fall within a BE limit, usually 80–125 % for the ratio of the product averages.

Although average BE is recommended for a comparison of BA measures in most BE studies, this chapter also describes two alternate approaches, termed population bio-equivalence (PBE) and individual bioequivalence (IBE). These alternate approaches may be useful, in some instances, for analyzing in vitro and in vivo BE studies. The average BE approach focuses only on the comparison of population averages of a BE measure of interest and not on the variances of the measure for the T and R products. The average BE method does not assess a subject-by-formulation interaction variance, that is, the variation in the average T and R difference among individuals.

In contrast, population and individual BE approaches include comparisons of both averages and variances of the measure. The population BE approach assesses total variability of the measure in the population. The individual BE approach assesses within-subject variability for the T and R products, as well as the subject-by-formulation interaction.

Figures 3.1, 3.2, and 3.3 illustrate the differences between equivalence in averages and in variabilities of bioavailability. Figure 3.1 illustrates the situation where distribution of the bioavailability measure, for example here ln(C-max), is equivalent in average and variability. Figure 3.2 illustrates that the two distributions can be equivalent in the average but not in the variability. Finally, Fig. 3.3 provides an illustration of how the two measures can be equivalent in variability but not on the average.

In Average BE, it is assumed that the Reference and the Test should give similar average exposure as in Figs. 3.1 and 3.2. In this case, the factors that might affect the BA measure are considered as noise and the study can be designed to minimize between individual variability. In Population BE, the statistical distribution of the drug exposures of the two formulations should be sufficiently similar as in Fig. 3.1

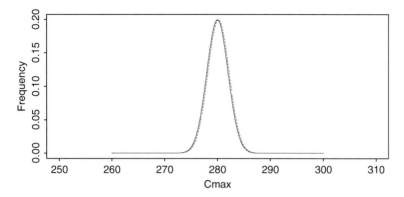

Fig. 3.1 Equivalence in both means and variabilities

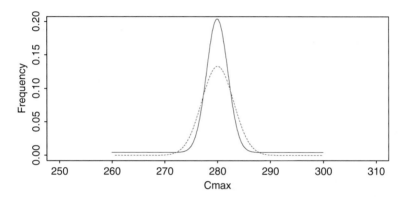

Fig. 3.2 Equivalence in means, but not in variabilities

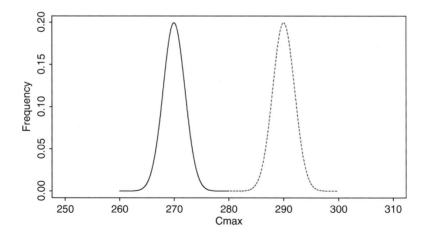

Fig. 3.3 Equivalence in variabilities, but not in means

in some appropriate population (Anderson and Hauck 1990). In this case, we see that the average and the variability of the bioavailability should be similar. Also, Population BE is referred to in the context of *prescribability* in the sense that if an individual is to take a product for the first time, the same therapeutic efficacy is expected no matter which formulation is prescribed. In Individual BE, the bioavailability of the Test is sufficiently close to that of the Reference in most individuals in some appropriate population (Anderson and Hauck 1990). In addition, Individual BE is referred to in the context of switchability, in the sense that an individual using one formulation should expect the same therapeutic effect after switching to the other formulation.

3.2   Statistical Model

Statistical analyses of BE data are typically based on a statistical model for the logarithm of the BA measures (e.g., AUC and $C_{\max}$). The model is a mixed-effects or two-stage linear model. Each subject, j, theoretically provides a mean for the log-transformed BA measure for each formulation, μ_{Tj} and μ_{Rj} for the T and R formulations, respectively. The model assumes that these subject-specific means come from a distribution with population means μ_T and μ_R, and between-subject variances σ_{BT}^2 and σ_{BR}^2, respectively. The model allows for a correlation, ρ, between μ_{Tj} and μ_{Rj}. The subject-by-formulation interaction variance component (Schall and Luus 1993), σ_D^2, is related to these parameters as follows:

$$
\begin{aligned}
\sigma_D^2 &= \text{Variance}\left(\mu_{Tj} - \mu_{Rj}\right) \\
&= \text{Variance}\left(\mu_{Tj}\right) + \text{Variance}\left(\mu_{Rj}\right) - 2\text{Covariance}\left(\mu_{Tj}, \mu_{Rj}\right) \\
&= \sigma_{BT}^2 + \sigma_{BR}^2 - 2\rho\sigma_{BT}\sigma_{BR} \\
&= \sigma_{BT}^2 + \sigma_{BR}^2 - 2\sigma_{BT}\sigma_{BR} + 2\sigma_{BT}\sigma_{BR} - 2\rho\sigma_{BT}\sigma_{BR} \\
&= \left(\sigma_{BT} - \sigma_{BR}\right)^2 + 2\left(1 - \rho\right)\sigma_{BT}\sigma_{BR}
\end{aligned}
\tag{3.1}
$$

For a given subject, the observed data for the log-transformed BA measure are assumed to be independent observations from distributions with means μ_{Tj} and μ_{Rj}, and within-subject variances σ_{WT}^2 and σ_{WR}^2. The total variances for each formulation are defined as the sum of the within- and between-subject components (i.e., $\sigma_{TT}^2 = \sigma_{WT}^2 + \sigma_{BT}^2$ and $\sigma_{TR}^2 = \sigma_{WR}^2 + \sigma_{BR}^2$). When BE is assessed through the analysis of crossover studies, the means are given additional structure by the inclusion of period and sequence effect terms.

3.3 Statistical Approaches for Bioequivalence

The general structure of a BE criterion is that a function (Θ) of population measures should be demonstrated to be no greater than a specified value (θ). Using the terminology of statistical hypothesis testing, this is accomplished by testing the hypothesis $H_0 : \Theta > \theta$ versus $H_A : \Theta \leq \theta$ at a desired level of significance, often 5 %. Rejection of the null hypothesis H_0 (i.e., demonstrating that the estimate of Θ is statistically significantly less than θ) results in a conclusion of BE. The choice of Θ and θ differs in average, population, and individual BE approaches. A general objective in assessing BE is to compare the log-transformed BA measure after administration of the T and R products. Population and individual approaches are based on the comparison of a measure of the difference (expected square distance) between the Test and Reference formulations to the same measure of difference between two administrations of the Reference formulation, denoted by R and R'. An acceptable Test (T) formulation is one where the difference between T and R' (i.e., $T - R'$) is not substantially greater than the difference between the two reference administrations (i.e., $R - R'$). In both population and individual BE approaches, this comparison appears as a comparison to the reference variance, which is referred to as *scaling to the reference variability*.

Population and individual BE approaches, but not the average BE approach, allow two types of scaling: reference-scaling and constant-scaling. Reference-scaling means that the criterion used is scaled to the variability of the R product, which effectively widens the BE limit for more variable reference products.

Although generally sufficient, use of reference-scaling alone could unnecessarily narrow the BE limit for drugs and/or drug products that have low variability but a wide therapeutic range. Hence, a mixed-scaling approach for the population and individual BE approaches is recommended by the guidance on *Statistical Approaches to Establishing Bioequivalence* (FDA 2001). Sections 3.3.2 and 3.3.3 describe such approaches in the population and IBE, respectively. With mixed scaling, the reference-scaled form of the criterion should be used if the reference product is highly variable; otherwise, the constant-scaled form should be used.

3.3.1 Average Bioequivalence

The average BE approach focuses on the comparison of population averages of a BE measure of interest for example log-transformed measure of AUC or $C_{\max}$.

The following criterion is recommended for average BE:

$$(\mu_T - \mu_R)^2 \leq \theta_A^2, \tag{3.2}$$

where μ_T is the population average response of the log-transformed measure for the Test formulation and μ_R is the population average response of the log-transformed measure for the Reference formulation as defined in Sect. 3.2.

This criterion is equivalent to:

$$-\theta_A \le (\mu_T - \mu_R) \le \theta_A \tag{3.3}$$

Typically, $\theta_A = \ln(1.25)$ this corresponds to the FDA bioequivalence limit of 80–125 % on the original scale. This criterion is not symmetric around 1 on the original scale for the ratio of the average bioavailability, but it is symmetric on the log scale around 0.

Average bioequivalence does not include a comparison of the variances of the measure for the T and R products and it does not assess a subject-by-formulation interaction variance. In the following sections, we will discuss two methods that will include such comparisons.

3.3.2   Population Bioequivalence

PBE is important in the context of drug interchangeability when prescribing new drug to a naïve subject. In terms of clinical setting, PBE allows the doctor to prescribe either the Test or the Reference drug product to the subject who has never used the drug with confidence. To assure PBE of the test and reference drugs, the distribution of pharmacokinetic measurements should be similar. This means that, in addition to the similarity of the population average of the pharmacokinetic measurements of the test and reference products, the population variability of the pharmacokinetic measurements should be similar as well.

In the guidance on *Statistical Approaches to Establishing Bioequivalence* (FDA 2001), the comparison of interest for population equivalence is presented in terms of the population difference ratio (PDR). PDR is defined as the ratio of the expected squared difference between T and R drugs administered to different subjects and the expected squared difference between two administrations of reference drugs (R and R') administered to different subjects. It is given by Eq. (3.4) as follows:

$$\mathrm{PDR} = \frac{\text{Difference between test and reference administrtered to different subjects}}{\text{Difference between two references administrtered to different subjects}}$$

$$\mathrm{PDR} = \sqrt{\frac{E\left(T - R'\right)^2}{E\left(R - R'\right)^2}} \tag{3.4}$$

Note that the notation $E(\cdot)$ denotes the expected value of a variable, which may be thought of as the theoretical mean value. It will be used throughout this chapter.

For two drug products to be considered population bioequivalent the PDR should be within acceptable limits. The 2001 Guidance proposed a mixed-scaling approach for population bioequivalence criteria (PBC). Under this approach, the guidance suggests the use of the reference-scaled method if the estimate of total standard deviation for the Reference drug (σ_{TR}) is greater than a specified constant for the total standard deviation (σ_{T0}); and the use of constant-scaled method if the estimate of total standard deviation for the Reference drug is less than or equal a specified constant for the total standard deviation.

The recommended criteria are:

- If $\sigma_{TR} > \sigma_{T0}$ then use reference-scaled

$$\frac{(\mu_T - \mu_R)^2 + (\sigma_{TT}^2 - \sigma_{TR}^2)}{\sigma_{TR}^2} \leq \theta_p \tag{3.5}$$

- If $\sigma_{TR} \leq \sigma_{T0}$ then use constant-scaled

$$\frac{(\mu_T - \mu_R)^2 + (\sigma_{TT}^2 - \sigma_{TR}^2)}{\sigma_{T0}^2} \leq \theta_p, \tag{3.6}$$

where μ_T is the population average response of the log-transformed measure for the Test formulation, μ_R is the population average response of the log-transformed measure for the Reference formulation, σ_{TT}^2 is the total variance of the Test formulation (i.e., sum of within- and between-subject variances), σ_{TR}^2 total variance of the Reference formulation (i.e., sum of within- and between-subject variances), and σ_{T0}^2 is a specified constant total variance and θ_p is the PBE limit. In other words, the PBC can be written as:

$$\text{PBC} = \frac{(\mu_T - \mu_R)^2 + (\sigma_{TT}^2 - \sigma_{TR}^2)}{\max(\sigma_{T0}^2, \sigma_{TR}^2)} \leq \theta_p \tag{3.7}$$

The above inequality (Eq. (3.7)) represents an aggregate approach where a single criterion on the left-hand side of the equation encompasses two major components. The first component addresses the difference between the T and R population averages ($\mu_T - \mu_R$) and the second component addresses the difference between the T and R total variances ($\sigma_{TT}^2 - \sigma_{TR}^2$). This aggregate measure is scaled to the total variance of the R product (σ_{TR}^2) or to a constant value (σ_{T0}^2), a standard that relates to a limit for the total variance), whichever is greater.

Under reference scaling, the PDR is monotonically related to the PBC. This can be shown as follows:

$$E(T - R)^2 = (\mu_T - \mu_R)^2 + \sigma_{TT}^2 + \sigma_{TR}^2 \tag{3.8}$$

$$E\left(R - R'\right)^2 = 2\sigma_{TR}^2 \tag{3.9}$$

$$\frac{E(T - R)^2}{E\left(R - R'\right)^2} = \frac{(\mu_T - \mu_R)^2 + \sigma_{TT}^2 + \sigma_{TR}^2}{2\sigma_{TR}^2}$$

$$= \frac{(\mu_T - \mu_R)^2 + \sigma_{TT}^2 + \sigma_{TR}^2 - \sigma_{TR}^2 + \sigma_{TR}^2}{2\sigma_{TR}^2}$$

$$= \frac{(\mu_T - \mu_R)^2 + \sigma_{TT}^2 - \sigma_{TR}^2 + 2\sigma_{TR}^2}{2\sigma_{TR}^2} \tag{3.10}$$

$$= \frac{(\mu_T - \mu_R)^2 + \sigma_{TT}^2 - \sigma_{TR}^2}{2\sigma_{TR}^2} + 1$$

$$= \frac{PBC}{2} + 1$$

This means that the PDR is related to the PBC through,

$$\begin{aligned} PDR &= \sqrt{\frac{E(T - R)^2}{E\left(R - R'\right)^2}} \\ &= \sqrt{\frac{PBC}{2} + 1} \end{aligned} \tag{3.11}$$

The determination of σ_{T0} is based on the maximum allowable PDR and the variance offset $(\sigma_{TT}^2 - \sigma_{TR}^2)$.

The 2001 Guidance states that the determination of θ_p should be based on the consideration of average BE criterion and the addition of variance terms to the population BE criterion, as expressed by the formula below:

$$\theta_p = \frac{\text{Average BE Limit} + \text{Variance factor}}{\text{Variance}}$$

$$\theta_p = \frac{(\ln 1.25)^2 + \varepsilon_p}{\sigma_{T0}^2} \tag{3.12}$$

The value of ε_p for population BE is guided by the consideration of the variance term $(\sigma_{TT}^2 - \sigma_{WR}^2)$. Per the 2001 guidance, sponsors or applicants wishing to use the population BE approach should contact the FDA for both ε_P and θ_P.

3.3.3 *Individual Bioequivalence*

In the previous section, we discussed the importance of PBE in the context of drug interchangeability when prescribing a new drug to a naïve subject. In this section, we present drug interchangeability in the context of switchability. In terms of the clinical setting, a finding of IBE allows the doctor to switch the prescribed drug from the Reference drug to the Test drug or vice versa for a subject who has been titrated to the most effective dose without compromising the efficacy and the safety of the drug. The basic idea in IBE is that most subjects will have similar bio-availabilities on the two formulations—Test and Reference products.

In the 2001 Guidance, the comparison of interest for individual equivalence is presented in terms of the ratio of the expected squared difference between T and R drugs administered to the same subject and the expected squared difference between two administrations of reference drugs (R and R') administered to the same subject. This ratio is the individual difference ratio (IDR) and is given by:

$$\text{IDR} = \frac{\text{Difference between test and reference administration on the same subject}}{\text{Difference between two reference administrations on the same subject}}$$

$$\text{IDR} = \sqrt{\frac{E(T-R)^2}{E(R-R')^2}}$$

For two drug products to be considered individual bioequivalent, the IDR should be within acceptable limits. The 2001 Guidance proposed a mixed-scaling approach for individual bioequivalence criteria (IBC). This approach uses the reference-scaled method if the estimate of the within-subject standard deviation of the Reference drug (σ_{WR}) is greater than a specified constant for the within-subject standard deviation (σ_{W0}) and the constant-scaled method if the estimate within-subject standard deviation of the Reference drug less or equal a specified constant for within subject standard deviation.

The recommended criteria are:

- If $\sigma_{WR} > \sigma_{W0}$ then use reference-scaled

$$\frac{(\mu_T - \mu_R)^2 + \sigma_D^2 + \left(\sigma_{WT}^2 - \sigma_{WR}^2\right)}{\sigma_{WR}^2} \leq \theta_I \tag{3.13}$$

- If $\sigma_{WR} \leq \sigma_{W0}$ then use constant-scaled

$$\frac{(\mu_T - \mu_R)^2 + \sigma_D^2 + \left(\sigma_{WT}^2 - \sigma_{WR}^2\right)}{\sigma_{W0}^2} \leq \theta_I, \tag{3.14}$$

where μ_T is the population average response of the log-transformed measure for the T formulation, μ_R is the population average response of the log-transformed measure for the R formulation, σ_D^2 is the subject-by-formulation interaction variance component, σ_{WT}^2 is the within-subject variance of the T formulation, σ_{WR}^2 within-subject variance of the R formulation, σ_{W0}^2 is a specified constant within-subject variance and is the individual BE limit.

In other words, the IBC can be written as:

$$\text{IBC} = \frac{(\mu_T - \mu_R)^2 + \sigma_D^2 + (\sigma_{WT}^2 - \sigma_{WR}^2)}{\max(\sigma_{W0}^2, \sigma_{WR}^2)} \leq \theta_I \qquad (3.15)$$

The above inequality [Eq. (3.15)] represents an aggregate approach where a single criterion on the left-hand side of the equation encompasses three major components. The first component addresses the difference between the T and R population averages $(\mu_T - \mu_R)$ which corresponds to the average bioequivalence criteria. The second component addresses the subject-by-formulation interaction (σ_D^2). From Eq. (3.1), we can see that σ_D^2 measures the extent to which the individual mean difference $(\mu_{Ti} - \mu_{Ri})$ are similar across the individuals. Finally, the third component addresses the difference between the T and R within-subject variances $(\sigma_{WT}^2 - \sigma_{WR}^2)$. This aggregate measure is scaled to the within-subject variance of the R σ_{WR}^2 product or to a constant value $(\sigma_{W0}^2$, a standard limit for the within-subject variance), whichever is greater.

The scaling approach in Eq. (3.13) is useful for drugs that exhibit high within-subject variability. Scaling to the reference variability will widen the bioequivalence limits. This is very important in this case because with high within-subject variability meeting the average bioequivalence limit of 80–125 % requires a large number of subjects. Even the reference drug might fail the average bioequivalence criterion against itself. For drugs with low within-subject variability, using the scaling approach as in Eq. (3.13) will unnecessarily tighten the limits. In this case scaling to constant variance (σ_{W0}^2) is recommended, as in Eq. (3.14). There are some exceptions to this mixed scaling criterion. This happens when the drug has a narrow therapeutic range and a reasonable public need to tighten the bioequivalence limits in this case even if $\sigma_{WR} \leq \sigma_{W0}$ scaling to the reference variability as expressed in Eq. (3.13) might be more appropriate.

Under reference scaling, the IDR is monotonically related to the IBC. This can be shown as follows:

$$E(T - R)^2 = (\mu_T - \mu_R)^2 + \sigma_D^2 + \sigma_{WT}^2 + \sigma_{WR}^2 \qquad (3.16)$$

$$E\left(R - R'\right)^2 = 2\sigma_{WR}^2 \qquad (3.17)$$

$$\frac{E(T-R)^2}{E(R-R')^2} = \frac{(\mu_T - \mu_R)^2 + \sigma_D^2 + \sigma_{WT}^2 + \sigma_{WR}^2}{2\sigma_{WR}^2}$$

$$= \frac{(\mu_T - \mu_R)^2 + \sigma_D^2 + \sigma_{WT}^2 + \sigma_{WR}^2 - \sigma_{WR}^2 + \sigma_{WR}^2}{2\sigma_{WR}^2}$$

$$= \frac{(\mu_T - \mu_R)^2 + \sigma_D^2 + \sigma_{WT}^2 - \sigma_{WR}^2 + 2\sigma_{WR}^2}{2\sigma_{WR}^2} \qquad (3.18)$$

$$= \frac{(\mu_T - \mu_R)^2 + \sigma_D^2 + \sigma_{WT}^2 - \sigma_{WR}^2}{2\sigma_{WR}^2} + 1$$

$$= \frac{IBC}{2} + 1$$

This means that the IDR is related to the IBC through,

$$IDR = \sqrt{\frac{E(T-R)^2}{E(R-R')^2}}$$

$$= \sqrt{\frac{IBC}{2} + 1} \qquad (3.19)$$

The determination of σ_{WO} is based on the maximum allowable IDR and the variance offset. For example, if the maximum allowable IDR is 1.25 and the variance offset is 0, and, based on the convention that the limits for average bioequivalence are 80–125 %, then the scaling standard deviation σ_{WO} is 0.2104.

The 2001 Guidance states that the determination of should be based on the consideration of average BE criterion and the addition of θ_I variance terms to the population BE criterion as expressed by the formula below:

$$\theta_I = \frac{average\ BE\ Limit + variance\ factor}{variance}$$

$$\theta_I = \frac{(\ln 1.25)^2 + \varepsilon_I}{\sigma_{WO}^2} \qquad (3.20)$$

The value ε_I for individual BE is guided by the consideration of the estimate of subject-by-formulation interaction variance (σ_D^2) as well as the difference in within-subject variability ($\sigma_{WT}^2 - \sigma_{WR}^2$). The magnitude of σ_D is associated with the percentage of individuals whose average T to R ratios lie outside 80–125 %. A large subject-by-formulation interaction corresponds to a substantial proportion of individuals with large individual mean differences. As we can see from Eq. (3.1) the value of $\sigma_D = 0$ occurs when all the individual-specific mean differences are equal to the overall mean difference ($\mu_T - \mu_R$). Also, from Eq. (3.1) $\sigma_D \neq 0$ occurs if either the between-subject variance of T and R are not equal or if the correlation is not

Fig. 3.4 Two-group
parallel design [Design 1]

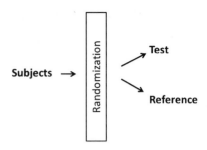

perfect, i.e., $\rho < 1$. In the 2001 Guidance, it is mentioned that if $\sigma_D = 0.1356$ approximately 10 % of the individuals would have their average ratios outside 80–125 % even if $\mu_T - \mu_R = 0$. Also, when $\sigma_D = 0.1741$, the probability is approximately 20 %. For more values of σ_D and their corresponding proportion of individuals outside the 80–125 %, see Hauck et al. 2000. The 2001 Guidance recommends that the allowance for the variance term $(\sigma_{WT}^2 - \sigma_{WR}^2)$ is 0.02 and that the allowance for (σ_D^2) is 0.03 (i.e., $\sigma_D = 0.1731$). This leads to the recommendation of ε_I to be equal to 0.05.

3.4 Study Design

The Code of Federal Regulations, 21CFR 320.25, indicates that the basic design of an in vivo bioavailability study is determined by the scientific questions to be answered, the nature of the reference material and the dosage form to be tested, the availability of analytical methods, and the benefit–risk considerations in regard to testing in humans. Also, 21CFR 320.26 and 21CFR 320.27 indicate that a single-dose or a multiple-dose bioequivalence study should be crossover in design, unless a parallel design or other design is more appropriate for valid scientific reasons. In the following sections, we will describe some of the experimental designs that are appropriate for bioequivalence studies; we will distinguish between nonreplicated and replicated designs and discuss each type in turn.

3.4.1 Nonreplicated Designs

A conventional nonreplicated design is an experimental design in which a treatment or a set of treatments is assigned to an experimental unit without replicating the treatment. Examples of such design are the parallel study design and the standard two-formulation, two-period, two-sequence crossover design.

In the parallel design [Design 1], each subject is randomized to only one treatment group. The simplest form of such a design is the two-group parallel design as shown in Fig. 3.4. Each subject is randomly assigned to one of the treatment groups and usually each treatment group has the same number of

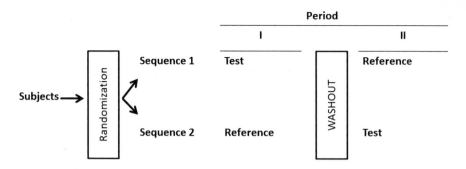

Fig. 3.5 Two-formulation, two-period, two-sequence crossover design [Design 2 TR/RT]

subjects. A parallel design is not commonly used in bioequivalence studies since it cannot distinguish between the intersubject variability and the intrasubject variability because each subject receives only one treatment. Therefore, the sample size is larger for a parallel design compared to other designs, such as the crossover designs. Under certain circumstances parallel designs should be used. For example, if the drug has a long half-life, a parallel design may be more appropriate than other designs due to the likelihood of dropouts during the required long washout period. Also, if the study is to be in patient population, a shorter study is recommended and hence a parallel design might be more appropriate. FDA sometimes recommends using a parallel design in determining bioequivalence. Topical antibacterial and antifungals ointments are examples of products for which three-arm parallel designs, with Test, Reference, and Vehicle arms, are useful.

Another example of the nonreplicated study designs is the standard two-formulation, two-period, two-sequence crossover design [Design 2 TR/RT]. This is a modified, randomized block design, where each block receives the test or reference drug at different periods as shown in Fig. 3.5. In this design, each subject is randomized either to sequence one, where the subject receives the Test drug in the first period then Reference drug in the second period, or to sequence two, where the subject receives the Reference drug in the first period and then the Test drug in the second period. The two periods are separated by a washout period. The length of the washout period should be sufficient for the drug received in the first period to be eliminated from the body. In this design, each subject serves as his/her own control; the design allows a within subject comparison between the test and the reference drugs.

Both the standard two-formulation, two-period, two-sequence crossover design and parallel can be used to generate data where an average or population approach is chosen for BE comparisons.

3.4.2 Replicated Crossover Designs

In a replicated crossover design, at least one treatment is repeated and there are usually more periods than there are treatments. In this section, we present five

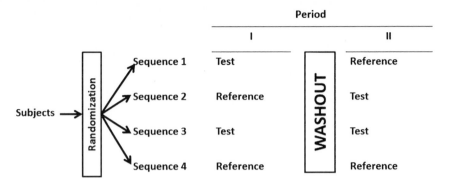

Fig. 3.6 Two-formulation, two-period, four-sequence crossover design [Design 3 TR/RT/TT/RR]

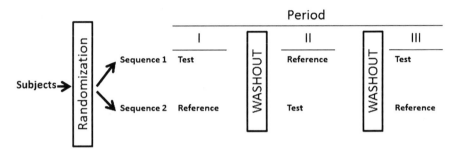

Fig. 3.7 Two-formulation, three-period, two-sequence crossover design [Design 4 TRT/RTR]

examples of replicated crossover designs. The first example of the replicated crossover design is the two-period replicated crossover design [Design 3, TR/RT/TT/RR] as shown in Fig. 3.6.

In Design 3, each subject is randomized to one of the four sequences. The subject receives the Test in the first period and the Reference in the second period in sequence one; in sequence two, the Reference in the first period and then Test in the second period; in sequence three, the Test in the first period and then Test in the second period; and, in sequence four, the Reference in the first period and then Reference in the second period. In each sequence, there is an adequate washout period between the two periods. Such a design is called the Balaam design.

The second and third examples of the replicated crossover designs (Design 4 and Design 5) in this section are two-sequence, three-period designs. In Design 4 [TRT/RTR], shown in Fig. 3.7, each subject is randomized to either sequence one, where the subject receives the Test in the first period, the Reference in the second period, then the Test in the third period or to sequence two where the subject receives the Reference in the first period, the Test in the second period, then the Reference again in the third period. Periods are separated with an adequate washout period.

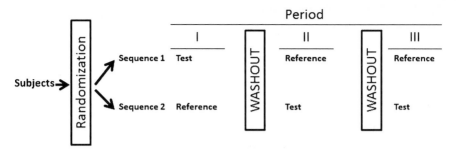

Fig. 3.8 Two-formulation, three-period, two-sequence crossover design [Design 5 TRR/RTT]

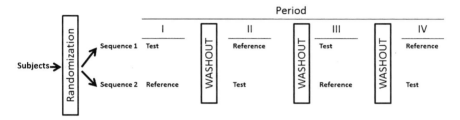

Fig. 3.9 Two-formulation, four-period, two-sequence crossover design [Design 6 TRTR/RTRT]

In Design 5 [TRR/RTT] as shown in Fig. 3.8, each subject is randomized to either sequence one, where the subject receives the Test in the first period, the Reference in the second period, then the Reference in the third period or sequence two, where the subject receives the Reference in the first period, the Test in the second period, then the Test in the third period. Periods are separated with an adequate washout period.

The fourth example is the two-sequence, four-period design [Design 6 TRTR/RTRT] as shown in Fig. 3.9. In Design 6, each subject is randomized to either sequence one, where the subject receives the Test in the first period, the Reference in the second period, then the Test in the third period, and finally the Reference in the fourth period or sequence two where the subject receives the Reference in the first period, the Test in the second period, then the Reference in the third period, and finally the Test in the fourth period. Periods are separated with an adequate washout period.

Our fifth and final example is the four-sequence, four-period designs [Design 7 TRRT/RTTR/TTRR/RRTT]. Subjects are randomized to a sequence and treatments are applied, alternating between Test and Reference as shown in Fig. 3.10.

3.4.2.1   Choosing Among These Designs: Statistical Considerations

Replicated crossover designs can be used irrespective of which approach is selected to establish BE, although they are not necessary when an average or population

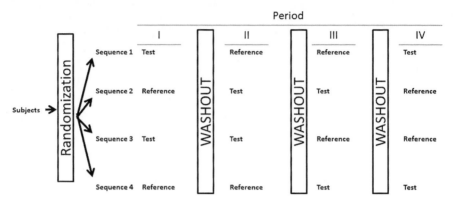

Fig. 3.10 Two-formulation, four-period, four-sequence crossover design [Design 7 TRRT/RTTR/TTRR/RRTT]

approach is used. Replicated crossover designs are critical when an individual BE approach is used. They allow a separate estimation of within-subject variances for the Test and Reference measures and the subject-by-formulation interaction variance component. An exception is the Balaam design [Design 3 TR/RT/TT/RR]. This design should be avoided for individual BE because subjects in the TT or RR sequence do not provide any information on subject-by-formulation interaction. However, the Balaam design may be useful for particular drug products (e.g., a long half-life drug for which a two-period study would be feasible but a three- or more-period study would not).

The replicated crossover design with only two sequences is the preferred design for the individual BE approach. In particular, the two-sequence, four-period design [Design 6, TRTR/RTRT], where half of the subjects receive the first sequence TRTR and the other half receive the sequence RTRT, is recommended by the 2001 Guidance. There are many reasons for this recommendation. To be able to explain them, we need first to mention that in a replicated crossover design, each unique combination of sequence and period can be called a *cell* of the design. For example, the two-sequence, four-period design has eight cells; the two-sequence, three-period design has six cells, etc. Also, the total number of degrees-of-freedom attributable to comparisons among the cells is just the number of cells minus one (unless there are cells with no observations). The fixed effects that are usually included in the statistical analysis are sequence, period, and treatment (i.e., formulation). The number of degrees-of-freedom attributable to each fixed effect is generally equal to the number of levels of the effect, minus one. Thus, in the case of Design 6 TRTR/RTRT, there would be 1 (one) degree-of-freedom due to sequence, 3 degrees-of-freedom due to period, and 1 degree-of-freedom due to treatment, for a total of 5 degrees-of-freedom due to the three fixed effects. These 5 degrees-of-freedom do not account for all 7 degrees-of-freedom attributable to the eight cells of the design; hence, we say the fixed effects model is not *saturated*. An effect for sequence-by-treatment interaction might be included in addition to the three *main effects*—sequence, period, and treatment. Alternatively, a sequence-by-period interaction effect might be included, which would fully saturate the fixed effects model.

If the replicated crossover design has only two sequences, use of only the three main effects (sequence, period, and treatment) in the fixed effects model or use of a more saturated model makes little difference to the results of the analysis, provided there are no missing observations and the study is carried out in one group of subjects (for example, a single site). The least squares estimate of $\mu_T - \mu_R$ will be the same for the main effects model and for the saturated model. Also, the method of moments (MM) estimators of the variance terms in the model used in some approaches to assessment of population and individual BE, which represent within-sequence comparisons, are generally fully efficient regardless of whether the main effects model or the saturated model is used. It is worth noting that the same lots of the T and R formulations should be used for the replicated administration. If the replicated crossover design has more than two sequences, the above advantages are no longer present. Main effects models will generally produce different estimates of $\mu_T - \mu_R$ than saturated models (unless the number of subjects in each sequence is equal), and there is no well-accepted basis for choosing between these different estimates. Also, MM estimators of variance terms will be fully efficient only for saturated models, while for main effects models fully efficient estimators would have to include some between-sequence components, complicating the analysis. Thus, use of designs with only two sequences minimizes or avoids certain ambiguities due to the method of estimating variances or due to specific choices of fixed effects to be included in the statistical model.

One of the reasons to use the four-sequence, four-period design as in Design 7 [TRRT/RTTR/TTRR/RRTT] above is that it is thought to be optimal if carryover effects are included in the model. Similarly, the two-sequence, three-period design as in Design 5 [TRT/RTT] is thought to be optimal among three-period replicated crossover designs. Both of these designs are *strongly balanced for carryover effects*, meaning that each treatment is preceded by each other treatment *and itself* an equal number of times.

With these designs, no efficiency is lost by including *simple* first-order carryover effects in the statistical model. *Simple* first-order carryover effect occurs if the formulation of one period affects the response to the formulation in the next period only. Since the washout period should be sufficient to eliminate carryover effect only the *simple* first-order carryover effect is of concern. However, if the possibility of carryover effects is to be considered in the statistical analysis of BE studies, then the possibility that the carryover is due to the formulation of the current treatment in addition to the previous treatment should also be considered. This is called a direct-by-carryover interaction. If direct-by-carryover interaction is present in the statistical model, these favored designs are no longer optimal. Design 5 [TRR/RTT] does not permit an unbiased within-subject estimate of $\mu_T - \mu_R$ in the presence of general direct-by-carryover interaction.

In the 2001 Guidance, it is mentioned that another design, the three-period, two-sequence design, Design 4 [TRT/RTR] can be used. Using this three-period design will require greater number of subjects compared to the recommended four-period design to achieve the same statistical power.

3.4.3   Sample Size, Study Power, and Dropouts

The assessment of average bioequivalence is based on the "Two One-Sided Test Procedure" as proposed by Schuirmann (1987). Sample size calculation is based on the power of this test procedure; power, here, is the pre-hoc chance of concluding BE when the two products are truly bioequivalent.

There are published formulas to calculate the sample size for average BE studies with various study designs, and references to these calculations will be given in the text below. Sample sizes for population and individual BE studies should be based on simulated data. The simulation is conducted using a default situation allowing the two formulations to vary as much as 5 % in average BA with equal variances and certain magnitudes of subject-by-formulation interaction. Usually, the study is designed, i.e., the sample size is calculated, to have 80 or 90 % power to conclude BE between these two formulations. Sample size also depends on the magnitude of variability and the design of the study. Variance estimates to determine the number of subjects for a specific drug can be obtained from the biomedical literature and/or pilot studies.

A minimum number of 12 evaluable subjects usually are included in any BE study, based on the FDA recommendations. Also, a sufficient number of subjects should enter the study to allow for dropouts. Replacement of subjects during the study could violate the assumptions of the statistical model and complicate the analysis; dropouts generally should not be replaced. If dropouts are to be replaced during the study, this intention has to be indicated in the protocol. The protocol should also state whether samples from replacement subjects, if not used, will be assayed. If the dropout rate is high and more subjects are added, a modification of the statistical analysis may be needed. Also, additional subjects should not be included after data analysis unless the trial was designed from the beginning as a sequential or group sequential design.

Tables 3.1, 3.2, 3.3, and 3.4 give sample sizes for 80 and 90 % power using the specified study design, given a selection of within-subject standard deviations (natural log scale), between-subject standard deviations (natural log scale), and subject-by-formulation interaction, as appropriate. These tables are taken from the 2001 Guidance.

Table 3.1 provides the estimated number of subjects needed for 80 and 90 % power for average bioequivalence in the two-sequence, two-period [RT/TR] and the two-sequenc, four-period [RTRT/TRTR] crossover designs. The calculations for the two-period designs use the method of Diletti et al. (1991) and the results for the four-period designs are based on the relative efficiency data of Liu (1995).

Table 3.1 Estimated numbers of subjects for average bioequivalence with $\Delta = 0.05$ for the two-sequence two-period [RT/TR] and two-sequence four-period [RTRT/TRTR] designs

$\sigma_{WT} = \sigma_{WR}$	σ_D	80 % Power		90 % Power	
		2P	4P	2P	4P
0.15	0.01	12	6	16	8
	0.10	14	10	18	12
	0.15	16	12	22	16
0.23	0.01	24	12	32	16
	0.10	26	16	36	20
	0.15	30	18	38	24
0.30	0.01	40	20	54	28
	0.10	42	24	56	30
	0.15	44	26	60	34
0.5	0.01	108	54	144	72
	0.10	110	58	148	76
	0.15	112	60	150	80

σ_D is the subject-by-formulation interaction standard deviation

σ_{WT} and σ_{WR} are the within-subject standard deviation for the Test and Reference formulation, respectively

Table 3.2 Estimated numbers of subjects for population bioequivalence with $\varepsilon_p = 0.02, \Delta = 0.05$ for the two-sequence four-period [RTRT/TRTR] design

$\sigma_{WT} = \sigma_{WR}$	$\sigma_{BT} = \sigma_{BR}$	80 % Power	90 % Power
0.15	0.15	18	22
	0.30	24	32
0.23	0.23	22	28
	0.46	24	32
0.30	0.30	22	28
	0.60	26	34
0.5	0.50	22	28
	1.00	26	34

σ_{WT} and σ_{WR} are the within-subject standard deviation for the Test and Reference formulation, respectively

σ_{BT} and σ_{BR} are the between-subject standard deviation for the Test and Reference formulation, respectively

The calculations were done for different values of within subject variability for the test and reference products (assuming that they are equal) and different values for the subject-by-formulation interaction.

Table 3.2 provides the estimated number of subjects needed for 80 and 90 % power for PBE in a balanced design across sequences, with the two-sequence and four-period crossover design RTRT/TRTR. Here, ε_p which is the difference between the Test and Reference total variances ($\sigma_{TT}^2 - \sigma_{TR}^2$), is set at 0.02, and $\Delta = 0.05$. The calculations were based on 1,540 simulations for each parameter combination (2001 Guidance).

Table 3.3 provides the estimated number of subjects needed for 80 and 90 % power for IBE in two designs that are balanced design across sequences. The first is the RTRT/TRTR and the second is the RTR/TRT. Here, $\varepsilon_I = 0.05$ where ε_I is the sum of the subject-by-formulation interaction variance and the difference between the Test and Reference total variances $\sigma_D^2 + (\sigma_{TT}^2 - \sigma_{TR}^2)$, and $\Delta = 0.05$. The calculations were based on 5,000 simulations for each parameter combination.

Table 3.3 Estimated numbers of subjects for individual bioequivalence with $\varepsilon_I = 0.05$, $\Delta = 0.05$ for the RTR/TRT and RTRT/TRTR designs

$\sigma_{WT} = \sigma_{WR}$	σ_D	80 % Power		90 % Power	
		3P	4P	3P	4P
0.15	0.01	14	10	18	12
	0.10	18	14	24	16
	0.15	28	22	36	26
0.23	0.01	42	22	54	30
	0.10	56	30	74	40
	0.15	76	42	100	56
0.30	0.01	52	28	70	36
	0.10	60	32	82	42
	0.15	76	42	100	56
0.5	0.01	52	28	70	36
	0.10	60	32	82	42
	0.15	76	42	100	56

σ_D is the subject-by-formulation interaction standard deviation
σ_{WT} and σ_{WR} are the within-subject standard deviation for the Test and Reference formulation, respectively

Table 3.4 Estimated numbers of subjects for individual bioequivalence with $\varepsilon_p = 0.05$, $\Delta = 0.10$ with constraint on $\hat{\Delta}$, $0.8 \leq \exp(\hat{\Delta}) \leq 1.25$ for the RTRT/TRTR design

$\sigma_{WT} = \sigma_{WR}$	σ_D	80 % Power	90 % Power
		4P	4P
0.30	0.01	30	40
	0.10	36	48
	0.15	42	56
0.5	0.01	34	46
	0.10	36	48
	0.15	42	56

σ_D is the subject-by-formulation interaction standard deviation
σ_{WT} and σ_{WR} are the within-subject standard deviation for the Test and Reference formulation, respectively

Tables 3.4 provides the estimated number of subjects needed for 80 and 90 % power for IBE for a two-sequence, four-period crossover design trial with a balanced design across sequences. The design used in the simulations is the RTRT/TRTR design with and $\Delta = 0.10$. A constraint on the point estimate of Δ, $\hat{\Delta}$ was also added so that $0.8 \leq \exp(\hat{\Delta}) \leq 1.25$. The calculations were based on 5,000 simulations for each parameter combination. Note that if $\Delta = 0.05$ is used in the simulation, sample sizes remain the same as given in Table 3.3. This is because the studies are already powered for variance estimation and inference, and therefore, a constraint on the point estimate of Δ has little influence on the sample size for small values of Δ.

While the above sample sizes assume equal within-subject standard deviations, simulation studies for three-period and four-period designs reveal that if $\Delta = 0$ and $\sigma_{WT}^2 - \sigma_{WR}^2 = 0.05$, the sample sizes given will provide either 80 or 90 % power for these studies. If sample sizes calculated are less than the recommended 12 subjects, the sample size should be increased to that minimum sample size, as per the 2001 Guidance.

3.5 Statistical Analysis

The pharmacokinetic parameters, such as AUC and C_{max}, and the clinical end-points, are analyzed statistically to determine if the test and reference products are equivalent. The following sections provide an overview of the statistical methodology for assessment of average, population, and individual BE.

3.5.1 *Logarithmic Transformation*

There are clinical and pharmacokinetic rationales to explain why the pharmacokinetic data (e.g., AUC and C_{max}) are usually log transformed using either common logarithms to the base 10 or natural logarithms. The limited sample size in a typical BE study precludes a reliable determination of the distribution of the data set. In the 2001 Guidance, FDA does not encourage the test for normality of error distribution after log transformation, and it also states that the normality of error distribution should not be a reason to carry out the statistical analysis on the original scale. In addition, if the investigator feels BE study data should be statistically analyzed on the original rather than on the log scale, justification should be provided.

The clinical rationale is based on the recommendation by the FDA Generic Drugs Advisory Committee in 1991. This committee concluded that the primary comparison of interest in a BE study is the ratio, rather than the difference, between average parameter data from the T and R formulations. Using logarithmic transformation, the general linear statistical model employed in the analysis of BE data allows inferences about the difference between the two means on the log scale, which can then be retransformed into inferences about the ratio of the two averages (means or medians) on the original scale. Logarithmic transformation thus achieves a general comparison based on the ratio rather than the difference.

The pharmacokinetic rational is based on Westlake (1973, 1988). Westlake observed that a multiplicative model is postulated for pharmacokinetic measures in BA and BE studies (i.e., AUC and C_{max}, but not T_{max}). Assuming that elimination of the drug is first order and only occurs from the central compartment, the following equation holds after an extravascular route of administration:

$$\text{AUC}_{0-\infty} = \frac{\text{FD}}{\text{CL}} \qquad (3.21)$$

$$\text{AUC}_{0-\infty} = \frac{\text{FD}}{(VK_e)}, \qquad (3.22)$$

where F is the fraction absorbed, D is the administered dose, and FD is the amount of drug absorbed. CL is the clearance of a given subject that is the product of the apparent volume of distribution (V) and the elimination rate constant (K_e). The use of AUC as a measure of the amount of drug absorbed involves a multiplicative term (CL) that might be regarded as a function of the subject. For this reason, Westlake contends that the subject effect is not additive if the data are analyzed on the original scale of measurement.

Logarithmic transformation of the AUC data will bring the CL (VK_e) term into the following equation in an additive fashion:

$$\ln AUC_{0-\infty} = \ln F + \ln D - \ln V - \ln K_e \qquad (3.23)$$

Note that a more general equation can be written for any multicompartmental model as:

$$AUC_{0-\infty} = \frac{FD}{V_{d\beta}\lambda_n}, \qquad (3.24)$$

where $V_{d\beta}$ is the volume of distribution relating drug concentration in plasma or blood to the amount of drug in the body during the terminal exponential phase and λ_n is the terminal slope of the concentration–time curve.

Similar arguments were given for C_{max}. The following equation applies for a drug exhibiting one compartmental characteristic:

$$C_{max} = \frac{FD}{V}xe^{-K_e \cdot T_{max}}, \qquad (3.25)$$

where again F, D and V are introduced into the model in a multiplicative manner. However, after logarithmic transformation, the equation becomes as:

$$\ln C_{max} = \ln F + \ln D - \ln V - K_e T_{max} \qquad (3.26)$$

Thus, log transformation of the C_{max} data also results in the additive treatment of the V term.

3.5.2 Data Analysis

3.5.2.1 Average Bioequivalence

3.5.2.1.1 Overview

The analysis of log-transformed BE measures is usually carried out using parametric (normal theory) methods. To show that two-drug products are equivalent under average BE using the criterion stated in Sect. 3.3.1, Eqs. (3.2) or (3.3), the statistical analysis can be done in the context of the following hypothesis framework as noted by Hauck and Anderson (1984). The null hypothesis of nonequivalence is

$$H_0 : \mu_T - \mu_R \le -\theta_A \text{ or } \mu_T - \mu_R \ge \theta_A$$

and the alternative hypothesis of bioequivalence is given as:

$$H_1 : -\theta_A < \mu_T - \mu_R < \theta_A$$

The general approach is to construct a $(1 - 2\alpha)100\%$ confidence interval with $\alpha = 0.05$, i.e., a 90 % confidence interval, for the quantity $(\mu_T - \mu_R)$, and to conclude average BE if this confidence interval is contained in the interval $[-\theta_A, \theta_A]$. Due to the nature of normal-theory confidence intervals, this corresponds to carrying out two one-sided tests of hypothesis at the 5 % level of significance (Schuirmann 1987) as follows:

$$H_{01} : \mu_T - \mu_R \le -\theta_A$$
$$H_{11} : \mu_T - \mu_R > -\theta_A$$

and

$$H_{02} : \mu_T - \mu_R \ge \theta_A$$
$$H_{12} : \mu_T - \mu_R < \theta_A$$

The 90 % confidence interval for the difference in the means of the log-transformed parameters can be calculated using methods appropriate to the experimental design as described below. The antilogs of the confidence limits obtained constitute the 90 % confidence interval for the ratio of the geometric means between the T and R products.

3.5.2.1.2   Nonreplicated Crossover Design

Parametric (normal-theory) procedures can be used to analyze log-transformed BA measures or the nonreplicated crossover designs. General linear model procedures available in PROC GLM in SAS or equivalent software are preferred, although linear mixed-effects model procedures can also be used for analysis of nonreplicated crossover studies. For example, for a conventional two-treatment, two-period, two-sequence (2×2) randomized crossover design, the statistical model typically includes factors accounting for the following sources of variation: sequence, subjects nested in sequences, period, and treatment. The *Estimate* statement in SAS PROC GLM, or an equivalent statement in other software, is used to obtain estimates for the adjusted differences between treatment means and the standard error associated with these differences.

The following SAS code illustrates an example of program statements to run the average BE analysis using PROC GLM in SAS version 9.3, with SEQ, SUBJ, PER, and TRT identifying sequence, subject, period, and treatment variables, respectively, and Y denoting the response measure (e.g., log(AUC) or log(C_{max})) being analyzed as follows:

```
PROC GLM data=;
  CLASS SUB PER SEQ TRT;
  MODEL Y = SEQ SUB(SEQ) PER
    TRT/SS1 SS3;
  TEST H=SEQ    E=SUB(SEQ);
  ESTIMATE 'TEST VS REFERENCE' TRT 1 -1;
  LSMEANS TRT/CL PDIFF ALPHA=.10;
RUN;
```

The ESTIMATE statement assumes that the code for T formulation precedes the code for R formulation in sort order (this would be the case, for example, if T were coded as 1 and R were coded as 2). If the R code precedes the T code in sort order, the coefficients in the ESTIMATE statement would be changed to -1 1. These statements assume that the study is carried out in one group of subjects. Modification can be made if the study is carried out in more than one group of subjects.

3.5.2.1.3   Replicated Crossover Designs

For replicated crossover designs, the parametric (normal-theory) procedures can be used to analyze log-transformed BA measures. Linear mixed-effects model procedures, available in PROC MIXED in SAS or equivalent software, can be used for the analysis of replicated crossover studies for average BE.

The following illustrates an example of program statements to run the average BE analysis using PROC MIXED in SAS version 9.3, with SEQ, SUBJ, PER, and TRT identifying sequence, subject, period, and treatment variables, respectively, and Y denoting the response measure [e.g., log(AUC) and log(C_{max})] being analyzed:

```
PROC MIXED data=;
  CLASS SEQ SUBJ PER TRT;
  MODEL  Y = SEQ PER TRT/ DDFM=SATTERTH;
  RANDOM TRT/TYPE=FA0(2)  SUB=SUBJ G;
  REPEATED/GRP=TRT SUB=SUBJ;
  ESTIMATE 'T vs. R' TRT 1 -1/CL ALPHA=0.1;
run;
```

The ESTIMATE statement assumes that the code for the T formulation precedes the code for the R formulation in sort order (this would be the case, for example, if T were coded as 1 and R were coded as 2). If the R code precedes the T code in sort order, the coefficients in the Estimate statement would be changed to -1 1. In the *Random* statement, TYPE = FA0(2) could possibly be replaced by TYPE = CSH. The use of TYPE = UN is not recommended, as it could result in an invalid (i.e., not nonnegative definite) estimated covariance matrix. Additions and modifications to these statements can be made if the study is carried out in more than one group of subjects.

3.5.2.1.4 Parallel Designs

For parallel designs, the confidence interval for the difference of means in the log scale can be computed using the total between-subject variance. As in the analysis for replicated designs above, equal variances should not be assumed.

3.5.2.2 Population Bioequivalence

3.5.2.2.1 Overview

To show that two drug products are equivalent under population BE using the criterion stated in Sect. 3.3.2, Eq. (3.5) or (3.6), the statistical analysis can be based on the following hypothesis framework:

$H_0 : \theta \geq \theta_P$ versus $H_1 : \theta < \theta_P$ is equivalent to testing the following hypotheses

$$H_0 : \eta \geq 0 \quad \text{versus} \quad H_1 : \eta < 0$$

where

$$\eta = (\mu_T - \mu_R)^2 + \left(\sigma_{TT}^2 - \sigma_{TR}^2\right) - \theta_p \max\left\{\sigma_{TR}^2, \sigma_{T0}^2\right\}$$

Analysis of BE data using the population approach should focus first on estimation of the mean difference between the T and R for the log-transformed BA measure and estimation of the total variance for each of the two formulations. This can be done using relatively simple unbiased estimators such as the method of moments (MM) (Chinchilli 1996, and Chinchilli and Esinhart 1996). After the estimation of the mean difference and the variances has been completed, a 95 % upper confidence bound for the population BE criterion can be obtained, or equivalently a 95 % upper confidence bound for a linearized form of the population BE criterion can be obtained. Population BE should be considered to be established for a particular log-transformed BA measure if the 95 % upper confidence bound for the criterion is less than or equal to the BE limit, or equivalently if the 95 % upper confidence bound for the linearized criterion is less than or equal to 0.

To obtain the 95 % upper confidence bound of the criterion, intervals based on validated approaches can be used. The following section describes an example of upper confidence bound determination using a population BE approach for a four-period crossover design as presented in the 2001 Guidance. The 2001 Guidance adopts the method in Hyslop, Hsuan, and Holder (2000). This method is based on a method first proposed by Howe (1974) which then was generalized by Graybill and Wang (1980) and Ting et al. (1990).

3.5.2.2.2   Method for Statistical Test of PBE Criterion for Four-Period Crossover Designs

This section describes a method for using the population BE criterion for four-period crossover design. The procedure involves the computation of a test statistic that is either positive (does not conclude population BE) or negative (concludes population BE). Consider the following statistical model which assumes a four-period design with equal replication of T and R in each of s sequences with an assumption of no (or equal) carryover effects (equal carryovers go into the period effects):

$$Y_{ijkl} = \mu_k + \gamma_{ijk} + \delta_{ijk} + \varepsilon_{ijkl}, \tag{3.27}$$

where $i = 1, \ldots, s$ indicates sequence, $j = 1, \ldots, n_i$ indicates subject within sequence i, $k = R, T$ indicates treatment, and $l = 1, 2$ indicates replicate on treatment k for subjects within sequence i.

Y_{ijkl} is the response of replicate l on treatment k for subject j in sequence i, γ_{ijk} represents the fixed effect of replicate l on treatment k in sequence i, δ_{ijk} is the random subject effect for subject j in sequence i on treatment k, and ε_{ijkl} is the random error for subject j within sequence i on replicate l of treatment k. The ε_{ijkl}'s are assumed to be mutually independent and identically distributed as:

$$\varepsilon_{ijkl} \sim N\left(0, \sigma_{Wk}^2\right)$$

for $i = 1, \ldots, s, j = 1, \ldots, n_i, k = R, T$, and $l = 1, 2$. Also, the random subject effects $\delta_{ij} = (\mu_R + \delta_{ijR}, \mu_T + \delta_{ijT})'$ are assumed to be mutually independent and distributed as:

$$\delta_{ij} \sim N_2\left[\begin{pmatrix} \mu_R \\ \mu_T \end{pmatrix}, \begin{pmatrix} \sigma_{BR}^2 & \rho\sigma_{BT}\sigma_{BR} \\ \rho\sigma_{BT}\sigma_{BR} & \sigma_{BT}^2 \end{pmatrix}\right].$$

The following constraint is applied to the nuisance parameters to avoid over parameterization of the model for $k = R, T$:

$$\sum_{i=1}^{s}\sum_{l=1}^{2} \gamma_{ikl} = 0$$

This statistical model proposed by Chinchilli and Esinhart assumes $s*p$ location parameters (where p is the number of periods) that can be partitioned into t treatment parameters and $sp-t$ nuisance parameters (Chinchilli and Esinhart 1996). This produces a saturated model. The various *nuisance* parameters are estimated in this model, but the focus is on the parameters needed for population BE. In some designs, the sequence and period effects can be estimated through a reparametrization of the nuisance effects. This model definition can be extended to other crossover designs.

The Linearized Criterion (from Sect. 3.3.2, Eqs. (3.5) and (3.6)) for the reference-scaled is given by:

$$\eta_1 = (\mu_T - \mu_R)^2 + (\sigma_{TT}^2 - \sigma_{TR}^2) - \theta_p \cdot \sigma_{TR}^2 < 0$$

And for the constant-scaled is given by:

$$\eta_2 = (\mu_T - \mu_R)^2 + (\sigma_{TT}^2 - \sigma_{TR}^2) - \theta_p \cdot \sigma_{T0}^2 < 0$$

The estimation of the linearized criterion depends on study designs. The remaining estimation and confidence interval procedures assume a four-period design with equal replication of T and R in each of s sequences. The reparametrizations are defined as:

$$U_{Tij} = \frac{1}{2}\left(Y_{ijT1} + Y_{ijT2}\right)$$

$$U_{Rij} = \frac{1}{2}\left(Y_{ijR1} + Y_{ijR2}\right)$$

$$V_{Tij} = \frac{1}{\sqrt{2}}\left(Y_{ijT1} - Y_{ijT2}\right)$$

$$V_{Rij} = \frac{1}{\sqrt{2}}\left(Y_{ijR1} - Y_{ijR2}\right)$$

$$I_{ij} = Y_{ijT.} - Y_{ijR.}$$

For $i = 1, \ldots, s$ and $j = 1, \ldots, n_i$, where

$$Y_{ijT.} = \frac{1}{2}\left(Y_{ijT1} + Y_{ijT2}\right) \quad \text{and} \quad Y_{ijR.} = \frac{1}{2}\left(Y_{ijR1} + Y_{ijR2}\right)$$

Compute the formulation means pooling across sequences we get:

$$\hat{\mu}_k = 1/s \sum_{i=1}^{s} \hat{Y}_{i.k.}, \ k = R, T \quad \text{and} \quad \hat{\Delta} = \hat{\mu}_T - \hat{\mu}_R$$

where

$$\overline{Y}_{i.k.} = \frac{1}{n_i}\sum_{j=1}^{n_i}\frac{1}{2}\sum_{l=1}^{2} Y_{ijkl.}$$

Table 3.5 Construction of a $(1-\alpha)$ level upper confidence for the reference-scaled criterion population bioequivalence

$H_q =$ confidence bound	$E_q =$ point estimate	$U_q = (H_q - E_q)^2$
$H_D = \left(\left\lvert \hat{\Delta} \right\rvert + t_{1-\alpha,n-s} \left(\dfrac{1}{s^2} \displaystyle\sum_{i=1}^{s} n_i^{-1} M_I \right)^{1/2} \right)^2$	$E_D = \hat{\Delta}^2$	U_D
$H_1 = \dfrac{(n-s)\cdot E_1}{\chi^2_{n-s,\alpha}}$	$E_1 = MU_T$	U_1
$H_2 = \dfrac{(n-s)\cdot E_2}{\chi^2_{n-s,\alpha}}$	$E_2 = 0.5 MV_T$	U_2
$H_3 = \dfrac{(n-s)\cdot E_3}{\chi^2_{n-s,1-\alpha}}$	$E_3 = -(1+\theta_p)\cdot MU_R$	U_3
$H_4 = \dfrac{(n-s)\cdot E_4}{\chi^2_{n-s,1-\alpha}}$	$E_4 = -(1+\theta_p)\cdot MV_R$	U_4
$H_{\eta_1} = \sum E_q + \left(\sum U_q \right)^{1/2}$		

Compute the variances of U_{Tij}, U_{Rij}, V_{Tij}, and V_{Rij}, pooling across sequences and denote these variance estimates by MU_T, MU_R, MV_T and MV_R, respectively. Specifically,

$$MU_T = \frac{1}{n_{U_T}} \sum_{i=1}^{s} \sum_{j=1}^{n_i} \left(U_{Tij} - \overline{U}_{Ti} \right)^2$$

$$MV_T = \frac{1}{n_{V_T}} \sum_{i=1}^{s} \sum_{j=1}^{n_i} \left(V_{Tij} - \overline{V}_{Ti} \right)^2$$

$$MU_R = \frac{1}{n_{U_R}} \sum_{i=1}^{s} \sum_{j=1}^{n_i} \left(U_{Rij} - \overline{U}_{Ri} \right)^2$$

$$MV_R = \frac{1}{n_{V_R}} \sum_{i=1}^{s} \sum_{j=1}^{n_i} \left(V_{Rij} - \overline{V}_{Ri} \right)^2$$

$$n_I = n_{U_T} = n_{U_R} = n_{V_T} = n_{V_R} = \left(\sum_{i=1}^{s} n_i \right) - s$$

Then, the linearized criterion is estimated by:

$\hat{\eta}_1 = \hat{\Delta}^2 + MU_T + 0.5 \cdot MV_T - (1+\theta_P) \cdot [MU_R + 0.5.MV_R]$ for the reference-scaled and $\hat{\eta}_2 = \hat{\Delta}^2 + MU_T + 0.5 \cdot MV_T - (1) \cdot [MU_R + 0.5MV_R] - \theta_p \cdot \sigma_{T0}$ for the constant-scaled.

The $(1-\alpha)\%$ Upper Confidence Bounds for the reference-scaled criterion (H_{η_1}) and the constant-scaled criterion (H_{η_2}) are estimated by finding an upper or lower confidence limit for each component of η_1 and η_2, respectively.

Table 3.5 illustrates the construction of a $(1-\alpha)$ level upper confidence bound based on the two-sequence, four-period design, for the reference-scaled criterion, η_1.

Table 3.6 Construction of a $(1 - \alpha)$ level upper confidence for the constant-scaled criterion in population bioequivalence

H_q = confidence bound	E_q = point estimate	$U_q = (H_q - E_q)^2$
$H_D = \left(\left(\left| \hat{\Delta} \right| + t_{1-\alpha,n-s} \left(\dfrac{1}{s^2} \displaystyle\sum_{i=1}^{s} n_i^{-1} M_I \right)^{1/2} \right)^2 \right)$	$E_D = \hat{\Delta}^2$	U_D
$H_1 = \dfrac{(n-s) \cdot E_1}{\chi^2_{n-s,\alpha}}$	$E_1 = MU_T$	U_1
$H_2 = \dfrac{(n-s) \cdot E_2}{\chi^2_{n-s,\alpha}}$	$E_2 = 0.5 MV_T$	U_2
$H_3 = \dfrac{(n-s) \cdot E_3}{\chi^2_{n-s,1-\alpha}}$	$E_3 = -1 \cdot MU_R$	U_3
$H_4 = \dfrac{(n-s) \cdot E_4}{\chi^2_{n-s,1-\alpha}}$	$E_4 = -(0.5) \cdot MV_R$	U_4
$H_{\eta_2} = \displaystyle\sum E_q - \theta_P \cdot \sigma_{T0}^2 + \left(\displaystyle\sum U_q \right)^{1/2}$		

Then, $H_{\eta_1} = \displaystyle\sum E_q + \left(\displaystyle\sum U_q \right)^{1/2}$ where the sum is over $q = D,\ 1,\ 2,\ 3,\ 4$, is the upper $(1 - \alpha)$ confidence bound for η_1. Use $\alpha = 0.05$ for a 95 % upper confidence bound. Note that, $n = \displaystyle\sum_{i=1}^{s} n_i$, where s is the number of sequences, n_i is the number of subjects per sequence, and $\chi^2_{n-s,\alpha}$ is from the cumulative distribution function of the chi-square distribution with $n - s$ degrees of freedom, i.e., $\Pr(\chi^2_{n-s} \leq \chi^2_{n-s,\alpha}) = \alpha$.

The confidence bound for the constant-scaled criterion η_2 is computed similarly, adjusting the constants associated with the variance components where appropriate (in particular, the constant associated with MU_R and MV_R). In this case, $H_{\eta_2} =$

$$\sum E_q - \theta_P \cdot \sigma_{T0}^2 + \left(\sum U_q \right)^{1/2}$$ where the sum is over $q = D,\ 1,\ 2,\ 3,\ 4$, is the upper $(1 - \alpha)$ confidence bound for η_2 (Table 3.6).

Using the mixed-scaling approach, to test for population BE, compute the 95 % upper confidence bound of either the reference-scaled or constant-scaled linearized criterion. The selection of either reference-scaled or constant-scaled approach depends on the study estimate of total standard deviation of the reference product (estimated by $[MU_R + 0.5MV_R]^{1/2}$ in the four-period design). If the study estimate of standard deviation is $\leq \sigma_{W0}$, the constant-scaled criterion and its associated confidence interval should be computed. Otherwise, the reference-scaled criterion and its confidence interval should be computed. The procedure for computing each of the confidence bounds is described above. If the upper confidence bound for the appropriate criterion is negative or zero, conclude population BE. If the upper bound is positive, do not conclude population BE.

For nonreplicated crossover studies, any available method (e.g., SAS PROC GLM or equivalent software) can be used to obtain an unbiased estimate of the mean difference in log-transformed BA measures between the T and R products. The total variance for each formulation should be estimated by the usual sample variance, computed separately in each sequence and then pooled across sequences.

For replicated crossover studies, the approach is the same as for nonreplicated crossover designs, but care should be taken to obtain proper estimates of the total variances. One approach is to estimate the within- and between-subject components separately, as for individual BE and then sum them to obtain the total variance. The method for the upper confidence bound should be consistent with the method used for estimating the variances.

3.5.2.2.3   Parallel Design

For parallel design studies, the estimate of the means and variances is the same as for nonreplicated crossover designs. The method for the upper confidence bound is modified to reflect independent rather than paired samples and to allow for unequal variances.

3.5.2.3   Individual Bioequivalence

3.5.2.3.1   Overview

To show that two drug products are equivalent under individual BE using the criterion stated in Sect. 3.3.3, Eq. (3.13) or (3.14), the statistical analysis can be incorporated into the following hypothesis:

$H_0 : \theta \geq \theta_I$ versus $H_1 : \theta < \theta_I$ is equivalent to testing the following hypotheses:

$$H_0 : \eta \geq 0 \quad \text{versus} \quad H_1 : \eta < 0$$

where

$$\eta = (\mu_T - \mu_R)^2 + \sigma_D^2 + (\sigma_{WT}^2 - \sigma_{WR}^2) - \theta_I \max\{\sigma_{WR}^2, \sigma_{W0}^2\}$$

Analysis of BE data using an individual BE approach (Sect. 3.3) should focus on estimation of the mean difference between T and R for the log-transformed BA measure, the subject-by-formulation interaction variance, and the within-subject variance for each of the two formulations. For this purpose, the MM approach is recommended. To obtain the 95 % upper confidence bound of a linearized form of the individual BE criterion, intervals based on validated approaches can be used. An example is described in the section below. After the estimation of the mean difference and the variances has been completed, a 95 % upper confidence bound for the individual BE criterion can be obtained, or equivalently a 95 % upper confidence bound for a linearized form of the individual BE criterion can be obtained. Individual BE should be considered to be established for a particular log-transformed BA measure if the 95 % upper confidence bound for the criterion is less than or equal to the BE limit, or equivalently if the 95 % upper confidence bound for the linearized criterion is less than or equal to 0. The restricted maximum likelihood (REML) method may be useful to estimate mean differences and variances when subjects with some missing data are included in the statistical analysis. A key distinction between the REML and MM methods relates to differences in estimating variance terms.

3.5.2.3.2   Method for Statistical Test of Individual Bioequivalence Criterion for Four-Period Crossover Designs

This section describes a method for using the individual BE criterion for four-period crossover design. The procedure involves the computation of a test statistic that is either positive (does not conclude individual BE) or negative (concludes individual BE). Consider the following statistical model which assumes a four-period design with equal replication of T and R in each of s sequences with an assumption of no (or equal) carryover effects (equal carryovers go into the period effects).

$$Y_{ijkl} = \mu_k + \gamma_{ijk} + \delta_{ijk} + \varepsilon_{ijkl} \tag{3.28}$$

where $i = 1, \ldots, s$ indicates sequence, $j = 1, \ldots, n_i$ indicates subject within sequence i, $k = R, T$ indicates treatment, and $l = 1, 2$ indicates replicate on treatment k for subjects within sequence i.

Y_{ijkl} is the response of replicate l on treatment k for subject j in sequence i, γ_{ijk} represents the fixed effect of replicate l on treatment k in sequence i, δ_{ijk} is the random subject effect for subject j in sequence i on treatment k, and ε_{ijkl} is the random error for subject j within sequence i on replicate l of treatment k. The ε_{ijkl}s are assumed to be mutually independent and identically distributed as:

$$\varepsilon_{ijkl} \sim N\left(0, \sigma_{Wk}^2\right)$$

for $i = 1, \ldots, s, j = 1, \ldots, n_i, k = R, T$, and $l = 1, 2$. Also, the random subject effects $\delta_{ij} = (\mu_R + \delta_{ijR}, \mu_T + \delta_{ijT})'$ are assumed to be mutually independent and distributed as:

$$\delta_{ij} \sim N_2\left[\begin{pmatrix} \mu_R \\ \mu_T \end{pmatrix}, \begin{pmatrix} \sigma_{BR}^2 & \rho\sigma_{BT}\sigma_{BR} \\ \rho\sigma_{BT}\sigma_{BR} & \sigma_{BT}^2 \end{pmatrix}\right].$$

The following constraint is applied to the nuisance parameters to avoid overparameterization of the model for $k = R, T$:

$$\sum_{i=1}^{s}\sum_{l=1}^{2}\gamma_{ikl} = 0$$

This statistical model proposed by Chinchilli and Esinhart assumes $s*p$ location parameters (where p is the number of periods) that can be partitioned into t treatment parameters and $sp\text{-}t$ nuisance parameters (Chinchilli and Esinhart 1996). This produces a saturated model. The various *nuisance* parameters are estimated in this model, but the focus is on the parameters needed for individual BE. In some designs, the sequence and period effects can be estimated through a reparametrization of the nuisance effects. This model definition can be extended to other crossover designs.

The Linearized Criteria (from Sect. 3.3.3, Eqs. (3.13) and (3.14)) for the reference-scaled is given by:

$$\eta_1 = (\mu_T - \mu_R)^2 + \sigma_D^2 + (\sigma_{WT}^2 - \sigma_{WR}^2) - \theta_I \cdot \sigma_{WR}^2$$

and, for the constant-scaled, by:

$$\eta_2 = (\mu_T - \mu_R)^2 + \sigma_D^2 + (\sigma_{WT}^2 - \sigma_{WR}^2) - \theta_I \cdot \sigma_{W0}^2$$

The estimation of the linearized criterion depends on study designs. The remaining estimation and confidence interval procedures assume a four-period design with equal replication of T and R in each of s sequences. The reparametrizations are defined as:

$$I_{ij} = Y_{ijT.} - Y_{ijR.}$$
$$T_{ij} = Y_{ijT1} - Y_{ijT2}$$
$$R_{ij} = Y_{ijR1} - Y_{ijR2}$$

For $i = 1,\ldots,s$ and $j = 1,\ldots,n_i$ where

$$Y_{ijT.} = \frac{1}{2}\left(Y_{ijT1} + Y_{ijT2}\right) \quad \text{and} \quad Y_{ijR.} = \frac{1}{2}\left(Y_{ijR1} + Y_{ijR2}\right)$$

Compute the formulation means, and the variances of I_{ij}, T_{ij}, and R_{ij}, pooling across sequences, and denote these variance estimates by M_I, M_T, and M_R, respectively, where

$$\hat{\mu}_k = \frac{1}{s}\sum_{i=1}^{s}\overline{Y}_{i.k..}, \quad k = R,T \quad \text{and} \quad \hat{\Delta} = \hat{\mu}_T - \hat{\mu}_R$$

$$\overline{Y}_{i.k.} = \frac{1}{n_i}\sum_{j=1}^{n_i}\frac{1}{2}\sum_{l=1}^{2}Y_{ijkl}$$

Table 3.7 Construction of a $(1-\alpha)$ level upper confidence for the reference-scaled criterion in individual bioequivalence

H_q = confidence bound	E_q = point estimate	$U_q = (H_q - E_q)^2$
$H_D = \left(\left\| \hat{\Delta} \right\| + t_{1-\alpha, n-s} \left(\frac{1}{s^2} \sum\limits_{i=1}^{s} n_i^{-1} M_I \right)^{1/2} \right)^2$	$E_D = \hat{\Delta}^2$	U_D
$H_I = \frac{(n-s) \cdot M_I}{\chi^2_{\alpha, n-s}}$	$E_I = M_I$	U_I
$H_T = \frac{0.5 \cdot (n-s) \cdot M_T}{\chi^2_{\alpha, n-s}}$	$E_T = 0.5 M_T$	U_T
$H_R = \frac{-(1.5+\theta_I) \cdot (n-s) \cdot M_R}{\chi^2_{1-\alpha, n-s}}$	$E_R = -(1.5+\theta_I) M_R$	U_R
$H_{\eta_1} = \sum E_q + \left(\sum U_q \right)^{1/2}$		

$$M_I = \hat{\sigma}_I^2 = \frac{1}{n_I} \sum_{i=1}^{s} \sum_{j=1}^{n_i} \left(I_{ij} - \bar{I}_i \right)^2$$

$$n_I = n_T = n_R = \left(\sum_{i=1}^{s} n_i \right) - s$$

$$M_T = \hat{\sigma}_{WT}^2 = \frac{1}{2n_T} \sum_{i=1}^{s} \sum_{j=1}^{n_i} \left(T_{ij} - \bar{T}_i \right)^2$$

$$M_R = \hat{\sigma}_{WR}^2 = \frac{1}{2n_R} \sum_{i=1}^{s} \sum_{j=1}^{n_i} \left(R_{ij} - \bar{R}_i \right)^2$$

Then, the linearized criterion is estimated by:
$\hat{\eta}_1 = \hat{\Delta}^2 + M_I + 0.5 \cdot M_T - (1.5 + \theta_I) \cdot M_R$ for the reference-scaled and $\hat{\eta}_2 = \hat{\Delta}^2 + M_I + 0.5 \cdot M_T - 1.5 \cdot M_R - \theta_I \cdot \sigma_{W0}^2$ for the constant-scaled.
The subject-by-formulation interaction variance component can be estimated by:

$$\hat{\sigma}_D^2 = \hat{\sigma}_I^2 - \frac{1}{2} \left(\hat{\sigma}_{WT}^2 + \hat{\sigma}_{WR}^2 \right)$$

The $(1-\alpha)\%$ Upper Confidence Bounds for the reference-scaled criterion (H_{η_1}) and the constant-scaled criterion (H_{η_2}) are estimated by finding an upper or lower confidence limit for each component of η_1 and η_2 respectively. Table 3.7 illustrates the construction of a $(1-\alpha)$ level upper confidence bound based on the two-sequence, four-period design, for the reference-scaled criterion, η_1. Then, $H_{\eta_1} = \sum E_q + \left(\sum U_q \right)^{1/2}$ where the sum is over $q = D, I, T, R$, is the upper $(1-\alpha)$ confidence bound for η_1. Use $\alpha = 0.05$ for a 95 % upper confidence bound. Note that,

Table 3.8 Construction of a $(1-\alpha)$ level upper confidence for the constant-scaled criterion in individual bioequivalence

H_q = confidence bound	E_q = point estimate	$U_q = (H_q - E_q)^2$
$H_D = \left(\left(\lvert\hat{\Delta}\rvert + t_{1-\alpha,n-s}\left(\frac{1}{s^2}\sum\limits_{i=1}^{s} n_i^{-1}M_1\right)^{1/2}\right)^2\right)$	$E_D = \hat{\Delta}^2$	U_D
$H_I = \frac{(n-s)\cdot M_I}{\chi^2_{a,n-s}}$	$E_I = M_I$	U_I
$H_T = \frac{0.5\cdot(n-s)\cdot M_T}{\chi^2_{a,n-s}}$	$E_T = 0.5M_T$	U_T
$H_R = \frac{-(1.5)\cdot(n-s)\cdot M_R}{\chi^2_{1-a,n-s}}$	$E_R = -(1.5)M_R$	U_R
$H_{\eta_2} = \sum E_q - \theta_I \sigma^2_{W0} + \left(\sum U_q\right)^{1/2}$		

$n = \sum\limits_{i=1}^{s} n_i$, s is the number of sequences, and $\chi^2_{a,n-s}$ is from the cumulative distribution function of the chi-square distribution with $n-s$ degrees of freedom, i.e., $\Pr(\chi^2_{n-s} \leq \chi^2_{a,n-s}) = \alpha$.

The confidence bound for the constant-scaled criterion η_2 is computed similarly, adjusting the constants associated with the variance components where appropriate (in particular, the constant associated with M_R). In this case, $H_{\eta_2} = \sum E_q - \theta_I \cdot \sigma^2_{W0} + \left(\sum U_q\right)^{1/2}$ where the sum is over $q=D,I,T,R$, is the upper $(1-\alpha)$ confidence bound for η_2. This is shown in Table 3.8.

Using the mixed-scaling approach, to test for individual BE, compute the 95 % upper confidence bound of either the reference-scaled or constant-scaled linearized criterion. The selection of either reference-scaled or constant-scaled criterion depends on the study estimate of within-subject standard deviation of the reference product. If the study estimate of standard deviation is $\leq \sigma_{W0}$, the constant-scaled criterion and its associated confidence interval should be computed. Otherwise, the reference-scaled criterion and its confidence interval should be computed. The procedure for computing each of the confidence bounds is described above. If the upper confidence bound for the appropriate criterion is negative or zero, conclude individual BE. If the upper bound is positive, do not conclude individual BE.

3.6 Other Considerations

3.6.1 Reference-Scaled Average Bioequivalence

In recent years, FDA has adopted other methods to establish bioequivalence for certain drug products, such as highly variable drugs (HVD) and narrow therapeutic index (NTI) drugs. HVD are drugs with within-subject variability (%CV) in BE measure of 30 % or greater; see Chap. 6 for more details on HVD. Also, NTI drugs are those drugs where a small difference in drug concentration may lead to serious therapeutic failure or adverse events. Chapter 8 describes NTI drugs.

Studies designed to show that generic HVDs are bioequivalent to their corresponding reference HVDs need to enroll a large number of subjects even when the two drugs have no significant differences in mean. Clinical data strongly support a conclusion that HV drugs have wide therapeutic indices. Otherwise, there would have been significant safety issues and lack of efficacy during the pivotal safety and efficacy clinical trials required for initial FDA marketing approval (Davit et al. 2012). Hence, wider BE intervals might be acceptable to avoid unnecessary human testing On the other hand, other drugs, such as NTI drugs, may pose serious health consequences and the BE limits need to be tightened. With these considerations in mind, a new method was developed by the FDA in the Office of Generic Drugs. This method is called reference-scaled average bioequivalence (RSABE). In this approach, the BE acceptance limits are scaled to the variability of the reference product.

For the two products to be considered RSABE, the following criterion is recommended for the log-transformed measures of BE measures such as AUC and $C_{\max}$.

$$\frac{(\mu_T - \mu_R)^2}{\sigma_{WR}^2} \leq \theta_S \tag{3.29}$$

where μ_T is the population average response of the log-transformed measure for the Test formulation, μ_R is the population average response of the log-transformed measure for the Reference formulation, σ_{WR}^2 is the population within-subject variance of the reference formulation, and $\theta_S = [\ln(\Delta)]^2/\sigma_{W0}^2$ is the BE limit; Δ and σ_{W0}^2 are predetermined constants set by FDA.

In this case, the null hypothesis of nonequivalence is

$$H_0 : \quad \frac{(\mu_T - \mu_R)^2}{\sigma_{WR}^2} > \theta_S$$

and the alternative, of bioequivalence, is given as:

$$H_1 : \quad \frac{(\mu_T - \mu_R)^2}{\sigma_{WR}^2} \leq \theta_S,$$

with testing usually at level $\alpha = 0.05$.

The alternative hypothesis H_1 may be rewritten as:

$$H_1 : \quad (\mu_T - \mu_R)^2 - \theta_S \sigma_{WR}^2 \leq 0$$

The strategy for testing this hypothesis is to obtain a $1 - \alpha$ (i.e., 95 %) upper confidence bound for the quantity $(\mu_T - \mu_R)^2 - \theta_S \sigma_{WR}^2$, and to reject H_0 in favor of H_1 if this confidence bound is less than or equal to zero. A method for obtaining the upper confidence bound is Howe's Approximation I (Howe 1974).

3.6.2 Multiple Groups

If a crossover study is carried out in two or more groups of subjects (e.g., if for logistical reasons only a limited number of subjects can be studied at one time), the statistical model should be modified to reflect the multigroup nature of the study. In particular, the model should reflect the fact that the periods for the first group are different from the periods for the second group. This applies to all of the approaches (average, population, and individual BE) described in this chapter. The multigroup analysis is beyond the scope of the material covered in this chapter.

A *sequential* design, in which the decision to study a second group of subjects is based on the results from the first group, calls for different statistical methods and is outside the scope of this chapter.

3.6.3 Carryover

Use of crossover designs for BE studies allows each subject to serve as his or her own control to improve the precision of the comparison. One of the assumptions underlying this principle is that *carryover effects* (also called *residual effects*) are either absent (the response to a formulation administered in a particular period of the design is unaffected by formulations administered in earlier periods) or equal for each formulation and preceding formulation. If carryover effects are present in a crossover study and are not equal, the usual crossover estimate of $(\mu_T - \mu_R)$ could be biased. One limitation of a conventional two-formulation, two-period, two-sequence crossover design is that the only statistical test available for the presence of unequal carryover effects is the sequence test in the analysis of variance (ANOVA) for the crossover design. This is a between-subject test, which would be expected to have poor discriminating power in a typical BE study. Furthermore, if the possibility of unequal carryover effects cannot be ruled out, no unbiased estimate of $(\mu_T - \mu_R)$ based on within-subject comparisons can be obtained with this design.

For replicated crossover studies, a within-subject test for unequal carryover effects can be obtained under certain assumptions. Typically only first-order carryover effects are considered of concern (i.e., the carryover effects, if they occur, only affect the response to the formulation administered in the next period of the design). Under this assumption, consideration of carryover effects could be more complicated for replicated crossover studies than for nonreplicated studies. The carryover effect could depend not only on the formulation that proceeded the current period but also on the formulation that is administered in the current period. This is called a *direct-by-carryover* interaction. The need to consider more than just *simple* first-order carryover effects has been emphasized (Fleiss 1989). With a replicated crossover design, a within-subject estimate of $\mu_T-\mu_R$ unbiased by general first-order carryover effects can be obtained, but such an estimate could be imprecise, reducing the power of the study to conclude BE.

In most cases, for both replicated and nonreplicated crossover designs, the possibility of unequal carryover effects is considered unlikely in a BE study under the following circumstances:

- It is a single-dose study.
- The drug is not an endogenous entity.
- More than an adequate washout period has been allowed between periods of the study and in the subsequent periods the predose biological matrix samples do not exhibit a detectable drug level in any of the subjects.
- The study meets all scientific criteria (e.g., it is based on an acceptable study protocol and it contains sufficient validated assay methodology).

The possibility of unequal carryover effects can also be discounted for multiple-dose studies and/or studies in patients, provided that the drug is not an endogenous entity and the studies meet all scientific criteria as described above. If a carryover effects are an issue a parallel design may be conducted for BE study.

3.6.4 Outlier Consideration

Outlier data in BE studies are defined as subject data for one or more BA measures that are discordant with corresponding data for that subject and/or for the rest of the subjects in a study. Because BE studies are usually carried out as crossover studies, the most important type of subject outlier is the within-subject outlier, where one subject or a few subjects differ notably from the rest of the subjects with respect to a within-subject *T–R* comparison. The existence of a subject outlier with no protocol violations could indicate either product failure or subject-by-formulation-interaction.

Product failure could occur, for example, when a subject exhibits an unusually high or low response to one or the other of the products because of a problem with the specific dosage unit administered. This could occur, for example, with a sustained and/or delayed-release dosage form exhibiting dose dumping or a dosage unit with a coating that inhibits dissolution.

A subject-by-formulation interaction could occur when an individual is representative of subjects present in the general population in low numbers, for whom the relative BA of the two products is markedly different than for the majority of the population, and for whom the two products are not bioequivalent, even though they might be bioequivalent in the majority of the population.

In the case of product failure, the unusual response could be present for either the *T* or *R* product. However, in the case of a subpopulation, even if the unusual response is observed on the *R* product, there could still be concern for lack of interchangeability of the two products. For these reasons, deletion of outlier values is generally discouraged, particularly for nonreplicated designs. With replicated crossover designs, the *retest* character of these designs should indicate whether to delete an outlier value or not.

3.6.5 Clinical Endpoints

Sometimes the nature of the drug products is such that a clinical endpoint study in patients is the only feasible approach to BE assessment. Most clinical endpoint BE studies use a parallel design. Clinical endpoint may be binary (e.g., cure/no cure and success/failure), categorical (e.g., a 4-point scale, 0="absent" to 3="Severe"), and essentially continuous (e.g., lesion counts and averaged scales over several assessments).

Typically, a placebo arm is included in the study, unless there is a compelling reason not to. To ensure that the study was capable of finding a difference if it was there, each active treatment (*T* and *R*) must be statistically significantly better than placebo in the study.

Equivalence criteria will depend on the nature of the endpoint. For binary endpoints (success/failure), typically the difference between the success probabilities must be shown to fall within the interval $[-0.2, 0.2]$. For essentially continuous endpoints, the criterion is usually similar to that used for PK BE studies. Analysis is dependent on the type of endpoint, nature of the measurement, and other study design features.

References

Anderson S, Hauck WW (1990) Consideration of individual bioequivalence. J Pharmacokinet Biopharm 18:259–273

Chinchilli VM (1996) The assessment of individual and population bioequivalence. J Biopharm Stat 6:1–14

Chinchilli VM, Esinhart JD (1996) Design and analysis of intra-subject variability in cross-over experiments. Stat Med 15:1619–1634

Davit B, Chen M-L, Connor D, Haider S, Kim S, Lee C, Lionberger R, Makhlouf F, Nwakama P, Patel D, Schuirmann D, Yu L (2012) Implementation of a reference-scaled average bioequivalence approach for highly variable generic drug products by the US Food and Drug Administration. AAPS J 14(4):915–924

Diletti E, Hauschke D, Steinijans VW (1991) A sample size determination for bioequivalence assessment by means of confidence intervals. Int J Clin Pharmacol Therap 29:1–8

FDA (2001) Guidance on statistical approaches to establishing bioequivalence. Center for Drug Evaluation and Research, U.S. Food and Drug Administration

Fleiss JL (1989) A critique of recent research on the two-treatment crossover design. Control Clin Trials 10:237–243

Graybill F, Wang CM (1980) Confidence intervals on nonnegative linear combinations of variances. J Am Stat Assoc 75:869–873

Hauck WW, Anderson S (1984) A new statistical procedure for testing equivalence in two-group comparative bioavailability trials. J Pharmacokinet Biopharm 12:83–91

Hauck WW, Hyslop T, Chen M-L, Patnaik R, Williams RL, and the FDA Population and Individual Bioequivalence Working Group (2000) Subject-by-formulation interaction in bioequivalence: conceptual and statistical issues. Pharm Res 17:375–380

Howe WG (1974) Approximate confidence limits on the mean of X + Y where X and Y are two tabled independent random variables. J Am Stat Assoc 69:789–794

Hyslop T, Hsuan F, Holder DJ (2000) A small-sample confidence interval approach to assess individual bioequivalence. Stat Med 19:2885–2897

Liu J-P (1995) Use of the repeated crossover designs in assessing bioequivalence. Stat Med 14:1067–1078

Schall R, Luus HG (1993) On population and individual bioequivalence. Stat Med 12:1109–1124

Schuirmann DJ (1987) A comparison of the two one-sided tests procedure and the power approach for assessing the equivalence of average bioavailability. J Pharmacokinet Biopharm 15:657–680

Schuirmann DJ (1989) Treatment of bioequivalence data: log transformation. In: Proceedings of bio-international '89—issues in the evaluation of bioavailability data, Toronto, Canada, October 1–4, pp 159–161

Ting N, Burdick RK, Graybill FA, Jeyaratnam S, Lu TFC (1990) Confidence intervals on linear combinations of variance components that are unrestricted in sign. J Stat Comput Sim 35:135–143

Westlake WJ (1973) The design and analysis of comparative blood-level trials. In: Warbrick J (ed) Current concepts in the pharmaceutical sciences, dosage form design and bioavailability. Lea and Febiger, Philadelphia, pp 149–179

Westlake WJ (1988) Bioavailability and bioequivalence of pharmaceutical formulations. In: Peace KE (ed) Biopharmaceutical statistics for drug development. Marcel Dekker, New York, pp 329–352

Chapter 4
The Effects of Food on Drug Bioavailability and Bioequivalence

Wayne I. DeHaven and Dale P. Conner

4.1 Mechanisms How Food Can Affect Drug Bioavailability

The absorption of an orally dosed drug product involves the dissolution, or release, of the active component from the drug product into the surrounding GI fluids. Once dissolved, the active drug substance is absorbed through the wall of the GI tract into the systemic circulation, where it reaches its target site of action. The rate and extent of the drug absorption is considered its bioavailability (BA), defined in more detail later in this chapter. While simplistic in concept, this process is quite involved. Several factors, including gastric and intestinal pH, gastric emptying, intestinal transit, formulation release (i.e., immediate- or controlled-release formulations), drug dissolution, and diffusion all come in to play (Fleisher et al. 1999). Of course, the presence of food in the GI tract can and does influence all of these factors (Welling 1996). This section discusses some of the ways food can affect drug absorption, giving specific examples along the way.

4.1.1 Gastrointestinal pH

The drug concentration in the lumen of the GI tract is a factor of the drug dissolution rate, which is influenced by the pH within the lumen of the GI tract. Some drugs are highly soluble and dissolve rapidly in the physiologically relevant pH range of 1–7.5. This chapter discusses these drugs further at the beginning of Sect. 4.3.

W.I. DeHaven (✉) • D.P. Conner
Center for Drug Evaluation and Research, U.S. Food and Drug Administration,
10903 New Hampshire Avenue, Silver Spring, MD 20993, USA
e-mail: Wayne.Dehaven@fda.hhs.gov

L.X. Yu and B.V. Li (eds.), *FDA Bioequivalence Standards*, AAPS Advances in the Pharmaceutical Sciences Series 13, DOI 10.1007/978-1-4939-1252-0_4, © The United States Government 2014

However, for many of today's drugs, the rate of dissolution is intimately dependent upon the GI pH.

This is illustrated by looking at the solubility of weak acid and weak base drug substances, which encompass a large portion of the currently marketed drugs in the USA. The solubility of these compounds depends on the ionization constant (K_a) of the particular drug substance and the pH of the GI fluids (Hörter and Dressman 2001). Weakly acidic drug substances generally increase in solubility in a linear relationship at pH values, which exceed $pH = pK_a + 1$ until the limiting solubility of the ionized form of drug is reached (Hörter and Dressman 2001). The opposite is true for weak bases. Therefore, weakly basic drug substances will often dissolve well in the acidic environment of the stomach, whereas weakly acidic drug substances will not dissolve until after they exit the stomach and reach the more alkaline environment of the small intestine (also usually where absorption occurs). Thus, the pH of the GI fluids plays an important role in the solubility of many drug substances. This, in turn, is a critical step in drug absorption and BA. Of course, how the drug substance is formulated into a final drug product will also be critical in determining its dissolution rates in various GI tract pH environments (e.g., enteric coating excipients, etc.), and subsequently the overall BA of the drug.

4.1.2   GI Tract pH Changes Which Occur After a Meal

In the fasting state, the gastric pH is held approximately at pH 1.5–2, whereas the duodenal pH has been reported at approximately pH 6.5 (Malagelada et al. 1976; Dressman et al. 1990; Russell et al. 1993; Charman et al. 1997; Hörter and Dressman 2001). The pH of gastric fluids rises dramatically in the duodenum due, in part, from the pancreatic bicarbonate added to the digestive mix of fluids and food components. The jejunum pH typically ranges between 6 and 7, whereas the ileum pH is reported as 6.5–8 (Evans et al. 1988; Charman et al. 1997; Hörter and Dressman 2001). These pH values vary from person to person and are influenced by a variety of factors such as age, physical activity, and overall health (Charman et al. 1997).

After eating a meal, signals to the parietal cells lining the stomach cause an increase in the secretion of acid into the stomach. Likewise, pancreatic bicarbonate fluid secretion elevates and the chyme in the duodenum is partly neutralized. Bile is also added to the chyme in order to help in the digestion of fats in the food.

Despite the increase in stomach acid output right after a meal, the gastric pH may actually elevate for a brief time (Hörter and Dressman 2001). This is likely caused by the ingested foods' ability to buffer and dilute the acid produced in the stomach. A few studies have been conducted looking at the early effects of food on gastric pH, and they generally come to a consensus that the gastric pH increases to approximately pH 5 shortly after ingestion of a meal (approximately 10 min after meal) and returns to the fasting state pH in approximately 1.5–2 h (Charman et al. 1997; Malagelada et al. 1976, 1977).

This change in the gastric pH shortly after a meal has the potential to influence the solubility, and thus BA, of a drug substance. Let us reconsider weak acid and weak base drug substances formulated as immediate-release drug products, but now let us consider these drugs as it pertains to gastric pH differences between the fasted and fed states. First, let us reconsider the weak bases, since in the fasted state these drugs generally dissolve well in the acidic environment of the stomach. What can the temporary elevation of gastric pH possibly do to the solubility and dissolution rate of a weak base? Well, the elevated gastric pH may reduce the dissolution of a weak base drug, leaving poorly water-soluble weak bases vulnerable to the pH-related changes in gastric pH. Further, upon gastric emptying, drug precipitation may occur from the combined effects of food on gastric pH and gastric emptying rates (which slows down when food is present. This is discussed later). Predictably, the overall food-effect on absorption would be decreased BA of the drug substance. For instance, ketoconazole is a weak basic drug shown to reduce BA when gastric pH is elevated (Charman et al. 1997).

Conversely, weak acids may show increased solubility in the stomach when coadministered with a meal due to the temporary elevation in gastric pH. This may lead to increased absorption.

Many drug products are formulated as modified-release products (e.g., delayed-release or extended-release), which often modify the release of the drug substance through a pH-dependent mechanism. For instance, enteric coatings are designed to protect the drug substance from the acidic environment of the stomach and release the drug once it reaches the more alkaline environment of the duodenum. The food-effects on gastric pH may also influence some of these enteric coatings, especially if the pH range in which it dissolves is around pH 5 (i.e., the approximate pH of the stomach contents right after meal ingestion), which could cause premature release of the active ingredient from the modified-release formulation. Conversely, lower duodenal pH (e.g., when pancreatic bicarbonate has not completely buffered the acidic chyme after a meal) could prevent dissolution because the pH threshold of the enteric coating is not reached (Charman et al. 1997). In either scenario, the oral BA may be compromised after dosing with a meal.

4.1.3   Gastric Emptying

Later in this chapter, gastric emptying will be discussed as it relates to its influence on the absorption of some BCS class 1 drug substances formulated as immediate-release drug products (BCS class 1 drug products contain highly soluble and highly permeable drug substance). Since BCS class 1 drugs rapidly dissolve independent of pH, generally the only effect of food on drug absorption of these drugs is from changes in gastric emptying (FDA's Guidance for industry: waiver of in vivo bioavailability and bioequivalence studies for immediate-release solid oral dosage forms based on a biopharmaceutics classification system, 2000). However, gastric emptying also impacts other drugs, which are not highly soluble and permeable.

Delayed gastric emptying can influence how long it will take a drug to dissolve and how long the dissolved drug will be present at high enough concentrations to be adequately absorbed. Multiple factors control the rate of gastric emptying. These factors include stomach content volume, pH, meal caloric breakdown (i.e., fats, proteins, and carbohydrates) and total calories, osmolarity, viscosity, and temperature of the stomach contents (Fleisher et al. 1999). The consensus from the scientific community is that solid meal contents empty more slowly than liquid meal contents, but the liquid caloric contents also slows gastric emptying (Collins et al. 1996; Camilleri et al. 1985). Therefore, if a drug product dissolves completely in the stomach when food is present, then it may leave the stomach earlier than a drug product which does not dissolve within the stomach. However, if the latter drug product disintegrates to a small enough particle size (data suggests less than 2 mm diameter; Meyer et al. 1985), then this drug will also leave the stomach earlier with the liquid component of the gastric contents.

After a meal, physiological changes in the GI tract occur which allow for proper breakdown, followed by absorption of the nutrients contained in the meal. In general, the delayed gastric emptying has the potential to increase absorption of some poorly water-soluble drugs due to increasing the potential dissolution time, thus increasing the amount of soluble drug available for BA. In contrast, absorption of unstable drugs might go down when dosed with a meal due to the increased amount of time the drug is within the harsh GI environment. An example of this is seen with the antiretroviral nucleoside analog didanosine (Davit and Conner 2008). Didanosine is acid labile, so it is formulated as buffered tablets and capsules with delayed-release beads, although these formulations clearly do not completely protect the drug substance from the food-effects on gastric emptying. Studies have shown that when the buffered tablet was given after a meal (up to 2 h after), the didanosine C_{max} and AUC both decreased by approximately 55 % compared to administration in the fasting state (Bristol-Myers Squibb, Videx® labeling 2006). When the delayed-release capsules with enteric coated beads were given with food, C_{max} and AUC decreased by approximately 46 % and 19 %, respectively, when compared to administration in the fasting state (Bristol-Myers Squibb, Videx EC® labeling 2011). Despite the formulations attempting to protect the drug substance from the gastric environment, the food-induced delay in gastric emptying caused a significant decrease in BA in the acid labile drug, didanosine. Of course, the FDA labeling for these products recommend taking on an empty stomach.

In contrast, the BA of the antibacterial drug nitrofurantoin is increased in the presence of food due to the delayed gastric emptying resulting in increased dissolution, and therefore, increased absorption (Maka and Murphy 2000). Bioavailability of nitrofurantoin is increased by approximately 40 % when dosed after a meal. The FDA-approved nitrofurantoin labeling states that it should be dosed with food to improve drug absorption (Procter & Gamble, Macrobid® labeling 2009).

Formulation scientists are trying to take advantage of the residence time a dosage form remains in the stomach in order to enhance the BA of certain drug substances. One case-in-point formulation is for gastro-floating tablets of cephalexin (Yin et al. 2013). Floating drug delivery systems are unique and promising in that they do not affect the motility of the GI tract. Rather, they simply float on top of the gastric contents.

Cephalexin is a broad-spectrum antibiotic which can be used to treat a wide range of bacterial infections (Shionogi, Inc., Keflex® labeling 2006). It is a lipophilic weak acid that is stable in gastric conditions but degrades within the more alkaline intestinal environment. It is well absorbed but has a very short half-life of approximately 1 h; therefore, immediate-release dosage forms must be dosed multiple times per day. Sustained-release formulations exist (not marketed in the USA), but the BA of these products is reduced due to the instability in the intestine and the narrow absorption window (Yin et al. 2013). Thus, the idea is a formulation which is maintained in the stomach and slowly released might provide improved BA when compared to the conventional sustained-release formulations.

Yin et al. tested this idea and demonstrated using hydroxypropyl methylcellulose (HPMC) as matrix and sodium bicarbonate as a gas forming agent, that the fasting relative BA of cephalexin (in beagle dogs) increased from 39.3 % in the conventional sustained-release formulation, up to 99 % in the gastro-floating tablet. Interestingly, when administered after a meal, conventional capsule C_{max} was reduced approximately 20 % and T_{max} was significantly prolonged. However, for the floating tablets, a very small increase in C_{max} and T_{max} was observed. Thus, because the tablets were floating on the gastric contents, food and gastric emptying had little to no significant impact on the rate and extent of absorption (as measured by C_{max}, T_{max} and AUC), while it had a significant effect on the absorption of the conventional capsules.

4.1.4   Intestinal Transit

Food has little of an effect on the small intestinal transit time (Yu et al. 1996; Fleisher et al. 1999). The small intestine transit takes approximately 4 h regardless of fasted or fed conditions. However, some drug substances or excipients can speed up the transit time of the intestine (Birkebaek et al. 1990; Adkin et al. 1995; Yuen 2010). Perhaps the more important differences between the fasted and fed environment of the small intestine, as they relate to drug absorption and BA, have to do with changes in pH, viscosity, enzymatic activity, complexation, chelation, and physical barriers to drug absorption, all of which might occur within the lumen of the small intestine.

4.1.5   Stimulate Bile Flow and Pancreatic Excretions

Bile is released from the gall bladder after a meal and enters the duodenum through the bile duct past the sphincter of Oddi. This results in elevated concentrations of bile salts in the small intestine. Besides functioning as a route of excretion for bilirubin, the function of bile in the digestive process is as a surfactant helping to emulsify the fats in food. Bile salts, containing both hydrophobic and hydrophilic sides, aggregate around the ingested fats forming micelles, greatly improving the body's ability to absorb these fats from the diet. Thus, for poorly soluble lipophilic drugs, dosing with a high-fat meal may enhance solubility and dissolution in the duodenum due to the presence of elevated bile salts compared to the fasting state (Charman et al. 1997; Fleisher et al. 1999). This, in turn, can increase the BA of the drug. Conversely, there is evidence that a high-fat meal-evoked increase in bile salt secretion reduces the solubility and dissolution for some hydrophilic compounds, thus reducing their BA.

Atenolol is a cardio-selective β-adrenoceptor antagonist used in treating hypertension and angina. Unlike the lipophilic β-blockers (e.g., propranolol or metoprolol), the hydrophilic atenolol does not go through extensive first-pass metabolism (Barnwell et al. 1993). Approximately only 50 % of atenolol becomes bioavailable due to poor absorption, and food intake can further reduce the absorption of atenolol (Tenormin® labeling, Astrazeneca 2012). In vivo data from Barnwell et al. suggests that bile acids can reduce the BA of atenolol by approximately 30 %. This reduction could not be explained by poor dissolution or degradation of the atenolol. Only C_{max} and AUC were affected with no significant differences in T_{max} and half-life observed (Barnwell et al. 1993).

4.1.6   Increase Splanchnic Blood Flow

Splanchnic blood flow can affect drugs, which are passively absorbed (e.g., paracellular absorption) by influencing the concentration gradient across the membrane (McLean et al. 1978; Toothaker and Welling 1980; Melander and McLean 1983; Fleisher et al. 1999). That is, if splanchnic blood flow is low, then the driving force for a dissolved drug substance, which is passively absorbed will be lower once that drug substance approaches equilibrium across the membrane. However, if splanchnic blood flow is increased, and the transluminal concentration gradient is steepened, then the driving force for absorption is greater, potentially enhancing BA.

High-protein meals have been shown to increase the rate of splanchnic blood flow. Therefore, dosing with high-protein meals may change the transluminal absorptive driving force for some drugs.

Presystemic metabolism (discussed in more detail below) must be taken into account when thinking about food-effect on drug absorption through changes in splanchnic blood flow. Altered splanchnic blood flow may also influence the absorption of drugs that undergo an extensive first-pass effect. That is, increased splanchnic blood flow may increase transluminal passive absorption, but it may also increase the first-pass effect, essentially countering the positive splanchnic effect on the overall BA (Toothaker and Welling 1980).

Drugs that are not absorbed in part by a passive or paracellular process should not be significantly impacted by any food-invoked changes in splanchnic blood flow. That is, for drugs that are absorbed through an active process, the rate-limiting step does not hinge upon a concentration gradient across the membrane.

4.1.7   Presystemic Elimination Pathways

Drug BA is controlled by the absorption processes (e.g., drug product dissolution, drug substance solubility, and permeability) *and* by presystemic clearance (Melander and McLean 1983; Fleisher et al. 1999). These are influenced by genetic and environmental factors, the latter including food ingestion. In BA and BE studies, the overall results reflect the sum of all metabolic transformations which occur to the drug substance. Food-evoked changes in drug presystemic clearance can cause significant increases or decreases in BA, as reflected in the BA or BE studies as changes in AUC and C_{max}.

There are many examples of drugs which are absorbed well; however, they show minimal BA due to presystemic metabolic transformation in the gut mucosa and/or during first-pass metabolism via the liver (first-pass effect) (Toothaker and Welling 1980; Melander and McLean 1983; Fleisher et al. 1999). First-pass hepatic metabolism is when a drug is absorbed across the GI tract and it enters the portal circulation prior to entering systemic circulation. In the liver, the drug is metabolized, leaving less drug substance to reach systemic circulation. The presence of food in the GI tract can significantly impact this effect on the overall BA. For instance, labetalol, a combined α- and β-adrenoceptor antagonist indicated for the management of hypertension, is subject to considerable presystemic metabolism, reducing the oral BA (Daneshmend and Roberts 1982; Melander and McLean 1983). However, when dosed after a meal, there is a food-induced increase in BA (38 %) which is attributed to, at least, in part, a reduction in presystemic clearance (Daneshmend and Roberts 1982). Hepatic metabolism (and gut mucosa metabolism) is often a saturable elimination process which can be influenced by rate of drug absorption.

Perhaps one of the best known examples of a food effect on presystemic clearance is with grapefruit juice (Deferme and Augustijns 2003), which is recognized to enhance BA of a number of drugs by inhibiting the efflux transporter P-glycoprotein (P-gp) and cytochrome P450 3A (CYP3A). The grapefruit effect appears more predominant in drugs that undergo extensive intestinal metabolism. Some drug products increase BA by as much as 300 % (Davit and Conner 2008).

4.1.8   Physical or Chemical Food–Drug Interactions

Sometimes the effect of food on the absorption of a particular drug is more of a direct interaction between the drug and something in the meal. For instance, drug binding to a component in the meal (e.g., pectin, or other dietary fiber) can reduce drug absorption and BA (Huupponen et al. 1984; Fleisher et al. 1999). The presence of the food can form a physical barrier between the GI lumen and the membrane where absorption occurs (Fleisher et al. 1999). The digestive process can form a viscous medium that can significantly diminish drug absorption through the reduction in drug diffusion (Fleisher et al. 1999). Chelation interactions can occur between metals present in food (such as dairy products and meats) and the drug substance. As previously alluded, some drug products (e.g., weak bases) may exhibit decreased BA with food due to the formation of complexes with bile acids.

Ciprofloxacin is a broad-spectrum quinolone antimicrobial agent. Multivalent cations (e.g., aluminum, iron, magnesium, and calcium) chelate with ciprofloxacin in the GI tract, resulting in reduced BA. Neuhofel et al. showed that both ciprofloxacin AUC and C_{max} were reduced by 21 % and 22 %, respectively, when dosed with calcium-fortified orange juice compared to nonfortified orange juice (Neuhofel et al. 2002). This reduced BA may result in loss of antibacterial effect. As a result, the FDA-approved labeling cautions that ciprofloxacin should not be taken with dairy products or calcium-fortified juices since BA may be significantly reduced (Bayer, Cipro® labeling 2013).

4.2   Clinical Relevance of the Effects of Food on Drug BA and BE

According to the US Department of Health and Human Services (HHS) Administration on Aging (AOA), people over the age of 65 represented 12.9 % of the US population in 2009 (roughly 39.6 million people). However, by 2030, it is expected that 19 % of the US population will be 65+, which represents nearly a doubling of this population to 72.1 million people (http://www.aoa.gov). Due in part to the advances made in medicine and other human health sciences, the US human population is living longer. With pharmaceutical advances and an increasing elderly population, patients chronically taking several prescribed oral drug products daily are becoming a routine way of life. Due to the convenience and compliance benefits of administering these medications at meal times (e.g., after breakfast or dinner), it is essential that there is a complete understanding of the food–drug (and drug–drug) effects on the BA of these drug products.

Likewise, generic prescription drugs make up the majority of the drug products prescribed in the USA today. Conservative estimates show that at over 80 % of the prescribed medications are filled with generics (http://www.FDA.gov). Therefore, any food-effect differences between generics and their matching RLDs should not exist in the US market, or else there could be safety and/or efficacy implications.

Food interactions with drug (substance or product) can be manifested as prolonged rate of absorption, decreased absorption, increased absorption, and unaffected absorption (Davit and Conner 2008). Such food-effects could potentially lower the BA of a particular drug below the therapeutically effective concentration range rendering the product less effective or ineffective. Likewise, the coadministration of a particular drug product with a high-fat, high-calorie meal (defined from the FDA's perspective in Sect. 4.4.3 of this chapter) could increase the BA to levels associated with higher prevalence of adverse events.

Human pharmacokinetic (PK) studies have demonstrated that drug plasma concentrations can be significantly altered by the administration of a particular drug product after a meal. Clinically significant effects of food on the absorption of a drug are typically evaluated by comparing the rate and extent of drug absorption when dosed with and without a high-fat, high-calorie meal. Food-effects on rate of drug absorption are reflected as changes in peak plasma concentrations (C_{max}) and time to reach peak plasma concentrations (T_{max}). Food-effects on extent of drug absorption are reflected as changes in the area under the drug plasma concentration versus time curve (AUC_{0-t} and AUC_{∞}). Therefore, in characterizing food-effects for regulatory purposes, applicants of NDAs to the FDA generally conduct human PK studies in which the investigator administers the drug to subjects with and without food and determines any changes in C_{max}, T_{max}, and AUC (FDA's Guidance for industry: waiver of in vivo bioavailability and bioequivalence studies for immediate-release solid oral dosage forms based on a biopharmaceutics classification system, 2000).

As discussed in Sect. 4.4, these NDA studies are often of a two-way crossover design in healthy subjects. By comparing the drug plasma concentration profiles, it can be ascertained whether the presence of food accelerated, delayed, increased, or decreased absorption when compared to same dosing but under the fasted state.

4.2.1   Oral Bioavailability Defined

Oral bioavailability (BA) of drugs is determined by the administered dose that is absorbed from the gastrointestinal (GI) tract. BA is defined in the FDA's regulations as "the *rate and extent* to which the active ingredient or active moiety is absorbed from a drug product and becomes available at the site of action" (21 CFR 320.1, 2011).

4.3   Food-Effect BA and Fed BE Studies

In practice, it is difficult to determine the exact mechanism by which food changes the BA of a drug without conducting specific mechanistic tests (Davit and Conner 2008). The physicochemical properties of the drug substance are important drug attributes which can give clues as to how the presence of food might affect drug absorption. Likewise, a strong understanding of the drug product excipients and how they might affect dissolution and absorption is important in predicting food-effects on BA.

The following section of this chapter focuses on the FDA perspective as it pertains to food-effects on drug substances (remember physicochemical properties) and drug products (remember excipients). Examples are given to highlight the scientific understanding behind some of the regulatory decision making.

4.3.1   Food and Drug Substance: BCS Class 1 Biowaivers

The coadministration of a particular drug product with a meal may change the BA by affecting either the drug substance (i.e., the active pharmaceutical ingredient, or API), or the drug product (including excipients). Certain drug substances, due to the intrinsic properties of that drug substance, are much less likely to encounter deleterious effects when dosed with a meal. These drug substances can generally be categorized as highly soluble and highly permeable drugs. When formulated as immediate-release drug products, these drugs dissolve independent of gastric pH.

There is a valuable tool used by the FDA and other regulatory agencies world-wide for the regulation of changes in oral drug products during scale-up and after postapproval, which is called the Biopharmaceutical Classification System (BCS; Amidon et al. 1995; Martinez and Amidon 2002). The BCS is thoroughly covered in Chap. 5. However, since it is relevant to the topic at hand, it is necessary that a brief description be given here. The BCS is a scientific framework for classifying drug substances based on their aqueous solubility and intestinal permeability. According to the BCS, class 1 drugs are both highly soluble and permeable, i.e., the highest dose strength is soluble in 250 mL or less of aqueous media over a pH range of 1–7.5, and the extent of absorption in humans is determined to be 90 % or more of the administered dose. BCS class 2 drug substances show low solubility, but are highly permeable like BCS class 1 drug substances. Therefore, dissolution is often rate-limiting for these drugs. BCS class 3 drugs are highly soluble, but show low permeability. Lastly, BCS class 4 drugs are both poorly soluble and poorly permeable. (Guidance for industry: waiver of in vivo bioavailability and bioequivalence studies for immediate-release solid oral dosage forms based on a biopharmaceutics classification system, 2000.)

BCS class 1 drug substances, which are highly soluble and permeable, when formulated as immediate-release drug products, generally are not significantly impacted by the presence of food in the GI tract. These products often can be dosed without regard to mealtime because the dissolution and absorption is often pH- and site-independent. Often, any physiological impact, which may occur when administered with food, is not clinically relevant. The primary physiological process, which does affect BA in these drug substances, is gastric emptying, which usually changes C_{max}; however, AUC remains unchanged. That is why for certain ANDA submissions, which reference a BCS class 1 compound, biowaivers are accepted by the FDA in place of in vivo BA or BE testing (as long as it is an immediate-release product in which the excipients do not affect the rate and extent of absorption). The applicant must submit data and supportive documentation showing high solubility, high permeability, GI stability, and rapid dissolution in the complete pH range relevant to the GI tract (pH 1–7.5). The final designation of BCS class I biowaiver eligible is made by the Agency. For further details, please refer to Chap. 5.

The BCS framework, a success by all accounts, does not address potential food-formulation effects on absorption, which might impact the BE between two generic products approved through the biowaiver process. Recently, the FDA's Office of Pharmaceutical Sciences (OPS) in collaboration with scientists from the University of Tennessee, conducted in vivo BE studies using two model BCS class 1 drugs and a single BCS class 3 drug, under fed conditions. Metoprolol and propranolol were selected as the highly permeable and soluble drugs (BCS class 1), whereas hydrochlorothiazide was tested as a BCS class 3 drug. The objective was to see if highly soluble and permeable drugs formulated in immediate-release solid oral dosage forms and that exhibit rapid in vitro dissolution are likely to be bioequivalent under fed conditions.

Two FDA approved products each for metoprolol tablets and propranolol/hydrochlorothiazide tablets (combination) were selected and a BE study was carried out in healthy volunteers under fed conditions (Yu et al. 2004). The studies were designed as single-dose two-way crossover studies with a 1-week washout between periods. The results showed that the two metoprolol products were bioequivalent (i.e., 90 % confidence intervals of the ratios for the parameters AUC and C_{max} were within 80.0–125.0 %) to each other when dosed with a high-fat, high-calorie meal. Likewise, both analytes of the propranolol/hydrochlorothiazide combination tablets were deemed bioequivalent. Based on these results, it is concluded that, generally, there is a low risk of nonbioequivalence with biowaivers for highly soluble, highly permeable, and rapidly dissolving immediate-release solid oral dosage forms.

The results of this study should not be interpreted as a lack of food-effect on these drugs. It simply suggests that whatever food effect occurs is unlikely to significantly impact the BE between two similarly formulated drug products with the same active ingredient. Actually, it has been published that propranolol (and metoprolol) BA is enhanced when dosed after a meal, possibly due to an increase in splanchnic blood flow (McLean et al. 1978); metabolic inhibition by amino acids (Semple and Fangming 1995); and/or saturation of hepatic first-pass metabolism (i.e., a reduction in presystemic clearance, but not improved absorption) (Melander et al. 1977; Tam 1993). The FDA-approved propranolol label notes a food-effect on propranolol BA and advises patients to take the drug product (at bedtime) consistently on either an empty stomach or with food (Reliant, InnoPran® XL label 2010).

It is important to understand that for BCS class 1 drugs, coadministration with a meal can and does affect the rate of absorption by delaying the gastric emptying time or delaying disintegration. For instance, drug substances which are absorbed in the small intestine may show an increase in T_{max} and a decrease in C_{max}. The contents of the meal can change this effect since gastric emptying can vary based on meal content. Often, the extent of absorption is not altered, i.e., the area under the curve (AUC) remains similar to the AUC in the fasted state.

Of course, there are always exceptions to the rule. Even for BCS class 1 drugs, it is possible to manufacture two immediate-release drug products that are not bioequivalent to each other when coadministered with food (Dressman et al. 2001). This can happen if excipients are added to one of the formulations (but not the other) which modifies the gastric emptying. For example, it has been shown that an acetaminophen formulation containing sodium bicarbonate is not bioequivalent to an acetaminophen formulation without bicarbonate under fed conditions, despite comparable dissolution rates (Grattan et al. 2000). It is for this reason that the FDA requires that the excipients must be shown not to affect absorption of the active ingredient in order for a BCS class 1 biowaiver to be approved.

For the other BCS class drugs (2–4) which are formulated as immediate-release drug products, and for all modified-release drug products (e.g., extended-release formulations, including BCS class 1 drug substances), any food-effects are most likely to result from a complex combination of factors that may influence dissolution and absorption of the drug substance. In these cases, the effect of food on absorption is difficult to predict. That is why for these products, the FDA recommends a fed BE or food-effect BA study.

4.3.2 Food and Drug Product: BA and BE Studies

Food-effects on the absorption of the drug substance should be distinguished from food-effects on drug absorption arising from interactions with the formulation (i.e., the excipients). It has already been alluded to that many excipients are used in order to get extended-release of the drug substance or protect the drug substance from the harsh environment within the GI tract, and these excipients' behavior depend on the environmental conditions of the GI tract to function as designed (e.g., gastro-floating tablets). Just like with the drug substance, food can interact with the excipients and therefore impact BA or BE through its interaction with the excipients.

A classic example of a food-effect on the drug product formulation, which is a big concern with the FDA, is "dose-dumping". Dose-dumping is a term that describes the rapid release of the active ingredient from the dosage form into the GI tract (FDA 2005). An example of food-effects on dose-dumping is of the extended-release formulations of theophylline (Davit and Conner 2008). The extended-release theophylline product (Theo-24®) was approved by the FDA in 1983 (Weinberger 1984). At that time, the drug product was approved without conducting fed BA studies. Only fasting studies were conducted to characterize the pharmacokinetics of this drug product. Therefore, any potential food-effect was unknown, yet the product was on the market. In 1985, Hendeles et al. published a manuscript showing that the theophylline C_{max} significantly increased when dosed after a high-fat meal (Hendeles et al. 1985). A similar study was conducted by Karim et al., and they reported similar findings when the extended-release theophylline tablet product Uniphyl® (discontinued) was given to healthy male subjects after a high-fat meal (Karim et al. 1985). Gai et al. compared the effects of fasting versus "normal", high-fat and high-fat/high-protein meals on theophylline rate (C_{max}) and extent (AUC) of absorption from two different extended-release tablet formulations, one based on a hydrophilic matrix, the other based on lipid matrix (Gai et al. 1997). Compared with fasting, any class of meal given with the hydrophilic matrix tablet produced a higher theophylline C_{max} but not AUC. By contrast, when given with the lipid matrix tablet, the high-fat and high-fat/high-protein meals increased both the theophylline AUC and C_{max}, whereas AUC and C_{max} following a normal meal were comparable to values in fasting subjects. The authors suggested that, for the hydrophilic matrix tablet, food increased the rate but not extent of theophylline absorption due to the delay in gastric emptying, whereas, for the lipid matrix tablet, the surface-active effect of bile salts together with erosion promoted by lipase action were responsible for the increases in both the rate and extent of theophylline absorption (Gai et al. 1997; summarized in Davit and Conner 2008).

Theophylline has a narrow therapeutic window. At the time of these studies, theophylline was an important drug product used to treat asthma (today, there are several alternatives available). Due to the narrow therapeutic window, any fluctuations in drug absorption and BA might impact the safety and/or efficacy profile. Yet, food caused significant differences in the absorption of theophylline. These findings highlight the importance of modified-release dosage forms releasing drug predictably as intended regardless if it is dosed with a meal. Or the label should clearly state that it should not be dosed after a meal.

Occasionally, the effect of food on BA and the possibility of "dose-dumping" can be predicted based on differences in in vitro dissolution results. For instance, Schug et al. observed formulation-dependent food-effects on relative BA from two extended-release nifedipine products approved for marketing in the European Union (EU; Schug et al. 2002a, 2002b). The two formulations had different dissolution properties in vitro which at least partly explained the in vivo differences when dosed with food (Grundy and Foster 1996).

However, often the in vitro dissolution results are not predictive. For instance, food had a pronounced effect on nifedipine BA from Nifedicron (another formulation of nifedipine), compared with fasted conditions, resulting in a pronounced increase in C_{max} values. However, the dissolution of Nifedicron was comparable to that of the nifedipine formulation which exhibited minimal food-effect on BA.

Thus, the observed potential for dose-dumping in the presence of food was not anticipated (Schug et al. 2002a, 2002b; Davit and Conner 2008).

These studies underscore the safety concerns regarding the potential for dose-dumping in similarly formulated drug products, and underscore that in vitro testing should not be used for modified-release drug products as a surrogate for in vivo testing. Because the result of the interaction between the food and the system used to control the liberation of the drug is difficult to predict, it becomes essential to test possible interactions with each new formulation. This leads us to the FDA's current thinking with regard to in vivo food-effect BA studies and fed BE studies.

4.3.3   FDA Recommendations for NDAs

The FDA's Guidance for industry: food effect bioavailability and fed bioequivalence studies document (2002) outlines the recommendations which are currently practiced regarding when and how industry should conduct these food-effect studies.

The FDA recommends NDA sponsors to conduct their food-effect BA studies early in the development process (i.e., during the IND period). The reasons for this are simple. First, knowing any potential effects of food on the BA of a drug can help direct formulation scientists toward the most safe and efficacious formulation possible. For instance, if it is determined that there is a food–drug interaction which occurs in the low pH environment of the stomach, then it would make sense for the formulators to add an enteric coating to protect the drug from this environment. Second, the food-effect studies should be conducted early so they can be used to steer the design of later clinical safety and efficacy studies. Third, the food-effect BA studies should be incorporated into the product labeling specifically as it pertains to dosing and administration.

The NDA sponsors should conduct food-effect BA studies for all new chemical entities (NCEs). A NCE means a drug that contains no active moiety that has been approved by FDA in any other application submitted under section 505(b) of the Federal Food, Drug and Cosmetic Act. Even if the active moiety is not an NCE, the FDA recommends food-effect BA studies conducted on all drugs formulated as modified-release drug products for reasons we just discussed. These studies should be designed to compare the BA when dosed in the fasting state versus when dosed after a standardized high-fat, high-calorie meal. If changes in formulation components or method of manufacture occur after these fed-study BA studies are carried out, then new fed studies are recommended to determine the impact of the formulation and/or manufacturing procedural changes on the BA of the drug substance. Due to the ever-growing complexities of modified-release formulations and the potential for dramatic food-effects on BA, these food-effect studies are recommended to ensure the safety and efficacy of drug products entering the US market.

4.3.4   FDA Recommendations for ANDAs

These potential formulation–dependent food interactions have implications for the generic drug industry. Under the FDA's regulations, inactive ingredients in a generic solid oral dosage form drug product can differ from the inactive ingredients used in the corresponding innovator drug product (or reference listed drug product) [21 CFR Section 320.1(c)]. In addition, generic modified-release products may be formulated with a different-release mechanism than their corresponding reference products (Pfizer v. Shalala 1998). Thus, because (1) each generic modified-release drug product can have a different formulation and release mechanism than its corresponding reference drug product and (2) the relative direction and magnitude of food-effects on modified-release formulations may be difficult to predict; the FDA asks generic drug applicants to conduct studies comparing the BE of the generic and corresponding reference drug products under fed conditions (Davit and Conner 2008). Such studies are called fed BE studies.

The FDA currently recommends a fed BE study for all orally administered drug products submitted as an ANDA, with only a few specific exceptions. This recommendation includes immediate-release and modified-release drug products, for reasons already discussed. The first exception we already addressed. This is when the drug substance is considered a BCS class 1 drug, and dissolution, solubility, and permeability data support a biowaiver of in vivo BE testing (for both fasting and fed BE studies). Remember that the excipients used cannot affect the rate and extent of absorption.

The second exception is when the reference listed drug (RLD) product's labeling (usually in the DOSAGE AND ADMINISTRATION section) clearly states that the product should be taken on an empty stomach. The RLD labeling will include such a comment when the NDA sponsor conducted a food-effect BA study and determined that a clinically relevant food effect occurred when dosed after a meal. Such an effect might be a decrease in BA below what is thought as the therapeutic window, or an increase in BA to potentially harmful systemic levels. Since the label clearly states do not take with food, a BA study conducted in subjects in the fed state is not necessary and could put the volunteers at risk. An example of this is represented in the package insert for Mycophenolic Acid Delayed-Release Tablets (Myfortic® labeling 2004). The insert states, "Compared to the fasting state, administration of Myfortic® 720 mg with a high-fat meal (55 g fat, 1,000 calories) had no effect on the systemic exposure (AUC) of mycophenolic acid (MPA). However, there was a 33 % decrease in the maximal concentration (C_{max}), a 3.5- hour delay in the T_{lag} (range, −6 to 18 h), and 5.0-hour delay in the T_{max} (range, −9 to 20 h) of MPA. To avoid the variability in MPA absorption between doses, Myfortic® should be taken on an empty stomach." The FDA does not recommend that sponsors conduct fed BE studies for generic products referencing Myfortic®.

Sometimes, a specific study population has difficulty in successfully ingesting a high-fat meal. For instance, certain drug products are not safe to use in healthy volunteers. Therefore, the recommended BE studies are actually conducted in patients, e.g., cancer patients. Due to the health of these patients, it may not be feasible that they eat a high-fat, high-calorie meal. In this case, the FDA recommends an alternative study design for the safety and well-being of the study population, due to meal tolerance issues. For instance, regarding Imatinib Mesylate Tablets, the FDA's Office of Generic Drugs (OGD) currently recommends only a fed study conducted in cancer patients already receiving a stable dose of the drug product. Since the patients may not be able to tolerate a high-fat meal, OGD recommends using a light, low-calorie and low-fat breakfast for the meal when conducting the fed BE study.

Conversely, the patients' health status may prevent the ability to fast for any prolonged duration of time, making a recommended fasting BE study difficult to complete. In these circumstances, the sponsor may provide a nonhigh-fat diet to the patients during the proposed study, provided that both study periods are conducted under the same conditions. The FDA recently made this type of study recommendation for sponsors proposing a fasting BE study in cancer patients dosed with Paclitaxel Suspension (injectable).

4.4   Study Considerations

Previous book chapters discuss in detail BA or BE study designs, which are recommended by the FDA. Of course, the traditional study design recommended by the FDA is a randomized, balanced, single-dose, two-treatment, two-period, two-sequence crossover design for studying the effects of food on the absorption of a particular drug product. For NDAs, where the sponsor is trying to determine the food-effects (if any), the treatments will be given under the fasted state in one period and under the fed state in the other period. For ANDAs, where the generic applicant wants to determine BE of its test product to the RLD under the fed state, both periods are under the fed condition, and in one period the subjects are dosed with test (generic), whereas in the other period they are dosed with the RLD.

Alternatives to the traditional two-way crossover can be used, especially if certain properties of the drug product make it more difficult to take the traditional approach. For instance, extremely long half-life drug products might benefit from carrying out a parallel study in place of a crossover study. Drug products showing high within-subject variability might benefit from conducting a partial or full replicate study in order to use the reference-scaled average BE approach (see Chap. 6). Often the applicant benefits from the submission of a study protocol to the FDA when the study design differs from the traditional approach.

4.4.1   Study Population

Unless there is a safety concern, the food-effect BA studies (NDAs) and fed BE studies (ANDAs) should be conducted using healthy volunteers drawn from the general population. Male and female volunteers should be equally enrolled unless there are specific safety concerns for one sex over the other (e.g., teratogenic effects), or the product is intended only for a single sex (e.g., females for oral contraceptives). The FDA recommends for drug products used predominantly in the older population that the sponsor include as many subjects older than 60 years of age as possible. A sufficient number of subjects should be enrolled to adequately power the study, but it is not expected that there will be sufficient power to draw conclusions for each subgroup. The FDA guidance sets a limit to the minimal number of subjects enrolled as 12 subjects.

4.4.2   Dosage Strength

The highest strength of the drug product should be dosed in the food-effect BA and fed BE studies, unless safety concerns warrant the use of a lower strength. Sometimes, due to the lack of an available sensitive analytical method, the dose is increased to more than one unit (e.g., $2\times$ or $3\times$). This is acceptable as long as the total dose does not exceed that which has been shown to be safe and is listed in the labeling as the maximum daily dose (MDD). Any dose higher than this will require an IND/protocol submission to the Agency for approval prior to conducting the study. For ANDA submissions, the applicant generally only needs to conduct the fed BE study on the highest strength. The lower strengths are often eligible for a biowaiver of in vivo BE testing (see Chap. 5 for details).

4.4.3   Meal Composition

According to the FDA's Guidance for Industry: Food Effect BA and Fed BE Studies document (December 2002), the food-effect BA and BE studies should be carried out using subjects who are dosed after eating a meal which provides the greatest effects on the GI physiology so that systemic drug availability is maximally affected. The meal should be standardized, and approximately 50 % of the total calories should come from fat. The overall caloric breakdown should be 150, 250, and 500–600 cal from protein, carbohydrate, and fat, respectively. Thus, the FDA's recommended meal includes around 1,000 cal, approximately half of the average adult person's total daily recommended caloric intake (2,000–2,500 cal).

   The guidance gives an example meal which includes two eggs fried in butter, two strips of bacon, two slices of toast with butter, four ounces of hash brown potatoes, and eight ounces of whole milk. Substitutions to this meal can be made as long as the caloric breakdown is comparable, and the meal volume and viscosity is comparable. While protein from an animal source is preferred, the FDA has accepted fed BA and BE studies where the volunteers were fed a vegetarian diet, so long as the caloric breakdown is the same as that described above, and the meal volume and viscosity is comparable. Perhaps, the most important thing to remember is that these meals should be standardized across the study and every subject must eat the entire meal as planned in the study protocol.

Sponsors of New Drug Applications (NDAs) can conduct additional fed BA studies using meals with different caloric breakdown than the high-fat, high-calorie meal. These studies are useful in exploring the mechanisms underlying a food–drug effect, and the information gained from these additional studies can be included on the product labeling. However, these studies are in addition to the fed study conducted in healthy volunteers fed a high-fat, high-calorie meal, not in place of this study. ANDAs generally only need to conduct a single fed BE study using the high-fat, high-calorie meal, unless the innovator's labeling dictates otherwise.

Mirabegron is a β3-adrenoceptor agonist indicated for the treatment of overactive bladder. The mirabegron drug product currently marketed in the USA is formulated as an extended-release tablet, which utilizes a hydrophilic gel-forming matrix to control release of the active drug substance along the GI tract (Myrbetriq[®] labeling 2012). Scientific evidence suggests mirabegron is a substrate for the efflux transporter P-glycoprotein (P-gp). Hence, BA increases from approximately 30 % for the 25 mg tablet to 45 % for the 100 mg tablet possibly due to saturation of P-gp (Lee et al. 2013).

In order to comply with the FDA, the innovator company conducted a fed in vivo BA study to support dosing recommendations in the labeling (Lee et al. 2013). The study was designed as a single-dose, randomized, open-label, crossover study in healthy adult subjects in the fasted state, or after a high- or low-fat breakfast. The high-fat breakfast consisted of two eggs, four sausage links, one slice of wheat bread with butter, hash brown potatoes with ketchup, cantaloupe, semi-skim milk, and orange juice (and met the FDA-recommended caloric breakdown for a high-fat, high-calorie meal). The low-fat breakfast consisted of cereal, two slices of whole wheat bread, ham, ketchup, and semi-skim milk. Subjects were dosed 50 or 100 mg tablets. The primary endpoints for determining food-effects were C_{max} and AUC (i.e., rate and extent of absorption).

Based on the results of the study, mirabegon BA was reduced when dosed after a meal, and the reduction was dependent on meal consumption. There was a greater reduction in plasma levels after the low-fat breakfast when compared to the high-fat breakfast. While the exact mechanism is unknown, the scientists who carried out this study hypothesized that the food-effects might be attributable to more efficient intestinal efflux under the fed state, which would limit the overall absorption of the drug. Since food delays gastric emptying, it might be that the P-gp is less likely to reach saturation when compared to the fasting state. Likewise, the differences seen between the high- and low-fat breakfasts might also be explained by P-gp efflux. Lipids have been reported to inhibit P-gp. Therefore, the authors postulate that at least part of the difference seen between the two fed states might be due to lipid inhibition of P-gp efflux. It is also possible that there are direct interactions with the food components, and this might be a part of the reason for the differences seen between the two meals.

As stated earlier, the FDA's guidance on food-effect BA and BE studies recommends eating a meal which provides the greatest effects on the GI physiology so that systemic drug availability is maximally affected. According to the guidance, a high-fat, high-calorie meal should provide the greatest effects. However, this case study shows that the high-fat, high-calorie meal is not always the meal which produces the greatest food-effect on drug absorption. That is, while the high-fat, high-calorie meal does probably cause the greatest effects on GI physiological processes, this does not always correlate with the greatest food-effect on the absorption of a drug. The reason why the FDA recommends the high-fat, high-calorie meal was represented in the previous theophylline example. The high-fat meal is the most stressful for certain types of formulations that can dose-dump.

4.4.4   Other Study Considerations

The Food-Effect BA and Fed BE Study Guidance clearly outlines additional considerations when planning a food-effect or fed BE study. The fasting-state and fed state should be standardized for these BA and BE studies. That is, all subjects should fast for the same amount of time (at least 10 h) prior to dosing in the fasting period of a food-effect BA study, and these subjects should fast for the same amount of time prior to eating a meal in the fed period of the same study. Likewise, for a fed BE study in both periods the subject should fast the same time prior to ingesting the meal. The meal should start 30 min prior to dosing, and the meal should be completed within that 30 min window. The drug products should be given with 240 mL of water. The subjects should not eat until 4 h after dosing. All other meals given during housing should also be standardized.

As already discussed, the effect of food in the BA study will be determined by comparing the rate and extent (C_{max} and AUC) of absorption when dosed without food to the rate and extent of absorption when dosed with a high-fat, high-calorie meal. Using the statistical approach described in Chap. 3, an absence of a food-effect is concluded when the 90 % confidence intervals for the ratio of population geometric means between the fed and fasted treatments (log-transformed data) is contained within the limits of 80–125 %. If not, then the sponsor should provide clinical relevance and add a description of food-effect in the labeling. A similar statistical approach is used in the fed BE studies submitted in support of ANDAs; however, now the study is comparing test versus reference, not fasted versus fed. Although no statistical criterion applies to T_{max}, the FDA's Guidance for Industry: Food Effect BA and Fed BE Studies document (2002) states that the FDA expects the T_{max} values for the test and reference products to be comparable based on clinical relevance.

In order to adequately capture C_{max}, sponsors must include adequate sampling time points in a food-effect BA study or fed BE study. It is sometimes difficult to predict what food will do to the absorption and BA of a particular product. For reasons already discussed, the time to reach C_{max} (T_{max}) may increase or decrease when dosed after a meal. Since these studies are not continuous sampling, it is essential to add enough time points to accurately capture C_{max} (and T_{max}). If insufficient sampling time points were used, then the FDA will not accept the study.

4.4.5 Sprinkle Study

In NDAs submitted to the FDA, the labeling of some drug products (e.g., controlled-release capsules containing beads) recommends that the product can be sprinkled on soft food (e.g., applesauce) and swallowed without chewing. If a similar comment is included in the labeling, then the innovator conducted a bridging BE study showing that the product when dosed sprinkled on a spoonful of soft food is bioequivalent to the same product swallowed whole. The labeling will specify which soft food was used in the study so that patients administer the drug on the same soft food.

For instance, the labeling of Nexium® (esomeprazole magnesium) delayed-release capsules states, "for patients who have difficulty swallowing capsules, one tablespoon of applesauce can be added to an empty bowl and the NEXIUM Delayed-Release Capsule can be opened, and the granules inside the capsule carefully emptied onto the applesauce. The granules should be mixed with the applesauce and then swallowed immediately: do not store for future use. The applesauce used should not be hot and should be soft enough to be swallowed without chewing. The granules should not be chewed or crushed. If the granules/applesauce mixture is not used in its entirety, the remaining mixture should be discarded immediately" (Nexium® labeling 2012).

Since generic drug products must show BE to the reference listed drug (RLD) products, and since the labeling is the same as the RLD, generic firms are recommended to conduct an additional BE study comparing the test drug product sprinkled on soft food to the RLD sprinkled on the same soft food under the fasting state (i.e., the only food ingested is the soft food used in the dosing). The data should be analyzed using the same rigorous statistical analysis as the other pivotal fasting and fed in vivo BE testing, i.e., using average BE approach and 90 % confidence interval in order to deem bioequivalent to the RLD.

Generic firms should keep in mind that bead size is of importance in developing a product in which the RLD labeling contains a sprinkle comment in the labeling. The beads should not be large enough to stimulate the urge to chew. Actually, the FDA published a draft guidance in 2012 (Guidance for industry: size of beads in drug products labeled for sprinkle, 2012), which gave specific recommendations on the size of beads allowed for these types of drug products. Based on chewing and swallowing particle size data in the literature, and on Agency experience with

NDAs and ANDAs, the FDA determined that the appropriate maximal bead size is 2.8 mm (10 % variation of the target, 2.5 mm). In addition, this guidance discusses recommended studies for drug products, which can be administered through a feeding tube.

4.5 Chapter Summary

The FDA's primary mission as it relates to the pharmaceutical industry is to provide safe and effective drug products (both innovator and generics) to the citizens of the USA. Traditionally, oral drug products were dosed at mealtimes. This was convenient for patients and helped in compliance. However, we now know that food can affect drug absorption and BA in a variety of ways, and the outcome depends on the drug substance (BCS class), drug product (i.e., the complete formulation including excipients), and the GI physiology. Now, among the many factors that the FDA examines to ensure that drug products are safe and effective is the effect of food on the BA of the active drug from the drug product. The FDA recommends food-effect BA studies for NDA submissions, which then are used in setting dosing recommendations that are included in the product labeling. Likewise, ANDA applicants are recommended to conduct fed BE studies showing that the test product is bioequivalent to the reference product when dosed after a high-fat meal. These studies are steps that the FDA has taken to ensure the safety and efficacy of the drug products it regulates.

These studies are considered the gold standard in determining potential effects of food on drug absorption, BA and BE. However, considerable research is looking at new, less costly approaches to predict food-effects based on the physicochemical properties of the drug substance, the drug product properties, and the modeled GI system. These in vitro approaches possibly will lead us to a further understanding of the complex effects of food on absorption and BA of solid oral dosing products. The goal of this modeling is to be able to make reliable, qualitative predictions during the preclinical phase of development based on biopharmaceutical properties (Jones et al. 2006; Dressman et al. 2007; Lentz 2008; Parrott et al. 2009; Klein 2010; Mathias and Crison 2012). The ultimate objective of the FDA regarding these food-effect studies is to provide a complete picture with regard to dosing recommendations which then can be conveyed to the prescribing physicians and patients alike through the labeling.

References

Adkin DA, Davis SS, Sparrow RA, Huckle PD, Phillips AJ, Wilding IR (1995) The effects of pharmaceutical excipients on small intestinal transit. Br J Clin Pharmacol 39(4):381–387

Amidon GL, Lennernas H, Shah VP, Crison JR (1995) A theoretical basis for a biopharmaceutic drug classification: the correlation of in vitro drug product dissolution and in vivo bioavailability. Pharm Res 12:413–420

Astellas Pharma US, Inc (2012) Myrbetriq® Tablets package insert

Astrazeneca Pharmaceuticals LP, Wilmington DE (2012) Tenormin® Tablets package insert

AstraZeneca Pharmaceuticals LP, Wilmington DE 19850 (2012) Nexium® Delayed-Release Capsules package insert

Barnwell SG, Laudanski T, Dwyer M, Story MJ, Guard P, Cole S, Attwood D (1993) Reduced bioavailability of atenolol in man: the role of bile acids. Int J Pharm 89(3):245–250

Bayer Pharmaceuticals Corporation, 400 Morgan Lane, West Haven CT 06516 (2005) Cipro® Tablets and Cipro® Oral Suspension package insert

Bayer Pharmaceuticals Corporation, 400 Morgan Lane, West Haven CT 06516 (2005) Cipro® XL Tablets package insert

Birkebaek NH, Memmert K, Mortensen J, Dirksen H, Christensen MF (1990) Fractional gastro-intestinal transit time: intra- and interindividual variation. Nucl Med Commun 11(3):247–252

Bristol-Myers Squibb Virology, Bristol-Myers Company, Princeton, NJ 08543 (2006) Videx® Buffered Tablets package insert

Bristol-Myers Squibb Virology, Bristol-Myers Company, Princeton, NJ 08543 (2011) Videx® EC package insert

Camilleri M, Malagelada JR, Brown ML, Becker G, Zinsmeister AR (1985) Relation between antral motility and gastric emptying of solids and liquids in humans. Am J Physiol 249(5 Pt 1): G580–G585

Charman WN, Porter CJH, Mithani S, Dressman JB (1997) Physicochemical and physiological mechanisms for the effects of food on drug absorption: the role of lipids and pH. J Pharm Sci 86 (3):269–282

Collins PJ, Horowitz M, Maddox A, Myers JC, Chatterton BE (1996) Effects of increasing solid component size of a mixed solid/liquid meal on solid and liquid gastric emptying. Am J Physiol 271(4 Pt 1):G549–G554

Daneshmend TK, Roberts CJC (1982) The influence of food on the oral and intravenous pharma-cokinetics of a high clearance drug: a study with labetalol. Br J Clin Pharmacol 14:73–78

Davit BM, Conner DP (2008) Food effects on drug bioavailability: implications for new and generic drug development. Biopharmaceutics Applications in Drug Development pp. 317–335

Deferme S, Augustijns P (2003) The effect of food components on the absorption of P-gp substrates: a review. J Pharm Pharmacol 55:153–162

Dressman JB, Berardi RR, Dermentzoglou LC, Russell TL, Schmaltz SP, Barnett JL, Jarvenpaa KM (1990) Upper gastrointestinal (GI) pH in young, healthy men and women. Pharm Res 7 (7):756–761

Dressman J, Butler J, Hempenstall J, Reppas C (2001) The BCS: where do we go from here? Pharm Technol 25:68–76

Dressman JB, Vertzoni M, Goumas K, Reppas C (2007) Estimating drug solubility in the gastrointestinal tract. Adv Drug Deliv Rev 59(7):591–602

Evans DF, Pye G, Bramley R, Clark AG, Dyson TJ, Hardcastle JD (1988) Measurement of gastrointestinal pH profiles in normal ambulant human subjects. Gut 29(8):1035–1041

FDA News (2005) FDA asks Purdue Pharma to withdraw Palladone® for safety reasons (July 12, 2005) http://www.fda.gov/NewsEvents/Newsroom/PressAnnouncements/2005/ucm108460.htm

Fleisher D, Li C, Zhou Y, Pao L, Karim A (1999) Drug, meal and formulation interactions influencing drug absorption after oral administration: clinical implications. Clin Pharmacokinet 36:233–254

Gai MN, Isla A, Andonaeugui MT, Thielemann AM, Seitz C (1997) Evaluation of the effect of three different diets on the bioavailability of two sustained release theophylline matrix tablets. Int J Clin Pharm Ther 35:565–571

Grattan T, Hickman R, Darby-Dowman A, Hayward M, Boyce M, Warrington S (2000) A five way crossover human volunteer study to compare the pharmacokinetics of paracetamol following oral administration of two commercially available paracetamol tablets and three development tablets containing paracetamol in combination with sodium bicarbonate or calcium carbonate. Eur J Pharm Biopharm 49(3):225–229

Grundy JS, Foster RT (1996) The nifedipine gastrointestinal therapeutic system (GITS). Evaluation of pharmaceutical, pharmacokinetic and pharmacologic properties. Clin Pharmacokinet 30:28–51

Hendeles L, Weingerger M, Milavetz G, Hill M, Vaughan L (1985) Food-induced "dose-dumping" from a once-a-day theophylline product as a cause of theophylline toxicity. Chest 87:758–765

Hörter D, Dressman JB (2001) Influence of physicochemical properties on dissolution of drugs in the gastrointestinal tract. Adv Drug Deliv Rev 46(1–3):75–87

Huupponen R, Seppälä P, Iisalo E (1984) Effect of guar gum, a fibre preparation, on digoxin and penicillin absorption in man. Eur J Clin Pharmacol 26(2):279–281

Jones HM, Parrott N, Ohlenbusch G, Lavé T (2006) Predicting pharmacokinetic food effects using biorelevant solubility media and physiologically based modelling. Clin Pharmacokinet 45 (12):1213–1226

Karim A, Burns T, Wearley L, Streicher J, Palmer M (1985) Food-induced changes in theophylline absorption from controlled-release formulations. I. Substantial increased and decreased absorption with Uniphyl tablets and Theo-Dur Sprinkle. Clin Pharmacol Ther 38:77–83

Klein S (2010) The use of biorelevant dissolution media to forecast the in vivo performance of a drug. AAPS J 12(3):397–406

Lee J, Zhang W, Moy S, Kowalski D, Kerbusch V, van Gelderen M, Sawamoto T, Grunenberg N, Keirns J (2013) Effects of food intake on the pharmacokinetic properties of mirabegron oral controlled-absorption system: a single-dose, randomized, crossover study in healthy adults. Clin Ther 35(3):333–341

Lentz KA (2008) Current methods for predicting human food effect. AAPS J 10(2):282–288

Maka DA, Murphy LK (2000) Drug-nutrient interactions: a review. AACN Clin Issues 11:580–589

Malagelada JR, Longstreth GF, Summerskill WH, Go VL (1976) Measurement of gastric functions during digestion of ordinary solid meals in man. Gastroenterology 70(2):203–210

Malagelada JR, Longstreth GF, Deering TB, Summerskill WH, Go VL (1977) Gastric secretion and emptying after ordinary meals in duodenal ulcer. Gastroenterology 73(5):989–994

Martinez MN, Amidon GL (2002) A mechanistic approach to understanding the factors affecting drug absorption: a review of fundamentals. J Clin Pharmacol 42:620–643

Mathias NR, Crison J (2012) The use of modeling tools to drive efficient oral product design. AAPS J 14(3):591–600

McLean AJ, McNamara PJ, du Souich P, Gibaldi M, Laika D (1978) Food, splanchnic blood flow, and bioavailability of drugs subject to first-pass metabolism. Clin Pharmacol Ther 24:5–10

Melander A, McLean A (1983) Influence of food intake on presystemic clearance of drugs. Clin Pharmacokinet 8(4):286–296

Melander A, Danielson K, Schersten B, Wahlin E (1977) Enhancement of the bioavailability of propranolol and metoprolol by food. Clin Pharmacol Ther 22(1):108–112

Meyer JH, Dressman J, Fink A, Amidon G (1985) Effect of size and density on canine gastric emptying of nondigestible solids. Gastroenterology 89(4):805–813

Neuhofel AL, Wilton JH, Victory JM, Hejmanowski LG, Amsden GW (2002) Lack of bioequivalence of ciprofloxacin when administered with calcium-fortified orange juice: a new twist on an old interaction. J Clin Pharmacol 42:461–466

Novartis Pharmaceuticals Corporation (2004) Myfortic® (mycophenolic acid) Delayed Release Tablet package insert

Parrott N, Lukacova V, Fraczkiewicz G, Bolger MB (2009) Predicting pharmacokinetics of drugs using physiologically based modeling—application to food effects. AAPS J 11(1):45–53

Pfizer, Inc. v. Shalala, 1 F.Supp.2d 38 (D.D.C. 1998)

Procter & Gamble Pharmaceuticals, Inc., Cincinnati, OH 45202 (2009) Macrobid® Capsules package insert

Purdue Pharmaceutical Products L.P., Stamford, CT 09601-3431 (2004) Uniphyl® Tablets package insert

Reliant Pharmaceuticals, Inc., Liberty Corner, NJ 07938 (2005) InnoPran® XL package insert

Russell TL, Berardi RR, Barnett JL, Dermentzoglou LC, Jarvenpaa KM, Schmaltz SP, Dressman JB (1993) Upper gastrointestinal pH in seventy-nine healthy, elderly, North American men and women. Pharm Res 10(2):187–196

Schug BS, Brendel E, Chantraine E, Wolf D, Martin W, Schall R, Blume HH (2002a) The effect of food on the pharmacokinetics of nifedipine in two slow release formulations: pronounced lag-time after a high fat breakfast. J Clin Pharmacol 53:582–588

Schug BS, Brendel E, Wonnemann M, Wolf D, Wargenau M, Dikngler A, Blume HH (2002b) Dosage form-related food interaction observed in a marketed once-daily nifedipine formulation after a high-fat American breakfast. Eur J Clin Pharmacol 58:119–125

Semple HA, Fangming X (1995) Interaction between propranolol and amino acids in the single-pass isolated, perfused rat liver. Drug Metab Dispos 23:794–798

Shionogi Inc (2013) Keflex® Capsules package insert

Tam YK (1993) Individual variation in first-pass metabolism. Clin Pharmacokinet 25:300–328

Toothaker RD, Welling PG (1980) The effect of food on drug bioavailability. Annu Rev Pharmacol Toxicol 20:173–199

U.S. Code of Federal Regulations, Title 21, Part 320—Bioavailability and Bioequivalence Requirements, Subpart A—General Provisions, Section 320.1 (2006) Definitions. U.S. Government Printing Office, Washington (April 2011), p 190

U.S. Department of Health and Human Services, Food and Drug Administration, Center for Drug Evaluation and Research (2000) Waiver of in vivo bioavailability and bioequivalence studies for immediate-release solid oral dosage forms based on a biopharmaceutics classification system (August 31, 2000)

U.S. Department of Health and Human Services, Food and Drug Administration, Center for Drug Evaluation and Research (2003) Guidance for industry: bioavailability and bioequivalence studies for orally administered drug products—general considerations (March 19, 2003)

U.S. Department of Health and Human Services, Food and Drug Administration, Center for Drug Evaluation and Research (2003) Guidance for industry: food-effect bioavailability and fed bioequivalence studies (2002)

U.S. Department of Health and Human Services, Food and Drug Administration, Center for Drug Evaluation and Research (2012) Guidance for industry: size of beads in drug products labeled for sprinkle (May 2012)

Weinberger MM (1984) Theophylline QID, BID and now QD? A report on 24-hour dosing with slow-release theophylline formulations with emphasis on analysis of data used to obtain Food and Drug approval for Theo-24. Pharmacotherapy 4:181–198

Welling PG (1996) Effects of food on drug absorption. Annu Rev Nutr 16:383–415

Yin L, Qin C, Chen K, Zhu C, Cao H, Zhou J, He W, Zhang Q (2013) Gastro-floating tablets of cephalexin: preparation and in vitro/in vivo evaluation. Int J Pharm 452(1–2):241–248

Yu LX, Lipka E, Crison JR, Amidon GL (1996) Transport approaches to the biopharmaceutical design of oral drug delivery systems: prediction of intestinal absorption. Adv Drug Deliv Rev 19(3):359–376

Yu LX, Straughn AB, Faustino PJ, Yang Y, Parekh A, Ciavarella AB, Asafu-Adjaye E, Mehta MU, Conner DP, Lesko LJ, Hussain AS (2004) The effect of food on the relative bioavailability of rapidly dissolving immediate-release solid oral products containing highly soluble drugs. Mol Pharm 1(5):357–362

Yuen KH (2010) The transit of dosage forms through the small intestine. Int J Pharm 395 (1–2):9–16

Chapter 5
Biowaiver and Biopharmaceutics Classification System

Ramana S. Uppoor, Jayabharathi Vaidyanathan, Mehul Mehta, and Lawrence X. Yu

5.1 Introduction

Bioavailability and bioequivalence are essential features that need to be assessed for a drug product, to evaluate the rate and extent to which the active ingredient or active moiety is absorbed from a drug product and becomes available at the site of action. CFR 320.22 gives FDA the authority under certain circumstances to waive the requirements for determining the in vivo bioavailability and bioequivalence (also called as biowaivers). In this chapter, we focus on the following biowaivers:

1. For drug products where bioavailability is self-evident, e.g., solutions.
2. Biowaivers in situations where one can rely on in vitro methods instead of in vivo for assessing bioavailability.
3. Biowaivers based on Biopharmaceutics Classification System (BCS).

5.2 General Biowaiver Considerations

5.2.1 For Certain Drug Products, the In Vivo Bioavailability or Bioequivalence of the Drug Product May Be Self-Evident (21CFR 320.22b)

As mentioned in the CFR, FDA waives the requirement for the in vivo bioavailability/bioequivalence studies when the bioavailability is considered self-evident based on data in the application if the product meets one of the following criteria:

R.S. Uppoor (✉) • J. Vaidyanathan • M. Mehta • L.X. Yu
Center for Drug Evaluation and Research, U.S. Food and Drug Administration,
10903 New Hampshire Avenue, Silver Spring, MD 20993, USA
e-mail: Ramana.Uppoor@fda.hhs.gov

L.X. Yu and B.V. Li (eds.), *FDA Bioequivalence Standards*, AAPS Advances in the Pharmaceutical Sciences Series 13, DOI 10.1007/978-1-4939-1252-0_5, © The United States Government 2014

1. The drug product is a parenteral solution administered by injection or an ophthalmic or otic solution and contains the same active and inactive ingredients in the same concentration as another approved drug product.
2. The drug product is administered by inhalation as a gas and contains the active ingredient in the same dosage form as another approved product.
3. The drug product is a solution (for application to skin, oral solution, elixir, syrup, tincture, solution for aerosolization or nebulization, nasal solution, etc.) and

    (a) Contains an active drug ingredient in the same concentration and dosage form as another approved drug product and
    (b) Contains no inactive ingredient that may significantly affect absorption of the active drug ingredient (systemic or local depending on the intended site of action).

So, for oral solutions, since the bioavailability is self-evident, in vivo bioequivalence studies are generally not necessary, unless the new solution contains excipients that affect absorption of the active ingredient.

5.2.2   Excipient Effect on Bioavailability

Some excipients could significantly impact the absorption of systemically administered drugs. For example, xylitol, sorbitol, and mannitol are commonly used formulation excipients for drug products (Fassihi et al. 1991; Fukahori et al. 1998). They are commonly used as artificial sweeteners in the food industry. These excipients are not well absorbed in the gastrointestinal (GI) tract. However, they increase the osmotic pressure in the intestine, which changes the flux of water in the GI tract. This osmotic stress can change the gastric emptying and the intestinal transit times at both the upper and lower parts of the intestine. The total amount of drug absorbed depends on the rate of absorption from the intestine and the total time that the drug is present in the intestine. Changes in the transit times in the GI tract may impact drug absorption.

When transit or emptying times are decreased, there is less time available for drug molecules in solution to be absorbed and thus the total absorption may decrease. Scintigraphic evidence suggests that osmotic agents can have minor effects on the residence time in the upper intestinal tract but significantly reduce the residence time in the lower intestinal tract (Adkin et al. 1995; Kruger et al. 1992). Using the GI transit time data from two literature studies, Chen et al. calculated the osmotic potential and plotted it against the small intestinal transit time (percent control) (Chen et al. 2013). A linear relationship was found between the small intestinal transit time and osmotic potential for both mannitol and PEG 400.

The osmotic pressure changes may also affect the rate of transport across the intestinal wall in addition to changing transit times, which could lead to changes in

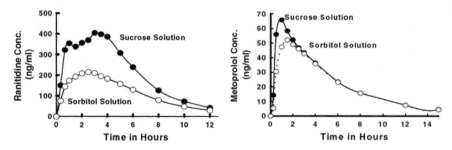

Fig. 5.1 Mean plasma concentrations of ranitidine (*left panel*) and metoprolol (*right panel*) in healthy volunteers after administration of 150 mg ranitidine solution and 50 mg metoprolol tartarate, respectively, with either 5 g of sorbitol (*open circles*) or 5 g of sucrose (*solid circles*). *Source*: Chen et al. (2007), pp. 75–76

absorption of low permeability drugs (Polli et al. 2004). As an example of this effect, Chen et al. (2007) studied the effect of two different sugars (sorbitol versus sucrose) on the bioavailability of ranitidine (low permeability) and metoprolol (high permeability). As shown in Fig. 5.1, the ranitidine C_{max} and AUC were significantly decreased in the presence of sorbitol as compared to sucrose. Similarly, for metoprolol, there was a reduction in C_{max} in the presence of sorbitol as compared to sucrose. Additionally, the T_{max} of metoprolol was delayed by ~30 min in the presence of sorbitol (Fig. 5.1).

In the same study, Chen et al. measured the pharmacokinetics of ranitidine in a four-way crossover study of ranitidine oral solution dosed with various amounts of sorbitol (Chen et al. 2007). The results indicated that doses of sorbitol greater than 1.25 g significantly reduced the bioavailability of ranitidine from an oral solution.

Chen et al (2013) also stated that mannitol exhibits similar osmotic effects as sorbitol and decreased the bioavailability of cimetidine, another low permeability drug substance (Chen et al. 2013). An apparent linear dose–response relationship was observed for different concentrations of sorbitol solutions with ranitidine. The mannitol/cimetidine data also were similar to that of sorbitol/ranitidine (Fig. 5.2). The authors concluded that better understanding of the dose–response relationship for such excipients on drug absorption and/or bioavailability will allow optimal use of these excipients during drug development. The authors also state that further research is needed to understand the mechanism of the effect of an excipient on drug absorption (Chen et al. 2013).

Another example of a pharmaceutical excipient with demonstrated effect on drug absorption is polyethylene glycol 400 (PEG 400). Several studies investigated the effect of PEG 400 on the absorption characteristics of ranitidine from the gastrointestinal tract (Basit et al. 2002; Schulze et al. 2003). These studies show that there is no significant effect of PEG 400 on gastric emptying; however, the presence of PEG 400 reduced the mean small intestinal transit times of the ranitidine solutions. This resulted in changes in drug absorption that depended upon the amount of PEG 400. Low concentrations of PEG 400 increased the

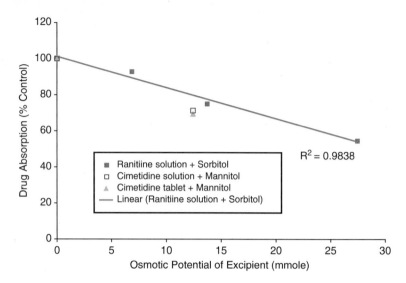

Fig. 5.2 Relationship between osmotic potential of sorbitol/mannitol and absorption of ranitidine/cimetidine compared to control. *Source*: Chen et al. (2013)

absorption of ranitidine, presumably due to changes in intestinal permeability of ranitidine, whereas high concentrations of PEG 400 reduced ranitidine absorption possibly due to shorter small intestinal transit time.

5.3 Biowaiver Based on Evidence Obtained In Vitro (21 CFR 320.22 (d))

For certain drug products, bioavailability may be measured or bioequivalence may be demonstrated by evidence obtained in vitro in lieu of in vivo data. Waiver of in vivo studies for different strengths of a drug product can be granted under 21 CFR 320.22(d)(2) when (1) the drug product is in the same dosage form but in a different strength; (2) this different strength is *proportionally similar* in its active and inactive ingredients to the strength of the product for which the same manufacturer has conducted an appropriate in vivo study; and (3) the new strength meets an appropriate in vitro dissolution test. The FDA guidance (2003) defines *proportionally similar* in the following ways:

- All active and inactive ingredients are in exactly the same proportion between different strengths (e.g., a tablet of 50-mg strength has all the inactive ingredients, exactly half that of a tablet of 100-mg strength and twice that of a tablet of 25-mg strength).
- Active and inactive ingredients are not in exactly the same proportion between different strengths as stated above, but the ratios of inactive ingredients to total

weight of the dosage form are within the limits defined by the SUPAC-IR (1995) and SUPAC-MR (1997) guidances up to and including Level II.

- For high-potency drug substances, where the amount of the active drug substance in the dosage form is relatively low, the total weight of the dosage form remains nearly the same for all strengths (within $\pm 10\,\%$ of the total weight of the strength on which a biostudy was performed), the same inactive ingredients are used for all strengths, and the change in any strength is obtained by altering the amount of the active ingredients and one or more of the inactive ingredients. The changes in the inactive ingredients are within the limits defined by the SUPAC-IR (1995) and SUPAC-MR (1997) guidances up to and including Level II.

Generally, the in vitro test used is a dissolution test. In vitro dissolution profiles need to be determined on at least 12 dosage units. Dissolution profiles of the test product should be compared to the reference product, using either model-independent or model-dependent approaches. The most common method used and recommended in the above guidances (1997) is using a model-independent approach, using similar factor f_2. An f_2 value between 50 and 100 suggests that the two dissolution profiles are similar.

$$f_2 = 50 \log\left\{ \left[1 + \frac{1}{n}\sum_{t=1}^{n}(R_t - T_t)^2 \right]^{-0.5} \times 100 \right\},$$

where $n =$ number of sampling time points, $R_t =$ dissolution at time point t of reference, $T_t =$ dissolution at time point t of test.

It is recommended that only one point past the plateau of the profiles be used in calculating the f value. Also, the average difference at any dissolution sampling time point should not be greater than 15 % between the test and reference.

- In addition, in cases where the SUPAC-IR or MR guidances recommend an in vivo study, in vitro assessments will be sufficient if there is a validated in vitro in vivo correlation established for the drug product.

5.4 BCS-Based Biowaiver

5.4.1 Biopharmaceutics Classification System

BCS is a scientific approach designed to predict drug absorption based on the aqueous solubility and intestinal permeation characteristics of the drug substance (Amidon et al. 1995). The BCS categorizes drug substances into one of the four BCS classes based on these characteristics:

Biopharmaceutics class	Solubility	Permeability
I	High	High
II	Low	High
III	High	Low
IV	Low	Low

Based on BCS, the FDA issued a guidance recommending biowaivers for high solubility and high permeability drugs provided that they are rapidly dissolving, stable in the gastrointestinal (GI) tract, and have a wide therapeutic index. The BCS-based biowaiver approach is used as a scientific tool to reduce unnecessary in vivo bioequivalence (BE) studies (2002). Within the BCS framework, when certain criteria (such as BCS Class I rapidly dissolving drug product) are met, this can be used as a drug development tool to help justify biowaiver requests. The underlying scientific basis is that observed in vivo differences in the rate and extent of absorption of a drug from two pharmaceutically equivalent solid oral products may be due to differences in drug dissolution in vivo. However, when the in vivo dissolution of an immediate release (IR) solid oral dosage form is rapid in relation to gastric emptying and the drug has high permeability, the rate and extent of drug absorption is unlikely to be dependent on drug dissolution and/or gastrointestinal transit time, and a biowaiver is appropriate.

Table 5.1 shows the current status regarding BCS-based biowaivers across various regulatory agencies.

5.4.2 Criteria for Determining BCS Class of a Drug Substance

5.4.2.1 Solubility

The solubility classification of a drug in the BCS is based on the highest strength in an IR product. A drug substance is considered *highly soluble* when the highest

Table 5.1 Current status of BCS-based biowaivers at Food and Drug Administration (FDA), European Medicines Agency (EMA), World Health Organization (WHO), and other regulatory agencies[a]

		FDA (2002)	Health Canada (2012)	EMA (2008)	WHO (2006)
Are Biowaivers allowed?	BCS Class I	Yes	Yes	Yes	Yes
	BCS Class II	No	No	No	Yes
	BCS Class III	No	Yes	Yes	Yes
	BCS Class IV	No	No	No	No

[a]At this time, ANVISA (Brazilian regulatory body) also allows biowaivers only for BCS Class 1 drug products (similar to the FDA), while the Japanese regulatory authority does not allow any BCS-based biowaivers (Gupta 2006)

strength is soluble in 250 mL or less of aqueous media over the pH range of 1.0–7.5. The volume estimate of 250 mL is derived from typical bioequivalence study protocols that prescribe administration of a drug product to fasting human volunteers with a glass (about 8 oz) of water. The highest strength instead of the highest dose is used because for a product with multiple strengths, the highest strength is usually recommended for use in a bioequivalence study (Haidar et al. 2008).

The FDA BCS guidance details the determination of the equilibrium solubility of a drug substance under physiological pH conditions. Specifically, the pH-solubility profile of the test drug substance is measured at $37 \pm 1\ ^{\circ}C$ in aqueous media with a pH in the range of 1–7.5 using a validated stability-indicating assay. A sufficient number of pH conditions is evaluated to accurately define the pH-solubility profile. The number of pH conditions for a solubility determination can be based on the ionization characteristics of the test drug substance. For example, when the pK_a of a drug is in the range of 3–5, solubility should be determined at $pH = pK_a$, $pH = pK_a + 1$, $pH = pK_a - 1$, and at $pH = 1$ and 7.5. A minimum of three replicate determinations of solubility in each pH condition is recommended. Standard buffer solutions described in the USP are considered appropriate for use in solubility studies. If these buffers are not suitable for physical or chemical reasons, other buffer solutions can be used. The final reported pH should be the pH determined after the equilibrium is reached.

5.4.2.2   Permeability

The permeability class of a drug substance can be determined by pharmacokinetic studies in human subjects using mass balance and absolute bioavailability methods or by intestinal perfusion approaches. Methods not involving human subjects include in vivo or in situ intestinal perfusion in a suitable animal model (e.g., rats), and/or in vitro permeability methods using excised intestinal tissues, or monolayers of suitable epithelial cells. However, human data are always preferable, and the results from animal perfusion or in vitro cell culture studies are generally considered as supportive. Specifically, we may consider the following methods to determine the permeability classification.

(a) Pharmacokinetic Studies in Humans

    (i) *Mass balance studies*. Pharmacokinetic mass balance studies using unlabeled, stable isotopes or a radiolabeled drug substance can be used to document the extent of absorption of a drug. Depending on the variability of the studies, a sufficient number of subjects should be enrolled to provide a reliable estimate of extent of absorption. Because this method can provide highly variable estimates of drug absorption for many drugs, other methods described below may be preferable.

    (ii) *Absolute bioavailability studies*. Oral bioavailability determination using intravenous administration as a reference can be used. Depending on the variability of the studies, a sufficient number of subjects should be enrolled in a study to provide a reliable estimate of the extent of absorption. When the absolute bioavailability of a drug is shown to be 90 % or more, additional data to document drug stability in the gastrointestinal fluid are not necessary.

(b) Intestinal Permeability Methods

The following methods can be used to determine the permeability of a drug substance from the gastrointestinal tract: (1) in vivo intestinal perfusion studies in humans; (2) in vivo or in situ intestinal perfusion studies using suitable animal models; (3) in vitro permeation studies using excised human or animal intestinal tissues; or (4) in vitro permeation studies across a monolayer of cultured epithelial cells.

In vivo or in situ animal models and in vitro methods, such as those using cultured monolayers of animal or human epithelial cells, are appropriate for passively transported drugs. The observed low permeability of some drug substances in humans could be caused by efflux of drugs via membrane transporters such as P-glycoprotein (P-gp). When the efflux transporters are absent in these models, or their degree of expression is low compared to that in humans, there may be a greater likelihood of misclassification of permeability class for a drug subject to efflux compared to a drug transported passively. Therefore, expression of known transporters in selected study systems should be characterized. Functional expression of efflux systems (e.g., P-gp) should be demonstrated with techniques such as bidirectional transport studies, demonstrating a higher rate of transport in the basolateral-to-apical direction as compared to apical-to-basolateral direction using selected model drugs or chemicals at concentrations that do not saturate the efflux system (e.g., cyclosporin A, vinblastine, rhodamine 123).

Pharmacokinetic studies on dose linearity or proportionality may provide useful information for evaluating the relevance of observed in vitro efflux of a drug. For example, there may be fewer concerns associated with the use of in vitro methods for a drug that has a higher rate of transport in the basolateral-to-apical direction at low drug concentrations but exhibits linear pharmacokinetics in humans. For application of the BCS, an apparent passive transport mechanism can be assumed when one of the following conditions is satisfied:

- A linear (pharmacokinetic) relationship between the dose (e.g., relevant clinical dose range) and measures of bioavailability (area under the concentration-time curve) of a drug is demonstrated in humans.
- Lack of dependence of the measured in vivo or in situ permeability is demonstrated in an animal model on initial drug concentration (e.g., 0.01, 0.1, and 1 times the highest dose strength dissolved in 250 mL) in the perfusion fluid.
- Lack of dependence of the measured in vitro permeability on initial drug concentration (e.g., 0.01, 0.1, and 1 times the highest strength dissolved in 250 mL) is demonstrated in donor fluid and transport direction (e.g., no statistically significant difference in the rate of transport between the apical-

Table 5.2 Model drugs suggested for use in establishing suitability of a permeability method. Potential internal standards (IS) and efflux pump substrates (ES) are also identified

Drug	Permeability class
Antipyrine	High (potential IS candidate)
Caffeine	High
Carbamazepine	High
Fluvastatin	High
Ketoprofen	High
Metoprolol	High (potential IS candidate)
Naproxen	High
Propranolol	High
Theophylline	High
Verapamil	High (potential ES candidate)
Amoxicillin	Low
Atenolol	Low
Furosemide	Low
Hydrochlorthiazide	Low
Mannitol	Low (potential IS candidate)
α-Methyldopa	Low
Polyethylene glycol (400)	Low
Polyethylene glycol (1,000)	Low
Polyethylene glycol (4,000)	Low (zero permeability marker)
Ranitidine	Low

Source: FDA BCS guidance, Attachment A; p. 13 (2002)

to-basolateral and basolateral-toapical direction for the drug concentrations selected) using a suitable in vitro cell culture method that has been shown to express known efflux transporters (e.g., P-gp).

To demonstrate suitability of a permeability method intended for application of the BCS, a rank-order relationship between test permeability values and the extent of drug absorption data in human subjects should be established using a sufficient number of model drugs. For in vivo intestinal perfusion studies in humans, six model drugs are recommended. For in vivo or in situ intestinal perfusion studies in animals and for in vitro cell culture methods, 20 model drugs are recommended. Depending on study variability, a sufficient number of subjects, animals, excised tissue samples, or cell monolayers should be used in a study to provide a reliable estimate of drug permeability. This relationship should allow precise differentiation between drug substances of low and high intestinal permeability attributes. Model drugs should represent a range of low (e.g., <50 %), moderate (e.g., 50–89 %), and high (≥90 %) absorption. Table 5.2 shows the recommended model drugs for in vitro permeability studies.

After demonstrating suitability of a method and maintaining the same study protocol, it is not necessary to retest all selected model drugs for subsequent studies intended to classify a drug substance. Instead, a low and a high permeability model drug should be used as internal standards (i.e., included in the perfusion fluid or donor fluid along with the test drug substance). These two internal standards are in addition to the fluid volume marker (or a zero permeability compound such as PEG 4000) that is included in certain types of perfusion techniques (e.g., closed loop techniques). The choice of internal standards should be based on compatibility with the test drug substance (i.e., they should not exhibit any significant physical, chemical, or permeation interactions). When it is not feasible to follow this protocol, the permeability of internal standards should be determined in the same subjects, animals, tissues, or monolayers, following evaluation of the test drug substance. The permeability values of the two internal standards should not differ significantly between different tests, including those conducted to demonstrate suitability of the method. At the end of an in situ or in vitro test, the amount of drug in the membrane should be determined.

For a given test method with set conditions, selection of a high permeability internal standard with permeability in close proximity to the low/high permeability class boundary may facilitate classification of a test drug substance. For instance, a test drug substance may be determined to be highly permeable when its permeability value is equal to or greater than that of the selected internal standard with high permeability. Typical internal standards for high permeability include metoprolol and labetalol although the use of labetalol has been questioned (Incecayir et al. 2013).

Numerous in vitro Caco-2 studies have suggested that transporters may enhance or limit the absorption of many drugs (Murakami and Takano 2008). The potential impact of transporters on absorption should be investigated.

(c) High Metabolism Method

Wu and Benet (2005) recognized that for drugs exhibiting high intestinal permeability rates, the major route of elimination in humans was via metabolism, whereas drugs exhibiting poor intestinal permeability rates were primarily eliminated in humans as unchanged drug in the urine and bile. They proposed a Biopharmaceutics Drug Disposition Classification System (BDDCS) that could serve as a basis for predicting the importance of transporters in determining drug disposition, as well as in predicting drug–drug interactions.

It was suggested that the extent of drug metabolism of ≥ 90 % metabolized could be used as an alternate method in defining Class 1 marketed drugs suitable for a waiver of in vivo studies of bioequivalence (Benet et al. 2008). That is, ≥ 90 % metabolized is an additional methodology that may be substituted for ≥ 90 % absorbed. Benet et al. (above) proposed that the following criteria can be used to define ≥ 90 % metabolized for marketed drugs: following a single oral dose to humans, administered at the highest dose strength, mass balance of the Phase 1 oxidative and Phase 2 conjugative drug metabolites in the urine and feces, measured as unlabeled, radioactive-labeled, or nonradioactive-labeled substances, account for ≥ 90 % of the drug dosed. For an orally administered drug to be ≥ 90 % metabolized by Phase 1 oxidative and Phase 2 conjugative processes, it is obvious that the drug must be absorbed.

This high metabolism method is currently not in the FDA guidance. However, it is a scientifically valid method.

(d) Instability in the Gastrointestinal Tract

Determining the extent of absorption in humans based on mass balance studies using total radioactivity in urine does not take into consideration the extent of degradation of a drug in the gastrointestinal fluid prior to intestinal membrane permeation. In addition, some methods for determining permeability could be based on loss or clearance of a drug from fluids perfused into the human and/or animal gastrointestinal tract either in vivo or in situ. Documenting the fact that drug loss from the gastrointestinal tract arises from intestinal membrane permeation, rather than a degradation process, will help establish permeability.

Stability in the gastrointestinal tract may be documented using gastric and intestinal fluids obtained from human subjects. Drug solutions in these fluids should be incubated at 37 °C for a period that is representative of in vivo drug contact with these fluids, e.g., 1 h in gastric fluid and 3 h in intestinal fluid. Drug concentrations should then be determined using a validated stability-indicating assay method. Significant degradation (>5 %) of a drug in this protocol could suggest potential instability. Obtaining gastrointestinal fluids from human subjects requires intubation and may be difficult in many cases. Use of gastrointestinal fluids from suitable animal models and/or simulated fluids such as gastric and intestinal fluids USP can be substituted when properly justified.

5.4.3 FDA BCS-Based Biowaiver

The FDA issued a guidance for industry on waivers of in vivo bioavailability and bioequivalence studies for immediate release solid oral dosage forms based on the BCS in August, 2000 (2002). This BCS guidance recommends that sponsors may request biowaivers for BCS Class I drugs in immediate release (IR) solid oral dosage forms that exhibit rapid in vitro dissolution, provided the following conditions are met: (a) excipients used in the IR solid oral dosage forms have no significant effect on the rate and extent of oral drug absorption; (b) the drug must not have a narrow therapeutic index; and (c) the product is designed not to be absorbed in the oral cavity.

The FDA guidance provides recommendations for sponsors of investigational new drug applications (IND), new drug applications (NDAs), abbreviated new drug applications (ANDAs), and supplements to these applications who wish to request a waiver of in vivo bioavailability (BA) and/or bioequivalence (BE) studies for immediate release (IR) solid dosage forms. The guidance provides the recommended methods for determining solubility, permeability, and in vitro dissolution and also provides criteria for granting a biowaiver based on BCS.

An IR drug product is characterized as a *rapid dissolution* product when not less than 85 % of the labeled amount of the drug substance dissolves within 30 min using USP Apparatus I at 100 rpm or USP Apparatus II at 50 rpm in a volume of

900 mL or less of each of the following media: (a) acidic media, such as 0.1 N HCl or USP simulated gastric fluid without enzymes (SGF); (b) a pH 4.5 buffer; and (c) a pH 6.8 buffer or USP simulated intestinal fluid without enzymes (SIF). Otherwise, the drug product is not considered to be a rapid dissolution product for the purpose of BCS-based biowaivers.

Based on the scientific principles of the BCS, observed in vivo differences in the rate and extent of absorption of a drug from two pharmaceutically equivalent solid oral products may be due to in vivo differences in drug dissolution. When the in vivo dissolution of an IR oral dosage form is rapid in relation to gastric emptying, the rate and extent of drug absorption is likely to be independent of drug dissolution. Therefore, similar to oral solutions, demonstration of in vivo bioequivalence may not be necessary as long as the inactive ingredients used in the dosage form do not significantly affect absorption of the active ingredient. Thus, for BCS Class I drug substances, demonstration of rapid in vitro dissolution using the recommended test methods would provide sufficient assurance of rapid in vivo dissolution, thereby ensuring human in vivo bioequivalence. The benefit of this FDA BCS guidance is not only lowering expenditures associated with bioavailability/bioequivalence studies but also more critically expediting the development of new chemical entities for the marketplace, which will ultimately be of benefit to the health of the American public.

5.4.4   Is There a Need to Redefine Solubility and Permeability?

(a) *Definition of high solubility.* The pH range requirement of 1.0–7.5 in the solubility classification is likely too conservative and may not be necessary. Under the fasted conditions, the pHs in the GI tract vary from 1.4 to 2.1 in the stomach, 4.9 to 6.4 in the duodenum, 4.4 to 6.6 in the jejunum, and 6.5 to 7.4 in the ileum (Oberle and Amidon 1987). Even under fed conditions, the pH is not expected to go above pH 6.8 in the stomach. Further, it generally takes approximately 85 min for a drug to reach the ileum (Yu et al. 1996). Thus, by the time the drug reaches the ileum, the dissolution of the drug product is likely complete if it meets the rapid dissolution criterion, i.e., no less than 85 % dissolved within 30 min. Therefore, it would be reasonable to redefine the pH range for BCS solubility class boundary from 1.0–7.5 to 1.0–6.8, in alignment with dissolution pHs, which are pH 1.0, 4.5, and 6.8 buffers. The new pH requirement of 1.0–6.8 has been adopted by many regulatory agencies worldwide (2006, 2008, 2012).

The dose volume of 250 mL seems also a conservative estimate of what actually is available in vivo for solubilization and dissolution. The physiological volume of the small intestine varies from 50 to 1,100 mL with an average of 500 mL under the fasted conditions (Lobenberg and Amidon 2000). When administered with a glass of water, the drug is immersed in approximately 250 mL of liquid in the stomach.

If the drug is not in solution in the stomach, gastric emptying would then expose it to the small intestine and the solid drug would then dissolve under the effect of additional small intestinal fluid. However, due to the large variability of the small intestinal volume, an appropriate definition of the volume for solubility class boundary would be difficult to set.

Another factor influencing in vivo solubility is bile salt/micelle solubilization (Fleisher et al. 1999). Intestinal solubility is perhaps the most important solubility since this is the absorbing region for most drugs. Many acidic drugs whose solubility is low at low pHs are well absorbed. For example, most nonsteroidal anti-inflammatory drugs (NSAIDs), such as flurbiprofen, ketoprofen, naproxen, and oxaprozin, are poorly soluble in the stomach but are highly soluble in the distal intestine, and their absolute human bioavailabilities are 90 % or higher, thus exhibiting behavior similar to those of BCS Class I drugs (Dressman 1997; Sheng et al. 2006; Yazdanian et al. 2004).

The solubility classification is based on the ability of a drug to dissolve in plain aqueous buffers. However, bile salts are present in the small intestine, even in the fasted state. Based on physiological factors, Dressman et al. (2008) designed two kinds of media: one to simulate the fasted-state conditions in the small intestine and the other to simulate the fed-state conditions in the small intestine (Dressman et al. 2008). These two media may be used in drug discovery and development processes to assess in vivo solubility and dissolution and have the potential to be utilized in drug regulation, i.e., dissolution methodology for bioequivalence demonstration using more physiologically relevant media, although more extensive research is needed.

Other criteria, such as intrinsic dissolution rate, may be useful in the classification of the biopharmaceutic properties of drugs. The intrinsic dissolution method has been widely used in pharmaceutical industries to characterize drug substances. Our recent data have shown that the intrinsic dissolution method is robust and easily determined. A good correlation between the intrinsic dissolution rate and BCS solubility classification was found for 15 BCS model drugs (Yu et al. 2004). Thus, the intrinsic dissolution rate may be used when the solubility of a drug cannot be accurately determined although more validation research needs to be conducted.

(b) *Definition of high permeability.* The permeability class boundary is based on the extent of intestinal absorption (fraction of dose absorbed) of a drug substance in humans or on measurements of the rate of mass transfer across intestinal membranes. Under the current BCS classification, a drug is considered to be highly permeable when the fraction of dose absorbed is equal to or greater than 90 %. The criterion of 90 % for the fraction of dose absorbed can be considered conservative since the experimentally determined fraction of dose absorbed is seen to be less than 90 % for many drugs that are generally considered completely or well absorbed. This raises a question as to whether an alternate permeability boundary (e.g., 85 %) should be considered for high permeability classification.

Benet and Larregieu (2010) have suggested that the FDA should eliminate the ambiguities in the current BCS biowaiver guidance and make public the drugs for which BCS biowaivers have been granted. They argue that although BCS Class I drugs are designated as high permeability drugs, in fact, the criterion utilized is high extent of absorption. They suggested that this ambiguity should be eliminated, and the FDA criterion should explicitly be stated as $\geq$90 % absorption based on absolute bioavailability or mass balance. Nevertheless, as stated in the previous section, the FDA BCS guidance does allow BCS classification based in situ or in vitro cell culture permeability, which, over the years, has granted numerous biowaivers. As in vivo absolute bioavailability or mass balance studies produce the extent of absorption while the in situ or in vitro methods yield permeability, there is really no universally suitable terminology. Nonetheless, for drugs with permeability-limited absorption, there exist an excellent relationship between the extent of absorption and permeability (Yu and Amidon 1999). As such, it really does not matter which term is used. For drugs with dissolution or solubility-limited absorption, the extent of absorption really does not reflect in vivo capability of permeation. In this regard, it is not appropriate to use the extent of absorption to classify drugs.

5.4.5   Biowaiver Extension Potential to BCS Class III Drugs

Drugs with high solubility and low permeability are classified as BCS Class III drugs. It has been suggested that biowaivers be extended to BCS Class III drugs with rapid dissolution property. It has been contended that there are equally compelling reasons to grant biowaivers to Class III drugs as there are for Class I drugs (Blume and Schug 1999). The absorption of a Class III drug is likely limited by its permeability and less dependent upon its formulation, and its bioavailability may be determined by its in vivo permeability pattern. If the dissolution of Class III products is very rapid under all physiologic pH conditions, it can be expected that they will behave like an oral solution in vivo. In vivo bioequivalence studies are generally waived for oral solution drug products since the release of the drug from an oral solution is self-evident (2003).

Nevertheless, the absorption kinetics from the small intestine are influenced by a combination of physiological factors and biopharmaceutical properties such as gastrointestinal motility, permeability, metabolism, dissolution, and the interaction/binding of drugs with excipients. A recent survey of the FDA data of over ten BCS Class III drugs shows that most commonly used excipients in solid dosage forms have no significant effect on absorption. If the excipients used in two pharmaceutically equivalent solid oral IR products do not affect drug absorption and the two products dissolve very rapidly in all physiologically relevant pHs (i.e., >85 % in 15 min), there would appear to be no reason to believe that these two products would not be bioequivalent (Yu et al. 2002).

5.4.5.1   Potential Excipient Effect on Motility and Permeability

Since Class III compounds often exhibit site-dependent absorption properties (Wu and Benet 2005), the transit time through specific regions of the upper intestine may be critical for bioequivalence, suggesting a more stringent dissolution criterion to ensure complete dissolution in the stomach. Certain excipients have been shown to influence gastrointestinal transit time. For example, scintigraphy has indicated that sodium acid pyrophosphate could reduce the small intestinal transit time by as much as 43 % compared to controls (Koch et al. 1993). Poorly absorbed sugar alcohols, such as sorbitol and mannitol, can also decrease the small intestinal transit time (Chen et al. 2013). Therefore, Class III oral drug products containing a significant amount of transit-influencing excipients should be excluded from consideration of biowaivers. While most commonly used excipients in solid dosage forms are unlikely to influence the gastrointestinal transit time significantly, the evidence by no means is conclusive.

The effects of excipients on permeability have been reviewed in the literature (Aungst 2000; Gupta et al. 2013). Excipients that can significantly affect permeability in vitro include surfactants, fatty acids, medium-chain glycerides, steroidal detergents, acyl carnitine and alkanoylcholines, N-acetylated non-α amino acids, and chitosans and other mucoadhesive polymers. Rege et al. (2001) investigated the effect of some formulation excipients on Caco-2 permeability and found that several commonly used IR formulation excipients did not modulate drug permeability across Caco-2 monolayers. When excipient effects are observed in Caco-2 systems, they are generally at much higher concentrations than those observed in vivo.

5.4.5.2   Dissolution

In vivo dissolution plays a more important role for Class III IR drug products than it does for Class I drug products. Dissolution tests with USP Apparatus I at 100 rpm (or USP Apparatus II at 50 rpm) in a volume of 900 mL of various pH media are recommended in the FDA BCS guidance to evaluate the product dissolution in vitro. For highly soluble and highly permeable drugs, rapid dissolution in vitro (no less than 85 % in 30 min) can most likely ensure rapid in vivo dissolution. However, demonstration of rapid in vitro dissolution of Class III drug products may not ensure rapid dissolution in vivo simply because sink conditions may not exist under in vivo conditions. In order to minimize the possibility of dissolution behavior anomalies, it was found in our simulation studies that it would be necessary to set a more rapid in vitro dissolution rate criterion of no less than 85 % within 15 min for Class III drugs (Yu et al. 2001).

5.5 Summary

Currently, biowaivers are being granted for:

- Certain drug products when the in vivo bioavailability or bioequivalence of the drug product may be self-evident.
- For certain drug products, bioavailability may be measured or bioequivalence may be demonstrated by evidence obtained in vitro in lieu of in vivo data. For example, when the drug product is in the same dosage form but in a different strength; this different strength is *proportionally similar* in its active and inactive ingredients to the strength of the product for which the same manufacturer has conducted an appropriate in vivo study, and the new strength meets an appropriate in vitro dissolution test.
- For certain drug products based on their BCS designation.

The current status regarding BCS-based biowaivers differs across various regulatory agencies. The current FDA BCS guidance allows for biowaivers based on conservative criteria with biowaivers applicable for only BCS Class I products. Possible new criteria and class boundaries are discussed for additional biowaivers based on the underlying physiology of the gastrointestinal tract. These changes in class boundaries for solubility and permeability are (a) to narrow the required solubility pH range from 1.0–7.5 to pH 1.0–6.8 and (b) to reduce the high permeability requirement from 90 to 85 %. Also, literature data suggest that it may be reasonable to extend biowaiver to BCS Class III drugs as long as any impact of excipients on absorption is controlled.

References

Adkin DA, Davis SS, Sparrow RA, Huckle PD, Phillips AJ, Wilding IR (1995) The effect of different concentrations of mannitol in solution on small intestinal transit: implications for drug absorption. Pharm Res 12(3):393–396

Amidon GL, Lennernas H, Shah VP, Crison JR (1995) A theoretical basis for a biopharmaceutic drug classification: the correlation of in vitro drug product dissolution and in vivo bioavailability. Pharm Res 12(3):413–420

Aungst BJ (2000) Intestinal permeation enhancers. J Pharm Sci 89(4):429–442. doi:10.1002/(SICI)1520-6017(200004)89:4<429::AID-JPS1>3.0.CO;2-J, [pii] 10.1002/(SICI)1520-6017(200004)89:4<429::AID-JPS1>3.0.CO;2-J

Basit AW, Podczeck F, Newton JM, Waddington WA, Ell PJ, Lacey LF (2002) Influence of polyethylene glycol 400 on the gastrointestinal absorption of ranitidine. Pharm Res 19(9):1368–1374

Benet LZ, Larregieu CA (2010) The FDA should eliminate the ambiguities in the current BCS biowaiver guidance and make public the drugs for which BCS biowaivers have been granted. Clin Pharmacol Ther 88(3):405–407. doi:10.1038/clpt.2010.149, clpt2010149 [pii]

Benet LZ, Amidon GL, Barends DM, Lennernas H, Polli JE, Shah VP, Stavchansky SA, Yu LX (2008) The use of BDDCS in classifying the permeability of marketed drugs. Pharm Res 25(3):483–488. doi:10.1007/s11095-007-9523-x

Blume HH, Schug BS (1999) The biopharmaceutics classification system (BCS): class III drugs—better candidates for BA/BE waiver? Eur J Pharm Sci 9(2):117–121, S0928-0987(99)00076-7 [pii]

Chen ML, Straughn AB, Sadrieh N, Meyer M, Faustino PJ, Ciavarella AB, Meibohm B, Yates CR, Hussain AS (2007) A modern view of excipient effects on bioequivalence: case study of sorbitol. Pharm Res 24(1):73–80. doi:10.1007/s11095-006-9120-4

Chen ML, Sadrieh N, Yu L (2013) Impact of osmotically active excipients on bioavailability and bioequivalence of BCS class III drugs. AAPS J 15(4):1043–1050. doi:10.1208/s12248-013-9509-z

Code of Federal Regulations 21 320.22. Bioavailability and Bioequivanece Requirements. http://www.accessdata.fda.gov/scripts/cdrh/cfdocs/cfCFR/CFRSearch.cfm?fr=320.22

Dressman JB (1997) Physiological aspects of the design of dissolution tests. In: Amidon GLR, Willimans RL (eds) Scientific foundations for regulating drug product quality. AAPS, Alexandria, pp 155–168

Dressman JB, Thelen K, Jantratid E (2008) Towards quantitative prediction of oral drug absorption. Clin Pharmacokinet 47(10):655–667. doi:10.2165/00003088-200847100-00003, 47103 [pii]

EMA (2008) European Medicines Agency: Guideline on the investigation of bioequivalence. http://www.ema.europa.eu/docs/en_GB/document_library/Scientific_guideline/2009/09/WC500003011.pdf

Fassihi AR, Dowse R, Robertson SSD (1991) Influence of sorbitol solution on the bioavailability of theophylline. Int J Pharm 72:175–178

FDA (1995) Food and Drug Administration. Guidance for industry: immediate release solid oral dosage forms. Scale-up and post-approval changes: chemistry, manufacturing, and controls, in vitro dissolution testing, and in vivo bioequivalence documentation. http://www.fda.gov/downloads/Drugs/GuidanceComplianceRegulatoryInformation/Guidances/UCM070636.pdf

FDA (1997) Food and Drug Administration. Guidance for industry: SUPAC-MR: modified release solid oral dosage forms—scale up and post-approval changes: chemistry, manufacturing, and controls; In vitro dissolution testing and in vivo bioequivalence documentation. http://www.fda.gov/downloads/Drugs/GuidanceComplianceRegulatoryInformation/Guidances/UCM070640.pdf

FDA (2002) Food and Drug Administration. Guidance to industry: waiver of in vivo bioavailability and bioequivalence studies for immediate-release solid oral dosage forms based on a biopharmaceutics classification system. http://www.fda.gov/downloads/Drugs/GuidanceComplianceRegulatoryInformation/Guidances/UCM070246.pdf

FDA (2003) Food and Drug Administration. Guidance for industry: bioavailability and bioequivalence studies for orally administered drug products—general considerations. http://www.fda.gov/downloads/Drugs/GuidanceComplianceRegulatoryInformation/Guidances/UCM070124.pdf

Fleisher D, Li C, Zhou Y, Pao LH, Karim A (1999) Drug, meal and formulation interactions influencing drug absorption after oral administration. Clinical implications. Clin Pharmacokinet 36(3):233–254. doi:10.2165/00003088-199936030-00004

Fukahori M, Sakurai H, Akatsu S, Negishi M, Sato H, Goda T, Takase S (1998) Enhanced absorption of calcium after oral administration of maltitol in the rat intestine. J Pharm Pharmacol 50(11):1227–1232

Gupta E, Barends DM, Yamashita E, Lentz KA, Harmsze AM, Shah VP, Dressman JB, Lipper RA (2006) Review of global regulations concerning biowaivers for immediate release solid oral dosage forms. Eur J Pharm Sci 29:315–324

Gupta V, Hwang BH, Doshi N, Mitragotri S (2013) A permeation enhancer for increasing transport of therapeutic macromolecules across the intestine. J Control Release. DOI: S0168-3659(13)00248-4 [pii]. 10.1016/j.jconrel.2013.05.002

Haidar S, Kwon H, Lionberger R, Yu LX (2008) Bioavailability and bioequivalence. Kluwer Academic, New York, pp 262–289

Health Canada (2012) Biopharmaceutics classification system based biowaiver: draft guidance document. http://www.hc-sc.gc.ca/dhp-mps/alt_formats/pdf/consultation/drug-medic/bcs_draft_guide_ebauche_ld_scb-eng.pdf

Incecayir T, Tsume Y, Amidon GL (2013) Comparison of the permeability of metoprolol and labetalol in rat, mouse, and Caco-2 cells: use as a reference standard for BCS classification. Mol Pharm 10(3):958–966. doi:10.1021/mp300410n

Koch KM, Parr AF, Tomlinson JJ, Sandefer EP, Digenis GA, Donn KH, Powell JR (1993) Effect of sodium acid pyrophosphate on ranitidine bioavailability and gastrointestinal transit time. Pharm Res 10(7):1027–1030

Kruger D, Grossklaus R, Herold M, Lorenz S, Klingebiel L (1992) Gastrointestinal transit and digestibility of maltitol, sucrose and sorbitol in rats: a multicompartmental model and recovery study. Experientia 48(8):733–740

Lobenberg R, Amidon GL (2000) Modern bioavailability, bioequivalence and biopharmaceutics classification system. New scientific approaches to international regulatory standards. Eur J Pharm Biopharm 50(1):3–12, S0939-6411(00)00091-6 [pii]

Murakami T, Takano M (2008) Intestinal efflux transporters and drug absorption. Expert Opin Drug Metab Toxicol 4(7):923–939. doi:10.1517/17425255.4.7.923

Oberle RL, Amidon GL (1987) The influence of variable gastric emptying and intestinal transit rates on the plasma level curve of cimetidine; an explanation for the double peak phenomenon. J Pharmacokinet Biopharm 15(5):529–544

Polli JE, Yu LX, Cook JA, Amidon GL, Borchardt RT, Burnside BA, Burton PS, Chen ML, Conner DP, Faustino PJ, Hawi AA, Hussain AS, Joshi HN, Kwei G, Lee VH, Lesko LJ, Lipper RA, Loper AE, Nerurkar SG, Polli JW, Sanvordeker DR, Taneja R, Uppoor RS, Vattikonda CS, Wilding I, Zhang G (2004) Summary workshop report: biopharmaceutics classification system—implementation challenges and extension opportunities. J Pharm Sci 93(6):1375–1381. doi:10.1002/jps.20064

Rege BD, Yu LX, Hussain AS, Polli JE (2001) Effect of common excipients on Caco-2 transport of low-permeability drugs. J Pharm Sci 90(11):1776–1786. doi:10.1002/jps.1127 [pii]

Schulze JD, Waddington WA, Eli PJ, Parsons GE, Coffin MD, Basit AW (2003) Concentration-dependent effects of polyethylene glycol 400 on gastrointestinal transit and drug absorption. Pharm Res 20(12):1984–1988

Sheng JJ, Kasim NA, Chandrasekharan R, Amidon GL (2006) Solubilization and dissolution of insoluble weak acid, ketoprofen: effects of pH combined with surfactant. Eur J Pharm Sci 29 (3–4):306–314. doi:10.1016/j.ejps.2006.06.006, S0928-0987(06)00169-2 [pii]

WHO (2006) World Health Organization: Proposal to waive in vivo bioequivalence requirements for WHO Model List of Essential Medicines immediate-release, solid oral dosage forms. World Health Technical Report Series 937

Wu CY, Benet LZ (2005) Predicting drug disposition via application of BCS: transport/absorption/elimination interplay and development of a biopharmaceutics drug disposition classification system. Pharm Res 22(1):11–23

Yazdanian M, Briggs K, Jankovsky C, Hawi A (2004) The "high solubility" definition of the current FDA Guidance on Biopharmaceutical Classification System may be too strict for acidic drugs. Pharm Res 21(2):293–299

Yu LX, Amidon GL (1999) A compartmental absorption and transit model for estimating oral drug absorption. Int J Pharm 186(2):119–125, doi: S0378517399001477 [pii]

Yu LX, Lipka E, Crison JR, Amidon GL (1996) Transport approaches to the biopharmaceutical design of oral drug delivery systems: prediction of intestinal absorption. Adv Drug Deliv Rev 19(3):359–376. doi:10.1016/0169-409X(96)00009-9 [pii]

Yu LX, Ellison CD, Conner DP, Lesko LJ, Hussain AS (2001) Influence of drug release properties of conventional solid dosage forms on the systemic exposure of highly soluble drugs. AAPS PharmSci 3(3):E24

Yu LX, Amidon GL, Polli JE, Zhao H, Mehta MU, Conner DP, Shah VP, Lesko LJ, Chen ML, Lee VH, Hussain AS (2002) Biopharmaceutics classification system: the scientific basis for biowaiver extensions. Pharm Res 19(7):921–925

Yu LX, Carlin AS, Amidon GL, Hussain AS (2004) Feasibility studies of utilizing disk intrinsic dissolution rate to classify drugs. Int J Pharm 270(1–2):221–227, doi: S037851730300574X [pii]

Chapter 6
Bioequivalence of Highly Variable Drugs

Barbara M. Davit and Devvrat T. Patel

6.1 Introduction: General Bioequivalence Principles

Bioequivalence (BE) is defined as the absence of a significant difference in the rate and extent to which an active ingredient or active moiety in pharmaceutical equivalents[1] or pharmaceutical alternatives[2] becomes available at the site of action when administered at the same molar dose under the same conditions in an appropriately designed study (Office of the Federal Register 2013). BE between a test and reference product is established in order to demonstrate therapeutic equivalence. Therapeutically equivalent drug products can be substituted with the full

---

[1] Pharmaceutical equivalents, as defined in the US FDA regulations (21 CFR Part 320.1(c)), means drug products in identical dosage forms that contain identical amounts of the identical active drug ingredient, i.e., the same salt or ester of the same therapeutic moiety, or, in the case of modified-release dosage forms that require a reservoir or overage or such forms as prefilled syringes where residual volume may vary, that deliver identical amounts of the active drug ingredient over the identical dosing period; do not necessarily contain the same inactive ingredients; and meet the identical compendial or other applicable standard of identity, strength, quality, and purity, including potency and, where applicable, content uniformity, disintegration times, and/or dissolution rates.

[2] Pharmaceutical alternatives, as defined in 21 CFR Part 320.1(d), mean drug products that contain the identical therapeutic moiety, or its precursor, but not necessarily in the same amount or dosage form or as the same salt or ester. Each such drug product individually meets either the identical or its own respective compendial or other applicable standard of identity, strength, quality, and purity, including potency and, where applicable, content uniformity, disintegration times, and/or dissolution rates.

B.M. Davit (✉)
Biopharmaceutics, Clinical Research, Merck, Sharp and Dohme Corp., One Merck Drive, P.O. Box 100, Whitehouse Station, NJ 08889, USA
e-mail: Barbara.Davit@merck.com

D.T. Patel
Office of Generic Drugs, Center for Drug Evaluation and Research, US Food and Drug Administration, 7500 Standish Place, Rockville, MD 20855, USA

L.X. Yu and B.V. Li (eds.), *FDA Bioequivalence Standards*, AAPS Advances in the Pharmaceutical Sciences Series 13, DOI 10.1007/978-1-4939-1252-0_6, © The United States Government 2014

expectation that the substituted (test) product will produce the same safety effect and safety profile as the originally prescribed (reference) product. Acceptable BE between a test and reference product is among the criteria required by the US Food and Drug Administration (FDA) for approval of new generic drug products (US Department of Health and Human Services et al. 2013a, b). With respect to new drug development, the FDA requires BE documentation to establish links between (1) early and late clinical efficacy trial formulations; (2) formulations used in clinical trial and stability studies, if different; and (3) clinical trial formulations and to-be-marketed formulations (US Department of Health and Human Services 2003). Thus, BE documentation play a pivotal role in new and generic drug development.

In a regulatory application for drug marketing approval, the pivotal studies demonstrating BE compare test and reference availability at the site of action in human subjects. For drugs which are systemically available, test and reference rate and extent of availability at the site of action is assessed by using pharmacokinetic (PK) profiles to determine rate and extent of systemic absorption. The PK parameter C_{max} (peak plasma drug concentration) is used to assess rate of absorption and the PK parameter AUC (area under the plasma concentration versus time profile) is used to assess extent of absorption. Most BE studies enroll healthy normal subjects who receive single doses of the test or reference product via a two-way crossover design. For the most part, the US FDA asks new and generic drug applicants to statistically compare generic and reference C_{max} and AUC values using the two one-sided tests procedure (Schuirmann 1987). Under the two one-sided tests procedure, the 90 % confidence interval (CI) around the geometric mean ratio (GMR) of the test and reference values of C_{max} and AUC is required to fit within BE limits, set from 80 to 125 % (Westlake 1981). The width of the 90 % CI depends upon the number of subjects in the study and the variability of the BE measure.

6.2 Definition of Highly Variable Drugs

Highly variable (HV) drugs are defined as drugs in which the within-subject variability (defined as the % coefficient of variation, %CV) in one or more of the BE measures is 30 % or greater (Blume and Midha 1993; Shah et al. 1996). The BE measures of interest are, as mentioned above, AUC, representing extent of drug absorption, and C_{max}, representing rate of drug absorption. A survey of some generic products reviewed by the FDA from 2003 to 2005 suggested that about 20 % of the generic drugs evaluated for marketing approval in the USA are HV due to their drug substance dispositional characteristics (Davit et al. 2008).

The FDA in 1992 began to routinely use the two one-sided tests procedure for analyzing BE study data (Davit et al. 2009). As stated above, for a test and reference product to be deemed bioequivalent, the 90 % CIs of the test/reference GMRs for both BE measures AUC and C_{max} must fall within the BE limits of 80–125 %. Determining the BE of HV generic drugs is challenging because the high within-subject variability

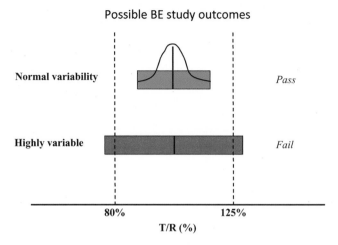

Fig. 6.1 The 80–125 % BE limits are represented along the *x*-axis as two "goal posts." The BE limits are compared to the hypothetical 90 % CIs of the test/reference BE measure GMRs for two drugs, a drug with "normal variability" (*green bar*) and an HV drug (*red bar*). The 90 % CIs of the two drugs are represented by colored bars. For the normal variability drug, the 90 % CI meets the BE limits. For the HV drug, the 90 % CI fails to meet the acceptance limits. As the width of the CI is influenced by the number of study subjects, in the case of this hypothetical HV drug, it is likely that the study would have met the BE limits if more subjects had been used

means that large numbers of subjects may be needed in order for the studies to meet the 80–125 % limits. Figure 6.1 shows the results of two hypothetical BE studies. The 90 % CIs of both products are represented by colored bars. Both products have GMRs near 1.00. The product represented by the green bar has low within-subject variability, and easily meets BE limits. The product represented by the red bar has high within-subject variability and fails to meet the BE limits. Notably, although this second product (red bar) has a test/reference GMR near 1.00, and appears to be well designed to perform the same as the reference product in vivo, it will be necessary to increase the number of study subjects—perhaps dramatically—in order for this product to meet the BE limits.

Several factors influence the sample size needed to meet the regulatory criteria for acceptable BE. First, each one-sided test (in the two one-sided tests procedure) is carried out at the 5 % level of significance, corresponding to the 90 % CI (US Department of Health and Human Services 2001). The 5 % level of significance represents the type I error rate (α), which is the probability of incorrectly deeming as bioequivalent two formulations whose population GMR fails to meet the BE limits. The second factor influencing sample size is study power, defined as the likelihood or chance of correctly demonstrating BE when it, in fact, exists (Patterson et al. 2001; Phillips 1990). A third factor influencing sample size is the test/reference BE measure ratios. If the true test/reference ratio differs from unity, the overall power to show BE is reduced at any given sample size, resulting in an increase in the number of study subjects needed. Other factors influencing sample

Table 6.1 The number of study subjects required to show BE with 80 % power is a function of within-subject variability and GMR (sample size estimations are for the case $\sigma_{WT} = \sigma_{WR}$ and $\sigma_D = 0$)

Within-subject %CV	GMR (%)	Sample size for a two-way crossover design	Sample size for a four-way crossover design
15	100	10	6
	105	12	8
	110	20	12
30	100	32	18
	105	38	20
	110	68	36
45	100	66	34
	105	80	42
	110	142	72
60	100	108	56
	105	132	66
	110	236	118
75	100	156	80
	105	190	96
	110	340	172

size include the study design and the expected within-subject variability. For example, a replicate four-way crossover BE study design, in which each subject receives the test and reference products twice, requires fewer subjects than a two-way crossover BE study design. As within-subject variability increases, the number of subjects needed in a crossover design will also increase, assuming that all other factors remain constant. Thus, BE study sample size is calculated based on a type I error rate of 5 % per test, the desired study power, and the best estimates of test/reference ratios and within-subject variability. Table 6.1 illustrates how these factors influence the number of subjects needed to provide an 80 % chance of an acceptable BE study.

As shown in Table 6.1, the number of study subjects needed to show BE increases dramatically for HV drugs. The FDA observed in a survey of generic drug product BE studies reviewed from 2003 to 2005 that studies of HV drugs generally used more subjects than studies of lower variability drugs (Davit et al. 2008).

One important observation is that clinical data strongly support a conclusion that HV drugs have wide therapeutic indices. Otherwise, there would have been significant safety issues and lack of efficacy during the pivotal safety and efficacy trials required for initial FDA marketing approval (Benet 2004). In other words, the reference product, when dosed on different occasions, was safe and efficacious, despite high PK variability.

6.3 Causes of Highly Variable Drugs

The majority of HV drugs appear to fall into Biopharmaceutics Classification System (BCS) Class II or IV, which are drugs having low aqueous solubility/high intestinal permeability or low aqueous solubility/low intestinal permeability, respectively (Cook et al. 2010; Amidon 2004). Dispositional characteristics of HV drugs include extensive presystemic metabolism, low bioavailability, high acid lability, and/or high lipophilicity (Davit et al. 2008). Consequently, plasma concentrations of these HV drugs are often very low. In such situations, it may not be possible to accurately characterize PK profiles, with the result that within-subject variability of BE measures can exceed 30 % (Conner 2009). As the FDA discourages unnecessary human testing, these observations raised questions about whether large numbers of subjects should be used in BE studies of drug products for which highly variable PK do not appear to impact safety and efficacy. It should be stressed that this concern pertains to high PK variability due to drug substance dispositional characteristics and has nothing to do with the formulation performance assessment that is the key question in BE comparisons.

An additional concern about imposing the 80–125 % BE limits to HV drugs is that the necessity of using a large sample size could serve to deter the development of generic drug products or reformulations of innovator drug products (DiLiberti 2004; Tothfalusi et al. 2009). For example, development of a generic drug product line could be halted because of a high failure rate of the in vivo BE studies necessary for approval; alternatively, it may be necessary to repeat the in vivo BE studies until a successful outcome is achieved (Endrenyi 2004). Not only does this situation lead to unnecessary human testing, but it also can contribute to increase the cost of drug development, which can be reflected in higher prices to consumers.

A final concern is that an HV drug reference product may not even be bioequivalent to itself with a relatively small number (i.e., 18–40) of subjects using a two-way crossover design with static BE limits of 80–125 % (Conner 2009). This situation was shown to occur in practice, as illustrated by results of BE studies of brand-name formulations of the HV drugs chlorpromazine and verapamil (Midha et al. 2005; Tsang et al. 1996). Such findings support the contention that a "one-size-fits-all" approach based on static BE limits is not suitable for drugs that are HV.

The issues surrounding BE evaluation of HV drugs and proposals for modifying the BE approach for such products were discussed over many years within the pharmaceutical sciences community, in the literature, and at various national and international venues (Davit et al. 2012). In 2004, FDA's Office of Generic Drugs (OGD) brought the issue to a meeting of the FDA Advisory Committee for Pharmaceutical Sciences and Clinical Pharmacology (Advisory Committee) (Executive Secretary 2004). Various proposals for optimizing study design and data analysis were discussed, including whether to expand BE limits to 70–143 %, or whether to use the within-subject PK variability of the reference product to scale the BE limits. The Advisory Committee recommended that FDA explore the use of reference scaling, to include a limit on the GMR. The concept of reference scaling evolved from the Individual Bioequivalence (IBE) approach, which the FDA worked toward implementing for a time, with the objective of improving formulation switchability (Hauck et al. 2000; Patnaik et al. 1997). The proposed criterion for acceptable IBE included the (a) comparison of test and reference means; (b) comparison of within-subject variances; (c) assessment of subject-by-formulation interaction; and (d) ability to scale the BE limits if within-subject variability following the administration of the reference product exceeded a predetermined value (Tothfalusi et al. 2009). Although the FDA halted its implementation of IBE in 2001, the Advisory Committee recommended that it consider applying reference scaling to BE studies of HV drugs (FDA Advisory Committee for Pharmaceutical Science and Clinical Pharmacology 2001).

Thus, to expand upon the 2004 Advisory Committee's recommendations and validate the usefulness of a reference-scaling approach, the FDA formed an interdisciplinary working group, charged with exploring these concepts via a series of simulation studies. The simulation studies investigated whether reference scaling of the BE limits was feasible, and which study designs and statistical analysis approaches were optimal to maintain a type I error rate of 5 % (Haidar et al. 2008a, b). In 2006, the OGD presented to the Advisory Committee the results of the working group's simulation studies (Executive Secretary 2006). The Advisory Committee made recommendations about the number of subjects and use of constraints on the reference-scaled BE limits. The FDA considered the Advisory Committee's recommendations in finalizing an RSABE approach for HV drugs.

6.4 US FDA's Recommendations for BE Studies of HV Drugs

Thus, since 2006, FDA has accepted a reference-scaled average bioequivalence (RSABE) approach for HV drugs (US Department of Health and Human Services 2012). Using the RSABE approach, the implied BE limits can widen to be larger than 80–125 % for drugs that are HV, provided that certain constraints are applied to this approach in order to maintain an acceptable type I error rate and satisfy any public health concerns (Davit et al. 2012).

The usual way of statistically analyzing BE study data is by the average bioequivalence (ABE) approach, based on the two one-sided tests procedure. The acceptance of BE is stated if the difference between logarithmic means is between preset regulatory limits, as shown below:

$$(\mu_{\mathrm{T}} - \mu_{\mathrm{R}})^2 \leq \theta_{\mathrm{A}}^2$$

where:

μ_T is the population average response of the log-transformed measure for the test (T) formulation

μ_R is the population average response of the log-transformed measure for the reference (R) formulation

θ_A is equal to $\ln(1.25)$

As $-\ln(1.25) = \ln(0.8)$, using the ABE approach, the BE acceptance limits are as follows:

$$\ln(0.8) \le (\mu_T - \mu_R) \le \ln(1.25)$$

Thus, via ABE, two products are deemed bioequivalent when the 90 % confidence intervals of the GMRs for AUC and C_{max} fall within the limits of 80–125 %.

By contrast to the ABE approach, using the RSABE approach, the BE acceptance limits are derived as shown in the following equation:

$$\frac{(\mu_T - \mu_R)^2}{\sigma_{WR}^2} \le \theta_s$$

where

σ_{WR}^2 is the population within-subject variance of the reference formulation

$\theta_s = \left(\frac{(\ln(1.25))^2}{\sigma_{W_0}^2} \right)$ is the BE limit

$\sigma_{W_0}^2$ is a predetermined constant set by the regulatory agency, in this case, the FDA

Under this model, the implied limits (which represent FDA's desired consumer risk model) on $\mu_T - \mu_R$ are as follows:

$$-\left[\ln(1.25) \frac{\sigma_{WR}}{\sigma_{W_0}} \right] \le \mu_T - \mu_R \le \frac{\ln(1.25)\sigma_{WR}}{\sigma_{W0}}$$

When $\sigma_{WR} = \sigma_{W0}$, the implied limits are equal to the standard unscaled BE limits of $\pm\ln(1.25)$ (0.8–1.25). If $\sigma_{WR} \ge \sigma_{W0}$, the implied limits are wider than the standard limits.

The FDA recommends using a mixed scaling approach to RSABE analysis. This is because the Agency determined that it is acceptable for the implied limits to be wider than the standard limits only when σ_{WR} is large (as for HV drugs).

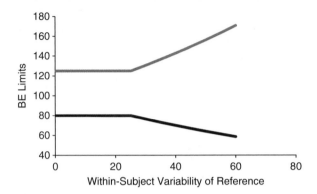

Fig. 6.2 Implied BE limits are plotted as a function of the population reference product within-subject variability of the BE measure. When $\sigma_{W_0} \leq 0.25$, for an acceptable BE study, the 90 % CI of the BE measure test/reference GMRs must fall within 80–125 % limits. When $\sigma_{W_0} > 0.25$, the implied limits scale as the reference product within-subject variability increases. The slope of this portion of the curve is determined by the value of σ_{W0}. The FDA does not permit scaling of the limits to be applicable until $\sigma_{WR} \geq 0.294$

Under the mixed scaling model, test and reference are considered bioequivalent if

$$\frac{(\mu_T - \mu_R)^2}{\sigma_{W_0}^2} \leq \frac{(\ln(1.25))^2}{\sigma_{W_0}^2}, \quad \text{when } \sigma_{WR} \leq \sigma_{W0}$$

and if

$$\frac{(\mu_T - \mu_R)^2}{\sigma_{WR}^2} \leq \frac{(\ln(1.25))^2}{\sigma_{W_0}^2}, \quad \text{when } \sigma_{WR} > \sigma_{W0}$$

FDA sets the value of σ_{W0} at 0.25 (Haidar et al. 2008a, b; US Department of Health and Human Services 2012). Under the mixed scaling model and with $\sigma_{W0} = 0.25$, the implied limits on $\mu_T - \mu_R$ are as depicted in Fig. 6.2.

Direct implementation of FDA's desired consumer risk model is impossible because σ_{WR} is a characteristic of the entire population and thus not directly measured in any particular study. Therefore, FDA proposes an implementation algorithm (US Department of Health and Human Services 2012). In FDA's implementation algorithm for mixed scaling studies, the observed within-subject variability of the reference s_{WR} (determined in the BE study) is compared to a cutoff value of 0.294, above which reference scaling is used. This implementation reduces the type I error (defined relative to FDA's desired consumer risk model) when the within-subject variability is near σ_{W0}.

6.4.1 Study Design

To use the RSABE approach, the reference product must be administered twice in order to determine its within-subject standard deviation (Haidar et al. 2008a, b). As such, the BE study can use either a partial replicate (three-way crossover, RTR, RRT, or TRR) or full replicate (four-way crossover, RTRT or TRTR) design, should enroll a minimum of 24 subjects (US Department of Health and Human Services 2012; Davit and Conner 2010). The FDA recommends a s_{WR} cutoff value of 0.294, at or above which reference scaling is permitted and below which the unscaled limits of 0.8–1.25 are applied (Haidar et al. 2008a, b; US Department of Health and Human Services 2012). The selection of 0.294 as the variation at which use of reference scaling of the limits is permissible is consistent with the general understanding that drugs are considered HV if the within-subject %CV observed in the study is ≥ 30 %, and, as such, is determined by using the conversion formula of $s^2 = \ln(CV^2 + 1)$.

FDA determined that there are advantages to choosing a s_{WR} cutoff somewhat larger than σ_{W0}. Agency scientists conducted a series of simulations of BE study results (1,000,000 per condition) to compare the effects of applying different values to σ_{W0} (Haidar et al. 2008a, b). Results of these simulation studies showed that using a σ_{W0} of 0.25 both (a) increased the study power compared to ABE without causing relatively large numbers of studies to pass when the GMR ≥ 1.2; and (b) resulted in a lower inflation of type I error compared to using a value of 0.294 (the same value as the cutoff) for σ_{W0}. Simulation studies conducted by others showed similar findings (Karalis et al. 2012).

The FDA recommends a secondary ("point estimate") constraint of 0.8–1.25 on the GMR (Haidar et al. 2008a, b; US Department of Health and Human Services 2012). As it is possible that, using RSABE, two products could be shown to be bioequivalent but have an estimated GMR outside of the 0.8–1.25 range, it is thought that use of the secondary GMR constraint will improve the confidence of clinicians and patients (Tothfalusi et al. 2009; Benet 2006). Several simulation studies investigating the relationship between FDA's recommended RSABE approach and study power have shown that, when within-subject variability exceeds 50–60 %, the GMR constraint becomes the dominant regulatory criterion rather than the scaling (Haidar et al. 2008a, b; Endrenyi and Tothfalusi 2009; Tothfalusi and Endrenyi 2011). It has also been shown that applying the GMR constraint can increase the sample size needed to show BE for drugs with within-subject variability >50–60 % (Tothfalusi and Endrenyi 2011). Thus, using a GMR constraint results in a situation where the benefits of using RSABE may be reduced when applied to drugs with very high within-subject variability.

However, it has also been argued that, without a GMR constraint in effect, the permissiveness of the RASBE approach can become excessively high (Benet 2006). Another argument favoring the incorporation of a GMR constraint is that it has a long history of successful application by regulatory agencies as a BE study acceptance criterion (Tothfalusi et al. 2009). For example, using a GMR constraint as a criterion for study outcome acceptance is used by the Health Canada Health Products and Food Branch for C_{max} in all BE studies (Health Canada 2012), and, until 2003, by the FDA for both AUC and C_{max} in BE studies that were conducted under fed conditions (Davit et al. 2012).

6.4.2   Data Processing

Using the RSABE approach recommended by the FDA, two products are bioequivalent when the 95 % upper confidence bound for $((\mu_T - \mu_R)^2/\sigma_{WR}^2) \leq \theta_s$, or, equivalently, a 95 % upper confidence bound for $(\mu_T - \mu_R)^2 - \theta_s\sigma_{WR}^2$ (US Department of Health and Human Services 2012). In addition, the GMR of the two products should fall between 0.8 and 1.25.

It should be noted that we do not know the values of the above population parameters and can never know their values. What we can do and do in actuality is calculate CIs around the BE measures, log(AUC) and/or log(C_{max}).

FDA posted a Guidance for Industry providing step-by-step instructions on how to statistically analyze BE study data using RSABE (US Department of Health and Human Services 2012). The instructions from the guidance are shown in an Appendix to this chapter.

The intention to use RSABE for an HV drug should be stated a priori in the study protocol. The first step in the analysis is to determine s_{WR}, the within-subject standard deviation (SD) of the reference product estimated from the study, for each of the BE measures AUC and C_{max}. If $s_{WR} < 0.294$, then the traditional two one-sided tests procedure should be used for data analysis, and BE acceptance limits of 80–125 % applied to the 90 % CIs of the test/reference BE measure GMRs. If $s_{WR} \geq 0.294$, then RSABE can be applied to the BE measure (either AUC, C_{max}, or both parameters). Figure 6.3 illustrates the decision-making process used by FDA when it determines whether the RSABE is applicable to BE study data.

Once it has been determined that the RSABE can be applied to a BE measure, the next step is to determine, using *Howe's Approximation I* (Howe 1974) the 95 % upper confidence bound for $\left(\overline{Y}_T - \overline{Y}_R\right)^2 - \theta_s\sigma_{WR}^2$, where

$\overline{Y}_T$ and $\overline{Y}_R$ are the means of the ln-transformed PK endpoints (AUC and/or C_{max}) obtained in the BE study for the test and reference product, respectively.

The test and reference products are concluded to be bioequivalent if

(a) The 95 % upper confidence bound for $\left(\overline{Y}_T - \overline{Y}_R\right)^2 - \theta_s\sigma_{WR}^2 \leq 0$
(b) The test/reference GMR in the study falls within [0.8, 1.25]

Two examples, provided in Table 6.2, show how the FDA applies the RSABE to studies of HV drugs. Both cases in the illustration are from BE submissions of generic products that were subsequently approved; thus, the RSABE was used to

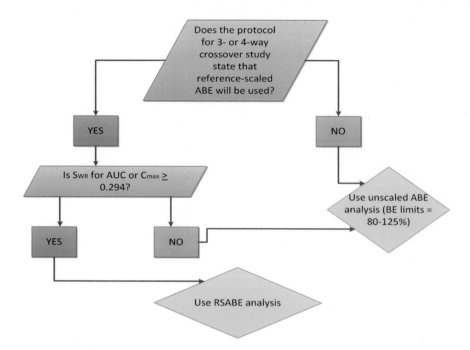

Fig. 6.3 Decision tree showing the process whereby the FDA scientific review staff decide whether it is possible to use RSABE or unscaled ABE. The first condition to be met is that the study protocol must state a priori that RSABE will be the method of statistical analysis

support the approvals in these two cases. In both cases, the investigators used three-way partial replicate study designs. The data in Table 6.2 are from FDA's calculations; it is standard review practice for FDA BE review staff to conduct calculations independently of the analysis performed by the applicant. SAS$^{®}$ Version 9.2 was used in performing the calculations illustrated.

As shown in Table 6.2, in the case of Drug A, the first step of the analysis determined that the values of s_{WR} exceeded the regulatory cutoff (0.294) for AUC_{0-t}, AUC_{∞}, and C_{max}. Thus, the BE application reviewer proceeded to use RSABE for all three BE measures. Table 6.2 shows that the calculated 95 % upper confidence bound for all three parameters <0. Thus, the product met the first BE study acceptance criterion. In addition, the point estimates for all three BE measures fell within 0.8 and 1.25. Thus, Drug A met the two RSABE acceptance criteria and was deemed bioequivalent to its corresponding reference.

The case of Drug B in Table 6.2 illustrates the decision-making approach in a situation where only one of the BE measures met the criteria for application of RSABE. In this case, s_{WR} values for AUC_{0-t} and AUC_{∞} were less than 0.294. Thus, the ABE two one-sided tests approach was applied for analyzing AUC_{0-t} and AUC_{∞}. With 90 % CIs of 89.82–109.53 % and 92.57–111.77 %, respectively, both of these two parameters met the traditional BE study [unscaled] limits of 80–125 %. However, the RSABE approach could be applied to the Drug B BE

Table 6.2 Summary BE statistics for two HV generic drugs, analyzed using RSABE or unscaled ABE, depending upon the value of s_{WR} (*PE* point estimate)

Parameter	T/R	90% CI	S_{WR}^2	S_{WR}	Criteria bound	Method used	Outcome
Drug A, fed bioequivalence study, N = 43							
AUC_{0-t}	1.17	93.91-132.91	0.5723	0.7565	-0.2892	Scaled/PE	Pass
$AUC\infty$	1.11	94.64-124.91	0.3452	0.5875	-0.1767	Scaled/PE	Pass
C_{max}	1.19	100.47-133.26	0.3768	0.6138	-0.1684	Scaled/PE	Pass
Drug B, fasting bioequivalence study, N = 36							
AUC_{0-t}	0.99	89.82-109.53	0.07425	0.2725	-0.03914	Unscaled	Pass
$AUC\infty$	1.02	92.57-111.77	0.06561	0.2561	-0.03126	Unscaled	Pass
C_{max}	0.98	85.70-110.54	0.1069	0.3270	-0.05294	Scaled/PE	Pass

The *green-colored blocks* designate which parameters were tested to conclude that bioequivalence was present

study C_{max} data, as its s_{WR} value of 0.327 exceeded the cutoff of 0.294. In this illustration, C_{max} met both acceptance criteria; its 95 % upper confidence bound, at -0.05294, was <0, and its point estimate, at 0.98, fell within the 0.8–1.25 limits. The Drug B case in Table 6.2 illustrates two important points in the analysis of BE study data from HV drugs. The first point is that both ABE and RSABE analysis can be applied to data from the same study depending on BE measure variability. The second point is that there is no penalty if the applicant uses a three- or four-way study design with the intent of using RSABE but s_{WR} fails to meet the cutoff of 0.294.

6.4.3 Application of RSABE to Generic Drug Development: Overview

From 2007 to 2012, the FDA evaluated 46 Abbreviated New Drug Applications (ANDAs) containing 64 BE studies in which RSABE was applied (Davit et al. 2012). Of these 64 studies, 62 met the RSABE criteria, and thus, the test (generic) and reference (corresponding innovator) HV drug products were deemed bioequivalent. Of the 62 studies, two were unacceptable. One of these two studies was unacceptable because s_{WR} did not meet the 0.294 cutoff (i.e., was <0.294) for one of the BE measures, and in addition this product did not meet average BE limits when the two one-sided tests procedure was applied. The second of these two studies was unacceptable because the BE study results did not meet the GMR constraint, i.e, the GMR exceeded 1.25 despite having a 90 % upper confidence bound <0. Interestingly, the applicant later conducted for this product a second successful RSABE study which met both the BE limits and the GMR constraint without reformulating or increasing the number of subjects. This example suggests that, due to uncertainty around the BE measures obtained when PK variability is high, chance plays a large role in determining the outcome of a BE study of an HV drug, particularly if the number of study subjects is not high enough for adequate study power.

From 2007 to 2012, RSABE supported four full approvals of new generic drug products, and one tentative approval. The tentative rather than full approval was because the patent had not yet expired on the reference product.

6.4.4 Application of RSABE to New Drug Development: Detailed Case Study

RSABE successfully supported the approval of a new delayed-release (DR) capsule formulation of mesalamine (Office of Clinical Pharmacology and Office of New Drug Quality Assurance Biopharmaceutics 2012). The reformulation, which was evaluated by the FDA under a New Drug Application (NDA), consisted of the mesalamine DR capsule (Delzicol®, NDA 204412), which replaced the mesalamine DR tablet (Asacol®, NDA 19651). The innovator formulated Delzicol® to replace Asacol® for safety reasons; Asacol® contained the plasticizer dibutyl phthalate in the enteric coating. In Delizcol®, dibutyl sebacate replaces dibutyl phthalate as the plasticizer. Due to the nature of the post-approval formulation change on this modified-release (MR) product, the FDA and NDA applicant agreed that it was necessary to conduct an in vivo clinical study with PK endpoints to show that Delzicol® was bioequivalent to Asacol®. Oral mesalamine, which in the DR capsule and DR tablet formulations is indicated to treat mild to moderately active ulcerative colitis, has high within-subject PK variability (>30 %). Thus, the FDA and the NDA applicant agreed that RSABE was an appropriate design for the in vivo study.

The primary PK endpoints for comparison in the BE study were C_{max}, AUC_{0-t}, and the partial (p) AUC_{8-48h}. The pAUC was recommended by the FDA as it was perceived to reflect drug absorption (and therefore drug availability) at the site of action, which is the colon. These mesalmine MR oral dosage forms were good candidates for RSABE analysis, due to extraordinarily high within-subject variability occurring for all PK parameters of interest. In the pivotal BE study submitted to support Delzicol® approval, within-subject variability values for mesalamine C_{max}, AUC_{0-t}, and AUC_{4-48h} were 170 %, 272 %, and 268 %, respectively, for the DR capsule, and 200 %, 306 %, 286 %, respectively, for the DR tablet. The DR tablet was the reference in the BE study. An important point to remember in RSABE for HV drugs is that only the within-subject variance of the reference

Table 6.3 The RSABE analysis below shows that Delzicol® (mesalamine DR capsule, Test product) is bioequivalent to Asacol® (mesalamine DR tablet, Reference product)

Parameter	Geometric mean		Reference product variability measures		Point estimate (test/reference)	95 % UCB of the linearized criterion
	Test	Reference	s_{WR}^2	s_{WR}		
C_{max} (ng/mL)	109.9	99.4	1.516	1.231	1.11	−1.030
AUC_{8-48h} (ng h/mL)	618.3	556.4	2.073	1.439	1.10	−1.265
AUC_{0-t} (ng h/mL)	719.6	648.4	1.849	1.360	1.08	−1.608

Calculations for this table were performed by FDA reviewers. PK parameters were determined for mesalamine. For all three key PK parameters (in this case C_{max}, AUC_{8-48h} and AUC_{0-t}), s_{WR} exceeded the regulatory cutoff of 0.294; thus, RSABE analysis could proceed. For the formulation comparisons to pass the RSABE analysis, the 95 % upper confidence bound (UCB) of the linearized criterion should be ≤0, and the point estimates of the Test/Reference GMR should be within 0.80–1.25 for the key PK parameters

product (the tablet in this case) PK parameters is used in the analysis to scale the BE limits; the within-subject variance of the test product (the capsule in this case) is not considered. The RSABE for the Delzicol® versus Asacol® comparison used 238 subjects who received single doses of the test and reference product via a fully replicated crossover design. Subjects were randomly assigned to one of the two following sequences: Reference—Test—Reference—Test (Sequence A), or Test—Reference—Test—Reference (Sequence B). PK was determined via noncompartmental analysis.

All statistical analysis was performed using PC/SAS®, Version 9.2, with an SAS® program code prepared specifically by FDA's OGD for the RSABE analysis. The GMR was calculated as $(T/R) = [(T1 \times T2)/(R1 \times R2)]1/2$. The test/reference GMRs for C_{max}, AUC_{8-48} and AUC_{0-t} were calculated. The within-subject standard deviation for each formulation and each PK parameter of interest was estimated from the analysis of variance of the log-transformed parameter using the RSABE procedure as described in the February 2011 Draft Guidance on Progesterone (US Department of Health and Human Services 2012). The same procedure was used to determine the 95 % (one-sided) upper-confidence bound (UCB) on the linearized criterion for these PK parameters.

As shown in Table 6.3, FDA's analysis of the study data demonstrated BE between Delzicol® and Asacol® using RSABE; similar results were obtained by the NDA applicant. The Delzicol® example is a particularly compelling case study for two reasons. First, due to the extraordinarily high within-subject variability of mesalamine PK parameters, it may not have been feasible to show BE for this product using traditional BE approaches. Second, although the RSABE was implemented to promote development of generic HV drugs by easing regulatory burdens, this was the first time that the approach was successfully used to support a major post-approval reformulation of an innovator's MR drug product.

6.5 Approaches Used by Other Regulatory Agencies in Designing BE Studies of HV Drugs

Regulatory agencies use a number of diverse approaches to reduce the number of subjects needed for an acceptable study of two bioequivalent HV drugs (Davit et al. 2013). Table 6.4 summarizes similarities and differences in such approaches. Several agencies, for example, permit widening BE limits for C_{max}, if previously established in the study protocol and if scientifically justified. Japan, the Association of South East Asian Nations (ASEAN), and the World Health Organization (WHO) suggest using a steady-state BE study design to reduce variability. Japan also recommends using studies with stable isotopes for HV drugs that may require large sample sizes in the BE studies. Like the FDA, the European Medicines Agency recommends an RSABE approach. In the EMA approach, the reference product is administered twice in the study (either a three-way or four-way design is acceptable), and the acceptance limits scale based on the within-subject variability of the reference product. Like the FDA, the EMA imposes a GMR constraint when using RSABE for HV drugs. However, as shown in Table 6.4, aside from the above similarities, the EMA and the USA implement RSABE in different ways. In addition, the EMA only permits RSABE to be used for C_{max}, whereas the FDA will accept RSABE for both AUC and C_{max}. Australia will consider using the EMA RSABE approach for BE studies of HV drugs provided that (1) the drug has highly variable PK due to incomplete or variable absorption or substantial (>40 %) first-pass metabolism; and (2) the reference product is one marketed in Australia. Finally, Health Canada states that there is no compelling need for a distinct category of HV drugs.

6.6 Conclusion

The RSABE approach is currently being used to support approvals of generic as well as new drugs. As of 2012, acceptable RSABE studies supported the approvals of ANDAs for four new HV generic drugs. Although originally developed for use with generic drug development, RSABE is now being applied to new drug development in some situations. Within the last year (2013), an NDA applicant was successful in using RSABE study design and analysis to support the approval of a major post-approval change in an MR formulation of the HV drug mesalamine (US Department of Health and Human Services et al. 2013a, b). Clearly, RSABE represents a significant advance in BE assessment of new and generic drug development. The availability of this science-driven approach has reduced unnecessary human testing and offered greater flexibility for designing BE studies of challenging drug products.

Table 6.4 Similarities and differences in how international regulatory agencies recommend designing BE studies of HV drugs

Highly variable drugs	
Similarities	All recommend
	• Crossover or parallel study designs
	• Non-compartmental analysis to determine PK parameters
	• ANOVA, performed at the 5 % level of significance, on the GMRs
Differences	• Australia: As follows:

- If the generic drug has highly variable PK due to incomplete or variable absorption or substantial (>40 %) first-pass metabolism, then the reference product must be one marketed in Australia
- If this criterion is met, then follow the EMA recommendations for HVD BE studies

• Brazil: A wider BE acceptance limit may be applied to C_{max}, if previously established in the study protocol and if scientifically justified

• Chinese Taipei, South Korea: Do not specify/mention

• Canada: No compelling need for a distinct category of HV drugs

• Singapore/ASEAN, WHO: One of the following approaches can be used:

- In rare cases, a wider BE limit acceptance range may be applied to AUC and C_{max}, if based on sound clinical justification
- A steady-state BE study can be conducted to reduce variability

• EMA: An RSABE approach may be applied to C_{max} only. A brief summary of study design and acceptance criteria is as follows:

- The reference product should be administered at least twice to determine within-subject variability
- Either a 3-period or 4-period replicate design study is acceptable
- BE limits are scaled to the within-subject variability of the reference product
- The C_{max} GMR in the study should fall within 0.80–1.25
- The within-subject %CV and corresponding BE limits are shown below

Within-subject CV (%)	BE limits
30	80.00 125.00
35	77.23 129.48
40	74.62 134.02
45	72.15 138.59
≥ 50	69.84 143.19

• Japan: One of the following approaches may be used to reduce variability:

- For drugs with a wide therapeutic index, it may be appropriate to set BE limits wider than 80–125 % for C_{max}
- A steady-state BE study
- A study with stable isotope

• USA: An RSABE approach may be applied to AUC and C_{max} (67). A brief summary of study design and acceptance criteria is as follows:

- The reference product should be administered at least twice to determine within-subject variability
- Either a 3-period or 4-period replicate design study is acceptable; The AUC and C_{max} GMRs in the study should fall within 0.80–1.25

(continued)

Table 6.4 (continued)

Highly variable drugs
– BE limits are scaled to the within-subject variability of the reference product, but not until the within-subject standard deviation of the reference product (s_{WR}) is ≥ 0.294
– A 95 % upper confidence bound for $(\mu_T - \mu_R)^2 - \theta s_{WR}^2$ must be ≤ 0, where μ_T = test product mean μ_R = reference product mean $\theta = \frac{(\ln\Delta)^2}{\sigma_{w0}^2}$ $\Delta = 1.25$ $\sigma_{w0} = 0.25$

The regulatory drug approval agencies surveyed include those of Australia, Brazil, Canada, Chinese Taipei, the European Medicines Agency (EMA), Japan, Switzerland (which follows EMA guidelines), Singapore (which follows the Association of South East Asian Nations guidelines), South Korea, the USA. The World Health Organization (WHO) is also included in the survey

Appendix. Method for Statistical Analysis Using the Reference-Scaled Average Bioequivalence Approach

Step 1. Determine s_{WR}, the within-subject standard deviation (SD) of the reference product, for the pharmacokinetic (PK) parameters AUC and C_{max}.

(a) If $s_{WR} < 0.294$, use the two one-sided tests procedure to determine bioequivalence (BE) for the individual PK parameter(s)

(b) If $s_{WR} \geq 0.294$, use the reference-scaled procedure to determine BE for the individual PK parameter(s)

Calculation for s_{WR} can be conducted as follows:

$$s_{WR}^2 = \frac{\sum_{i=1}^{m} \sum_{j=1}^{n_i} \left(D_{ij} - \bar{D}_i\right)^2}{2(n-m)}$$

where:

i = number of sequences m used in the study

[$m = 3$ for partially replicated design: TRR, RTR, and RRT; $m = 2$ for fully replicated design: TRTR and RTRT]

j = number of subjects within each sequence
T = Test product
R = Reference product
$D_{ij} = R_{ij1} - R_{ij2}$ (where 1 and 2 represent replicate reference treatments)

$$\overline{D}_i = \frac{\sum_{j=1}^{n_i} D_{ij}}{n_i}$$

$n = \sum_{i=1}^{m} n_i$ (i.e., total number of subjects used in the study, while n_i is number of subjects used in sequence i)

> *Mixed scaling*: AUC (AUC_{0-t} and $AUC_{0-\infty}$, as applicable) and C_{max} may have different s_{WR} values. Only use the reference-scaled procedure for the specific PK parameter that has a $s_{WR} \geq 0.294$. The two one-sided tests procedure must be used for PK parameters with $s_{WR} < 0.294$.

Continue with steps 2 and 3 for PK parameters that have a $s_{WR} \geq 0.294$.
Continue with steps 2 and 3 for PK parameters that have a $s_{WR} \geq 0.294$.

Step 2. Determine the 95 % upper confidence bound for:

$$\left(\bar{Y}_T - \bar{Y}_R\right)^2 - \theta s_{WR}^2$$

where:

$\bar{Y}_T$ and $\bar{Y}_R$ are the means of the ln-transformed PK endpoint (AUC and/or C_{max}) obtained from the BE study for the test and reference products, respectively

$\theta \equiv \left(\frac{\ln(1.25)}{\sigma_{w0}}\right)^2$ (scaled average BE limit)

$\sigma_{w0} = 0.25$ (regulatory limit)

The method of obtaining the upper confidence bound is based on *Howe's Approximation I*, which is described in the following chapter:

W. G. Howe (1974), Approximate Confidence Limits on the Mean of X+Y Where X and Y are Two Tabled Independent Random Variables, *Journal of the American Statistical Association*, 69 (347): 789–794.

Step 3. For the test product to be bioequivalent to the reference product, *both* of the following conditions must be satisfied for each PK parameter tested:

(a) The 95 % upper confidence bound for $\left(\bar{Y}_T - \bar{Y}_R\right)^2 - \theta s_{WR}^2$ must be ≤ 0

(b) The point estimate of the Test/Reference GMR must fall within [0.80, 1.25]

> If SAS® is used for statistical analysis (not necessary to use SAS® if other software accomplishes same objectives)
>
> - PROC MIXED should be used for fully replicated (four-way) BE studies
> - PROC GLM should be used for partially replicated (three-way) BE studies

Example SAS Codes: Partial Reference-Replicated Three-Way Design

For a bioequivalence study with the following sequence assignments in a partial reference-replicated three-way crossover design:

	Period 1	Period 2	Period 3
Sequence 1	T	R	R
Sequence 2	R	T	R
Sequence 3	R	R	T

The following codes are an example of the determination of reference-scaled ABE for LAUCT.

Dataset containing TEST observations:

```
data test;
  set pk;
  if trt='T';
  latt=lauct;
run;
```

Dataset containing REFERENCE 1 observations:

```
data ref1;
  set ref;
  if (seq=1 and per=2) or (seq=2 and per=1) or (seq=3 and per=1);
  lat1r=lauct;
run;
```

Dataset containing REFERENCE 2 observations:

```
data ref2;
  set ref;
  if (seq=1 and per=3) or (seq=2 and per=3) or (seq=3 and per=2);
  lat2r=lauct;
run;
```

Define the following quantities:

T_{ij} = the observation on T for subject j within sequence i
R_{ijk} = kth observation ($k = 1$ or 2) on R for subject j within sequence i

$$I_{ij} = T_{ij} - \frac{R_{ij1} + R_{ij2}}{2}$$

and
$$D_{ij} = R_{ij1} - R_{ij2}$$

I_{ij} is the difference between a subject's (specifically subject j within sequence i) observation on T and the mean of the subject's two observations on R, while D_{ij} is the difference between a subject's two observations on R.

Determine I_{ij} and D_{ij}

```
data scavbe;
  merge test ref1 ref2;
  by seq subj;
  ilat=latt-(0.5*(lat1r+lat2r));
  dlat=lat1r-lat2r;
run;
```

Intermediate analysis—ilat

```
proc glm data=scavbe;
  class seq;
  model ilat=seq/clparm alpha=0.1;
  estimate 'average' intercept 1 seq 0.3333333333 0.3333333333
0.3333333333;
  ods output overallanova=iglm1;
  ods output Estimates=iglm2;
  ods output NObs=iglm3;
  title1 'scaled average BE';
run;
```

From the dataset IGLM2, calculate the following:

IGLM2:   pointest=exp(estimate);
            x=estimate**2-stderr**2;
            boundx=(max((abs(LowerCL)),(abs(UpperCL))))**2;

Intermediate analysis—dlat

```
proc glm data=scavbe;
  class seq;
  model dlat=seq;
  ods output overallanova=dglm1;
  ods output NObs=dglm3;
  title1 'scaled average BE';
run;
```

From the dataset DGLM1, calculate the following:

DGLM1:   dfd=df;
            s2wr=ms/2;

From the above parameters, calculate the final 95 % upper confidence bound:

```
theta=((log(1.25))/0.25)**2;
y=-theta*s2wr;
boundy=y*dfd/cinv(0.95,dfd);
sWR=sqrt(s2wr);
critbound=(x+y)+sqrt(((boundx-x)**2)+((boundy-y)**2));
```

Example SAS Codes: Fully Replicated Four-Way Design

For a bioequivalence study with the following sequence assignments in a partial reference-replicated four-way crossover design:

	Period 1	Period 2	Period 3	Period 4
Sequence 1	T	R	T	R
Sequence 2	R	T	R	T

The following codes are an example of the determination of reference-scaled ABE for LAUCT.

Dataset containing TEST 1 observations:

```
data test1;
  set test;
  if (seq=1 and per=1) or (seq=2 and per=2);
  lat1t=lauct;
run;
```

Dataset containing TEST 2 observations:

```
data test2;
  set test;
  if (seq=1 and per=3) or (seq=2 and per=4);
  lat2t=lauct;
run;
```

Dataset containing REFERENCE 1 observations:

```
data ref1;
  set ref;
  if (seq=1 and per=2) or (seq=2 and per=1);
  lat1r=lauct;
run;
```

Dataset containing REFERENCE 2 observations:

```
data ref2;
  set ref;
  if (seq=1 and per=4) or (seq=2 and per=3);
  lat2r=lauct;
run;
```

Further assume that there are no missing observations. All subjects provide two observations on T and two observations on R. The number of subjects in each sequence is n_1 and n_2 for sequences 1 and 2, respectively.

Define the following quantities:

$T_{ijk} = k$th observation ($k = 1$ or 2) on T for subject j within sequence i
$R_{ijk} = k$th observation ($k = 1$ or 2) on R for subject j within sequence i

$$I_{ij} = \frac{T_{ij1} + T_{ij2}}{2} - \frac{R_{ij1} + R_{ij2}}{2}$$

and
$$D_{ij} = R_{ij1} - R_{ij2}$$

I_{ij} is the difference between the mean of a subject's (specifically subject j within sequence i) two observations on T and the mean of the subject's two observations on R, while D_{ij} is the difference between a subject's two observations on R.

Determine I_{ij} and D_{ij}

```
data scavbe;
  merge test1 test2 ref1 ref2;
  by seq subj;
  ilat=0.5*(lat1t+lat2t-lat1r-lat2r);
  dlat=lat1r-lat2r;
run;
```

Intermediate analysis—ilat

```
proc mixed data=scavbe;
   class seq;
   model ilat =seq/ddfm=satterth;
   estimate 'average' intercept 1 seq 0.5 0.5/e cl alpha=0.1;
   ods output CovParms=iout1;
   ods output Estimates=iout2;
   ods output NObs=iout3;
   title1 'scaled average BE';
   title2 'intermediate analysi - ilat, mixed';s
run;
```

From the dataset IOUT2, calculate the following:

```
IOUT2:      pointest=exp(estimate);
            x=estimate**2-stderr**2;
            boundx=(max((abs(lower)),(abs(upper))))**2;
```

Intermediate analysis—dlat

```
proc mixed data=scavbe;
   class seq;
   model dlat=seq/ddfm=satterth;
   estimate 'average' intercept 1 seq 0.5 0.5/e cl alpha=0.1;
   ods output CovParms=dout1;
   ods output Estimates=dout2;
   ods output NObs=dout3;
   title1 'scaled average BE';
   title2 'intermediate analysis - dlat, mixed';
run;
```

From the dataset DOUT1, calculate the following:

```
DOUT1:      s2wr=estimate/2;
```

From the dataset DOUT2, calculate the following:

```
DOUT2:      dfd=df;
```

From the above parameters, calculate the final 95 % upper confidence bound:

```
theta=((log(1.25))/0.25)**2;
y=-theta*s2wr;
boundy=y*dfd/cinv(0.95,dfd);
sWR=sqrt(s2wr);
critbound=(x+y)+sqrt(((boundx-x)**2)+((boundy-y)**2))
```

For PK parameters with an $s_{WR} < 0.294$, use the unscaled ABE approach:

Calculation of unscaled 90 % bioequivalence confidence intervals:

```
PROC MIXED
  data=pk;
  CLASSES SEQ SUBJ PER TRT;
  MODEL LAUCT = SEQ PER TRT/ DDFM=SATTERTH;
  RANDOM TRT/TYPE=FA0(2) SUB=SUBJ G;
  REPEATED/GRP=TRT SUB=SUBJ;
  ESTIMATE 'T vs. R' TRT 1 -1/CL ALPHA=0.1;
  ods output Estimates=unsc1;
  title1 'unscaled BE 90% CI - guidance version';
  title2 'AUCt';
run;

data unsc1;
  set unsc1;
  unscabe_lower=exp(lower);
  unscabe_upper=exp(upper);
run;
```

References

Amidon G (2004) Sources of variability: physicochemical and gastrointestinal, FDA Advisory Committee for Pharmaceutical Sciences and Clinical Pharmacology meeting transcript. Available via US Food and Drug Administration Dockets. http://www.fda.gov/ohmrs/dockets/ac/04/transcripts/4034T2.htm. Accessed 19 Feb 2014

Benet L (2004) Why highly variable drugs are safer, FDA Advisory Committee for Pharmaceutical Sciences and Clinical Pharmacology meeting transcript. Available via US Food and Drug Administration Dockets. http://www.fda.gov/ohrms/dockets/ac/04/transcripts/4034T2.htm. Accessed 19 Feb 2014

Benet L (2006) Therapeutic considerations of highly variable drugs, FDA Advisory Committee for Pharmaceutical Sciences and Clinical Pharmacology meeting transcript. Available via US Food and Drug Administration Dockets. http://www.fda.gov/ohrms/dockets/ac/06/minutes/2006-4241t2-01.pdf. Accessed 19 Feb 2014

Blume H, Midha K (1993) Bio-International 92, conference on bioavailability, bioequivalence, and pharmacokinetic studies. J Pharm Sci 82:1186–1189

Conner D (2009) Bioequivalence methods for highly variabile drugs and drug products, FDA Advisory Committee for Pharmaceutical Sciences and Clinical Pharmacology meeting transcript. Available via US Food and Drug Administration Dockets. http://www.fda.gov/downloads/ADvisoryCommittees/CommitteesMeetingMaterials/Drugs/AdvisoryCommitteeforPharmaceuticalScienceandClinicalPharmacology/UCM179891.pdf. Accessed 19 Feb 2014

Cook J, Davit B, Polli J (2010) Impact of biopharmaceutics classification system-based biowaivers. Mol Pharm 7:1539–1544

Davit B, Braddy A, Conner D, Yu L (2013) International guidelines for bioequivalence of systemically-available orally-administered generic drug products: a survey of similarities and differences. AAPS J 15:974–990

Davit B, Chen M-L, Conner DP et al (2012) Implementation of a reference-scaled average bioequivalence approach for highly variable generic drug products by the US Food and Drug Admininstration. AAPS J 14:915–924

Davit BM, Conner DP (2010) The United States of America. In: Kanfer I, Shargel L (eds) Generic drug product development: international regulatory requirements for bioequivalence. Informa Healthcare, New York, pp 254–281

Davit BM, Conner DP, Fabian-Fritsch B et al (2008) Highly variable drugs: observations from bioequivalence data submitted to the FDA for new generic drug applications. AAPS J 10:148–156

Davit BM, Nwakama PE, Buehler GJ et al (2009) Comparing generic and innovator drugs: a review of 12 years of bioequivalence data from the United States Food and Drug Administration. Ann Pharmacother 43:1583–1597

DiLiberti C (2004) Why bioequivalence of highly variable drugs is an issue, FDA Advisory Committee for Pharmaceutical Sciences and Clinical Pharmacology meeting transcript. Available via US Food and Drug Administration Dockets. http://www.fda.gov/ohrm/dockets/ac/04/transcripts/4034T2.htm. Accessed 19 Feb 2014

Endrenyi L (2004) Bioequivalence methods for highly variable drugs, FDA Advisory Committee for Pharmaceutical Science and Clinical Pharmacology meeting transcript. Available via US Food and Drug Administration Dockets. http://www.fda.gov/ohrms/dockets/ac/04/transcripts/4034T2.htm. Accessed 19 Feb 2014

Endrenyi L, Tothfalusi L (2009) Regulatory conditions for the determination of bioequivalence of highly variable drugs. J Pharm Pharmaceut Sci 12:138–149

Executive Secretary (2004) Minutes of the meeting of the FDA Advisory Committee for Pharmaceutical Science and Clinical Pharmacology. Available via US Food and Drug Administration Dockets. http://www.fda.gov/ohrms/dockets/ac/04/minutes/4034M1.htm. Accessed 19 Feb 2014

Executive Secretary (2006) Minutes of the meeting of the FDA Advisory Committee for Pharmaceutical Science and Clinical Pharmacology. Available via US Food and Drug Administration Dockets. http://www.fda.gov/ohrms/dockets/ac/06/minutes/2006-4241m2.pdf. Accessed 19 Feb 2014

FDA Advisory Committee for Pharmaceutical Science and Clinical Pharmacology (2001) Discussion of Individual Bioequivalence Issues. Available via US Food and Drug Administration Dockets. http://www.fda.gov/ohrms/dockets/ac/01/transcripts/3804t2_03_Afternoon_Session.pdf. Accessed 19 Feb 2014

Haidar SH, Davit BM, Chen M-L et al (2008a) Bioequivalence approaches for highly variable drugs and drug products. Pharm Res 25:237–241

Haidar SH, Makhlouf F, Schuirmann DJ et al (2008b) Evaluation of a scaling approach for the bioequivalence of highly variable drugs. AAPS J 10:450–454

Hauck WH, Hyslop T, Chen M-L et al (2000) Subject-by-formulation interaction in bioequivalence: conceptual and statistical terms. Pharm Res 17:375–380

Health Canada (2012) Guidance document—comparative bioavailability standards: formulations used for systemic effects. Available via the Health Products and Food Branch of Canada. http://www.hc-sc.gc.ca/dhp-mps/prodpharma/applic-demande/guide-ld/bio/gd_standards_ld_normes-eng.php. Accessed 19 Feb 2013

Howe W (1974) Approximate confidence limits on the mean of X+Y where X and Y are two independent random variables. J Am Stat Assoc 69:789–794

Karalis V, Sylmillides M, Macheras P (2012) Bioequivalence of highly variable drugs: a comparison of the newly-proposed regulatory approaches by FDA and EMA. Pharm Res 29:1066–1077

Midha K, Rawson MJ, Hubbard JW (2005) The bioequivalence of highly variable drugs and drug products. Int J Clin Pharmacol 43:495–498

Office of Clinical Pharmacology and Office of New Drug Quality Assurance Biopharmaceutics (2012) Clinical pharmacology and biopharmaceutics review, NDA 204412. Available via Drugs@FDA. http://www.accessdata.fda.gov/drugsatfda_docs/nda/2013/204412Orig1s000 ClinPharmR.pdf. Accessed 19 Feb 2014

Office of the Federal Register (2013) Section 320.1(e) Bioequivalence. Code of Federal Regulations, p 190

Patnaik R, Lesko L, Chen M-L, Williams R (1997) Individual bioequivalence: new concepts in the statistical assessment of bioequivalence metrics. Clin Pharmacokinet 33:1–6

Patterson S, Zariffa N, Montague T, Howland K (2001) Non-traditional study designs to demomnsrate average bioequivalence for highly variable drug products. Eur J Clin Pharmacol 57:663–670

Phillips K (1990) Power of the two one-sided tests procedure in bioequivalence. J Pharmacokinet Biopharm 18:137–144

Schuirmann D (1987) A comparison of the two one-sided tests procedure and the power approach for assessing the equivalence of average bioavailability. J Pharmacokinet Biopharm 15:657–680

Shah VP, Yacobi A, Barr WH et al (1996) Evaluation of orally administered highly variable drugs and drug formulations. Pharm Res 13:1590–1594

Tothfalusi L, Endrenyi L (2011) Sample size for designing bioequivalence studies for highly variable drugs. J Pharm Pharmaceut Sci 15:73–84

Tothfalusi L, Endrenyi L, Arieta A (2009) Evaluation of bioequivalence for highly variable drugs with scaled average bioequivalence. Clin Pharmacokinet 48:725–743

Tsang YC, Pop R, Gordon P et al (1996) High variability in drug pharmacokinetics complicates determination of bioequivalence: experience with verapamil. Pharm Res 13:846–850

US Department of Health and Human Services (2001) Statistical approaches to establishing bioequivalence. Available via Food and Drug Administration, Center for Drug Evaluation and Research. http://www.fda.gov/downloads/Drugs/GuidanceCompliance RegulatoryInformation/Guidances/UCM070244.pdf. Accessed 19 Feb 2014

US Department of Health and Human Services (2003) Guidance for industry: bioavailability and bioequivalence studies for orally-administered drug products, general considerations. Available via Food and Drug Administration, Center for Drug Evaluation and Research. http://www.fda.gov/downloads/Drugs/GuidanceComplianceRegulatoryInformation/Guidances/UCM070124.pdf. Accessed 19 Feb 2014

US Department of Health and Human Services (2012) Draft guidance for industry, bioequivalence of progesterone capsules. Available via Food and Drug Administration, Center for Drug Evaluation and Research. http://www.fda.gov/downloads/Drugs/GuidanceCompliance RegulatoryInformation/Guidances/UCM209294.pdf. Accessed 19 Feb 2014

US Department of Health and Human Services, Food and Drug Administration, Center for Drug Evaluation and Research (2013) Approval package, Delzicol, NDA 204412. Available via Drugs@FDA. http://www.accessdata.fda.gov/drugsatfda_docs/nda/2013/204412_delzicol_toc.cfm. Accessed 19 Feb 2014

US Department of Health and Human Services, Food and Drug Administration, Center for Drug Evaluation and Research (2013) Approved drug products with therapeutic equivalence evaluations (Orange Book). Available via Electronic Orange Book. http://www.fda.gov/downloads/Drugs/DevelopmentApprovalProcess/UCM071436.pdf. Accessed 19 Feb 2014

Westlake W (1981) Bioequivalence testing: a need to rethink. Biometrics 37:589–594

Chapter 7
Partial Area Under the Curve: An Additional Pharmacokinetic Metric for Bioavailability and Bioequivalence Assessments

Hao Zhu, Ramana S. Uppoor, Mehul Mehta, and Lawrence X. Yu

7.1 General Background

Bioavailability and bioequivalence assessments are routinely performed during drug development. Bioavailability is defined in the Code of Federal Regulations, Title 21, Section 320, Part1 (21 CFR 320.1) as "the rate and extent to which the active ingredient or active moiety is absorbed from a drug product and becomes available at the site of action." Bioavailability data for a given formulation provides an estimate of the relative fraction of the administered dose that is absorbed into the systemic circulation when compared to that of an optimally available formulation, such as an intravenous dosage form (100 % bioavailability). When an intravenous dosage form is not available, an oral solution or suspension could be used as the reference formulation. Absolute bioavailability is termed to quantify the relative fraction of systemic exposure (e.g., dose normalized $AUC_{0-\infty}$) for a given non-intravenous formulation when an intravenous dosage form is used as the reference formulation, whereas relative bioavailability is used when the reference formulation is another non-intravenous formulation (e.g., oral solution, oral suspension, or another tablet). Bioavailability of a drug is mainly determined by the properties of the drug substance (e.g., solubility and intestinal permeability) and the properties of the formulation (e.g., dissolution). Bioavailability value may also provide information on potential presystemic metabolism (Bylund and Bueters 2013) and/or efflux transporters (e.g., p-gp) (Banerjee et al. 2000). When the performance of different formulations needs to be compared, the comparison of their bioavailability will be one of the most critical aspects. A bioequivalence

---

Opinions expressed in this chapter are those of the authors and may not necessarily reflect the position of the Food and Drug Administration.

H. Zhu (✉) • R.S. Uppoor • M. Mehta • L.X. Yu
Center for Drug Evaluation and Research, U.S. Food and Drug Administration,
10903 New Hampshire Avenue, Silver Spring, MD 20993, USA
e-mail: Hao.Zhu@fda.hhs.gov

L.X. Yu and B.V. Li (eds.), *FDA Bioequivalence Standards*, AAPS Advances
in the Pharmaceutical Sciences Series 13, DOI 10.1007/978-1-4939-1252-0_7,
© The United States Government 2014

study is intended to demonstrate an absence of a meaningful difference in the bioavailability between two formulations (i.e., test product and reference product). In § 320.1 (21CFR320.1), bioequivalence is defined as "the absence of a significant difference in the rate and extent to which the active ingredient or active moiety in pharmaceutical equivalents or pharmaceutical alternatives becomes available at the site of drug action when administered at the same molar dose under similar conditions in an appropriately designed study." Once bioequivalence is established, clinical efficacy and safety information obtained from one formulation may be linked to the other formulation. For example, a bioequivalence study is useful to link a clinical trial formulation to a to-be-marketed formulation during a new drug development. Whenever a major change in the manufacturing process of an approved product occurs, a bioequivalence study may be required to ensure no significant changes in drug release and absorption with the new process (FDA 1997a, b, 1999). Established bioequivalence can be the basis for approval of a generic product via 505 (j) or a new product (e.g., different formulations or salt forms) via 505 b (2) route.

Peak drug concentration (C_{max}) and area under the plasma/serum/blood concentration–time curve (AUC) are quantitative peak and systemic exposure variables associated with the rate and extent of drug absorption. They are commonly used for bioavailability and bioequivalence assessments. C_{max} represents the highest exposure in a dosing interval. Three AUC metrics, AUC_{0-t}, $AUC_{0-\infty}$, and $AUC_{0-\tau}$, are used to quantify total exposure. Both AUC_{0-t} (where t is the last pharmacokinetic observation) and $AUC_{0-\infty}$ are used to measure total exposure in a single dose study, whereas $AUC_{0-\tau}$ (where τ is the dosing interval at steady state) is used to describe total exposure in a multiple-dose study (Shargel and Yu 1999). A typical bioequivalence study compares the C_{max} and AUC values obtained from subjects treated with the test and reference products. An average bioequivalence test is generally used to determine whether the 90 % confidence interval of the geometric mean ratio of C_{max} (or AUC) between the test and reference products falls into the bioequivalence range of 80 % (non-inferiority margin) to 125 % (non-superiority margin). The test product is considered bioequivalent to the reference product only when C_{max} and AUC of the test product are neither superior nor inferior to the reference product. Bioequivalence can generally be demonstrated by measurements of peak and total exposure; however, some exceptions exist and are discussed below.

As shown in the following example, quantitative comparisons of peak and total exposure alone may be inadequate for bioavailability and bioequivalence assessments of products with multimodal release mechanisms intended to generate desirable clinical responses. For example, methylphenidate is a central nervous system stimulant indicated for the treatment of attention deficit hyperactivity disorder (ADHD) (Pringsheim and Steeves 2011). Several methylphenidate formulations have been developed with various combinations of fast-release and slow-release components in order to target different clinical responses over a typical school day. Metadate CD® and Concerta® are methylphenidate products with different combinations of fast-release and slow-release components (Endrenyi and Tothfalusi 2010) (Sect. 7.5.1). The geometric mean ratios with 90 % confidence intervals of dose normalized C_{max} and $AUC_{0-\infty}$ for the two formulations fall into the bioequivalence range of 80–125 %. Following traditional criteria, Metadate $CD^{\circledR}$ and Concerta$^{\circledR}$ would be considered bioequivalent, which implies that the two products are therapeutically equivalent. Nevertheless, the two products yield distinctively different pharmacokinetic profiles (Sect. 7.5.1). Metadate $CD^{\circledR}$ shows two peaks. The first peak at 1.5 h post dose appears to be a shoulder of the second peak, which is the main peak occurring at 4.5 h post dose. Methylphenidate concentrations following the treatment of Concerta$^{\circledR}$ reaches the first peak at about 1 h post dose, followed by a gradual increase in the concentrations over the 5–9 h. The mean T_{max} for the second peak is around 6–10 h post dose. In addition, it appears that Metadate $CD^{\circledR}$ and Concerta$^{\circledR}$ work well at different time intervals on a typical school day. The clinical outcomes seem to be consistent with the pharmacokinetic profiles of the two products, which suggest that these products are not therapeutically equivalent. Therefore, additional pharmacokinetic metrics may be necessary to distinguish the potential therapeutic differences for products like Concerta$^{\circledR}$ and Metadate $CD^{\circledR}$.

The above case illustrates the need for additional pharmacokinetic metrics to characterize the rate and extent of absorption for different products. A different pharmacokinetic metric, partial AUC, focusing on the extent of exposure over a time interval of interest, has been proposed. This chapter will provide the readers with an overview of clinical and pharmacokinetic aspects associated with partial AUC.

7.2 Definitions

Partial AUC ($pAUC_{t_0-t_p}$) is defined as the area under the plasma concentration (C_t) versus time profile over two specified time points (Eq. (7.1), t_0 and t_p are two time points of interest) (FDA 2010b). For example, to quantify an early exposure, a partial AUC can be defined as the AUC between the time of dosing to the median time of the maximum concentration of a reference product ($pAUC_{0-t_{max}(R)}$). In addition to the description of an early exposure, the concept of partial AUC can be applied to different time intervals of interest. If there is a need to ensure equivalent exposures from 6 to 8 h post dose between the reference and test products, a partial AUC between 6 and 8 h ($pAUC_{6-8}$) can be used (FDA 2010b).

$$pAUC_{t_0-t_p} = \int_{t_0}^{t_p} c_t \cdot dt \qquad (7.1)$$

Partial AUC represents the drug exposure during the time interval of interest. Partial AUC is generally calculated by using trapezoidal rule. Each AUC segment between two adjacent time points (i.e., from t_i to t_{i+1}) is calculated by using

Eq. (7.2), where C_i and C_{i+1} are the concentrations at the ith and $i+1$th time points, respectively. Δt_i is the time interval between t_i and t_{i+1} (i.e., $\Delta t_i = t_{i+1} - t_i$).

$$AUC_i = \frac{(C_i + C_{i+1}) \cdot \Delta t_i}{2} \qquad (7.2)$$

Partial AUC ($pAUC_{0-p}$) can then be calculated by summing up all these AUC segments within the time interval of interest (Eq. (7.3)).

$$pAUC_{t_0 - t_p} = \sum_{i=1}^{n} \frac{(C_i + C_{i+1}) \cdot \Delta t_i}{2} \qquad (7.3)$$

To simplify this equation, let's assume all time intervals are the same (i.e., $\Delta t_i = \Delta t$ for all i's). Equation (7.3) can thus be converted into Eq. (7.4), where $\overline{C}_i$ is the average concentration between t_i and t_{i+1} and $\overline{C}$ is the average concentration during the entire time interval of interest.

$$pAUC_{t_0 - t_p} = \Delta t \sum_{i=1}^{n} \overline{C}_i = \Delta t \cdot n \cdot \overline{C} \qquad (7.4)$$

As shown in Eq. (7.4), partial AUC is equivalent to the entire time interval of interest (i.e., $t_p - t_0 = n \cdot \Delta t$), which is the sum of a series of constant time interval, multiplied by the average concentration within the same time interval (i.e., $\overline{C}$). In other words, partial AUC divided by its time interval will equal to the average concentration during this interval (Wang 2009).

Because each partial AUC represents the drug exposure for each time interval of interest, all pAUCs together contain the information on the shape of the pharmacokinetic profile. For example, the pharmacokinetic profile following the administration of Ritalin LA®, a methylphenidate product, can be divided into two parts at the cut off of 5 h post dose. Partial AUCs between 0 and 5 h, and 5 h to the last pharmacokinetic observation can be calculated separately. The former partial AUC represents the early exposure and the latter partial AUC represents the late exposure. To match partial AUCs with the same time interval from two pharmacokinetic profiles following the administration of reference or test products ensures similar drug exposure in the given time interval. Equivalence of partial AUCs across all time intervals of interest between two pharmacokinetic profiles indicates similarity in the shape of the two pharmacokinetic profiles. Because of this feature, partial AUCs are especially useful in describing pharmacokinetic profiles following the administration of products with complicated release mechanisms.

7.3 Formulations with Various Release Mechanisms

Pharmacokinetic profiles of various products are determined by their designed release mechanisms. Regular solid dosage forms, like tablets and capsules, undergo disintegration, dissolution, and absorption before the active compound becomes available in the systemic circulation (Qiu et al. 2009). The release rate of a drug may differ across different formulations. An immediate-release formulation is designed to disintegrate rapidly and to allow a quick release of the active moiety into gastrointestinal fluid. A delayed-release formulation delays the starting point of the drug release in the gastrointestinal tract. A typical technique to produce a delayed-release formulation is to coat the formulations with pH-sensitive membranes (i.e., enteric coat) so that the drug release is prohibited until the product passes through the stomach. The pharmacokinetic profile for a delayed-release product is featured by a time interval after dosing (e.g., approximately 1 h) with observations of no detectable drug concentrations in the systemic circulation. An extended-release formulation is intended to ensure the drug is available over a prolonged time interval. This formulation is usually applied to a compound with short half-life in order to reduce dosing frequencies. In general, there are two types of extended-release formulations—conventional extended-release formulations and formulations with combined fast-release and slow-release components. A typical conventional extended-release formulation creates barriers for drug diffusion, controls drug release through pores with designed sizes, or drives drug release through osmotic pressure so that the drug release rate is controlled or reduced. Pharmacokinetic profiles for conventional extended-release formulations are relatively flat (i.e., low fluctuation) with increased T_{max} and apparent half-lives. For practical purpose, immediate-release, delayed-release, and conventional extended-release formulations are considered as formulations with monomodal release mechanisms. In recent years, extended-release formulations with combinations of fast-release and slow-release components, known as multimodal release products, have become available, which yield more complicated pharmacokinetic profiles typically engineered to meet specific clinical needs.

A bioavailability or bioequivalence study may be conducted between products with various release mechanisms (FDA 2003). In most cases, the test and reference products share the same conventional immediate-release or extended-release mechanisms. Sometimes, comparisons can be made between products both shown to possess multimodal release mechanisms. There are also cases where comparisons are performed between products with different release patterns. For example, a bioavailability or bioequivalence study may be conducted between an immediate-release and an extended-release product, or between extended-release products with or without delayed-release features. The needs and utility for partial AUCs for products with various release mechanisms may vary and are discussed in detail in the following sections.

7.4 Utility of Partial AUCs

Partial AUC has only recently been considered as a metric for bioavailability and bioequivalence studies. Instead, the maximum concentration observed in a dosing interval (i.e., C_{max}) has been used as the main metric to characterize the rate of absorption in a bioavailability or bioequivalence study. The geometric mean ratio and the 90 % confidence interval of C_{max} between test and reference products are used as part of the bioequivalence assessment. However, using C_{max} to characterize the rate of absorption does not appear to be ideal because C_{max} is not solely driven by the absorption rate. Indeed, C_{max} is also affected by the extent of drug absorption, distribution, and drug elimination. Therefore, in the early 1990s, the selection of alternative pharmacokinetic metrics to further assess rate of absorption attracted a lot of attention in the scientific community. Multiple metrics were proposed including AUC normalized C_{max}, and ratio of intercepts extrapolated from logarithmic concentration/time values of the two products (Endrenyi et al. 1998; Lacey et al. 1994; Reppas et al. 1995; Tozer et al. 1996; Rostami-Hodjegan et al. 1994; Duquesnoy et al. 1998) etc. In 1992, Chen proposed using the ratio of partial AUC between reference and test products as an additional metric for bioequivalence testing (Chen 1992). In 1996, Tozer et al. interpreted partial AUC as a measurement of early exposure and further recommended the use of partial AUC in bioequivalence studies (Tozer et al. 1996). In the past decades, partial AUC has been further applied in bioequivalence studies for products with various release mechanisms.

7.4.1 Formulations with the Same Release Mechanisms (Monomodal Release Mechanisms)

In most cases, bioavailability or bioequivalence studies are to compare products with the same immediate-release or extended-release mechanisms, known as monomodal release mechanisms. The release rate (amount of drug released/time) from an immediate-release product is relatively fast, whereas that from an extended-release product is slow. Despite the difference in release rate among different products, the drug release pattern from the same product is expected to be consistent over time. From a modeling perspective, drug absorption from these products is generally assumed to follow first-order or zero-order release characteristics. If a product follows zero-order release pattern, the release rate of the compound from the product should be constant over time. Whereas, if a product follows first-order release pattern, the release rate of the compound is proportional to the amount of the compound left in the formulation with the first-order rate constant unchanged over time. Therefore, it is reasonable to assume that the rate or rate constant for a drug released from a product with monomodal release mechanism stays the same over time.

Under hypothetical situations with no variability, the pharmacokinetic profiles for products with the same monomodal release mechanism (e.g., first-order or zero-order absorption alone) would be the same if C_{max} and AUC values from different products are *identical* and there is no lag time for absorption in any product. As an example, for two formulations with first-order absorption and first-order elimination given to the same subject in a crossover manner, their plasma concentration (C) over time (t) profiles can be described by:

$$C = \frac{F \cdot D \cdot K_a}{V_d \cdot (K_a - K_e)} \cdot \left(e^{-K_e \cdot t} - e^{-K_a \cdot t}\right) \tag{7.5}$$

where K_a and K_e are absorption and elimination rate constants, respectively; D is the dose to the patient; V_d is volume of distribution; and F is absolute bioavailability. Among all the pharmacokinetic parameters (e.g., K_e, V_d, F, and K_a), only F and K_a are related to the performance of formulations. As shown below, AUC is proportional to F (Eq. (7.6)) and C_{max} is a function of K_a (Eq. (7.7)).

$$\text{AUC} = \frac{F \cdot D}{CL} \tag{7.6}$$

$$C_{max} = \frac{F \cdot D \cdot K_a}{V_d \cdot (K_a - K_e)} \cdot \left(e^{-K_e \cdot \frac{\ln\left(\frac{K_a}{K_e}\right)}{(K_a - K_e)}} - e^{-K_a \cdot \frac{\ln\left(\frac{K_a}{K_e}\right)}{(K_a - K_e)}}\right) \tag{7.7}$$

To ensure identical AUC and C_{max}, values of F and K_a for the two products must be identical if there is no lag time involved in absorption for either product. In this case, the entire pharmacokinetic profiles for the two products would be the same, and there is no need to further examine partial AUCs.

However, in reality, all pharmacokinetic profiles are associated with inherent intrasubject and intersubject variability. Hence, a bioequivalence study is designed *to show sufficiently similar, but not to show identical*, C_{max} and AUC obtained from the pharmacokinetic profiles following the administration of different products. As such, the pharmacokinetic profiles can still be different when C_{max} and AUC are determined to meet the bioequivalence criteria. Nevertheless, the difference in the shape of the pharmacokinetic profiles should be limited. Figure 7.1 demonstrates the simulated pharmacokinetic profiles of different products sharing identical AUC but with different C_{max} values. Profile 2 was generated by assuming a standard C_{max}. Profile 1 and 3 were simulated with C_{max} values 25 % higher or 20 % lower than the standard C_{max}. Any pharmacokinetic profiles with C_{max} values meeting bioequivalence criteria may have a different shape from Profile 2, but are anticipated to be within the range of Profile 1 and 3. If Profile 1 and 3 are associated with meaningfully different clinical responses, pharmacokinetic parameter(s) in addition to C_{max} might be necessary to distinguish the difference in early exposures. Under this scenario, partial AUC can be useful for a product when (1) control of early drug exposure is important for efficacy or safety reason and (2) partial AUC is more sensitive than C_{max} in identifying differences in early exposures.

Fig. 7.1 Simulated
pharmacokinetic profiles
with identical AUC but with
different C_{max} values. *Note*:
Profile 1, 2, and 3 share the
same AUC. Profile 2 has a
standard C_{max}. Profile 1 and
3 were simulated with C_{max}
values 25 % higher or 20 %
lower than the standard
C_{max}, respectively

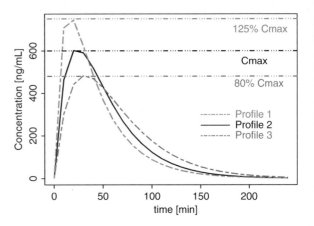

Partial AUC calculated up to the T_{max} seems to correlate well with pharmacodynamic parameters applied for describing early onset (e.g., time to reach 50 % of
the maximal effect). Dokoumetzidis et al. conducted pharmacokinetic and pharmacodynamic simulations to explore the relationships between partial AUC and
pharmacodynamic parameters (Dokoumetzidis and Macheras 2000). Early exposures (i.e., concentration–time profiles) up to the T_{max} were simulated by using a
first-order absorption and elimination, one compartment model with different
release characteristics. K_e was assumed to be constant (i.e., 0.355), whereas K_a
increased about tenfold, ranging from 2 to 0.2. Pharmacodynamic effect was
described by using an E_{max} model (Eq. (7.8)):

$$E = \frac{E_{max} \cdot C}{EC_{50} + C} \tag{7.8}$$

where E_{max} and EC_{50} were the maximal effect and the concentration to evoke 50 %
of the maximal effect. E was the corresponding effect at the concentration level of
C (Gabrielesson and Weiner 2001). The simulated concentration–time profile was
linked to pharmacodynamic effect through an effect compartment model to com-
pensate a delayed effect. Two types of pharmacodynamic effects were assessed.
One is the effect at T_{max} ($E_{t_{max}}$), which represents the maximal possible effect during
a defined dosing interval if the delayed effect is not apparent. The other one is time
to reach 50 % of the maximal effect ($T_{EC_{50}}$), which is more relevant to early onset.

In a crossover bioequivalence study, the ratio of partial AUC between the test
and reference products was typically used. The simulations were performed to
compare ratios of partial AUC and ratios of major pharmacodynamic parameters
between the test and reference products. The correlation between the ratio of partial
AUC ($pAUC_{t_{max},Ref}T/R$) and the ratio of effect at T_{max} ($E_{t_{max},Ref}T/R$) was evaluated
at first. The simulation indicated that this correlation was affected by EC_{50}. When
EC_{50} was small, the correlation was poor. However, when EC_{50} was increased, the
correlation was improved but still not ideal. However, the author found that the

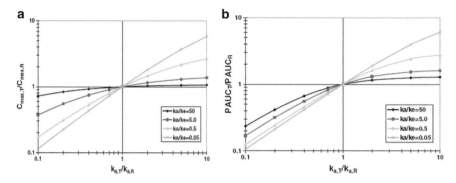

Fig. 7.2 Relationship of absorption rate constant (K_a) and maximum concentration (**a**), or partial AUC (**b**) between reference and test products (Chen et al. 2011)

correlation between the ratio of partial AUC and ratio of $T_{EC_{50}}$ between the reference and test products was much better than that for $E_{t_{max}, Ref}$, and was less affected by EC_{50}. Therefore, the partial AUC calculated up to T_{max} can be considered as a reasonable variable to describe early onset effect.

Partial AUC appears to be more sensitive than C_{max} in detecting difference in drug release from an immediate-release formulation, whereas the sensitivities for partial AUC and C_{max} appear similar for a conventional extended-release formulation. Chen et al. performed simulations to compare partial AUC and C_{max} in detecting early exposure differences (i.e., sensitivity) (Chen et al. 2011). The simulation was based on a one-compartment model with first-order absorption and elimination. Partial AUC was calculated between the time of dosing to the time of maximal concentration for the reference product in each subject. The results are shown in Fig. 7.2. Figure 7.2a shows the relationship of C_{max} (C_{max}, T/R) and K_a ratios (K_a, T/R) between test (T) and reference (R) product, stratified by K_a/K_e values. The log-transformed values were plotted. Likewise, Fig. 7.2b demonstrates the relationship of partial AUC and K_a ratios (pAUC, T/R versus K_a, T/R). It appeared that both C_{max} and partial AUC ratios correlated well with K_a ratio when modified release products were administered (i.e., under "flip-flop" situation). When K_a/K_e was less than 1, both C_{max} and partial AUC ratios changed with K_a ratio. The smaller the value of K_a/K_e, the slope between C_{max} or partial AUC ratios and K_a ratio was closer to 1. There was no apparent different pattern when C_{max} or partial AUC ratio was used. The results suggested no added benefit of including partial AUC when C_{max} has been used for bioequivalence assessment for conventional extended-release products. For K_a/K_e values greater than 1, the slope of the curve between partial AUC ratio and K_a ratio was steeper than that for C_{max} ratio versus K_a ratio curve, suggesting that partial AUC can be a more sensitive metric in detecting K_a difference between a test and a reference product. This is particularly true in the region when K_a of the test product (K_a, T) is smaller than that for the reference product (K_a, R). The results indicated that for immediate-release formulation, partial AUC seemed to be more sensitive than C_{max} in detecting the difference of K_a, especially when the drug release from the test product was slower than that from the reference product.

To appropriately describe early exposure, various cutoff time points for partial AUC of an immediate-release formulation has been assessed through simulations. It has been shown that the sensitivity of partial AUC in detecting different release characteristics decreased significantly when the cutoff time points were 1.5–2-fold longer than T_{max} (Rostami-Hodjegan et al. 1994). These findings support the use of T_{max} as a cutoff time point for partial AUC intended to characterize early exposure.

One potential hurdle of using partial AUC as a measurement of early exposure in a bioequivalence study for an immediate-release formulation is large variability typically associated in the early phase of pharmacokinetic profiles. Chen et al. conducted post hoc analyses to compare the performance of partial AUC relative to C_{max} and AUC of several approved drugs in bioequivalence studies. It was shown that for immediate-release formulations, intrasubject variability for partial AUC (%C.V. = 21–67 %) is larger than that for C_{max} (%C.V. = 6–23 %) or AUC (%C.V. = 4–24 %). In addition, intrasubject variability for partial AUC obtained from immediate-release formulations (%C.V. = 21–67 %) tended to be higher than that from extended-release formulations (%C.V. = 10–25 %). Multiple reasons may contribute to the observed differences in variability. Partial AUC calculated up to T_{max} is mainly driven by the absorption phase of a pharmacokinetic profile, which can be variable when an immediate-release formulation goes through the stomach and different sections of the gastrointestinal tract. In addition, the sampling frequency prior to T_{max} is generally limited. The collected data might not be optimal for partial AUC characterization (Chen et al. 2011). As a consequence, the sample size required to demonstrate bioequivalence with partial AUC might be higher than a regular bioequivalence study with the focus on C_{max} and AUC only. Sometimes, the need for a large sample size can make a trial impractical. Even though universal agreement on the determination of bioequivalence for highly variable drugs has not been achieved, several approaches were proposed and some could be applied to partial AUC analyses, such as scaled bioequivalence approach. This approach may be used to expand the bioequivalence limit based on the variability obtained from the reference product (Haidar et al. 2008). Appropriate designs and statistical analyses are required to ensure successful employment of this approach.

Scientific findings discussed above have shown that for an immediate-release formulation with indications critical with early onset, partial AUC shows its value because this metric is more sensitive than existing measurement such as C_{max}. As a reflection to these findings, partial AUC has been included in the current FDA guidance as a measure for early exposure in bioavailability or bioequivalence study using an immediate-release formulation. As stated in the guidance, partial AUC is only critical when early exposure shown to be important in terms of clinical response at early time points (e.g., to ensure adequate early onset or to avoid adverse events associated with early drug release). As recommended by the guidance, partial AUC is truncated at the median time to the maximum concentration for the reference product (i.e., T_{max} (Ref)) (FDA 2003).

Simulation results also have shown that compared to C_{max}, partial AUC calculated up to T_{max} does not provide added benefit in detecting difference in drug release for conventional extended-release formulations with the same dosing interval. The utility of using partial AUC as an additional bioequivalence metric in conventional extended-release formulations was discussed in a workshop sponsored by the American Association of Pharmaceutical Scientists (AAPS) held in Baltimore, Maryland in 2009. Stakeholders from industry, academia, and regulatory agency met and shared their positions. It was concluded in the workshop that "the current regulatory approaches criteria for bioequivalence evaluation were considered adequate for the assessment of therapeutic equivalence and interchangeability of conventional monophasic extended-release formulations." Hence, the experts that attended the workshop did not recommend partial AUC be included for bioequivalence assessment in studies comparing test and reference products which are conventional extended-release formulations (Chen et al. 2010a, b, c).

7.4.2 Formulations with Combined Fast-Release and Slow-Release Components (Multimodal Release Mechanisms)

Products with multimodal release mechanisms are extended-release formulations with combined fast-release and slow-release components. The need for a product with the joint components should be determined by the intended pharmacological effect. For a product with slow onset and the effect is intended to be maintained over a long time, the dosing pattern and time course of plasma concentration over a day might not be critical. Treatments including antidepressants such as Brintellix® (FDA 2013c) and antipsychotics (McEvoy et al. 2007), take days or weeks before the optimal effect can be achieved, and thus it does not appear to be necessary to design a product with multimodal release components for these indications. The products requiring a combination of fast-release and slow-release components are for those compounds that exhibit rapid onset-offset, and need maintenance pharmacological effect like Ambien CR® for sleep aid (FDA 2013b) and Concerta® for ADHD treatment (FDA 2013d). Products with multimodal release mechanisms are designed to generate pharmacokinetic profiles suitable for these specific clinical needs. Different components of the formulation are supposed to deliver drug at predetermined sections of the gastrointestinal tract over designated time intervals. Although both the fast-release and slow-release components may follow the same release pattern (e.g., first-order release/absorption process), the release characteristics (e.g., release/absorption constants) differ from one component to the other. Immediately after the formulation is taken, the pharmacokinetic profile may be controlled by the fast-release component. Over time, a transition is expected from an immediate-release profile into an extended-release profile. Ultimately, the pharmacokinetic profile is mainly driven by the slow-release component. Thus, the

Fig. 7.3 Pharmacokinetic
profiles simulated from
products with different
combinations of release
mechanisms with identical
$AUC_{0-\infty}$ and C_{max}. *Note*:
Profile 1 and 2 have
identical $AUC_{0-\infty}$ and C_{max}

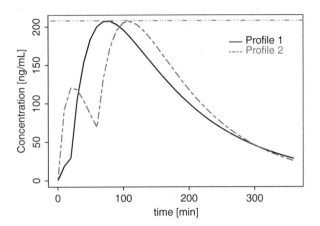

overall pharmacokinetic profile following the administration of a product with multimodal release mechanisms can be complicated.

Formulations with multimodal release mechanisms bring challenges in bioavailability and bioequivalence evaluation. Figure 7.3 shows pharmacokinetic profiles simulated from two products with two combinations of fast-release and slow-release components. Drug release from each component was assumed to follow first-order release/absorption. Two one-compartment models with first-order elimination were applied for simulation—one model represented the fast-release component and the other represented the slow release component. The amount of drug in the two components, lag time of drug release from the two components, and different release and absorption kinetics (as reflected in different absorption constants) were adjusted to yield distinctively different shapes of plasma concentration–time profiles, however with *identical* C_{max} and $AUC_{0-\infty}$. As shown in Fig. 7.3, one profile shows two peaks whereas the other profile shows one peak.

In reality, the case of Concerta® and Metadate CD® discussed under "General Background" section further illustrate that the traditional bioequivalence criteria with the focus on C_{max} and $AUC_{0-\infty}$ alone may not be sufficient to ensure therapeutic equivalence of products with multimodal release mechanisms. As Endrenyi et al. pointed out that dose normalized partial AUCs calculated between dosing to 4 or 6 h post dose (i.e., $pAUC_{0-4}$ or $pAUC_{0-6}$) clearly demonstrated the difference between Concerta® and Metadate CD®, whereas $AUC_{0-\infty}$ and C_{max} met bioequivalence standard. For example, the dose normalized (normalized to 20 mg of Metadate CD® or Concerta®) geometric mean ratios of $pAUC_{0-4}$ and $pAUC_{0-6}$ were 69.9 % and 77.9 %, respectively. The lower 90 % confidence intervals were 66.2 and 74.2 %. Hence, to further characterize the shape difference in pharmacokinetic profiles, it is valuable to include partial AUC as an additional metric for bioequivalence and bioavailability assessment (Endrenyi and Tothfalusi 2010).

As shown earlier, the shape of pharmacokinetic profiles may change and value of partial AUC can be affected by a combination of various factors associated with

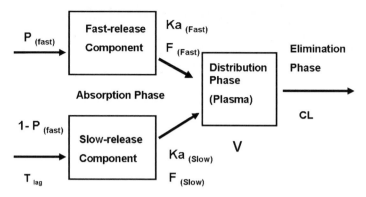

Fig. 7.4 One typical pharmacokinetic model applied for products with multimodal release mechanisms

fast-release and slow-release components. These factors may include release mechanisms of multiple components, release duration of each component, time delay of drug release from different components, and proportion of the total dose in each component. A typical pharmacokinetic model following the administration of a product with multimodal release mechanisms can be illustrated in Fig. 7.4. The absorption phase is described by two compartments, one for fast-release component and one for slow-release component, with dual parallel inputs into systemic circulation. $P_{(fast)}$ and $1 - P_{(fast)}$ are the proportions of the total dose allocated to the fast-release and slow-release components, respectively. T_{lag} is the lag time before the drug releases from the slow-release component. $F_{(fast)}$ and $F_{(slow)}$ are the absolute bioavailabilities for the fast-release and slow-release components, whereas $K_{a(fast)}$ and $K_{a(slow)}$ are the release/absorption rate constants for the fast-release and slow-release components. The duration of drug release from each component should be a function of the amount of drug in each component and release/absorption rate constant of each component. The value of partial AUC is a function of $P_{(fast)}$, T_{lag}, $F_{(fast)}$, $F_{(slow)}$, $K_{a(fast)}$, and $K_{a(slow)}$. All the pharmacokinetic factors are affected by formulation design and can be changed individually/independently or jointly. Any change in the values of the above-mentioned parameters can potentially affect the value of the partial AUC of interest. Simulations were conducted to investigate the relationship between the changes of individual pharmacokinetic parameter and partial AUC, as well as the power and sensitivity of using partial AUC as additional criteria for bioequivalence assessment.

Wang et al. presented at the AAPS workshop (Wang 2009) the function of probability of claiming bioequivalence, by using partial AUC as an additional metric in addition to C_{max} and AUC, versus changes in the underlying formulation-related individual pharmacokinetic parameters. His simulation was based on the pharmacokinetic model similar to the model presented in Fig. 7.4. About 200 simulations of a crossover bioequivalence study were performed to generate the power function (i.e., the probability of claiming bioequivalence

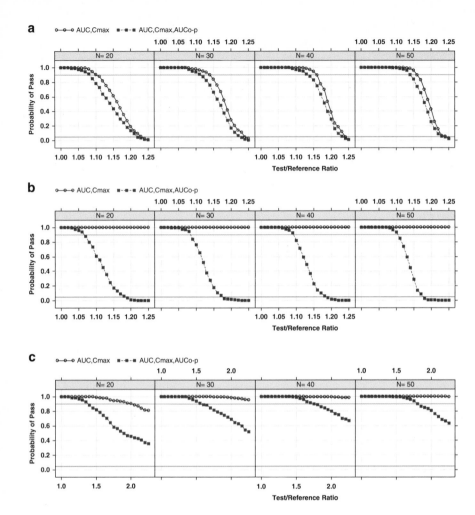

Fig. 7.5 The probability of claiming bioequivalence by using $C_{\max}$, AUC or by using $C_{\max}$, AUC, and partial AUC versus the individual underlying pharmacokinetic parameter changes (Wang 2009). *Note*: (**a**) the absolute bioavailabilities for both $F_{(\text{fast})}$ and $F_{(\text{slow})}$ increases from 1.05 to 1.25. (**b**) The absolute bioavailability for $F_{(\text{fast})}$ increases from 1.05 to 1.25. (**c**) The $K_{a(\text{fast})}$ increases from 1.25 to 2.25

when two products are bioequivalent). The results are shown in Fig. 7.5. The line with open cycles represents the probability of claiming bioequivalence by using traditional pharmacokinetic variables for bioequivalence testing (i.e., $C_{\max}$ and AUC) and the line with solid squares represents the probability of claiming bioequivalence by including partial AUC as an additional metric. The power functions with different sample sizes under four scenarios were generated. Figure 7.5a represents the scenario where the ratio of the true absolute bioavailability of both fast-release and slow-release components ($F_{(\text{fast})}$ and $F_{(\text{slow})}$)

between the test and reference products varies between 1.05, 1.1, 1.15, 1.2, and 1.25. Figure 7.5b shows the results when the ratio of the underlying absolute bioavailability of the fast-release component between the test and reference products changes from 1.05 to 1.25. Figure 7.5c demonstrates the power function when the ratio of the underlying first-order release/absorption rate constant between the test and reference products increases from 1.25 to 2.25. The results suggested if the pharmacokinetic profiles are reasonably similar between the reference and test products, a drastic increase in sample size is not necessary when partial AUC is included as an additional metric for bioequivalence testing.

Zirkelbach et al. investigated the sensitivity of using partial AUC ratios between the test and reference products to the changes in the release/absorption rate constant of the fast-release component (i.e., $K_{a(fast)}$) (Fourie Zirkelbach et al. 2013). The pharmacokinetic model applied for simulation was similar to the model presented in Fig. 7.4, except that the release/absorption from the fast-release component is a zero-order process. The effect of changes in the ratio of $K_{a(fast)}$ between the reference and the test products (i.e., $K_{a(fast),test}/K_{a(fast),reference}$) on the ratio of partial AUC truncated at 4 and 6 h (i.e., $pAUC_{(0-4),test}/pAUC_{(0-4),reference}$ or $pAUC_{(0-6),test}/pAUC_{(0-6),reference}$) between the two products were assessed, respectively. It was shown that partial AUCs truncated at different time intervals appear to be associated with different sensitivities in the detection of the difference in drug release rate from the fast-release component. Further simulations were performed to show the relationships between changes in T_{lag}, $K_{a(slow)}$, and release/absorption duration on the mean ratios of $pAUC_{test}/pAUC_{reference}$, where partial AUCs were truncated at 4 or 6 h, respectively. Partial AUC truncated at 6 h (i.e., $pAUC_{(0-6)}$) covers longer time intervals than partial AUC truncated at 4 h (i.e., $pAUC_{(0-4)}$). Hence, compared with $pAUC_{(0-4)}$, $pAUC_{(0-6)}$ was affected more by changes in T_{lag}, $K_{a(slow)}$, and release/absorption duration. If there is an increase in T_{lag}, $K_{a(slow)}$, and release/absorption duration of the test product, the ratio of $pAUC_{(0-6)}$ between the test and reference products was affected more than the ratio of $pAUC_{(0-4)}$. Hence different partial AUC might perform differently in detecting the underlying difference in drug release.

The selection of appropriate cutoff time point for partial AUC is challenging. Typically, two approaches have been taken to determine the cutoff time points. The first approach is based on the characteristics of pharmacokinetic profiles. Taking a pharmacokinetic profile with two distinctive peaks, for example, the first and second peaks are considered to be related to the drug release from the fast-release and slow-release components, respectively. The cutoff time point for partial AUC can be selected based on the T_{max} of the first peak so that the early exposure can be separated from the late exposure. Even if when the first peak is not apparent (e.g., as a shoulder), the T_{max} from an immediate-release formulation with the same active ingredient can be applied as the cutoff point with the assumption that the release characteristics from the fast-release component of the product with multimodal release mechanism would be similar to that from an immediate-release formulation. The second approach is based on pharmacokinetic/pharmacodynamic (PK/PD) relationship for a product of interest. In general, a direct link PK/PD relationship is expected for a product which requires multimodal release mechanism. This relationship links the drug concentration and its pharmacodynamic effect (e.g., clinical outcomes). An established PK/PD relationship can be applied to identify concentration range associated with a specific pharmacodynamic effect of interest (e.g., onset effect). This approach, by principle, is highly recommended by industry, academia, and regulatory experts because the determined cutoff time points can be more clinically relevant (Chen et al. 2010a, b, c). However, no generally agreed approach and criteria across various products have been established at present.

In the year of 2010, an advisory committee meeting was held by the U.S. Food and Drug Administration. One meeting objective was to discuss the need of using partial AUC to characterize products with multimodal release mechanisms and applying partial AUC for bioequivalence testing. The committee endorsed the use of partial AUC as an additional pharmacokinetic metric for bioequivalence testing (FDA 2010b). Since then, several product-related bioequivalence guidance documents on the use of partial AUC have been published (FDA 2010a, 2011, 2012a). Some of these examples are discussed in Sect. 7.5.

7.4.3   Formulations with Different Release Mechanisms

Bioavailability and bioequivalence studies are sometimes conducted to compare two products with different release mechanisms. Partial AUCs can be applied to identify unique drug release feature among products with different release mechanisms. In other cases, partial AUC may demonstrate that products with different release mechanisms yield similar pharmacokinetic profiles.

Partial AUCs may be applied to distinguish drug release features that result in unique clinical responses for products with different release mechanisms. For example, mesalamine has been approved for treatment of patients with active ulcerative colitis, which is one form of inflammatory bowl disease featured by open sores or ulcers in colon. Even though the pharmacological mechanism is unknown, it is thought that mesalamine inhibits prostaglandin production in the colon by blocking cyclooxygenase, and hence controls inflammation locally (FDA 2013g). Drug delivery to the local inflammatory site in colon, rather than drug entry into the systemic circulation, appears to be important to ensure effectiveness (Klotz 2012). Hence, the delayed-release feature of a formulation is critical because less mesalamine is anticipated to reach the colon in products without this delayed-release feature. It is likely that two products, with or without delayed-release features, may be falsely considered therapeutically equivalent by using C_{max} and AUC alone as the pharmacokinetic parameters for bioequivalence determination. With the understanding of the mechanism, the formulation with no delayed-release feature is likely to be less effective. To identify the delayed-release feature of a product, $pAUC_{0-3}$ and $pAUC_{3-t}$ are recommended as additional bioequivalence criteria by the US FDA (FDA 2012a).

Partial AUCs can be used as an additional metric to ensure similar pharmacokinetic profiles are produced by formulations with different release mechanisms. In a drug development program, it is common that an extended-release formulation is developed after an immediate-release formulation is available to reduce dosing frequency. Generally, the shape of pharmacokinetic profile for an extended-release formulation is anticipated to be different from an immediate-release formulation. If the concern exists that change in pharmacokinetic profile might lead to different clinical responses, a clinical efficacy and safety study is required to support the approval of the extended-release formulation. This principle has been applied to the development of most extended-release formulations of antiepileptic agents. A different approach was used in a recent development program for topiramate extended-release formulation. Topiramate is an antiepileptic agent initially approved in 1996 under the trade name of Topamax® for the treatment of epilepsy (FDA 2012b). The compound is formulated as an immediate-release tablet or capsule, and administered to patients twice daily. Topiramate extended-release formulation was subsequently developed to target a once-daily dosing. One developmental approach is to demonstrate that the new formulation, even with the changes in the pharmacokinetic profile, is safe and effective by using clinical efficacy and safety trials. This approach is apparently costly and time consuming. As an alternative, a developer chose to prove that the difference in pharmacokinetic profiles between the immediate-release and extended-release formulation is sufficiently small, hence no difference in clinical responses is anticipated. In this program, a bioequivalence study was conducted in subjects receiving multiple doses of either topiramate immediate-release formulation or extended-release formulation in a crossover manner. Pharmacokinetic samples were collected at steady state. In addition to AUC_{0-24} and C_{max} at steady state, all observed concentration values and partial AUCs from time zero to each time point post dose were also compared. It has been shown that the 90 % confidence intervals for all the partial AUCs are within the defined bioequivalence criteria of 80–125 %. As shown in Sect. 1.2, partial AUC represents the average exposure within the defined time intervals. Equivalence of partial AUCs between time zero and all subsequent time points suggests equivalence in the average exposures within any time points of interest and thus no clinically meaningful difference is anticipated between the two formulations. This approach has been accepted by the agency and the product was approved in 2013 under the trade name of Trokendi XR® (FDA 2013l).

7.5 Case Studies

In addition to the cases presented in the previous sections, partial AUC has been implemented as an additional metric for bioequivalence testing of several other products. Examples including methylphenidate extended-release products and zolpidem extended-release products are discussed in detail in Sect. 7.5.

7.5.1 Methylphenidate Extended-Release Products

As discussed earlier, methylphenidate has been indicated for the treatment of ADHD. Even though the underlying mechanism of action is unclear, it is generally believed that the pharmacological effect of methylphenidate is related to the reuptake inhibition of norepinephrine and dopamine. The original methylphenidate product was approved in 1955 and marketed under the trade name of Ritalin® in the USA. This is an immediate-release formulation and can be administered twice or three times a day. Practically, it is inconvenient to store the product at school and to dose patients multiple times on a school day because methylphenidate is a controlled substance (FDA 2013i).

Methylphenidate products with prolonged release characteristics were developed to reduce dosing frequency. Ritalin SR® is an extended-release product given once daily with an intended treatment duration of 8 h (FDA 2013j). However, the treatment effect of Ritalin SR® does not appear to be optimal. Therefore, various methylphenidate formulations with combinations of fast-release and slow-release components were further developed subsequently. Ritalin LA® is available in capsules containing 50 % of fast-release beads and 50 % of enteric-coated, delayed-release beads (FDA 2013k). Ritalin LA® shows double peaks in the pharmacokinetic profile with the first peak around 1–3 h and the second peak around 7 h post dose. The two peaks are consistent with the peaks in the pharmacokinetic profiles following the administration of Ritalin® administered 4 h apart (FDA 2013k). Focalin XR® is a product of dextromethylphenidate, the enantiomer of racemic methylphenidate (2013e). The formulation of Focalin XR® is similar to Ritalin LA®, hence their pharmacokinetic profiles are similar (FDA 2013e). Metadate CD® is another capsule dosage form with a mixture of 30 % fast-release and 70 % slow-release beads. The concentration–time profile of Metadate CD® shows two distinctive peaks. The first peak occurring around 1.5 h appears to be sharp and second peak around 4.5 h is relatively flat (FDA 2013f). Concerta® is available in tablets including a fast-release overcoat accounting for 22 % of the total dose and a slow-release core containing the remaining 78 % of the dose. The slow-release core has two drug layers and one osmotically active layer. In aqueous solution, the osmotic layer expands and pushes methylphenidate out through a laser-drilled orifice in a controlled manner. The concentration–time profile of Concerta® shows two peaks with the first peak of methylphenidate similar to a shoulder (FDA 2013d). Quillivant XR® is a methylphenidate oral suspension including 20 % of fast-release and 80 % of slow-release powder. The concentration–time profile of Quillivant XR® has one apparent peak (FDA 2013h). All the five products of methylphenidate have been approved by the US FDA under New Drug Application (NDA) with different plasma concentration versus time profiles generated by adjusting various components of the formulations and with unique pharmacodynamic time profiles to meet special clinical needs.

Partial AUC is useful in the bioavailability and bioequivalence studies for methylphenidate products with multimodal release mechanisms. The selection of appropriate time interval for the calculation of partial AUC is critical to ensure therapeutic equivalence of methylphenidate products. Taking Ritalin LA®, for example, the pharmacokinetic profile of Ritalin LA® is generated to ensure adequate symptom improvement in the morning and sustained performance through a typical school/work day. The cutoff point of partial AUC was determined based on the knowledge of the composition of the formulation (Stier et al. 2012). The fast-release component of Ritalin LA® works similar to an immediate-release product of methylphenidate. The maximum concentration of methylphenidate obtained from various immediate-release methylphenidate products is achieved at around 2 ($\pm$0.5) and 3 ($\pm$0.5) hours post dose under fast and fed conditions. The estimated T_{max} values under fast and fed conditions are expanded by two times of the standard error in order to ensure sufficient coverage of the true T_{max}'s. Hence, 3 and 4 h are used as the cutoff time points for partial AUC calculation in fasted and fed bioequivalence studies. Clinically, appropriate early exposure during the 3–4 h period post dose will be important, as it is typically essential for school-age pediatric patients to stay focused in classroom in the morning (which is about 3–4 h within the morning dose). The variability in partial AUCs truncated at 3–4 h post dose may be large but is still reasonable for bioequivalence assessment. Therefore, per US FDA's guidance, $pAUC_{0-3}$ and $pAUC_{0-4}$ have been chosen as additional parameters to assess bioequivalence under fast and fed conditions, respectively (FDA 2010a).

7.5.2 Zolpidem Extended-Release Products

Zolpidem tartrate immediate-release formulation was approved in 1992 for the treatment of insomnia characterized by difficulties with sleep initiation. It is marketed in the USA. under the trade name of Ambien® (FDA 2013a). Zolpidem is a gamma-aminobutyric-acid (GABA) A agonist mainly interfering with GABA-BZ1 receptor. Following a single dose, the maximum concentration is achieved at around 1.5 h (median) post dose. Zolpidem is converted into inactive metabolites and eliminated through kidney rapidly with a mean half-life of 2.5 h. The pharmacokinetic feature of Ambien® ensures its rapid effect on sleep initiation in insomnia patients. However, because the plasma concentration declines rapidly after the peak concentration, the sleep maintenance effect during a typical sleep cycle does not appear to be optimal.

Ambien CR® was approved in the USA in 2005 with an indication of treating insomnia patients with difficulty of sleep onset and sleep maintenance (FDA 2013b). Ambien CR® shows unique features in rapid sleep onset, sufficient sleep maintenance, and limited residual effect. The pharmacodynamic effect following the treatment of Ambien CR® is a consequence of its unique pharmacokinetic profile generated by the formulation. The Ambien CR® tablets contain two layers—a fast-release layer allowing quick release of zolpidem and a slow-release layer ensuring adequate zolpidem concentration over a sustained time interval to cover the entire sleep cycle. The pharmacokinetic profiles following a single dose

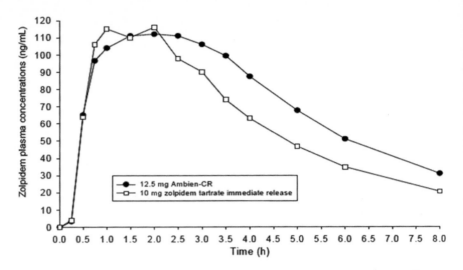

Fig. 7.6 Zolpidem plasma concentration–time profile following a single dose of Ambien® (*open square*) and Ambien CR® (*close circle*) over the time course of 8 h (FDA 2013b)

Table 7.1 Different developed formulation for zolpidem

Formulation	Fast-release component (%)	Slow-release component (%)
C	80	20
E	60	40
G	40	60

of Ambien® and Ambien CR® are shown in Fig. 7.6. The maximum concentration following the administration of a single dose of Ambien® or Ambien CR® is reached at about 1.5 h. The mean half-lives are 2.6 h and 2.8 h for Ambien® and Ambien CR®, respectively. There is no apparent double peak following the administration of Ambien CR®. Even though the T_{max} values of Ambien® and Ambien CR® are similar, the small difference in elimination half-lives ensures higher zolpidem concentration following the dosing of Ambien CR® between 2 and 8 h (Fig. 7.6).

Partial AUC is recommended in bioequivalence studies using Ambien CR® as the reference product because of the multimodal release nature of the product. For the selection of appropriate formulation of Ambien CR®, the developer conducted a pharmacodynamic study to assess the effect of different combination of fast-release and slow-release components on sleep induction, maintenance, and residual effect (Lionberger et al. 2012). Three formulations were developed with different ratios of fast-release and slow-release components (Table 7.1). Although all formulations were designed to ensure the completion of zolpidem release in 4 h, formulations C, E, and G demonstrated different pharmacodynamic effects. Formulations C and G showed either lack of sufficient sleep duration or unacceptable residual effect.

Formulation E appeared to maintain a balance of adequate sleep duration and minimal residual effect after awakened and was further developed into Ambien CR®. It has been shown that most patients fall into sleep after receiving Ambien CR® within 1.5 h post dose, which is consistent with the T_{max} of both Ambien and Ambien CR®. This time point is considered as time of onset, and thus exposure obtained between dosing to 1.5 h post dose represents the early exposure. To better quantify the onset of effect, $pAUC_{0-1.5}$ has been proposed as a metric for bioequivalence testing.

It appears that the onset of sleep between Ambien and Ambien CR® is similar. A comparison of cumulative absorption showed similar profiles of AMBIEN and AMBIEN CR® up to 2 h post dose. The results further confirm that the selection of $pAUC_{0-1.5}$ based on the pharmacodynamic effect (i.e., sleep onset) is consistent with drug release profiles.

Additional modeling and simulations have been conducted to assess the suitability of using $pAUC_{0-1.5}$ for bioequivalence testing when Ambien CR® is used as the reference product. Pharmacokinetic profiles following administration of various formulations were simulated based on in vitro dissolution profiles using IVIVC, deconvolution, and CAT (compartment and absorption transit) models. Using the traditional bioequivalence variables, C_{max} and AUC, all three prototype formulations met bioequivalence criteria, whereas the designated pharmacodynamic study showed the three formulations are therapeutically inequivalent. The use of $AUC_{0-1.5}$ allows detection of the difference in pharmacokinetic profile among formulations C, E, and G.

Using $pAUC_{0-1.5}$ appears to be sufficient to characterize early exposure, which is critical to the onset of the pharmacodynamic effect following the treatment of Ambien CR®. Another pharmacodynamic feature for Ambien CR® is sleep maintenance and low residual effect, which appears to be more relevant to the zolpidem middle and late exposures generated by the slow-release component of the formulation. To illustrate the needs of including additional partial AUCs (e.g., $pAUC_{3-6}$) to characterize middle and late exposures for bioequivalence testing, pharmacokinetic simulations were performed. The pharmacokinetic profile following the treatment of Ambien CR® was generated by using CAT model and zolpidem in vitro dissolution profiles described by a Weibull model. Hypothetical test and reference products are differed by altering the shape parameters of Weibull model. Pharmacokinetic variables applied for standard bioequivalence testing, such as C_{max} and AUC, between the hypothetical test and reference products were generated. In addition, multiple partial AUCs including, $pAUC_{0-1.5}$ for early exposure, $pAUC_{3-6}$ for middle exposure, and $pAUC_{6-\infty}$ for late exposure, were calculated. The bioequivalence criteria were considered to be met when the ratios between the hypothetical test and reference products fall within 0.90–1.11, since the simulation focused on the mean values without accounting for variability. The results have revealed that the allowable range of release rate is narrow for products to be bioequivalent to Ambien CR® based on C_{max}, AUC, and $pAUC_{0-1.5}$. It has further been shown that once C_{max}, AUC, and $pAUC_{0-1.5}$ meet bioequivalence criteria (i.e., 0.90–1.11 for mean ratio between hypothetical test and reference products), pAUC$_{3-6}$ and pAUC$_{6-\infty}$ are likely to meet the bioequivalence standard as well. Hence, there is no need to include pAUC$_{3-6}$ and pAUC$_{6-\infty}$ as additional pharmacokinetic variables for bioequivalence testing.

In summary, pAUC$_{0-1.5}$ has been recommended by the US FDA guidance to be included as an additional pharmacokinetic metric for bioequivalence test in studies using Ambien CR$^{\circledR}$ as the reference product (FDA 2010c, 2011). This partial AUC characterizes early zolpidem exposure which is consistent with early drug absorption and critical to the onset of pharmacodynamic effect. This additional parameter distinguishes the performances of different formulations that are likely to be considered as bioequivalent products using traditional bioequivalence approaches (i.e., based on C_{max} and AUC). Once pAUC$_{0-1.5}$ is added to AUC and C_{max} for bioequivalence testing, additional partial AUCs to characterize the middle and late exposures do not appear to be necessary. The variability of pAUC$_{0-1.5}$ appears to be within an acceptable range so that a bioequivalence study with a regular sample size should be sufficient.

7.6 Summary and Future Perspectives

Bioavailability and bioequivalence studies are routinely conducted at various stages of product development. To facilitate specific product development, AUCs obtained over time intervals of interest, known as partial AUC, can be applied for bioavailability and bioequivalence assessments. Partial AUC is to describe shapes of pharmacokinetic profiles with the focus on quantification of exposures over specific time intervals. Partial AUC is most useful in products with multimodal release mechanisms, where traditional bioequivalence metrics such as C_{max} and AUC are not sufficient to distinguish performance of products and ensure therapeutic equivalence. Typical products requiring multimodal release components are products associated with rapid onset and offset of the pharmacodynamic effect (e.g., the treatment of insomnia and ADHD). The appropriate approach for the selection of cutoff time points for partial AUC calculation is still under active investigation. Traditionally, features from pharmacokinetic profile (e.g., T_{max}) are the basis for the selection of cutoff time points. In recent years, pharmacokinetic–pharmacodynamic relationship has been proposed as an alternative source for the determination of cutoff time points. Partial AUC has been applied as an additional metric for bioequivalence testing in several products, including methylphenidate and zolpidem products. In addition, partial AUC can be applied to quantify early exposures in immediate-release formulations and to compare performance of products with different release mechanisms.

The concept and application of partial AUC is still evolving. The development of novel techniques and formulations allows more complicated pharmacokinetic profiles to ensure clinical needs. With the pharmacokinetic profiles demonstrating unique features, it may be critical to quantify exposures in defined ranges of pharmacokinetic profiles. In a bioavailability study, quantification of exposure over a time interval allows a direct comparison between a novel formulation and an existing formulation. In a bioequivalence study, the demonstration of similar exposures over a time interval of clinical interest further ensures therapeutic equivalence between the test and reference products. Appropriate application of partial AUC facilitates product development and hence promotes safe and effective use of various formulations. Additional proposals for the application of partial AUC are available from different researchers. For example, Gehring et al. discussed the potential use of partial AUC in the characterization of pharmacokinetic profiles generated by long acting formulations (Gehring and Martinez 2012). For convenience and improved compliance, some products may be developed with an intended dosing interval over weeks or months. Mechanistically, the prolonged dosing interval can be achieved by combinations of various release mechanisms. The active compound can be released from the formulation by sporadic bursts, combination of a loading release and a zero-order release, or sequential changes in drug release rate. Complex pharmacokinetic profiles can be generated by a formulation with multiple release mechanisms and prolonged release interval. For these formulations, using C_{max} and AUC alone may not be sufficient to characterize the rate or extent of drug absorption at any time interval. Under such circumstances, multiple segmented AUCs (i.e., partial AUC) over several time intervals may be applied to further characterize the pharmacokinetic profiles. However, the use of partial AUC in any novel formulations will still require careful examination. Practical and/or statistical issues may arise depending on the formulations under study, which will need to be resolved before the metric can be used in the bioavailability or bioequivalence evaluation.

References

Banerjee SK, Jagannath C, Hunter RL, Dasgupta A (2000) Bioavailability of tobramycin after oral delivery in FVB mice using CRL-1605 copolymer, an inhibitor of P-glycoprotein. Life Sci 67 (16):2011–2016

Bylund J, Bueters T (2013) Presystemic metabolism of AZ'0908, a novel mPGES-1 inhibitor: an in vitro and in vivo cross-species comparison. J Pharm Sci 102(3):1106–1115

21CFR320.1 Bioavailability and bioequivalence requirements [Online]. http://www.accessdata. fda.gov/scripts/cdrh/cfdocs/cfcfr/CFRSearch.cfm?fr=320.1. Accessed 8 Dec 2013

Chen ML (1992) An alternative approach for assessment of rate of absorption in bioequivalence studies. Pharm Res 9(11):1380–1385

Chen ML, Shah VP, Ganes D, Midha KK, Caro J, Nambiar P, Rocci ML Jr, Thombre AG, Abrahamsson B, Conner D, Davit B, Fackler P, Farrell C, Gupta S, Katz R, Mehta M, Preskorn SH, Sanderink G, Stavchansky S, Temple R, Wang Y, Winkle H, Yu L (2010a) Challenges and opportunities in establishing scientific and regulatory standards for determining therapeutic equivalence of modified-release products: workshop summary report. Clin Ther 32 (10):1704–1712

Chen ML, Shah VP, Ganes D, Midha KK, Caro J, Nambiar P, Rocci ML Jr, Thombre AG, Abrahamsson B, Conner D, Davit B, Fackler P, Farrell C, Gupta S, Katz R, Mehta M, Preskorn SH, Sanderink G, Stavchansky S, Temple R, Wang Y, Winkle H, Yu L (2010b) Challenges and opportunities in establishing scientific and regulatory standards for assuring therapeutic equivalence of modified release products: workshop summary report. AAPS J 12(3):371–377

Chen ML, Shah VP, Ganes D, Midha KK, Caro J, Nambiar P, Rocci ML Jr, Thombre AG, Abrahamsson B, Conner D, Davit B, Fackler P, Farrell C, Gupta S, Katz R, Mehta M, Preskorn SH, Sanderink G, Stavchansky S, Temple R, Wang Y, Winkle H, Yu L (2010c) Challenges and opportunities in establishing scientific and regulatory standards for assuring therapeutic equivalence of modified-release products: workshop summary report. Eur J Pharm Sci 40 (2):148–153

Chen ML, Davit B, Lionberger R, Wahba Z, Ahn HY, Yu LX (2011) Using partial area for evaluation of bioavailability and bioequivalence. Pharm Res 28(8):1939–1947

Dokoumetzidis A, Macheras P (2000) On the use of partial AUC as an early exposure metric. Eur J Pharm Sci 10(2):91–95

Duquesnoy C, Lacey LF, Keene ON, Bye A (1998) Evaluation of different partial AUCs as indirect measures of rate of drug absorption in comparative pharmacokinetic studies. Eur J Pharm Sci 6(4):259–264

Endrenyi L, Tothfalusi L (2010) Do regulatory bioequivalence requirements adequately reflect the therapeutic equivalence of modified-release drug products? J Pharm Pharmaceut Sci 13 (1):107–113

Endrenyi L, Csizmadia F, Tothfalusi L, Chen ML (1998) Metrics comparing simulated early concentration profiles for the determination of bioequivalence. Pharm Res 15(8):1292–1299

FDA (1995) Guidance for industry: immediate release solid oral dosage forms. http://www.fda.gov/downloads/Drugs/Guidances/UCM070636.pdf. Accessed 9 Dec 2013

FDA (1997a) Guidance for industry: SUPAC-MR: modified release solid oral dosage forms. http://www.fda.gov/downloads/Drugs/Guidances/UCM070640.pdf. Accessed 9 Dec 2013

FDA (1997b) Guidance for industry: SUPAC-SS: nonsterile semisolid dosage forms. http://www.fda.gov/downloads/Drugs/Guidances/UCM070930.pdf. Accessed 9 Dec 2013

FDA (1999) Guidance for industry: SUPAC-IR/MR: immediate release and modified release solid oral dosage forms manufacturing equipment addendum. http://www.fda.gov/downloads/Drugs/Guidances/UCM070637.pdf. Accessed 9 Dec 2013

FDA (2003) Guidance for industry: bioavailability and bioequivalence studies for orally administered drug products—general considerations. http://www.fda.gov/downloads/Drugs/.../Guidances/ucm070124.pdf. Accessed 8 Dec 2013

FDA (2010a) Draft guidance on methylphenidate hydrochloride. http://www.fda.gov/downloads/Drugs/GuidanceComplianceRegulatoryInformation/Guidances/UCM281454.pdf. Accessed 8 Dec 2013

FDA (2010b) Meeting of the advisory committee for pharmaceutical science and clinical pharmacology. http://www.fda.gov/downloads/AdvisoryCommittees/CommitteesMeetingMaterials/Drugs/AdvisoryCommitteeforPharmaceuticalScienceandClinicalPharmacology/UCM207955.pdf. Accessed 8 Dec 2013

FDA (2010c) Response to citizen's petition. Docket No, FDA-2007-P-0182. http://www.regulations.gov/#!documentDetail;D=FDA-2007-P-0182-0017. Accessed 9 Dec 2013

FDA (2011) Draft guidance on zolpidem. http://www.fda.gov/downloads/Drugs/GuidanceComplianceRegulatoryInformation/Guidances/UCM175029.pdf. Accessed 9 Dec 2013

FDA (2012a) Draft guidance on mesalamine. http://www.fda.gov/downloads/Drugs/GuidanceComplianceRegulatoryInformation/Guidances/UCM319999.pdf. Accessed 8 Dec 2013

FDA (2012b) U.S. package insert for Topamax®. http://www.accessdata.fda.gov/drugsatfda_docs/label/2012/020505s050lbl.pdf. Accessed 8 Dec 2013

FDA (2013a) U.S. package insert for Ambien®. http://www.accessdata.fda.gov/drugsatfda_docs/label/2013/019908s032s034,021774s013s015lbl.pdf. Accessed 9 Dec 2013

FDA (2013b) U.S. package insert for Ambien CR®. http://www.accessdata.fda.gov/drugsatfda_docs/label/2013/019908s032s034,021774s013s015lbl.pdf. Accessed 9 Dec 2013

FDA (2013c) U.S. package insert for Brintellix®. http://www.accessdata.fda.gov/drugsatfda_docs/label/2013/204447s000lbl.pdf. Accessed 9 Dec 2013

FDA (2013d) U.S. package insert for Concerta®. http://www.accessdata.fda.gov/drugsatfda_docs/label/2013/021121s031lbl.pdf. Accessed 8 Dec 2013

FDA (2013e) U.S. package insert for Focalin XR®. http://www.accessdata.fda.gov/drugsatfda_docs/label/2013/021802s025lbl.pdf. Accessed 8 Dec 2013

FDA (2013f) U.S. package insert for Metadate CD®. http://www.accessdata.fda.gov/drugsatfda_docs/label/2013/021259s026lbl.pdf. Accessed 8 Dec 2013

FDA (2013g) U.S. package insert for Pentasa®. http://www.accessdata.fda.gov/drugsatfda_docs/label/2013/020049s025lbl.pdf. Accessed 8 Dec 2013

FDA (2013h) U.S. package insert for Quillivant XR®. http://www.accessdata.fda.gov/drugsatfda_docs/label/2013/202100s002lbl.pdf. Accessed 8 Dec 2013

FDA (2013i) U.S. package insert for Ritalin®. http://www.accessdata.fda.gov/drugsatfda_docs/label/2013/010187s074,018029s044lbl.pdf. Accessed 8 Dec 2013

FDA (2013j) U.S. package insert for Ritalin SR®. http://www.accessdata.fda.gov/drugsatfda_docs/label/2013/010187s074,018029s044lbl.pdf. Accessed 8 Dec 2013

FDA (2013k) U.S. package insert for Ritalin LA®. http://www.accessdata.fda.gov/drugsatfda_docs/label/2013/021284s019lbl.pdf. Accessed 8 Dec 2013

FDA (2013l) U.S. package insert for Trokendi XR®. http://www.accessdata.fda.gov/drugsatfda_docs/label/2013/201635s000lbl.pdf. Accessed 8 Dec 2013

Fourie Zirkelbach J, Jackson AJ, Wang Y, Schuirmann DJ (2013) Use of partial AUC (PAUC) to evaluate bioequivalence—a case study with complex absorption: methylphenidate. Pharm Res 30(1):191–202

Gabrielesson J, Weiner D (2001) Pharmacokinetic and pharmacodynamic data analysis: concepts and applications, 3rd edn. Apokekarsocieteten, Stockholm

Gehring R, Martinez M (2012) Assessing product bioequivalence for extended-release formulations and drugs with long half-lives. J Vet Pharmacol 35(Suppl 1):3–9

Haidar SH, Makhlouf F, Schuirmann DJ, Hyslop T, Davit B, Conner D, Yu LX (2008) Evaluation of a scaling approach for the bioequivalence of highly variable drugs. AAPS J 10(3):450–454

Klotz U (2012) The pharmacological profile and clinical use of mesalazine (5-aminosalicylic acid). Arzneimittelforschung 62(2):53–58

Lacey LF, Keene ON, Duquesnoy C, Bye A (1994) Evaluation of different indirect measures of rate of drug absorption in comparative pharmacokinetic studies. J Pharm Sci 83(2):212–215

Lionberger RA, Raw AS, Kim SH, Zhang X, Yu LX (2012) Use of partial AUC to demonstrate bioequivalence of zolpidem tartrate extended release formulations. Pharm Res 29:1110–1120

McEvoy JP, Daniel DG, Carson WH Jr, McQuade RD, Marcus RN (2007) A randomized, double-blind, placebo-controlled, study of the efficacy and safety of aripiprazole 10, 15 or 20 mg/day for the treatment of patients with acute exacerbations of schizophrenia. J Psychiatr Res 41 (11):895–905

Pringsheim T, Steeves T (2011) Pharmacological treatment for attention deficit hyperactivity disorder (ADHD) in children with comorbid tic disorders. Cochrane Database Syst Rev. (4): 1–27

Qiu Y, Chen Y, Zhang G (2009) Developing solid oral dosage forms: pharmaceutical theory and practice. Elsevier, Amsterdam

Reppas C, Lacey LF, Keene ON, Macheras P, Bye A (1995) Evaluation of different metrics as indirect measures of rate of drug absorption from extended release dosage forms at steady-state. Pharm Res 12(1):103–107

Rostami-Hodjegan A, Jackson PR, Tucker GT (1994) Sensitivity of indirect metrics for assessing "rate" in bioequivalence studies—moving the "goalposts" or changing the "game". J Pharm Sci 83(11):1554–1557

Shargel L, Yu A (1999) Applied biopharmaceutics and pharmacokinetics, 4th edn. Appleton & Lange, Stamford, CT

Stier EM, Davit BM, Chandaroy P, Chen ML, Fourie-Zirkelbach J, Jackson A, Kim S, Lionberger R, Mehta M, Uppoor RS, Wang Y, Yu L, Conner DP (2012) Use of partial area under the curve metrics to assess bioequivalence of methylphenidate multiphasic modified release formulations. AAPS J 14(4):925–926

Tozer TN, Bois FY, Hauck WW, Chen ML, Williams RL (1996) Absorption rate vs. exposure: which is more useful for bioequivalence testing? Pharm Res 13(3):453–456

Wang Y (2009) Determining critical PK measures for prediction of therapeutic equivalence of MR products: in silico approaches. AAPS Workshop

Chapter 8
Bioequivalence for Narrow Therapeutic Index Drugs

Wenlei Jiang and Lawrence X. Yu

8.1 Introduction

Bioequivalence (BE) is defined as the absence of a significant difference in the rate and extent to which the active ingredient or active moiety in pharmaceutical equivalents or pharmaceutical alternatives becomes available at the site of drug action when administered at the same molar dose under similar conditions in an appropriately designed study. BE studies of systemically absorbed drug products are generally conducted by determining pharmacokinetic endpoints to compare the in vivo rate and extent of drug absorption of a test and a reference drug product in healthy subjects. A test product is considered bioequivalent to a reference product if the 90 % confidence intervals for the geometric mean test/reference ratios of the area under the drug's plasma concentration versus time curve (AUC) and peak plasma concentration (C_{max}) both fall within the predefined BE limits of 80.00–125.00 %.

Although this BE limit has been successfully used to approve thousands of generic drugs, questions persist about whether it is appropriate for narrow therapeutic index (NTI) drugs, for which small changes in blood concentration could potentially cause serious therapeutic failures and/or serious adverse drug reactions in patients. While health care professionals, pharmaceutical scientists, regulatory agencies, and consumer advocates agree that more stringent criteria for BE should be considered for regulatory approval of NTI drugs, they disagree about how much assurance is needed about the similarity of a generic and its original innovator (reference) product for NTI drugs to be considered therapeutically equivalent.

FDA recently reconsidered the BE approach for NTI drugs and recommends a new approach to demonstrate bioequivalence of NTI drugs (US Food and Drug

W. Jiang (✉) • L.X. Yu
Center for Drug Evaluation and Research, U.S. Food and Drug Administration,
10903 New Hampshire Avenue, Silver Spring, MD 20993, USA
e-mail: Wenlei.Jiang@fda.hhs.gov

L.X. Yu and B.V. Li (eds.), *FDA Bioequivalence Standards*, AAPS Advances
in the Pharmaceutical Sciences Series 13, DOI 10.1007/978-1-4939-1252-0_8,
© The United States Government 2014

Administration 2012). This chapter discusses various public perspectives on the interchangeability of NTI drugs, reviews definitions and regulatory BE approaches by various international regulatory bodies, and examines the key characteristics of NTI drugs. The discussion will focus on the FDA's approach for NTI drugs with illustration of case studies.

8.2 Public Perspectives

There exist numerous anecdotal post-market reports that claim therapeutic failure or increased adverse events when switching from reference to generic drug products. There is, however, no well-controlled clinical study that demonstrates these events are related to switching between generic and reference drug products. Current spontaneous adverse event reporting systems are limited in their ability to compare safety signal between one drug product and another. As a result, the bulk of clinical evidence related to interchangeability of generic drugs is found in case reports and observational studies, which are difficult to prove causality.

Surveys of pharmacists and other health care professionals show that some believe that generic versions of certain drugs should not be dispensed (Kirking et al. 2001; Vasquez and Min 1999). Medical associations have issued various official positions on this issue. A 2006 joint position statement from the American Association of Clinical Endocrinologists, the Endocrine Society, and the American Thyroid Association raised concerns about FDA's approach for evaluating BE of generic levothyroxine products and recommended that physicians not prescribe generic levothyroxine drug products. The American Medical Association (AMA) issued a report in 2007 (American Medical Association 2007; https://www.aace.com/files/position-statements/aace-tes-ata-thyroxineproducts.pdf) generally backing the use of generic drugs, but recommending continued research into the best approach to determine individual product BE and specifically advocating FDA to reexamine its BE criteria for levothyroxine. The 2006 position statement from the American Academy of Neurology opposed generic substitution of anticonvulsant drugs for the treatment of epilepsy without the attending physician's approval (Liow et al. 2007). The American Society of Transplantation Conference report indicated that physicians supported the use of generic immunosuppressive agents in low-risk transplant recipients (Alloway 2003), however maintained that data is inadequate to make recommendations on the use of generic immunosuppressant medications in potentially at-risk patient populations (e.g., African Americans and pediatrics). The report recommended that demonstration of BE in at-risk patient populations be incorporated into the generic drug approval process. Although not all drugs in these categories are necessarily NTI drugs, the medical associations' concerns about interchangeability point to areas for investigation.

Finally, in the United States, policies related to NTI drug substitution differ among states. Currently 13 states list specific NTI drugs that are considered nonsubstitutable (National Association of Boards of Pharmacy 2006). The pharmacy laws of North Carolina require that a prescription for an NTI drug be refilled "using only the same drug product by the same manufacturer that the pharmacist last dispensed under the prescription, unless the prescriber is notified by the pharmacist prior to the dispensing of another manufacturer's product and the prescriber and the patient give documented consent to the dispensing of the other manufacturer's product" (Pope 2009). Many states currently have mandatory generic substitution laws, although these laws may vary significantly. In Oklahoma, a pharmacist must obtain approval from the patient or prescriber before substituting with a generic product while in Vermont a physician must provide a statement of generic ineffectiveness to prevent generic substitution. The different approach states take to the regulation on generic substitution of NTI drugs underscores the continued uncertainties in the community.

8.3 Regulatory Definition of Narrow Therapeutic Index Drugs

Several terms are used to describe the drugs in which comparatively small differences in dose or concentration may lead to serious therapeutic failures and/or serious adverse drug reactions in patients, including *narrow therapeutic index*, *narrow therapeutic range*, *narrow therapeutic ratio*, *narrow therapeutic window*, and *critical-dose drugs*. Table 8.1 summarizes the terms for this type of drugs used by different regulatory bodies, as well as the drug list in regulatory guidance if provided.

Health Canada has long documented this category of drugs that required greater degree of assurance in bioequivalence studies and named them critical-dose drugs. Critical-dose drugs are defined as those drugs where comparatively small differences in dose or concentration lead to dose- and concentration-dependent, serious therapeutic failures and/or serious adverse drug reactions which may be persistent, irreversible, slowly reversible, or life-threatening, which could result in inpatient hospitalization or prolongation of existing hospitalization, persistent or significant disability or incapacity, or death (Health Canada 2012). Critical-dose drugs apply to products including, but not limited to, those containing cyclosporine, digoxin, flecainide, lithium, phenytoin, sirolimus, tacrolimus, theophylline, and warfarin.

European Medicines Agency (EMA) does not define a set of criteria to categorize drugs as NTI drugs and they decide it case-by-case by Committee for Human Medicinal Products (CHMP) whether an active substance is an NTI drug based on clinical considerations. For instance, in the "Questions & Answers: Positions on specific questions addressed to the pharmacokinetics working party" document, they specify that tacrolimus and cyclosporine are NTI drugs in their individual product bioequivalence guidance (European Medicines Agency).

Japan Pharmaceutical and Food Safety Bureau uses the term narrow therapeutic range drug but has no definition on it in relevant guidelines. Nonetheless, a long list

Table 8.1 Terms, regulatory definitions, and list of narrow therapeutic index (NTI) drugs

Regulatory agencies	Term used and regulatory definition of NTI drugs	NTI drug list
Health Canada	Critical-dose drugs	Cyclosporine, digoxin, flecainide, lithium, phenytoin, sirolimus, tacrolimus, theophylline, and warfarin
	Critical-dose drugs are defined as those drugs where comparatively small differences in dose or concentration lead to dose- and concentration-dependent, serious therapeutic failures and/or serious adverse drug reactions which may be persistent, irreversible, slowly reversible, or life-threatening, which could result in inpatient hospitalization or prolongation of existing hospitalization, persistent or significant disability or incapacity, or death	
Europe Medicines Agency (EMA)	Narrow therapeutic index drugs	No list
	No definition	
US Food and Drug Administration (FDA)	Narrow therapeutic index drugs; NTI drugs are those drugs where small differences in dose or blood concentration may lead to serious therapeutic failures and/or adverse drug reactions that are life-threatening or result in persistent or significant disability or incapacity	Warfarin, Tacrolimus. Additional drugs are being identified as NTI drugs
Japan Pharmaceutical and Food Safety Bureau	Narrow therapeutic range drugs	Aprindine, carbamazepine, clindamycin, clonazepam, clonidine, cyclosporine, digitoxin, digoxin, disopyramide, ethinyl estradiol, ethosuximide, guanethidine, isoprenaline, lithium, methotrexate, phenobarbital, phenytoin, prazosin, primidone, procainamide, quinidine, sulfonylurea antidiabetic drugs compounds, tacrolimus, theophylline compounds, valproic acid, warfarin, zonisamide, glybuzole
	No definition	
South Africa Medicine Control Council	Narrow therapeutic range drugs having steep dose response curve	No drug list
	No definition	

of narrow therapeutic range drug is provided including mostly antiepileptic drugs, antidiabetic compounds, immunosuppressants, and others (Japan Pharmaceutical and Food Safety Bureau 2012a).

For United States Food and Drug Administration, the Code of Federal Regulations (21CFR320.33) uses *narrow therapeutic ratio* and defines it as follows:

(a) There is less than a twofold difference in median lethal dose (LD50) and median effective dose (ED50) values
(b) There is less than a twofold difference in the minimum toxic concentrations (MTC) and minimum effective concentrations (MEC) in the blood
(c) Safe and effective use of the drug products requires careful titration and patient monitoring

The CFR definition about narrow therapeutic ratio highlights the importance of careful dosage titration and patient monitoring. However, it may not be clinically practical to assess it as the values of LD50, ED50, MTC, or MEC are frequently not available during drug development or even after approval. In its guidances to industry, FDA also identified *narrow therapeutic range* drug products as those containing certain drug substances that are "subject to therapeutic drug concentration or pharmacodynamic monitoring, and/or where product labeling indicates a narrow therapeutic range designation" (US Food and Drug Administration 2003, 2000).

The 2010 and 2011 FDA advisory committee meeting discussed the definitions of NTI drugs (US Food and Drug Administration 2010b, 2011). Following discussions in conjunction with the committee's recommendations, FDA is using the term NTI and defining NTI drugs as those drugs where small differences in dose or blood concentration may lead to serious therapeutic failures and/or adverse drug reactions that are life-threatening or result in persistent or significant disability or incapacity (US Food and Drug Administration 2012).

8.4   Characteristic of Narrow Therapeutic Index Drugs

For NTI drugs, small differences in dose or concentration may lead to serious therapeutic failures and/or serious adverse drug reactions in patients. This section describes the general characteristics of NTI drugs, which can be used to classify certain drugs as NTI drugs.

First, we need to determine what are considered serious therapeutic failure or serious adverse drug reactions. If drug concentrations are below therapeutic concentrations for these indications, e.g., epilepsy, depression, schizophrenia, immunosuppression, cardiovascular disease, heart failure and atrial fibrillation, asthma and bronchospasm, anticoagulation, the therapeutic failures may be rated severe. Drug product black box warnings are generally considered as suggestion of severe toxicities. The severe toxicities can be hematological, cardiovascular, and neurological related such as bleeding, QT prolongation, arrhythmia, tachycardia,

Table 8.2 Therapeutic drug ranges, concentrations associated with serious toxicity, and estimated toxic/effective concentration ratios (http://www.nlm.nih.gov/medlineplus/ency/article/003430.htm)

Drugs	Therapeutic range	Plasma concentration associated with serious toxicity	Estimated toxic/ effective concentration ratio
Phenytoin (http://www.clinicalpharmacology-ip.com/Forms/drugoptions.aspx?cpnum=484&n=Phenytoin)[a]	10–20 mcg/ml	>40 mcg/ml	2.7
Digoxin (http://www.clinicalpharmacology-ip.com/Forms/Monograph/monograph.aspx?cpnum=190&sec=monmp)	0.8–1.5 ng/ml (CHF) 1.5–2.0 ng/ml (arrhythmia)	>2.5 ng/ml	1.4
Lithium (http://www.clinicalpharmacology-ip.com/Forms/Monograph/monograph.aspx?cpnum=351&sec=monmp)	0.6–2 meq/L 0.8–1.2 meq/L	>1.5 meq/L >2 meq/L	2.5 2.5
Theophylline (http://www.clinicalpharmacology-ip.com/Forms/Monograph/monograph.aspx?cpnum=599&sec=monmp)	5–15 mcg/ml (bronchodilator) 6–13 mcg/ml (premature apnea)	>20 mcg/ml	2

[a]For drugs with both therapeutic range data and concentrations associated with serious toxicity available, the toxic/effective concentration ratio is estimated by toxic concentration/middle value of therapeutic range

bradycardia, heart palpitations, hypertension, strokes, coma, seizures, and others. However, only severe toxicities relevant to drug substance are included to support the determination of NTI. For example, the cremophor vehicle for Taxol is thought to be responsible for most of the hypersensitive reactions seen with paclitaxel (Liebmann et al. 1993). The toxicities induced by cremophor in the drug product should not be considered in the determination of the drug substance as NTI. Further, the degree of adverse events or toxic effects should be evaluated based on the relative severity of the disease under investigation. For example, most clinicians will not treat a mild disease at the risk of serious side effects. Yet, one may tolerate more serious side effects to treat a life-threatening disease. Severe allergic reactions such as anaphylaxis are not considered in NTI determination since they are only pertaining to a small specific patient population and are not dose-/concentration-dependent.

Second, NTI drugs often have close therapeutic and toxic doses (or the associated blood/plasma concentrations). Adverse events can either possess their own dose-/concentration-response relationships or reflect extensions of therapeutic effects. Due to limitations in clinical studies, complete dose-/concentration-response curves are seldom developed. Therefore, therapeutic range data, blood concentration data associated with serious toxicity, and/or drug–drug interaction data can be used to estimate the ratio of toxic concentration to effective concentration. Table 8.2 lists some drugs' therapeutic ranges and estimated toxic/effective concentration ratios. The estimated toxic/effective concentration ratios provide quantitative information

Table 8.3 Summary of residual variability (%CV)[a] from Abbreviated New Drug Applications (ANDAs)

Drug products	# of BE Studies	AUC0-t		C_{max}	
		Mean	Range	Mean	Range
Warfarin	29	5.7	3.3, 11.0	12.7	7.7, 20.1
Lithium carbonate	16	7.8	4.5, 14.0	13.5	6.4, 24.4
Digoxin	5	21.7	13.1, 32.2	21.0	14.3, 26.1
Phenytoin	12	9.2	4.1, 18.6	14.9	7.4, 20.0
Theophylline	3	17.9	12.8, 24.2	18.2	11.8, 25.8
Tacrolimus	6	21.9	16.8, 26.6	19.0	15.0, 24.4

[a]The residual variability is the derived ANOVA root mean square error (RMSE) from two-way crossover BE studies, comparing test and reference products. The RMSE, as it is calculated from combined test and reference data, is an estimate of the residual variability in the pharmacokinetic measures of each individual drug substance

about how close the effective and toxic concentrations are. It should be noted that not every drug would have therapeutic range data available. Further, the therapeutic range data are usually the mean estimates for a population, which may not reflect therapeutic range in an individual patient. In addition, the drug concentrations associated with serious toxicities are often not available, which adds challenges to define a clear range between effective and toxic dose/concentration.

Third, as small variations in drug exposure can have significant clinical impact, many NTI drugs are subject to therapeutic drug monitoring based on pharmacokinetic or pharmacodynamics measures. Nevertheless, not all drugs subject to therapeutic monitoring are NTI drugs. For example, clinicians may conduct therapeutic monitoring because patients have potential compliance problems or clinical observation alone could not optimize the drug dose.

Fourth, NTI drugs generally have small to medium within-subject variability (WSV). WSV is estimated via root mean square error (RMSE) values of the bioequivalence parameters C_{max} and AUC0-t (Davit et al. 2008). Here, WSV refers to a measure of variability in blood concentration within the same subject, when the subject is administered two doses of the same formulation on two different occasions (Van Peer 2010). This variability may be intrinsic to the drug substance and/or the formulation, but also includes analytical variability, drug product quality variability, and unexplained random variation. WSV is of particular importance for NTI drugs because variations in plasma concentrations may have severe consequences. Approved drugs with narrow therapeutic indices should have exhibited small WSV. Otherwise, patients would routinely experience cycles of toxicity and lack of efficacy, and therapeutic monitoring would be useless (Benet 2006). A drug is considered highly variable if its WSV for C_{max} and/or area under the curve (AUC) is greater than 30 % (Haidar et al. 2008). Table 8.3 summarizes the residual variability of PK parameters of six drugs from single-dose, two-way, crossover BE studies with mean residual coefficient of variation (CV) ranging from 5.7 to 21.7 %. The mean residual CV includes WSV as well as variations caused by differences between test and references formulations. Therefore, the actual WSV would be even smaller. All drugs in Table 8.3 possess low-to-moderate WSV. In some cases, the clinical use of NTI drug often involves small dose adjustments of less than 20 % (Parks 2006). There is the implicit assumption that product variation and specifically variation introduced by product substitution be less than the size of dose adjustments.

In summary, the following characteristics generally apply to NTI drugs: (a) sub-therapeutic concentrations may lead to serious therapeutic failure; (b) there is little separation between therapeutic and toxic doses (or the associated blood/plasma concentrations); (c) they are subject to therapeutic monitoring based on pharmacokinetic (PK) or pharmacodynamic (PD) measures; (d) they possess low-to-moderate (i.e., no more than 30 %) WSV; and (e) in clinical practice, doses are often adjusted in very small increments (less than 20 %). These characteristics can help the classification of drugs as NTI drugs.

8.5 Bioequivalence Approaches for Narrow Therapeutic Index Drugs

Bioequivalence (BE) studies are an integral component of the drug development and approval process. BE studies are designed to determine if there is a significant difference in the rate and extent to which the active drug ingredient, or active moiety, becomes available at the site of drug action. The conventional bioequivalence study is usually conducted in healthy subjects with pharmacokinetic (PK) endpoints using a single-dose, two-way, crossover study design. Samples of an accessible biologic fluid such as blood or urine are analyzed for drug concentrations, and pharmacokinetic measures such as AUC and peak concentration (C_{max}), are obtained from the resulting concentration-time profiles. The BE parameters, AUC and C_{max}, are statistically analyzed using a two one-sided test procedure to determine whether the average values for the measures estimated after administration of the test and reference products are comparable. Two products are generally judged bioequivalent if the 90 % confidence interval of the geometric mean ratio (GMR) of AUC and C_{max} fall within 80–125 % (US Food and Drug Administration 2003).

The BE limit of 80.00–125.00 % is based on the premise that a 20 % difference between test and reference product is not clinically significant. The two one-sided test procedure for evaluating BE simultaneously controls the average difference between the test and reference product and the precision with which the population averages are estimated. The precision is determined by the WSV of BE measures and the number of subjects in the study. A drug product with large WSV may need a large number of subjects to pass bioequivalence standards while a product with very low variability may pass with a larger difference in mean response, as shown in Fig. 8.1. The assumption that 20 % difference between test and reference product is not clinically significant may not hold for NTI drugs. Thus, the large difference in

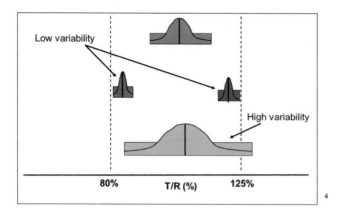

Fig. 8.1 Effect of variability on BE studies, where T is the test product and R is the reference product

mean C_{max} or AUC of two generic NTI products may potentially cause large plasma drug concentration fluctuation when patients switch between two products, potentially resulting in therapeutic failure or serious adverse events. As such, the conventional BE limits may be appropriate for most systemically available drug products, but not necessarily sufficient for NTI drugs.

Over the years various regulatory bodies have taken different bioequivalence approaches for NTI drugs (Table 8.4). Essentially, there are two approaches: Direct tightening of average bioequivalence limits and scaled average bioequivalence based on the WSV of the RLD.

8.5.1 Direct Tightening of Average Bioequivalence Limits

Considering that the bioequivalence limits of 80.00–125.00 % for the standard 90 % confidence interval may be too relaxed, some regulatory agencies take the approach of direct tightening of bioequivalence limits to a narrower range.

Health Canada requires the applicant to conduct a single-dose, two-way crossover or parallel study in healthy subjects or patients to demonstrate bioequivalence of NTI drugs. The criterion for the 90 % confidence interval of the relative mean AUC of the test to reference formulation is tightened to 90.0–112.0 % inclusive, whereas the criterion for the 90 % confidence interval of the relative mean C_{max} of the test to reference formulation remains to be 80.0 and 125.0 % (Health Canada 2012). These requirements are to be met in both the fasted and fed states. Steady-state studies are not required for critical-dose drugs unless warranted by exceptional circumstances. If a steady-state study is required, the 90 % confidence interval of the relative mean C_{min} of the test to reference formulation should also be between 80.0 and 125.0 % inclusive. If the bioequivalence study is conducted in patients

Table 8.4 Bioequivalence study design and criteria for narrow therapeutic index drugs

Regulatory agencies	Bioequivalence study design	Bioequivalence criteria
Health Canada	Single-dose two-way crossover or parallel study in healthy subjects	The 90 % confidence interval of the relative mean AUC and C_{max} of the test to reference formulation should be within 90.0–112.0 % and 80.0–125.0 %, respectively
European Medicine Agency	Single-dose two-way crossover or parallel study in healthy subjects	The 90 % confidence interval for AUC should be tightened to 90.00–111.11 %. Where C_{max} is of particular importance for safety, efficacy, or drug level monitoring, the 90.00–111.11 % acceptance interval should also be applied to C_{max}
South Africa Medicine Control Council	Single-dose two-way crossover or parallel study in healthy subjects	The 90 % confidence interval of the relative mean AUC and C_{max} of the test to reference formulation should be within 80.0–125.0 %
Japan Pharmaceutical and Food Safety Bureau	Single-dose two-way crossover or parallel study in healthy subjects	The 90 % confidence interval of the relative mean AUC and C_{max} of the test to reference formulation should be within 80.0–125.0 %
US FDA	Single-dose, fully replicated, four-way crossover study in healthy subjects	The 90 % confidence interval of the relative mean AUC and C_{max} of the test to reference formulation must pass both the reference-scaled limits and the unscaled average bioequivalence limits of 80.00–125.00 %. In addition, the upper limit of the 90 % confidence interval of the ratio of the within-subject standard deviation of the test to reference product (σ_{WT}/σ_{WR}) is less than or equal to 2.5

who are already receiving the drug as part of treatment, Health Canada highly recommends that the study group be as homogeneous as possible with respect to predictable sources of variation in drug disposition.

EMA recommends the acceptance interval for AUC of NTI drugs be tightened to 90.00–111.11 % (European Medicines Agency 2010). Where C_{max} is of particular importance for safety, efficacy, or drug level monitoring, the 90.00–111.11 % acceptance interval should also be applied for C_{max}. Therapeutic Goods Administration (TGA) of Australia follows the EMA guideline about NTI drugs (Therapeutic Goods Administration of Australia). Certain countries within the European Union have more specific policies and guidance related to NTI drugs. For example, the Danish Medicines Agency requires that the 90 % confidence interval for the ratio of the test and reference products for AUC and C_{max} must be within the 90.00–111.11 % (Danish Medicines Agency). Furthermore, the confidence interval must include 100 %.

As of South Africa Medicines Control Council, for general products, the 90 % confidence interval for the test/reference ratio of AUC and C_{max} should lie within the acceptance interval of 80–125 % and 75–133 %, respectively (South Africa Medicines Control Council 2011). For NTI products, the 90 % confidence interval for the test/reference ratio of C_{max} is tightened to 80–125 %.

In Japan, if the 90 % confidence interval of the difference in the average values of logarithmic C_{max} and AUC between test and reference products is within log (0.80)–log(1.25), products are considered to be bioequivalent. In some cases the confidence interval is not within the above range, however the test products can still be accepted as bioequivalent if (1) the total sample size of the initial bioequivalence study is not less than 20 ($n = 10$/group) or pooled sample size of the initial and add-on subject studies is not less than 30, (2) the differences in average values of logarithmic C_{max} and AUC between two products are within log(0.90)–log(1.11), and (3) dissolution rates of test and reference products are evaluated to be similar. These bioequivalence criteria apply to both conventional drug products and narrow therapeutic range products in Japan (Japan Pharmaceutical and Food Safety Bureau 2012b). However, for NTI drugs, a stricter requirement is used when applying biowaiver among different product strengths. For example, in the case of immediate release (IR) and enteric coated products containing NTI drugs, the test and reference are considered equivalent only if their dissolution profiles meet equivalence criteria and their average dissolution at 30 min are not less than 85 % under multiple testing conditions (Japan Pharmaceutical and Food Safety Bureau 2012a). For conventional drug products, they only need to meet the former dissolution criteria.

8.5.2 Reference-Scaled Average Bioequivalence Approach for NTI Drugs

The long time debate in the United States whether the BE limits of 80.00–125.00 % for the 90 % confidence interval is sufficient for NTI drugs was intensely discussed at the April 2010, FDA advisory committee meeting on NTI drugs (US Food and Drug Administration 2010a). The committee voted 11–2 that the BE criterion for the 90 % confidence interval to be within 80.00–125.00 % is insufficient for NTI drugs. Based on the input from the advisory committee, FDA conducted simulations to investigate the application of different BE approaches for NTI drugs, including the use of (1) direct tightening of BE limits and (2) tightening BE limits based on reference variability (the *reference-scaled average BE approach*). Variables evaluated in the simulations included WSV, sample size, and point estimate limit. The powers of a given study design using the reference-scaled average BE versus average BE approach were compared. Given the variation of WSV in NTI drugs (Yu 2011), the fixed average BE limits of 90–111 % can be too strict for truly equivalent generic drugs (i.e. GMR = 0.95–1.05) with medium WSV. The simulation results indicated that an approach that tightens BE limits based on reference variability is the preferred approach for evaluating BE of NTI drugs, i.e., the reference-scaled average bioequivalence approach. Based on these efforts, FDA is now recommending a four-way, fully replicated, crossover study design to demonstrate bioequivalence for NTI drugs. This study design permits not only the comparison of the mean of the test and reference drug products, but also the WSV of the test and reference drug products.

8.5.2.1 Mean Comparison

Because both test and reference drug products are given twice in each subject, the four-way, crossover, fully replicated study design enables the scaling of the acceptance BE limits to the WSV of the reference product. Scaled average BE for both AUC and $C_{\max}$ is evaluated by testing the following null hypothesis (US Food and Drug Administration 2012):

$$H_0 : \frac{(\mu_T - \mu_R)^2}{\sigma_{WR}^2} > \theta \qquad (8.1)$$

(given $\theta > 0$) versus the alternative hypothesis:

$$H_1 : \frac{(\mu_T - \mu_R)^2}{\sigma_{WR}^2} \leq \theta \qquad (8.2)$$

where μ_T and μ_R are the averages of the log-transformed measure ($C_{\max}$, AUC) for the test and reference drug products, respectively; usually testing is done at level $\alpha = 0.05$; and θ is the scaled average BE limits. Furthermore,

$$\theta = \frac{[\ln(\Delta)]^2}{\sigma_{W0}^2} \qquad (8.3)$$

where Δ is $1/0.9$, the upper BE limit for Test/Reference ratio of geometric means, and $\sigma_{W0} = 0.10$. Note that rejection of the null hypothesis, H_0, supports the conclusion of equivalence.

The baseline BE limits for NTI drugs are set at 90.00–111.11 % using the reference WSV (CV) of 10 %, but these limits would be scaled, based on the observed WSV of the reference product in the study. If reference WSV is less than or equal to 10%, then the reference-scaled BE limits are narrower than 90–111.11 %. If reference WSV is greater than 10 %, then the reference-scaled BE limits are wider than 90–111.11 %. These limits expand as the variability increases. However, since it is considered not desirable clinically to have these limits exceeded 80.00–125.00 %, FDA recommends that all BE studies on NTI drugs must pass both the reference-scaled approach and the unscaled average bioequivalence limits of 80.00–125.00 %. Because of these two criteria, the BE limits for these drug products would be tightened as shown in Fig. 8.2.

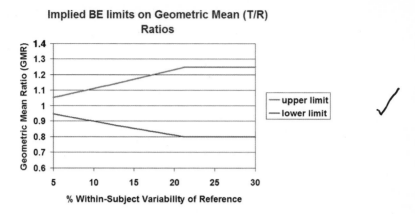

Fig. 8.2 Implied BE limits of geometric mean ratios for NTI drugs

8.5.2.2   Within-Subject Variability Comparison

WSV is of particular importance for NTI drugs because variations in plasma concentrations may have serious consequences. If an NTI test drug product has much higher WSV than the reference drug product in a BE study, the larger variation in blood concentration may result in higher likelihood of serious therapeutic failures and/or adverse reactions. Therefore, the test/reference ratio of within-subject standard deviation is evaluated. WSV comparison of the test and reference drug products is carried out by a one-sided F test. The null hypothesis for this test is the following.

$$H_0 : \sigma_{WT}/\sigma_{WR} > \delta \tag{8.4}$$

And the alternative hypothesis is:

$$H_1 : \sigma_{WT}/\sigma_{WR} \leq \delta \tag{8.5}$$

where δ is the regulatory limit to declare the WSV of the test drug product is not greater than that of the reference drug product. The 90 % confidence interval of the ratio of the within-subject standard deviation of the test to reference drug product, σ_{WT}/σ_{WR}, is given by $\left(\dfrac{s_{WT}/s_{WR}}{\sqrt{F_{\alpha/2}(v_1,v_2)}}, \dfrac{s_{WT}/s_{WR}}{\sqrt{F_{1-\alpha/2}(v_1,v_2)}} \right)$, where s_{WT} is the estimate of σ_{WT} with v_1 as the degrees of freedom, s_{WR} is the estimate of σ_{WR} with v_2 as the degrees of freedom, $F_{\alpha/2}(v_1, v_2)$ is the value of the F-distribution with v_1 (numerator) and v_2 (denominator) degrees of freedom that has a probability of $\alpha/2$ to its right, and $F_{1-\alpha/2}(v_1, v_2)$ is the value of the F-distribution with v_1 (numerator) and v_2 (denominator) degrees of freedom that has a probability of $1 - \alpha/2$ to its right. Here α is equal to 0.1. Equivalent WSV is declared when the upper limit of the 90 %

confidence interval for σ_{WT}/σ_{WR} is less than or equal to 2.5—i.e. the test statistic is based on the upper limit of the 90 % confidence interval (Jiang et al. 2012; US Food and Drug Administration 2012).

The reference-scaled average BE approach has been used successfully to demonstrate the BE of highly variable drugs and drug products (Davit et al. 2012). Highly variable drugs and drug products are those having greater than 30 % of WSV in pharmacokinetic measures (AUC and/or C_{max}), and generally exhibit a wide therapeutic window. Using the reference-scaled approach for highly variable drugs and drug products, the sample size required for a BE study is significantly reduced while avoiding the risk of allowing therapeutically inequivalent products to reach the market. Application of the reference-scaled approach for NTI drugs will tighten the BE limits of these drug products and circumvent the possibility of approving a generic product with a large mean difference from its reference drug product. Additional variability comparison will further reduce the risk of approving a generic drug product with a large variability difference from its reference drug product.

8.6 Case Studies

8.6.1 Warfarin (FDA) (http://www.clinicalpharmacology-ip. com/Forms/Monograph/monograph.aspx?cpnum= 650&sec=moninte; http://www.thomsonhc.com/ micromedex2/librarian/ND_T/evidencexpert/ND_PR/ evidencexpert/CS/D343B0/ND_AppProduct/ evidencexpert/DUPLICATIONSHIELDSYNC/175679/ ND_PG/evidencexpert/ND_B/evidencexpert/ND_P/ evidencexpert/PFActionId/evidencexpert. DisplayDrugpointDocument?docId=671285& contentSetId=100&title=Warfarin+Sodium& servicesTitle=Warfarin+Sodium&topicId= administrationMonitoringSection&subtopicId=null)

Warfarin is generally recognized as a NTI drug. Warfarin was first selected as a model drug to undergo a stepwise analysis to determine whether it satisfies all four general characteristics of NTI drugs: (1) sub-therapeutic concentrations may lead to serious therapeutic failure; (2) there is little separation between therapeutic and toxic doses (or the associated blood/plasma concentrations); (3) they are subject to therapeutic monitoring based on pharmacokinetic (PK) or pharmacodynamic (PD) measures; (4) they possess low-to-moderate WSV.

Warfarin is used for the following indication including (1) prophylaxis and/or treatment of venous thrombosis and its extension, and pulmonary embolism;

(2) prophylaxis and/or treatment of the thromboembolic complications associated with atrial fibrillation and/or cardiac valve replacement; and (3) reducing the risk of death, recurrent myocardial infarction, and thromboembolic events such as stroke or systemic embolization after myocardial infarction (FDA). The dosage and administration of warfarin must be individualized for each patient according to the particular patient's prothrombin (PT)/international normalized ratio (INR) response to the drug. If underdosed, failed treatment for the above indications may result in acute or recurrent thromboembolic episodes which are considered severe therapeutic failure. There is a black box warning in the warfarin label. If overdosed warfarin sodium can cause major or fatal bleeding, which are considered serious toxicity.

Warfarin's dose–response relationship in an individual patient is unpredictable based on population data and therefore a new patient's maintenance dose is difficult to predict. The label states that "It cannot be emphasized too strongly that treatment of each patient is a highly individualized matter. COUMADIN (Warfarin Sodium), a narrow therapeutic range (index) drug, may be affected by factors such as other drugs and dietary vitamin K. Dosage should be controlled by periodic determinations of prothrombin time (PT)/International Normalized Ratio (INR)." The relationship between warfarin dose and INR response is steep, which may lead to serious therapeutic failures and/or adverse drug reactions and make the selection of a maintenance dose challenging (Dalere et al. 1999) Based on the dose–response curve, one can estimate the toxic and effective doses for patients. INR < 2 or INR > 4 is considered likely to cause therapeutic failure or serious toxicity respectively (US Food and Drug Administration), the corresponding effective and toxic doses would be about 5 and 7 mg, respectively. Therefore, the ratio of toxic dose/effective dose is 1.4, which is very tight. In addition, some drug–drug interaction data also suggest that there is little separation between effective and toxic warfarin doses in patients. For example, studies have shown that rifampin increased the clearance of R-warfarin and S-warfarin 3.5-fold and 2-fold, respectively. Clinicians may need to increase warfarin's daily dose by two- to threefolds within the 1st week of starting rifampin. Upon discontinuation of rifampin, warfarin doses need to be reduced by half (FDA).

Warfarin undergoes pharmacodynamic monitoring. The biomarker that is used to measure warfarin's efficacy is the international normalized ratio (INR) and prothrombin time (PT). The INR is a good indicator of effectiveness and risk of bleeding during warfarin therapy. It is recommended to monitor INR levels in warfarin naïve patients starting after the initial two or three doses and at least once per month in patients receiving a stable dose regimen of warfarin (Ansell et al. 2008). Dose adjustment should be individualized to patient's INR to ensure efficacy and prevent adverse reactions (e.g., excessive bleeding). Patients at a higher risk of bleeding may benefit from more frequent INR monitoring, careful dose adjustment to desired INR, and a shorter duration of therapy.

In addition, the ANOVA RMSE was calculated based on the results of 2-period 2-sequence crossover bioequivalence studies in healthy subjects. Table 8.3 provides a summary of RMSE from approved warfarin ANDAs reviewed between 1996 and 2008. The analysis suggested that warfarin has mean within-subject CV of 5.7 % and 12.7 % for AUC and C_{max}, respectively. RMSE includes the variability between generic and reference drug products. Therefore, actual WSV would be even smaller.

In summary, warfarin therapeutic failure has serious consequences and overdose will cause severe toxicity. The dose that is minimally effective is relatively close to the minimum dose that leads to serious toxicity. Warfarin is subject to regular therapeutic monitoring based on INR, and has a small to medium (<30 %) WSV. Therefore, warfarin is classified as an NTI drug.

Since FDA has concluded that Warfarin sodium is a NTI drug, a fully replicated crossover design was recommended to demonstrate bioequivalence of generic warfarin sodium tablet in both fasting and fed states. The detailed statistical procedure and SAS code were provided in FDA individual bioequivalence guidance database (US Food and Drug Administration 2012) (see "Appendix").

8.6.2 Tacrolimus (FDA)

Tacrolimus capsule is a calcineurin-inhibitor immunosuppressant indicated for the prophylaxis of organ rejection in patients receiving allogeneic kidney transplants, allogeneic liver transplants, or allogeneic heart transplants. It is often used concomitantly with azathioprine or mycophenolate mofetil (MMF) and adrenal corticosteroids. The consequences of underdosing including morbidity/mortality associated with graft rejection are of major clinical importance and can substantially affect clinical outcome.

Tacrolimus can cause serious toxicities including malignancies, infection, nephrotoxicity, neurotoxicity, and hypertension. The black box warnings in the tacrolimus label include malignancies and serious infections. Patients receiving immunosuppressants, including Prograf, are at increased risk of developing lymphomas and other malignancies, particularly of the skin. The risk appears to be related to the intensity and duration of immunosuppression rather than to the use of any specific agent. Patients receiving immunosuppressants, including Prograf, are at increased risk of developing bacterial, viral, fungal, and protozoal infections, including opportunistic infections. These infections may lead to serious, including fatal outcomes. Tacrolimus can cause acute or chronic nephrotoxicity, particularly when used in high doses. Acute nephrotoxicity is most often related to vasoconstriction of the afferent renal arteriole, which is characterized by increasing serum creatinine, hyperkalemia, and/or a decrease in urine output, and is typically reversible. Chronic CNI nephrotoxicity is associated with mostly irreversible histologic damage to all compartments of the kidneys, including glomeruli, arterioles, and tubulo-interstitium.

The toxic tacrolimus concentration is not well defined. Acute oral overdose has been associated with tacrolimus levels of 19–97 ng/ml. The initial oral dosage recommendations for adult patients with kidney, liver, or heart transplants along with recommendations for whole blood trough concentrations in the package insert are shown in Table 8.5. In the case of heart transplantation, the observed whole blood trough concentrations ranged from 10 to 20 ng/mL. The observed whole blood trough concentration range suggested that tacrolimus has a close effective trough concentration and trough concentration associated with serious toxicity. Masuda and Inui et al. reported that surveys of tacrolimus trough concentrations at the steady-

Table 8.5 Summary of initial oral dosage recommendations and observed whole blood trough tacrolimus concentrations in adults

Patient population	Recommended Prograf initial oral dosage. Note: daily doses should be administered as two divided doses, every 12 h	Observed whole blood trough concentrations
Adult Kidney transplant In combination with azathioprine	0.2 mg/kg/day	Month 1–3: 7–20 ng/mL Month 4–12: 5–15 ng/mL
In combination with MMF/IL-2 receptor antagonist[a]	0.1 mg/kg/day	Month 1–12: 4–11 ng/mL
Adult liver transplant pediatric liver transplant	0.10–0.15 mg/kg/day	Month 1–12: 5–20 ng/mL
Adult heart transplant	0.075 mg/kg/day	Month 1–3: 10–20 ng/mL Month ≥4: 5–15 ng/mL

[a]In a second smaller trial, the initial dose of tacrolimus was 0.15–0.2 mg/kg/day and observed tacrolimus trough concentrations were 6–16 ng/mL during months 1–3 and 5–12 ng/mL during months 4–12

state and clinical events revealed that patients with a trough concentration of between 10 and 20 ng/mL avoided acute cellular rejection, infections, and side effects and most of the adverse effects occurred at a blood concentration higher than 20 ng/mL. Since surveyed patients were safely discharged from hospital without complications, the trough blood concentration of tacrolimus ranging between 10 and 20 (ng/mL) was suggested to be the therapeutic range (Masuda and Inui 2006).

In addition, available drug–drug interaction data also suggest that tacrolimus has a close effective drug concentration and drug concentration associated with serious toxicity. Tacrolimus is metabolized mainly by CYP3A enzymes, drug substances known to inhibit these enzymes may increase tacrolimus whole blood concentrations. Drugs known to induce CYP3A enzymes may decrease tacrolimus whole blood concentration. For example, sirolimus (2–5 mg/day) decreases tacrolimus blood concentrations (mean AUC0-12 and C_{min} by 30 %) vs tacrolimus alone. Sirolimus (1 mg/day) led to decrease in mean AUC0-12 and C_{min} by ~3 % and 11 %, respectively. This extent of tacrolimus pharmacokinetic parameter changes was considered major. Thus, use of sirolimus, in combination with tacrolimus, for prevention of graft rejection is not recommended. However, if concurrent use is deemed necessary, monitoring patients closely for loss of tacrolimus efficacy is required. In summary, therapeutic range data and drug–drug interaction data provide quantitative estimate about the closeness of effective tacrolimus concentration and concentration associated with serious toxicity.

Monitoring of tacrolimus blood concentrations in conjunction with other laboratory and clinical parameters is considered an essential aid to patient management for the evaluation of rejection, toxicity, dose adjustments, and compliance. The relative risks of toxicity and efficacy failure are related to tacrolimus whole blood trough concentrations. Therefore, monitoring of whole blood trough concentrations is recommended to assist in the clinical evaluation of toxicity and efficacy failure.

Factors influencing frequency of monitoring include but are not limited to hepatic or renal dysfunction, the addition or discontinuation of potentially interacting drugs and the post-transplant time.

In addition, the ANOVA RMSE from tacrolimus bioequivalence statistical analyses (Table 8.3) suggest that the WSV of tacrolimus is moderate.

In summary, for tacrolimus, therapeutic failure caused by underdose has serious consequences and overdose will cause severe toxicity. The minimum effective drug concentration is relatively close to the minimum drug concentration that leads to serious toxicity. Tacrolimus is subject to therapeutic drug monitoring based on trough whole blood concentration, and has medium (<30 %) WSV. Therefore, tacrolimus meets proposed NTI classification criteria and is an NTI drug.

Tacrolimus is also considered as a critical-dose drug by Health Canada based on the following (Health Canada 2012): (1) Tacrolimus may cause neurotoxicity and nephrotoxicity and the likelihood increases with higher blood levels; (2) Monitoring of tacrolimus blood levels in conjunction with other laboratory and clinical parameters is considered an essential aid to patient management. (3) In kidney transplant patients a significant correlation was found between tacrolimus levels and the incidence of both toxicity and rejection.

EMA also considers tacrolimus as a drug with a NTI (EMA 2012): (1) Tacrolimus is a drug that requires individual dose titration to achieve a satisfactory balance between maximizing efficacy and minimizing serious dose-related toxicity. Plasma level monitoring is routinely employed to facilitate dose titration; (2) Recommended Therapeutic Drug Monitoring schemes often set desirable levels close to the upper or lower limit of the therapeutic window (5 or 20 ng/ml); (3) The consequences of overdosing and of underdosing (including morbidity/mortality associated with graft rejection) are of major clinical importance and can substantially affect clinical outcome.

8.7 Future Perspectives

The adaptation of the BA/BE concept has enabled the approval of quality generic drug products. To provide enhanced assurance of the therapeutic equivalence of NTI drugs, FDA and other regulatory agencies have tightened their bioequivalence limits. As of Oct 2013, FDA has updated two product-specific bioequivalence recommendations and recommended reference-scaled bioequivalence approach for NTI drugs including warfarin sodium tablet and tacrolimus capsule. Broad implementation of this new bioequivalence approach is challenging because some drugs do not have an established NTI classification. It is imperative to establish a systematic process to identify and classify drugs as an NTI. Dose adjustment and therapeutic monitoring data in clinical practice may provide insight about the drug dose/concentration and response relationship. In 2013, FDA has initiated research projects to integrate clinical practice data with statistical tools to characterize the drug dose/concentration–response relationship and classify drugs with NTI (US Food and Drug Administration 2013).

Further, differences do exist in the determination and approval standards for NTI drugs among major regulatory bodies. Generic applicants have to conduct different types of bioequivalence studies for marketing the same generic NTI products in different regions of the world. A global NTI drug list and harmonized bioequivalence criteria are essential for creating lasting standards and will speed up the development and approval processes of generic NTI drugs.

Disclaimer   The views presented in this article by the authors do not necessarily reflect those of the US FDA.

Appendix: Method for Statistical Analysis Using the Reference-Scaled Average Bioequivalence Approach for Narrow Therapeutic Index Drugs

Step 1. Determine s_{WR}, the estimate of within-subject standard deviation (SD) of the reference product, for the pharmacokinetic (PK) parameters AUC and C_{max}. Calculation for s_{WR} can be conducted as follows:

$$s_{WR}^2 = \frac{\sum_{i=1}^{m} \sum_{j=1}^{n_j} \left(D_{ij} - \overline{D_{i.}} \right)^2}{2(n-m)}$$

where: $i =$ number of sequences m used in the study; [$m = 2$ for fully replicated design: TRTR and RTRT]; $j =$ number of subjects within each sequence; $T =$ Test product; $R =$ Reference product
$D_{ij} = R_{ij1} - R_{ij2}$ (where 1 and 2 represent replicate reference treatments)

$$\overline{D_{i.}} = \frac{\sum_{j=1}^{n_i} D_{ij}}{n_i}$$

$n = \sum_{i=1}^{m} n_j$ (i.e. total number of subjects used in the study, while n_i is the number of subjects used in sequence i)

Step 2. Use the reference-scaled procedure to determine BE for individual PK parameter(s).
Determine the 95 % upper confidence bound for:

$$\left(\overline{Y_T} - \overline{Y_R} \right)^2 - \theta s_{WR}^2$$

Where:

- $\overline{Y}_T$ and $\overline{Y}_R$ are the means of the ln-transformed PK endpoint (AUC and/or C_{max}) obtained from the BE study for the test and reference products, respectively
- $\theta \equiv \left(\frac{\ln(\Delta)}{\sigma_{w0}}\right)^2$ (scaled average BE limit)
- and $\sigma_{w0} = 0.10$ (regulatory constant), $\Delta = 1.11111$ ($= 1/0.9$, the upper BE limit)

The method of obtaining the upper confidence bound is based on Howe's Approximation I, which is described in the following paper:

W.G. Howe (1974), Approximate Confidence Limits on the Mean of X + Y Where X and Y are Two Tabled Independent Random Variables, Journal of the American Statistical Association, 69 (347): 789–794.

Step 3. Use the unscaled average bioequivalence procedure to determine BE for individual PK parameter(s). Every study should pass the scaled average bioequivalence limits and also unscaled average bioequivalence limits of 80.00–125.00 %.

Step 4. Calculate the 90 % confidence interval of the ratio of the within-subject standard deviation of test product to reference product σ_{WT}/σ_{WR}. The upper limit of the 90 % confidence interval for σ_{WT}/σ_{WR} will be evaluated to determine if σ_{WT} and σ_{WR} are comparable. The proposed acceptance criteria for the upper limit of the 90 % equal-tails confidence interval for σ_{WT}/σ_{WR} is less than or equal to 2.5.

The $(1 - \alpha)100\%$ CI for $\frac{\sigma_{WT}}{\sigma_{WR}}$ is given by

$$\left(\frac{s_{WT}/s_{WR}}{\sqrt{F_{\alpha/2}(v_1, v_2)}}, \frac{s_{WT}/s_{WR}}{\sqrt{F_{1-\alpha/2}(v_1, v_2)}}\right)$$

where

- s_{WT} is the estimate of σ_{WT} with v_1 as the degree of freedom
- s_{WR} is the estimate of σ_{WR} with v_2 as the degree of freedom
- $F_{\alpha/2, v_1, v_2}$ is the value of the F-distribution with v_1 (numerator) and v_2 (denominator) degrees of freedom that has probability of $\alpha/2$ to its right.
- $F_{1-\alpha/2, v_1, v_2}$ is the value of the F-distribution with v_1 (numerator) and v_2 (denominator) degrees of freedom that has probability of $1 - \alpha/2$ to its right.
- here $\alpha = 0.1$.

If SAS® is used for statistical analysis*
PROC MIXED should be used for fully replicated (4-period, 2-sequence replicated crossover 4-way) BE studies
*not necessary to use SAS® if other software accomplishes same objectives

Example SAS Codes: 4-Period, 2-Sequence Replicated Crossover Study

For a bioequivalence study with the following sequence assignments in a fully replicated 4-way crossover design:

	Period 1	Period 2	Period 3	Period 4
Sequence 1	T	R	T	R
Sequence 2	R	T	R	T

The following codes are an example of the determination of reference-scaled average bioequivalence for LAUCT. Assume that the datasets TEST and REF, have already been created, with TEST having all of the test observations and REF having all of the reference observations.

Dataset containing TEST 1 observations:

```
data test1;
 set test;
 if (seq=1 and per=1) or (seq=2 and per=2);
 lat1t=lauct;
run;
```

Dataset containing TEST 2 observations:

```
data test2;
  set test;if (seq=1 and per=3) or (seq=2 and per=4);
  lat2t=lauct;
run;
```

Dataset containing REFERENCE 1 observations:

```
data ref1;
  set ref;if (seq=1 and per=2) or (seq=2 and per=1);
  lat1r=lauct;
run;
```

Dataset containing REFERENCE 2 observations:

```
data ref2;

  set ref;

  if (seq=1 and per=4) or (seq=2 and per=3);

  lat2r=lauct;

run;
```

The number of subjects in each sequence is n1 and n2 for sequences 1 and 2, respectively.

Define the following quantities:

$T_{ijk} = k$th observation ($k = 1$ or 2) on T for subject j within sequence i
$R_{ijk} = k$th observation ($k = 1$ or 2) on R for subject j within sequence i

$$I_{ij} = \frac{T_{ij1} + T_{ij2}}{2} - \frac{R_{ij1} + R_{ij2}}{2}$$

and

$$D_{ij} = R_{ij1} - R_{ij2}$$

I_{ij} is the difference between the mean of a subject's (specifically subject j within sequence i) two observations on T and the mean of the subject's two observations on R, while D_{ij} is the difference between a subject's two observations on R.

Determine I_{ij} and D_{ij}

```
data scavbe;
  merge test1 test2 ref1 ref2;
  by seq subj;
  ilat=0.5*(lat1t+lat2t-lat1r-lat2r);
  dlat=lat1r-lat2r;
run;
```

Intermediate analysis—ilat

```
proc mixed data=scavbe;
  class seq;
  model ilat =seq/ddfm=satterth;
  estimate 'average' intercept 1 seq 0.5 0.5/e cl alpha=0.1;
  ods output CovParms=iout1;
  ods output Estimates=iout2;
  ods output NObs=iout3;
  title1 'scaled average BE';
  title2 'intermediate analysis - ilat, mixed';
run;
```

From the dataset IOUT2, calculate the following:
IOUT2:

```
pointest=exp(estimate);
x=estimate**2-stderr**2;
boundx=(max((abs(lower)),(abs(upper))))**2;
```

Intermediate analysis—dlat

```
proc mixed data=scavbe;
  class seq;
  model dlat=seq/ddfm=satterth;
  estimate 'average' intercept 1 seq 0.5 0.5/e cl alpha=0.1;
  ods output CovParms=dout1;
  ods output Estimates=dout2;
  ods output NObs=dout3;
  title1 'scaled average BE';
  title2 'intermediate analysis - dlat, mixed';
run;
```

From the dataset DOUT1, calculate the following:
DOUT1: s2wr=estimate/2;

From the dataset DOUT2, calculate the following:
DOUT2: dfd=df;

From the above parameters, calculate the final 95 % upper confidence bound:

```
theta=((log(1.11111))/0.1)**2;

y=-theta*s2wr;

boundy=y*dfd/cinv(0.95,dfd);

sWR=sqrt(s2wr);

critbound=(x+y)+sqrt(((boundx-x)**2)+((boundy-y)**2));
```

Calculate the unscaled average bioequivalence limits:
Calculation of unscaled 90 % bioequivalence confidence intervals:

```
PROC MIXED
data=pk;
CLASSES SEQ SUBJ PER TRT;
MODEL LAUCT = SEQ PER TRT/ DDFM=SATTERTH;
RANDOM TRT/TYPE=FA0(2) SUB=SUBJ G;
REPEATED/GRP=TRT SUB=SUBJ;ESTIMATE 'T vs. R' TRT 1 -1/CL ALPHA=0.1;ods output
Estimates=unscl;
title1 'unscaled BE 90% CI - guidance version';title2 'AUCt';
run;

data unscl;
  set unscl;
  unscabe_lower=exp(lower);
  unscabe_upper=exp(upper);
run;
```

References

21CFR320.33.

American Association of Clinical Endocrinologists (AACE), The Endocrine Society (TES), and American Thyroid Association (ATA). Joint position statement on the use and interchangeability of thyroxine products [Online]. https://www.aace.com/files/position-statements/aacetes-ata-thyroxineproducts.pdf. Accessed

Digoxin [Online]. http://www.clinicalpharmacology-ip.com/Forms/Monograph/monograph.aspx?cpnum=190&sec=monmp. Accessed

Lithium [Online]. http://www.clinicalpharmacology-ip.com/Forms/Monograph/monograph.aspx?cpnum=351&sec=monmp. Accessed

Phenytoin [Online]. http://www.clinicalpharmacology-ip.com/Forms/drugoptions.aspx?cpnum=484&n=Phenytoin. Accessed

Theophylline [Online]. http://www.clinicalpharmacology-ip.com/Forms/Monograph/monograph.aspx?cpnum=599&sec=monmp. Accessed

Therapeutic Drug Levels [Online]. http://www.nlm.nih.gov/medlineplus/ency/article/003430.htm. Accessed

Warfarin [Online]. http://www.clinicalpharmacology-ip.com/Forms/Monograph/monograph.aspx?cpnum=650&sec=moninte. Accessed

Warfarin Sodium [Online]. http://www.thomsonhc.com/micromedex2/librarian/ND_T/evidencexpert/ND_PR/evidencexpert/CS/D343B0/ND_AppProduct/evidencexpert/DUPLICATIONSHIELDSYNC/175679/ND_PG/evidencexpert/ND_B/evidencexpert/ND_P/evidencexpert/PFActionId/evidencexpert.DisplayDrugpointDocument?docId=671285&contentSetId=100&title=Warfarin+Sodium&servicesTitle=Warfarin+Sodium&topicId=administrationMonitoringSection&subtopicId=null. Accessed

Alloway RREA (2003) Report of the American Society of Transplantation conference on immunosuppressive drugs and the use of generic immunosuppressants. Am J Transplant 3:1211–1215

American Medical Association (2007) Generic substitution of narrow therapeutic index drugs. Report 2 of the council on science and public health [Online]. http://www.ama-assn.org/resources/doc/csaph/csaph2a07-fulltext.pdf. Accessed

Ansell J, Hirsh J, Hylek E, Jacobson A, Crowther M, Palareti G (2008) Pharmacology and management of the vitamin K antagonists: American College of Chest Physicians Evidence-Based Clinical Practice Guidelines (8th edition). Chest 133:160S–198S

Benet LZ (2006) Why highly variable drugs are safer? www.fda.gov/ohrms/dockets/ac/06/slides/2006-4241s2_2.ppt. Accessed

Dalere GM, Coleman RW, Lum BL (1999) A graphic nomogram for warfarin dosage adjustment. Pharmacotherapy 19:461–467

Danish Medicines Agency. http://laegemiddelstyrelsen.dk/en/topics/authorisation-and-supervision/licensing-of-medicines/marketing-authorisation/application-for-marketing-authorisation/bioequivalence-and-labelling-of-medicine–bstitution.aspx. Accessed

Davit BM, Conner DP, Fabian-Fritsch B, Haidar SH, Jiang X, Patel DT, Seo PR, Suh K, Thompson CL, Yu LX (2008) Highly variable drugs: observations from bioequivalence data submitted to the FDA for new generic drug applications. AAPS J 10:148–156

Davit BM, Chen ML, Conner DP, Haidar SH, Kim S, Lee CH, Lionberger RA, Makhlouf FT, Nwakama PE, Patel DT, Schuirmann DJ, Yu LX (2012) Implementation of a reference-scaled average bioequivalence approach for highly variable generic drug products by the US Food and Drug Administration. AAPS J 14:915–924

US Food and Drug Administration (2000) Guidance for industry: waiver of in vivo bioavailability and bioequivalence studies for immediate-release solid oral dosage forms based on a biopharmaceutics classification system [Online]. http://www.fda.gov/downloads/Drugs/.../Guidances/ucm070246.pdf. Accessed 16 June 2013

US Food and Drug Administration (2003) Guidance for industry: bioavailability and bioequivalence studies for orally administered drug products—general considerations [Online]. http://www.fda.gov/downloads/Drugs/.../Guidances/ucm070124.pdf. Accessed 16 June 2013

US Food and Drug Administration (2010a) April 13, 2010 Meeting of the Pharmaceutical Science and Clinical Pharmacology Advisory Committee: topic 1, revising the BE approaches for critical dose drugs [Online]. http://www.fda.gov/AdvisoryCommittees/Calendar/ucm203405.htm. Accessed 16 June 2013

US Food and Drug Administration (2010b) US FDA Pharmaceutical Science and Clinical Pharmacology Advisory Committee Meeting, April 13, 2010 [Online]. http://www.fda.gov/AdvisoryCommittees/Calendar/ucm203405.htm. Accessed

US Food and Drug Administration (2011) US FDA Pharmaceutical Science and Clinical Pharmacology Advisory Committee Meeting, July 26, 2011 [Online]. http://www.fda.gov/AdvisoryCommittees/Calendar/ucm261780.htm. Accessed

US Food and Drug Administration (2012) Draft guidance on warfarin sodium [Online]. http://www.fda.gov/downloads/Drugs/GuidanceComplianceRegulatoryInformation/Guidances/UCM201283.pdf. Accessed

US Food and Drug Administration (2013) Collection of dose adjustment and therapeutic monitoring data to aid narrow therapeutic index drug classification (U01) [Online]. http://grants.nih.gov/grants/guide/rfa-files/RFA-FD-13-020.html. Accessed

EMA (2014) Committee for Human Medicinal Products (CHMP) questions & answers: positions on specific questions addressed to the Pharmacokinetics Working Party. EMA/618604/2008 Rev. 9

European Medicines Agency (2010) Guideline on the investigation of bioequivalence [Online]. http://www.ema.europa.eu/docs/en_GB/document_library/Scientific_guideline/2010/01/WC500070039.pdf. Accessed

European Medicines Agency. Clinical efficacy and safety: clinical pharmacology and pharmacokinetics [Online]. http://www.ema.europa.eu/ema/index.jsp?curl=pages/regulation/general/general_content_000370.jsp&mid=WC0b01ac0580032ec5. Accessed

FDA. Warfarin sodium label [Online]. http://www.accessdata.fda.gov/scripts/cder/drugsatfda/index.cfm?fuseaction=Search.Overview&DrugName=WARFARIN%20SODIUM. Accessed

Japan Pharmaceutical and Food Safety Bureau (2012a) Guideline for bioequivalence studies for different strengths of oral solid dosage forms [Online]. http://www.nihs.go.jp/drug/be-guide(e)/strength/GL-E_120229_ganryo.pdf. Accessed

Japan Pharmaceutical and Food Safety Bureau (2012b) Guideline for bioequivalence studies of generic products [Online]. http://www.nihs.go.jp/drug/be-guide(e)/Generic/GL-E_120229_BE.pdf. Accessed

Haidar SH, Davit B, Chen ML, Conner D, Lee L, Li QH, Lionberger R, Makhlouf F, Patel D, Schuirmann DJ, Yu LX (2008) Bioequivalence approaches for highly variable drugs and drug products. Pharm Res 25:237–241

Health Canada (2012) Comparative bioavailability standards: formulations used for systemic effects [Online]. http://www.hc-sc.gc.ca/dhp-mps/alt_formats/pdf/prodpharma/applic-demande/guide-ld/bio/gd_standards_ld_normes-eng.pdf. Accessed

Jiang W, Schirman D, Makhlouf F, Lionberger R, Zhang X, Patel D, Subramaniam S, Connor D, Davit B, Grosser S, Yu L (2012) Within-subject variability comparison of narrow therapeutic index drug products. In: American association of pharmaceutical scientist, Chicago, IL

Kirking DM, Gaither CA, Ascione FJ, Welage LS (2001) Pharmacists' individual and organizational views on generic medications. J Am Pharm Assoc 41:723–728

Liebmann J, Cook JA, Mitchell JB (1993) Cremophor EL, solvent for paclitaxel, and toxicity. Lancet 342:1428

Liow K, Barkley GL, Pollard JR, Harden CL, Bazil CW (2007) Position statement on the coverage of anticonvulsant drugs for the treatment of epilepsy. Neurology 68:1249–1250

Masuda S, Inui K (2006) An up-date review on individualized dosage adjustment of calcineurin inhibitors in organ transplant patients. Pharmacol Ther 112:184–198

National Association of Boards of Pharmacy (2006) Survey of pharmacy laws. XIX: drug product selection laws http://www.nabp.net/publications/survey-of-pharmacy-law/

Parks MH (2006) Clinical perspectives on levothyroxine sodium products [Online]. http://www.fda.gov/ohrms/dockets/ac/06/slides/2006-4228S1-01-03-Parks%20Clinical.pdf. Accessed

Pope ND (2009) Generic substitution of narrow therapeutic index drugs. US Pharm (Generic Drug Review Suppl) 34:12–19

South Africa Medicines Control Council (2011) Biostudies [Online]. http://www.mccza.com/dynamism/default_dynamic.asp?grpID=30&doc=dynamic_generated_page.asp&categID=177&groupID=30. Accessed

Therapeutic Goods Administration of Australia. http://www.tga.gov.au/pdf/euguide/ewp140198rev1.pdf. Accessed

US Food and Drug Administration. Warfarin sodium label [Online]. http://www.accessdata.fda.gov/drugsatfda_docs/label/2011/009218s107lbl.pdf. Accessed

Van Peer A (2010) Variability and impact on design of bioequivalence studies. Basic Clin Pharmacol Toxicol 106:146–153

Vasquez EM, Min DI (1999) Transplant pharmacists' opinions on generic product selection of critical-dose drugs. Am J Health Syst Pharm 56:615–621

Yu LX (2011) Quality and bioequivalence standards for narrow therapeutic index drugs [Online]. http://www.fda.gov/downloads/Drugs/DevelopmentApprovalProcess/HowDrugsareDevelopedandApproved/ApprovalApplications/AbbreviatedNewDrugApplicationANDAGenerics/UCM292676.pdf. Accessed Dec 2013

Chapter 9
Pharmacodynamic Endpoint Bioequivalence Studies

Peng Zou and Lawrence X. Yu

9.1 Introduction

Pharmacokinetics (PK) is defined as the study of the time course of drug absorption, distribution, metabolism, and excretion. In contrast, pharmacodynamics (PD) refers to the relationship between drug concentration at the site of action and the pharmacological effects or adverse effects (Macdonald et al. 2004). PD is intrinsically linked to PK. A PD endpoint is defined as an indicator of pharmacologic response to a therapeutic intervention, which can be quantitatively measured and evaluated.

Figure 9.1 demonstrates that in vitro dissolution, PK and PD measurements can be used to establish bioequivalence of orally administered drug products. For instance, in vitro dissolution assays conducted at physiologically relevant pHs are used to establish bioequivalence of highly permeable, highly soluble drug substances formulated into rapidly dissolving drug products (FDA 2000). PK measurements, such as drug and/or metabolite concentrations in blood, urine, bile, or tissues, are used to demonstrate the equivalent exposure of two drug products. Alternatively, quantitative PD measurements of a dose-related pharmacological response can be used to establish bioequivalence of two drug products. When all the above three approaches are not applicable, clinical study may be selected to assess bioequivalence.

In descending order of preference, the US FDA recommends PK, PD, clinical, and in vitro studies for assessing bioequivalence of both innovative drug products and generic drug products (Chen et al. 2001). Comparison of PK profiles in human subjects is the most widely used method to demonstrate bioequivalence. PK endpoint-based bioequivalence evaluation is based on the assumption that the therapeutic effect of a drug product is a function of the systemic exposure or

P. Zou (✉) • L.X. Yu
Center for Drug Evaluation and Research, U.S. Food and Drug Administration,
10903 New Hampshire Avenue, Silver Spring, MD 20993, USA
e-mail: Peng.Zou@fda.hhs.gov

L.X. Yu and B.V. Li (eds.), *FDA Bioequivalence Standards*, AAPS Advances
in the Pharmaceutical Sciences Series 13, DOI 10.1007/978-1-4939-1252-0_9,
© The United States Government 2014

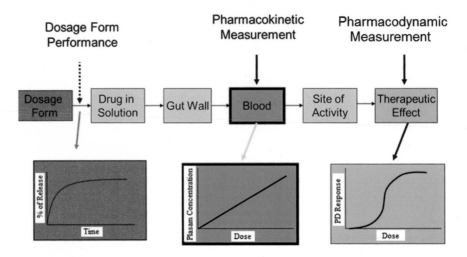

Fig. 9.1 Schematic representation of typical approaches for bioequivalence study of an orally administrated drug product

excretion profile of the active ingredient. However, PK endpoint-based approach is not applicable when (1) the drug and/or metabolite concentrations in plasma and/or urine are negligible; (2) drug and/or metabolite concentrations cannot be reliably measured based on currently available analytical methods; or (3) the measured drug concentration is not an indicator of efficacy and safety of a particular drug product. For these drug products, PD, clinical, and/or in vitro studies may be used to demonstrate bioequivalence.

For certain drug products, comparative clinical endpoint studies are currently the only acceptable approach to assess bioequivalence. However, the high costs, long duration, risk of failure, and insensitivity to detect formulation difference in comparative clinical endpoint studies form obstacles for generic drug development. In contrast, PD endpoint studies are relatively easier to perform, fairly reproducible, and shorter in duration. Consequently, PD studies result in limited drug exposure to the human subjects and require a relatively small number of subjects. On average, a PD endpoint bioequivalence study can save $2–6 million compared with a clinical endpoint bioequivalence study (Lionberger 2008). Hence, PD endpoint-based approach is highly recommended for bioequivalence studies when PK or in vitro studies are not applicable.

9.2   Guidelines Related to PD Endpoint-Based Bioequivalence Study

The regulatory authorities of many countries and World Health Organization (WHO) have published regulatory or scientific guidelines for bioequivalence studies. PD endpoint-based approaches have been recommended in these guidelines.

Table 9.1 Guidelines related to PD endpoint-based bioequivalence studies

Guidance title	Organizations or regulatory agencies	Recommendations for PD endpoint bioequivalence study	References
WHO expert committee on specifications for pharmaceutical preparation	WHO	The guideline adopts US FDA, European Regulatory Authority (EMEA) and other regulatory authorities' recommendations for the use of PD endpoint-based bioequivalence studies	WHO (2006)
Bioequivalence studies with pharmacokinetic endpoints for drugs submitted under an ANDA	US FDA	PD endpoint studies can be used to demonstrate bioequivalence but not recommended for drug products that are absorbed into the systemic circulation and for which a pharmacokinetic approach can be used	FDA (2013a)
Bioavailability and bioequivalence studies for nasal aerosols and nasal sprays for local action	US FDA	A PD study combined with in vitro and PK studies are recommended to establish bioequivalence of suspension formulations of locally acting nasal aerosol and spray drug products	FDA (2003a)
Topical dermatologic corticosteroids: in vivo bioequivalence	US FDA	A PD endpoint study (i.e., the vasoconstrictor bioassay) is recommended to establish bioequivalence of topical corticosteroids	FDA (1995)
Note for guidance on the investigation of bioavailability and bioequivalence	EMEA	PD endpoint studies are recommended for locally acting drugs without systemic absorption	EMEA (2000), FDA (1995)
Guideline on the requirements for clinical documentation for orally inhaled products (OIP) including the requirements for demonstration of therapeutic equivalence between two inhaled products for use in the treatment of Asthma and Chronic Obstructive Pulmonary Disease (COPD) in adults and for use in the treatment of asthma in children and adolescents	EMEA	PD safety and efficacy studies are the last step in cases where in vitro and PK data are unable to demonstrate equivalence. Bronchodilatation and bronchoprovocation PD models are recommended for bioequivalence study of orally inhaled short-acting $\beta 2$ adrenoceptor agonists (SABAs), long-acting $\beta 2$ adrenoceptor agonists (LABAs), anticholinergics, and glucocorticosteroids	EMEA (2009), EMEA (2000)

(continued)

Table 9.1 (continued)

Guidance title	Organizations or regulatory agencies	Recommendations for PD endpoint bioequivalence study	References
Data requirements for safety and effectiveness of subsequent market entry inhaled corticosteroid products for use in the treatment of asthma	Health Canada	A PD study by assessing the effect on the hypothalamic–pituitary–adrenal axis (HPA) is recommended for demonstrating bioequivalence of inhaled corticosteroid products if drug blood or plasma levels are too low to allow for reliable analytical measurement	Health-Canada (2011), EMEA (2009)
Guidance to establish equivalence or relative potency of safety and efficacy of a second entry short-acting beta2-agonist metered dose inhaler	Health Canada	Bronchodilation and bronchoprotection PD models are recommended for bioequivalence study of short-acting beta2-agonist metered dose inhaler products	Health-Canada (1999), Health-Canada (2011)
Guidance for bioavailability and bioequivalence studies	CDSCO, India	The guidance specifies justification and requirements for PD endpoint-based bioequivalence studies	CDSCO-India (2005), Health-Canada (1999)
Bioequivalence requirements guidelines	Saudi FDA	The guidance provides justification and requirements for PD endpoint-based bioequivalence studies	Saudi-FDA (2005), CDSCO-India (2005)

The guidelines related to recommendations for PD endpoint-based bioequivalence studies are summarized in Table 9.1. Meanwhile, the US FDA continually publishes bioequivalence recommendations for specific drug products to support generic drug product applications (FDA 2010b). The bioequivalence recommendation documents are updated periodically. One thousand one hundred twenty-seven documents for specific drug products have been published on the internet by the end of 2013 (http://www.fda.gov/drugs/guidancecomplianceregulatoryinformation/guidances/ucm075207.htm).

The WHO's guideline has requested the following justifications for the use of PD endpoint-based approach to demonstrate bioequivalence (WHO 2006).

- PD studies are usually not recommended for orally administered drug products when the drug is absorbed into the systemic circulation and a PK approach can be used to assess systemic exposure and establish bioequivalence. This is because variability in PD measures is generally greater than that in PK measures. However, in the instances where a PK approach is not possible, suitably validated PD endpoint methods can be used to demonstrate bioequivalence of orally administered drug products.

- PD endpoint bioequivalence studies are recommended if quantitative analysis of the active pharmaceutical ingredient (API) and/or metabolite(s) in plasma or urine cannot be performed with sufficient accuracy and sensitivity based on the currently available analytical methods.
- PD endpoint bioequivalence studies are required if measurements of API concentrations cannot be used as surrogate endpoints for the demonstration of efficacy and safety of the particular drug product.
- PD endpoint bioequivalence studies are especially appropriate for locally acting drug products such as human gastrointestinal (GI) tract locally acting drug products, topically applied dermatologic drug products, and oral inhalation drug products.

9.3 General Considerations for PD Endpoint-Based Bioequivalence Study

A PD endpoint-based approach is highly recommended by regulatory authorities when PK endpoint and in vitro approaches are not applicable. The PD endpoint-based bioequivalence study can only be applied to a limited number of drug products. To be eligible for a PD endpoint-based bioequivalence study, the drug product must meet the following criteria (Mastan et al. 2011).

- A dose–response relationship is demonstrated when possible.
- The PD effect at the selected dose(s) should be at the rising phase of the dose–response curve (i.e., below the maximum response which usually corresponds to a plateau level).
- Sufficient measurements should be taken to provide an appropriate PD response profile.
- All PD measurement methods should be validated for specificity, sensitivity, accuracy, and precision when possible.

When a PD endpoint-based approach is used to assess bioequivalence, the steepness of the dose–response curve (slope b) and the variability of PD endpoint measurements (s) are critical to the success of the bioequivalence study. The s/b ratio is inversely related to the ability of the PD endpoint assay to detect the difference between formulations (Ahrens et al. 2001). The smaller the s/b ratio, the more powerful the PD endpoint assay is. The s/b ratio is influenced by a number of factors including the drug product, selection of dose, selection of PD endpoint, disease severity, patient selection criteria, study design (crossover or parallel), period of treatment, and precision and sensitivity of PD endpoint measurements (FDA 2009; Ahrens et al. 2001). To improve the sensitivity of the PD endpoint assay and reduce the risk of failure of PD endpoint bioequivalence study, some general considerations should be taken into the study design, study conduct, and data analysis.

9.3.1 Selection of PD Endpoint

The selection of PD endpoint to be measured is extremely important. To establish a PD endpoint model with sufficient sensitivity to detect formulation differences between drug products, the selected PD endpoint should have a steep dose–response curve and be reproducible with low response variability.

Generally, the selected PD endpoint is a pharmacological or therapeutic effect which is relevant to the efficacy and/or safety of the drug product such as blood glucose level for anti-diabetic drug products. However, the PD endpoint can differ from the desired clinical effect(s) of a drug product. An example of this is the use of skin blanching as a PD endpoint for topical delivery of dermatologic formulations of corticosteroid. Following absorption into the skin, the corticosteroid causes a local vasoconstrictive effect. Although the blanching of the skin is not a desired clinical effect of these drug products, it does reflect drug exposure at the anti-inflammatory action site of these drug products. Quantitation of this skin blanching has been successfully used as a PD endpoint for assessing bioequivalence of test and reference dermatologic formulations of corticosteroid (Wiedersberg et al. 2008).

9.3.2 Dose Selection

A pilot study with the reference drug product is usually recommended to investigate the dose–response relationship and determine the dose and number of subjects to be used in the pivotal bioequivalence study. The doses that produce response in the steeply rising region of the dose–response curve should be selected for the pivotal study, resulting into a sensitive PD endpoint assay for detecting the formulation difference (EMEA 2009). The drug products should not produce a maximal response in the course of the study as the differences between formulations may not be detected at maximum or near-maximum response, which is usually at the saturated part of the dose–response curve. If a dose at the top or the bottom of the dose response curve is selected, the sensitivity of PD response to detect the difference of formulations will be compromised.

9.3.3 The Number of Study Subjects

The sample size of the study subjects must be selected with adequate statistical power, which is influenced by both the slope of and variability in the dose–response curve. Appropriate PD endpoint and study population selection can effectively reduce the sample size. Besides pilot studies, PD modeling and simulation (M&S) approach can also be used to estimate the number of subjects required for a PD endpoint study.

9.3.4 Inclusion and Exclusion Criteria of Subjects

To decrease the s/b ratio, participants should be screened prior to the PD study to exclude nonresponders (FDA 1995). The variability of PD measurements at baseline and following drug treatment can be minimized (small **s**) by careful selection of all responders. Meanwhile, responders are expected to exhibit steeper dose–response slopes (large **b**) than nonresponders. The criteria by which responders are distinguished from nonresponders must be stated in the study protocol.

The baseline status of the study subjects, e.g., diet, exercise, smoking, alcohol intake, and other habits possibly modifying drug dose–response curve, must be well defined (Zhi et al. 1995). When the PD study is conducted in patients, the characteristics of the disease, such as disease severity, secondary complications, other systems affected, and the use of other drugs, must be taken into account. For example, the disease severity of asthma patients can dramatically influence the steepness of dose–response curve of orally inhaled corticosteroids (Swainston Harrison and Scott 2004). Hence, if this PD endpoint is to be investigated, this study must be performed in patients with stable and appropriate severity.

9.3.5 Study Design

Many factors such as placebo effects (availability and application), treatment duration, dosing interval, and study design (parallel vs. crossover, double dummy, triple dummy, etc.) can affect the statistical power of dose–response analysis. The ideal PD study design should be double-blinded, subjecting each patient to four variables: baseline, placebo, formulation A, and formulation B (Zhi et al. 1995). The double-blinded design can minimize the variability of PD endpoint measurement. Baseline measurement and placebo treatment in the design of the study can decrease the variability of PD endpoint at baseline. In addition, adjustment of dosing intervals can shift PD response to the linear portion of the dose–response curve, thereby increasing statistical power of the PD response analysis. A crossover study design, which allows each subject to be used as "his or her own control,"

Table 9.2 Points for consideration in study designs and exposure–response analysis (FDA 2003b)

Study design	Points to consider in study design and exposure–response analysis
Crossover, fixed dose, dose response	• For immediate, acute, reversible responses • Provide both population mean and individual exposure–response information • Safety information obscured by time effects, tolerance, etc. • Treatment by period interactions and carryover effects are possible; dropouts are difficult to deal with • Changes in baseline-comparability between periods can be problematic
Parallel, fixed dose, dose response	• For long-term, chronic responses, or responses that are not quickly reversible • Provides only population mean, no individual dose response • Should have a relatively large number of subjects (1 dose per patient) • Gives good information on safety
Titration	• Provides population mean and individual exposure–response curves, if appropriately analyzed • Confounds time and dose effects, a particular problem for safety assessment
Concentration controlled, fixed dose, parallel, or crossover	• Directly provides group concentration-response curves (and individual curves, if crossover) and handles inter-subject variability in pharmacokinetics at the study design level rather than data analysis level • Requires real-time assay availability

would decrease the variability of PD endpoint baseline, substantially increasing statistical power (Ahrens et al. 2001). However, crossover design may not be appropriate when carryover effect is expected. The general considerations for study design and exposure–response analysis are summarized in Table 9.2 (FDA 2003b).

9.3.6 PD Sampling Time Points

In some PD endpoint studies, the PD measurements can be continuously recorded. The time course of the PD response intensity of the drug product can be plotted in ways similar to the PK profiles. Similarly, the PD profile parameters can be derived including the area under the effect–time curve (AUEC), the maximum response (E_{max}) and the time at which the maximum response occurred (T_{max}). For these PD endpoint studies, a rational design of the sampling scheme would improve the performance of the PD assay to detect the difference between two formulations.

For example, acarbose, a glycosidase inhibitor, is a GI tract locally acting anti-diabetic drug. The blood glucose level is selected as a PD endpoint to assess the bioequivalence of orally administrated acarbose tablets. The maximum reduction of blood glucose occurs within the first hour following acarbose administration upon sucrose challenge. Therefore, intensive sampling during the first hour postdose is required.

9.3.7   PD Endpoint Measurements and Method Validation

The PD response should be measured quantitatively, preferably under double-blind conditions, and be recordable by an instrument with low variability. If instrumental measurements are not possible, recordings on visual analogue scales may be acceptable. If only qualitative measurements are available, special statistical analysis will be required.

PD endpoint-based bioequivalence study should cover the sufficient time course, the initial baseline values in each period should be comparable, and the PD effect at the selected dose(s) should be within the rising phase of the dose–response curve (i.e., below the maximum response). The methodology must be validated for precision, accuracy, and specificity.

9.3.8   Data Analysis and PD Modeling

The nonlinear character of the dose/response relationship should be taken into account and baseline corrections should be considered during data analysis. The nonlinear dose–response relationship adds complexity in assessing bioequivalence by PD endpoint method. As shown in Fig. 9.2a, due to the linear dose–PK relationship for most drug products, the difference in PK measurement directly reflects the difference in the delivered dose amount. In contrast, the observed PD response difference may not proportionally reflect the dose difference because of nonlinear dose–PD response relationships. With the same PD response difference between the test and reference product ($T/R = 90$ % on PD response scale), the dose of the test product could be 82 % (Fig. 9.2b) or 46 % (Fig. 9.2c) of the reference product's dose. Therefore, the dose difference between the test and reference products associated with PD response difference is dependent on the position of the dose(s) on the nonlinear dose–response curve. Hence, assessing bioequivalence on the "PD response scale" can be misleading since it may not directly reflect the differences in relative bioavailability of two formulations. To address this issue, dose-scale analysis is introduced to translate nonlinear PD endpoint measurements to linear dose measurements. The relative bioavailability "F" of the test product relative to that of the reference product can be calculated using the dose scale analysis. The dose-scale analysis has been recommended for

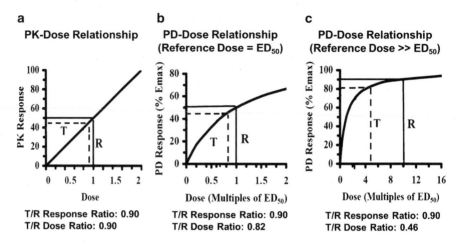

Fig. 9.2 Nonlinear dose–response relationships in PD endpoint studies. (**a**) Linear dose–PK relationship; (**b**) nonlinear dose–PD response curve when the reference drug product dose is ED50; (**c**) nonlinear dose–PD response curve when the reference drug product dose is >>ED50. ED50 is the dose required to produce 50 % of the maximum PD response

PD endpoint studies of orally inhaled short-acting beta-2 agonists (Treffel and Gabard 1993). The recommended bioequivalence acceptance range for the relative bioavailability (F) is within 67–150 % with 90 % confidence interval (CI).

The acceptance range for bioequivalence assessment may not be appropriate in certain cases and should be justified on a case-by-case basis. For example, a PK/PD modeling analysis revealed that the currently accepted PK bioequivalence criteria (0.80–1.25 with 90 % CI) is very restrictive for the ratio of E_{max} but is still appropriate for the ratio of AUEC between test and reference products (Navidi et al. 2008).

9.4 PD Endpoint-Based Bioequivalence Recommendations for Specific Drug Products

Although PD endpoint approaches are highly recommended when PK or in vitro studies are not applicable, the PD endpoint-based bioequivalence studies currently accepted by the FDA are still very limited. As shown in Table 9.3, the FDA has recommended PD endpoint studies to assess bioequivalence of eight drug products. Among them, acarbose tablet, lanthanum tablet, orlistat capsule are GI tract locally acting drug products. Fluticasone propionate cream is a topically applied dermatologic drug product. Albuterol sulfate metered dose aerosol is an orally inhaled corticosteroid. Fluticasone propionate/salmeterol xinafoate inhalation powder is an orally inhaled combination formulation of corticosteroid and long-acting beta₂ agonist (LABA). Enoxaparin sodium injection and dalteparin sodium injection

Table 9.3 PD bioequivalence studies recommended by the FDA

Drug products	Dosage form/route	Indication	Mechanisms of action	PD endpoints	BE criteria
Acarbose	Tablet/oral	Type 2 diabetes	Inhibition of pancreatic alpha-amylase and intestinal alpha-glucosidase hydrolase enzymes	Blood glucose[a]	90 % CIs for the T/R ratios of AUEC and C_{max} within 80–125 %[b]
Lanthanum carbonate	Chewable tablet/oral	End stage renal disease	Reducing the absorption of phosphate by forming insoluble lanthanum phosphate complexes	Change in urinary phosphate excretion	90 % CIs for the T/R ratios of reduced urinary phosphate excretion within 80–125 %
Orlistat	Capsule/oral	Obesity	Inhibiting the absorption of dietary fats	Ratio of amount of fecal fat excreted at steady-state to daily ingested fat	Dose-scale analysis; 90 % CI for the relative bioavailability (F) within 80–125 %
Fluticasone propionate (corticosteroid)	Cream/topical	Corticosteroid-responsive dermatoses	Anti-inflammatory, antipruritic, and vasoconstrictive properties	Degree of skin blanching	90 % CI for the T/R ratios of AUEC within 80–125 %
Albuterol sulfate	Aerosol/inhalation	Bronchospasm	Activation of beta2-adrenergic receptors on airway smooth muscle	Bronchoprovocation[c]: PC_{20} or PD_{20}; or Bronchodilatation: $AUEC_{0-4h}$, $AUEC_{0-6h}$ and maximum FEV_1	Dose-scale analysis; 90 % CI for the relative bioavailability (F) within 67–150 %
Fluticasone propionate/salmeterol xinafoate	Powder/inhalation	Asthma and chronic obstructive pulmonary disease	Bronchodilation (salmeterol) and anti-inflammatory effects (fluticasone)	AUC_{0-12h} and FEV_1	90 % CI for the T/R ratios of AUC_{0-12h} and FEV_1 within 80–125 %
Enoxaparin sodium	Injectable/subcutaneous	Deep vein thrombosis	Enhancing the inhibition of Factor Xa and thrombin by antithrombin	Anti-Xa and anti-IIa in plasma[d]	90 % CIs for the T/R ratios of anti-Xa AUEC and E_{max} within 80–125 %[d]

(continued)

Table 9.3 (continued)

Drug products	Dosage form/route	Indication	Mechanisms of action	PD endpoints	BE criteria
Dalteparin sodium	Injectable/subcutaneous	Deep vein thrombosis	Enhancing the inhibition of Factor Xa and thrombin by antithrombin	Anti-Xa and anti-IIa in plasma[e]	90 % CIs for the T/R ratios of anti-Xa AUEC and E_{max} within 80–125 %[d]

Note:

[a] In case that test product and reference product are not Q1/Q2

[b] T and R stand for test and reference drug products, respectively

[c] PC_{20} and PD_{20} are the provocative concentration and dose, respectively, of the methacholine challenge agent required to reduce the forced expiratory volume in one second (FEV_1) by 20 % following administration of differing doses of albuterol (or placebo) by inhalation

[d] PD study is used to support active ingredient sameness between test and reference products

[e] Anti-IIa data of test product and reference product are required as supportive evidence of active ingredient sameness between test and reference products

are anticoagulants. The dose, study population, study design and PD response measurements recommended by the FDA for individual drug products are summarized in Table 9.4. It is worthwhile to note that PD endpoint study may not be the only option for assessing bioequivalence of these drug products. For example, for acarbose tablets and lanthanum carbonate tablets, in vitro study is the other option to demonstrate bioequivalence. In addition, for complex drug products such as albuterol sulfate metered dose inhaler, enoxaparin sodium injection and dalteparin sodium injection, PD endpoint study alone is not sufficient to establish bioequivalence. It should be combined with in vitro study and/or in vivo PK study to establish bioequivalence.

9.4.1 Acarbose Tablet

Acarbose is used to treat type 2 diabetes. Acarbose inhibits intestinal α-glycosidase, which decreases the digestion of ingested starch and disaccharides in the small intestine and reduces their oral absorption, resulting in decreased post-meal glucose levels (Lee et al. 2012). PK endpoint method cannot be used to assess bioequivalence of acarbose tablet since its systemic exposure is negligible and the clinical efficacy is not correlated with its PK profile. Therefore, blood glucose concentration is recommended as a PD endpoint to assess bioequivalence of acarbose tablets in healthy subjects (FDA 2009). A pilot study is recommended to determine the appropriate dose for the pivotal bioequivalence study and to determine the appropriate number of study subjects needed to provide adequate statistical power. In the pilot study, after a challenge dose of 75 g of sucrose, a range of single dose of reference acarbose tablets are given to identify the lowest possible dose that will yield a PD response significantly different from the baseline. This is to make sure the selected dose falls into the steeply rising region of the dose–response curve. In the pilot and pivotal studies, blood samples are collected prior to treatment and throughout 4 h following acarbose and sucrose administration. Blood glucose levels are adjusted for pretreatment (i.e., baseline) levels. Bioequivalence assessment is based on the reduction of blood glucose levels following treatment with acarbose and sucrose together compared with only sucrose challenge. The parameters used for bioequivalence assessment are the maximum reduction in blood glucose concentration ($C_{\max}$) and the area under the blood glucose concentration reduction versus time curve through 4 h (AUEC_{0-4h}).

Recently, a set of new bioequivalence criteria have been investigated in healthy Chinese volunteers by following the FDA recommended study design (Zhang et al. 2012). In this study, consistent with the FDA guidance, serum glucose concentration was used as PD endpoint. However, it was found that a large percentage of subjects had negative values of AUEC_{0-4h} for both the reference and test formulations (35 % versus 45 % respectively). Furthermore, the variability in AUEC_{0-4h} for both test and reference formulations was very large, resulting in non-normal distributions. To overcome the limitation of AUEC_{0-4h}, the authors

Table 9.4 The pharmacodynamic study design recommended by the FDA for specific drug products

Drug products	Dose	Selection of subjects	Study design	PD response measurements
Acarbose	To be determined in pilot study	Healthy subject excluding obesity	Single-dose, two-way crossover, fasting study; 1-week washout between treatments	blood samples; sampling during 4 h postdose and intensive sampling in first hour postdose
Lanthanum carbonate	To be determined in pilot study	Healthy subject	Fed study	Urine samples
Orlistat	60 mg tid or 2× 60 mg tid	Healthy subject	Multiple-dose, three-way crossover, fed study (controlled diet); washout period ≥4 day	fecal samples; sampling over at least 24 h at steady state
Fluticasone propionate	To be determined in pilot study	Healthy subjects and sensitive responders	Multiple-dose, 40–60 subjects, eight testing sites per arm; T and R are assigned to two arms complementarily	Skin chromameter measurement through at least 24 h
Albuterol sulfate (bronchoprovocation)	0 mg, 0.09 mg of T and R, 0.18 mg of R	Patients with stable mild asthma	Single-dose, double-blind, double dummy, randomized, crossover study; ≥24 h washout period	To measure the concentration or dose of methacholine required to decrease FEV_1 by 20 % following administration of differing doses of albuterol
Albuterol sulfate (bronchodilatation)	0 mg, 0.09 mg of T and R, 0.18 mg of R	Patients with moderate-to-severe asthma	Single-dose, double-blind, double-dummy, randomized, crossover study; ≥24 h washout period	To measure $AUEC_{0-4h}$, $AUEC_{0-6h}$ and maximum FEV_1; FEV_1 is measured through 6 h postdose
Fluticasone propionate/ salmeterol xinafoate	100 µg/50 µg, twice daily	Males and non-pregnant females with asthma	Multiple-dose, randomized, parallel study consisting of a 2 week run-in period followed by a 4-week treatment period of the placebo, T or R product	To measure area under FEV_1-time curve from time zero to 12 h on the first day of the treatment and FEV_1 prior to dosing on the last day of 4-week treatment

Note: T and R stand for test and reference drug products, respectively. FEV_1 stands for forced expiratory volume in one second. $AUEC_{0-4h}$ and $AUEC_{0-6h}$ stand for areas under the effect curve calculated from the zero time to 4 and 6 h, respectively proposed three new bioequivalence parameters: glucose excursion (GE), GE' (glucose excursion without the effect of the homeostatic glucose control), and fAUC (the degree of fluctuation in serum glucose concentration based on AUC_{0-4h}). GE is calculated as the difference between the peak (C_{max}) and trough (C_{min}) serum glucose concentrations in the 4 h study period.

$$GE = C_{max} - C_{min}$$

GE' is calculated as:

$$GE' = C_{max} - C'_{min}$$

Where C'_{min} is the minimum serum glucose concentration in the time interval 0–T_{max}.

fAUC is calculated as:

$$fAUC = AUC(C \geq C_{ss}) + AUC(C \leq C_{ss})$$

where C_{ss} is the plateau concentration of glucose and calculated as $C_{ss} = (AUC_{0-4h})/4$

The authors found that the variability of GE, GE', and fAUC is less than that of $AUEC_{0-4h}$ and concluded that the combination of C_{ss} and one of three new glucose fluctuation parameters, GE, GE', and fAUC is preferable than $AUEC_{0-4h}$ for bioequivalence study of acarbose tablets. Although the proposed new bioequivalence parameters show advantages, more clinical studies are required to validate the new criteria.

9.4.2   Lanthanum Carbonate Chewable Tablet

Lanthanum carbonate chewable tablets are used to reduce serum phosphate in patients with end stage renal disease. The product is recommended to be administered with food. Lanthanum carbonate inhibits absorption of phosphate by forming highly insoluble lanthanum phosphate complexes in GI tract, consequently reducing serum phosphate level (Swainston Harrison and Scott 2004). Both Lanthanum carbonate and Lanthanum phosphate are poorly absorbed in the GI tract. The oral bioavailability of Lanthanum carbonate is <0.002 %. Because of the negligible systemic exposure and poor correlation between efficacy and systemic exposure, the reduction in urinary phosphate excretion is used as the PD endpoint to assess bioequivalence of Lanthanum carbonate products (FDA 2011c). Due to the relatively large variability of urinary phosphate excretion baseline in patients with end stage renal disease, the PD endpoint assay may not be sensitive to the difference of formulations if this study is conducted in patients with end stage renal disease. Therefore, healthy subjects are recommended to be recruited in this PD endpoint study.

9.4.3 Orlistat Capsule

Orlistat, an intestinal lipase inhibitor, is used in the treatment of obesity. By inhibiting the hydrolysis of dietary fat into free fatty acids and monoacylglycerols in GI tract, orlistat reduces the systemic absorption of dietary fat and increases fecal fat excretion (Drent and van der Veen 1993). Since systemic absorption of orlistat is negligible (Zhi et al. 1995), PK endpoint-based bioequivalence study is not applicable. Therefore, the ratio of the amount of fat excretion in feces over a 24-h period at steady-state to the amount of daily ingested fat is selected as the PD endpoint for bioequivalence assessment (FDA 2010a). To minimize the variability of fecal fat excretion (PD response baseline), a standardized and well-controlled diet containing 30 % of calories from fat, as per labeling, is recommended throughout the study and the controlled diet should start at least 5 days prior to drug treatment to achieve a steady-state of fecal fat excretion. A multiple-dose, three-way crossover study consisting of two doses of reference product and at least one dose of the test product in healthy subjects is recommended. To measure the steady-state fecal fat excretion, each treatment period needs to last for at least 9 days. Each treatment period should be separated by a washout period of at least 4 days. The fecal samples are collected over at least 24 h postdose. Due to the nonlinear relationship between orlistat dose and PD response, a dose-scale method incorporating E_{max} model is recommended to calculate the relative bioavailability of the test product. The 90 % CI of the relative bioavailability must be within 80–125 % in order to establish bioequivalence.

9.4.4 Fluticasone Propionate Cream

Fluticasone propionate is a synthetic corticosteroid and the topically applied fluticasone propionate cream is used to help relieve skin itching and inflammation. Due to the lack of systemic exposure, traditional PK bioequivalence study is not applicable to topically applied dermatologic products. For instance, for topical creams and gels, although PK approaches (such as skin stripping, microdialysis, and noninvasive imaging or spectroscopic detection) have shown promise in published references (N'Dri-Stempfer et al. 2009; Navidi et al. 2008), these methods are currently not adopted by regulatory authorities. For topical solutions, bioequivalence can be established based on qualitatively and quantitatively equivalent composition of the test and reference products. As such, PD skin blanching method is recommended for bioequivalence studies for topical corticosteroids (FDA 1995). Similarly, clinical endpoint studies are recommended to establish bioequivalence for most other topical products because no alternative methods have been accepted by regulatory authorities (Lionberger 2008).

The degree of skin blanching is used as a PD endpoint to assess bioequivalence of fluticasone propionate cream products (FDA 2011b). The application of corticosteroids to skin causes vasoconstriction of the microvasculature in the skin, leading to skin blanching (whitening) at the site of application. The intensity of skin blanching is correlated with the amount of the corticosteroids delivered through the stratum corneum and the clinical efficacy (Wiedersberg et al. 2008). According to the FDA guidance (FDA 1995), the skin blanching method can be applied to all the topical dermatologic corticosteroid products. A pilot study is performed to establish the dose duration–pharmacological response relationship of a reference product. The formulation is applied to skin for various durations up to 6 h. At the end of the treatment period, the skin blanching response is measured using a chromameter over the next 24–28 h. After baseline correction, the skin blanching response data are plotted against chromameter measurement time and the AUEC is calculated using the trapezoidal rule. The AUEC values are then plotted as a function of dose duration to obtain the dose duration–response curve. From these profiles, the maximum AUEC (E_{max}), the dose duration corresponding to half-maximal response (ED_{50}), the lowest dose duration (D_1) and the highest dose duration (D_2) for use in the pivotal bioequivalence study are determined.

The pivotal bioequivalence study then compares the in vivo response of the test product with that of the reference product using appropriate statistical tools. To be bioequivalent, the test to reference ratios of AUEC should fall within 80–125 % with 90 % CI (FDA 1995).

9.4.5 Albuterol Sulfate Inhalation

Albuterol sulfate is a beta$_2$-adrenergic agonist. Albuterol relaxes the smooth muscle of airways, thus protecting against all bronchoconstrictor challenges. Due to the limited relevance of PK data to drug delivery to the target sites, PK data alone are insufficient for establishing bioequivalence of orally inhaled drug products (OIPs) (Adams et al. 2010). Additionally, the lung deposition data from gamma scintigraphy methods are not considered to be reliable to demonstrate equivalent distribution patterns of two formulations of an OIP (Daley-Yates and Parkins 2011). Currently, the FDA uses a "Weight-of-Evidence" approach to determine bioequivalence of OIPs. This approach includes assessment of qualitative (Q1) and quantitative (Q2) sameness in formulations, device similarity, equivalent in vitro performance, equivalent systemic exposure (safety) demonstrated by PK or PD data, and equivalent local delivery demonstrated by PD or clinical endpoint data (efficacy) (Adams et al. 2010).

The FDA has recommended either bronchodilatation or bronchoprovocation studies for documentation of bioequivalence of albuterol sulfate metered dose inhalers as well as other orally inhaled short-acting beta$_2$-adrenergic agonists

(SABA) (FDA 2013b). The applicant can choose either PD endpoint approach. In the bronchodilatation study, the lung function parameter forced expiratory volume in 1 s (FEV_1) is selected as the PD endpoint, which is measured by a spirometry test. A single-dose, double-blind, double dummy, randomized, crossover study with a washout period of at least 24 h in patients with moderate-to-severe asthma (FEV_1 within 40–70 % of predicted) is recommended. A pilot study which identifies the sensitive responders and estimates the number of subjects is critical for the success of the bronchodilatation study.

In the bronchoprovocation study, methacholine is used to induce bronchoconstriction which is characteristic of asthma, resulting in a decreased FEV_1 (Creticos et al. 2002). In brief, methacholine is inhaled at a very low dose at the beginning and FEV_1 is subsequently measured. The concentration of methacholine is then progressively increased until the FEV_1 falls by >20 % from baseline value. The concentration of methacholine that causes exactly 20 % decrease is estimated, which is $PC_{20}FEV_1$ and serves as the PD endpoint in the bronchoprovocation study. The lower the $PC_{20}FEV_1$, the greater the airway responsiveness is. The $PC_{20}FEV_1$ baseline values in asthma patients are 10–1,000 times lower than that in healthy subjects. Therefore, patients with mild asthma are recommended as the study subjects. Clinically relevant doses of inhaled albuterol can increase the $PC_{20}FEV_1$ by 10–20-fold. This sensitive response is highly correlated with the dose administered, allowing the bronchoprovocation study to detect the difference between two albuterol metered dose inhaler products (EMEA 2009).

Compared with bronchoprovocation study, bronchodilatation PD study is technically easier to be conducted but is more likely to fail if patients are not well selected (Evans et al. 2012). In contrast, bronchoprovocation study design is more complicated but the study has higher probability of success since the PD endpoint $PC_{20}FEV_1$ is very sensitive to the dose change. Because both PD studies exhibit nonlinear dose–response relationships, the dose scale analysis approach is used, and therefore equivalence is established based on the "dose" scale rather than on the "response" scale. The FDA uses an acceptable limit of 67.00–150.00 % to determine bioequivalence (FDA 2013b). Methacholine is the currently favored bronchoprovocation agent and is recommended by the FDA. Adenosine monophosphate (AMP), mannitol or histamine also might be used as bronchoprovocation agents.

9.4.6 Fluticasone Propionate/Salmeterol Xinafoate Inhalation Powder

Fluticasone propionate/salmeterol xinafoate inhalation powder is a combination formulation containing fluticasone propionate and salmeterol xinafoate, indicated for asthma and chronic obstructive pulmonary disease (COPD). Fluticasone propionate is an anti-inflammatory corticosteroid and inhibits the production or secretion mediators involved in the asthmatic response. Salmeterol is a selective, long-acting $beta_2$-adrenergic agonist which can relax bronchial smooth muscle. Similar to albuterol sulfate inhalation drug products, PK data alone are insufficient for establishing bioequivalence of orally inhaled fluticasone propionate/salmeterol xinafoate powder. The FDA recommends a combination of in vitro studies, PK study, PD endpoint study, formulation sameness, and device assessment to demonstrate bioequivalence of fluticasone propionate/salmeterol xinafoate inhalation powder (FDA 2013c).

The FDA has recommended bronchodilatation test as the PD endpoint study for Fluticasone propionate/salmeterol xinafoate inhalation powder products. In this study, asthma patients are randomly assigned to three parallel groups. After a 2-week run-in period, the three groups receive a 4-week and twice daily treatment of placebo, test product or reference product. On the first day of the 4-week treatment, FEV_1 is determined at 0, 0.5, 1, 2, 3, 4, 6, 8, 10, and 12 h postdose. The area under the serial FEV_1-time curve calculated from time zero to 12 h ($AUEC_{0-12h}$) is used as one bioequivalence study endpoint. In addition, FEV_1 is measured in the morning prior to the dosing of inhaled medications on the last day of a 4-week treatment, which serves as the other bioequivalence study endpoint. Both the endpoints should be baseline adjusted. To be bioequivalent, the test to reference ratios of $AUEC_{0-12h}$ and FEV_1 should fall within 80–125 % with 90 % CI.

9.4.7 Low Molecular Weight Heparin Injections

Low molecular weight heparin (LMWH) products, such as enoxaparin sodium injection and dalteparin sodium injection, are mixtures of thousands of oligosaccharides. LMWH products act as anticoagulants by inactivating both factor Xa and factor IIa in the coagulation cascade (FDA 2011b). Due to the complexity of the active ingredient of LMWH injections, characterization of their dose–pharmacokinetics relationship in human is impossible. It is well established that the different LMWH products have different PD profiles based on their in vivo anti-Xa and anti-IIa activities. These differing PD profiles might be due in part to differences in anti-Xa/anti-IIa ratio or molecular weight distribution of oligosaccharide chains, among other reasons. Therefore, the in vivo anti-Xa and anti-IIa activities are selected as the PD endpoints to assess the active ingredient sameness between two LMWH products (FDA 2012). However, due to the poor correlation between the clinical outcome of LMWH products and the measured in vivo anti-Xa and anti-IIa activities, the PD endpoint study can only provide supporting evidence of active ingredient sameness and bioequivalence (Lee et al. 2013). The FDA recommends additional comprehensive characterization studies to demonstrate bioequivalence of enoxaparin sodium injection (FDA 2011a) and dalteparin sodium injection (FDA 2012).

9.5 Other Pharmacodynamic Endpoints for Bioequivalence Study

Besides the FDA recommended PD endpoint methods in Table 9.3, some other PD endpoint methods have been investigated for bioequivalence study, which are summarized in Table 9.5. Although these PD endpoint methods have not been adopted by the FDA, they appear promising.

9.5.1 Orally Inhaled Corticosteroids

Most PD endpoint models in Table 9.5 are developed to establish bioequivalence of orally inhaled corticosteroids (ICS). Establishing bioequivalence of ICS is difficult because ICS have relatively flat dose–response curves. The challenge for ICS bioequivalence study is to identify a PD model that provides sufficient statistical power. The proposed PD models include measurement of exhaled NO, the asthma

Table 9.5 Previously reported PD methods for bioequivalence assessment

Drug	Indication	PD model	PD endpoint	References
Orally inhaled prednisone	Asthma	Asthma stability model	Peak expiratory flow	Ahrens et al. (2001)
Orally inhaled ciclesonide	Asthma	AMP challenge model	AMP PC_{20}	Taylor et al. (1999)
Orally inhaled budesonide	Asthma	Nitric oxide (NO) model	Exhaled nitric oxide	Jatakanon et al. (1999)
Orally inhaled budesonide	Asthma	Allergen challenge	Allergen PC15	Swystun et al. (1998)
Orally inhaled budesonide	Asthma	Sputum eosinophilia model	The number of eosinophils in sputum	Jatakanon et al. (1999)
Topical tretinoin	Acne vulgaris	(1) Erythema; (2) exfoliation, and (3) increased transepidermal water loss (TEWL) models	(1) Score of erythema; (2) score of scaling/ peeling, (3) TEWL reading	Lehman and Franz (2012)
Topical ibuprofen	Anti-inflammatory	Erythema model	Reduction of methyl nicotinate-induced erythema	Treffel and Gabard (1993)
Epoetin alfa injection	Anemia	Hemoglobin model	Hemoglobin concentration	Lissy et al. (2011)

Allergen PC_{15}: the provocative concentration of allergen required to produce a 15 % fall in FEV_1
AMP PC_{20}: the provocative concentration of adenosine monophosphate (AMP) required to produce a 20 % fall in FEV_1

Table 9.6 PD models for establishing bioequivalence of orally inhaled corticosteroids (Evans et al. 2012)

PD model	Comments
Improvement in lung function	• Advantage: clearly clinically relevant • Disadvantage: concern about carryover between treatment arms prevents crossover design
Asthma stability model	• Advantage: clearly clinically relevant; sufficiently steep dose–response • Disadvantage: technically difficult to conduct; large screen failure rate to identify suitable subjects
Exhaled nitric oxide model	• Advantage: easy to measure • Disadvantage: clinical relevance less clear than asthma stability model or classic study of improvement in lung function. Several confounding factors including smoking, intercurrent infections, nitrate content of food, etc. To ensure adequate dose–response, careful subject selection required
AMP, methacholine or mannitol challenge model	• Advantage: easier to conduct than asthma stability • Disadvantage: clinical relevance uncertain. To ensure adequate dose–response, careful subject selection required
Sputum eosinophilia model	• Advantage: easier to conduct than asthma stability • Disadvantage: in some cases, clinical relevance is questionable. Practical concerns include inability of some patients to generate sputum, and the requirement for manual processing of the sample, which makes standardization across centers challenging. To ensure adequate dose–response, careful subject selection required

stability model, eosinophilia measurement in induced sputum samples (as recommended by Health Canada), and bronchoprovocation with adenosine, methacholine, mannitol, or allergen. These PD models were widely discussed at the 2010 ISAM/IPAC-RS European Workshop (Evans et al. 2012). The advantages and disadvantages of each model are summarized in Table 9.6. Among these models, the exhaled NO model and asthma stability model appeared promising even though they have not been adopted by regulatory authorities (Adams et al. 2010; Evans et al. 2012).

9.5.2 Long-Acting Beta₂ Agonists

Bronchodilatation and bronchoprovocation models have been accepted by the FDA to establish bioequivalence of orally inhaled short-acting $beta_2$ agonists (SABAs) such as orally inhaled albuterol sulfate (Table 9.3). These two PD models may be applied to assess bioequivalence of long-acting $beta_2$ agonists (LABAs) after minor modifications in study design, such as adjustment of the measurement time points and washout periods for crossover studies (Evans et al. 2012).

9.5.3   Topical Dermatologic Products and Others

Currently, the skin blanching assay is the only PD endpoint study adopted by the FDA for documenting bioequivalence of topical dermatologic products. However, the assay is applicable to topical corticosteroids only. In a recent publication (Lehman and Franz 2012), three PD endpoints, erythema score, exfoliation (score of scaling/peeling), and increased transepidermal water loss (TEWL) readings were successfully used to assess bioequivalence of topical retinoid products. The treatment with topical retinoid products is correlated with the appearance of an inflammatory response characterized by erythema and scaling. The resulting retinoid dermatitis is drug, concentration, and vehicle dependent. Therefore, the measurement of the signs of inflammation (erythema/scaling) can serve as potential PD endpoints to establish bioequivalence of topical retinoid products. In addition, chronic treatment with retinoids alters the structure of stratum corneum barrier, resulting in increased TEWL. The change in TEWL is another potential PD endpoint for bioequivalence study. The method validation showed that the developed PD endpoint approaches have sufficient sensitivity, specificity, and reproducibility to distinguish (1) three concentrations of tretinoin in a commercial cream product line, (2) two concentrations of tretinoin in a commercial gel product line, (3) different vehicles (gel vs. cream) containing the same concentration of tretinoin, and (4) tretinoin and adapalene at the same concentration.

Similarly, erythema score has been used as a PD endpoint to demonstrate bioequivalence of topical ibuprofen products (Treffel and Gabard 1993). In this study, erythema score is a PD endpoint reflecting the anti-inflammation activity of ibuprofen. The investigators simultaneously compared the inhibition of an inflammation induced by a methyl nicotinate assay following topical application of two 10 % ibuprofen formulations. A correlation ($r = 0.9603$, $p < 0.001$) between the amount of drug in the epidermis and the corresponding erythema score was observed, indicating that erythema score can be a potential PD endpoint to assess bioavailability and bioequivalence of topical nonsteroidal anti-inflammatory drug (NSAID) products.

9.5.4   Topical Dermatologic Products and Others

Besides locally acting drug products, PD endpoint studies have also been employed to assess bioequivalence of therapeutic protein product of epoetin alfa injections (Lissy et al. 2011). The EMEA requests both PK and PD endpoint studies to document bioequivalence of epoetin alfa products (EMEA 2007). Hemoglobin concentration is selected as the PD endpoint to assess the bioequivalence of three epoetin alfa products marketed in either US or Europe. Both PK and PD studies consistently demonstrated bioequivalence of the three products despite different formulations.

9.6 Summary

Overall, PD endpoint studies are very useful to establish bioequivalence of drug products when PK endpoint and in vitro approaches are not applicable. An ideal PD endpoint for establishing bioequivalence needs to (1) be sensitive (steep dose–response curve); (2) be reproducible; (3) have low variability of PD response at baseline and following drug treatment; and (4) have adequate statistical power with feasible sample size. Besides the PD endpoint selection, the study design, pilot study, and study population are critical for the success of the study.

Currently, the PD models accepted by regulatory authorities for bioequivalence assessment are still very limited, which has impeded the development of some generic drug products such as orally inhaled formulations and topically applied dermatologic formulations. Identification and validation of new PD endpoints for bioequivalence assessment is one scientific challenge that needs to be addressed through the collaboration of pharmaceutical industry, academia, and regulatory authorities.

References

Adams WP, Ahrens RC, Chen ML, Christopher D, Chowdhury BA, Conner DP, Dalby R, Fitzgerald K, Hendeles L, Hickey AJ, Hochhaus G, Laube BL, Lucas P, Lee SL, Lyapustina S, Li B, O'Connor D, Parikh N, Parkins DA, Peri P, Pitcairn GR, Riebe M, Roy P, Shah T, Singh GJ, Sharp SS, Suman JD, Weda M, Woodcock J, Yu L (2010) Demonstrating bioequivalence of locally acting orally inhaled drug products (OIPs): workshop summary report. J Aerosol Med Pulm Drug Deliv 23:1–29

Ahrens RC, Teresi ME, Han SH, Donnell D, Vanden Burgt JA, Lux CR (2001) Asthma stability after oral prednisone: a clinical model for comparing inhaled steroid potency. Am J Respir Crit Care Med 164:1138–1145

CDSCO-India (2005) Guidance for bioavailability and bioequivalence studies. http://cdsco.nic.in/html/be%20guidelines%20draft%20ver10%20march%2016,%2005.pdf

Chen ML, Shah V, Patnaik R, Adams W, Hussain A, Conner D, Mehta M, Malinowski H, Lazor J, Huang SM, Hare D, Lesko L, Sporn D, Williams R (2001) Bioavailability and bioequivalence: an FDA regulatory overview. Pharm Res 18:1645–1650

Creticos PS, Adams WP, Petty BG, Lewis LD, Singh GJ, Khattignavong AP, Molzon JA, Martinez MN, Lietman PS, Williams RL (2002) A methacholine challenge dose-response study for development of a pharmacodynamic bioequivalence methodology for albuterol metered- dose inhalers. J Allergy Clin Immunol 110:713–720

Daley-Yates PT, Parkins DA (2011) Establishing bioequivalence for inhaled drugs; weighing the evidence. Expert Opin Drug Deliv 8:1297–1308

Drent ML, Van Der Veen EA (1993) Lipase inhibition: a novel concept in the treatment of obesity. Int J Obes Relat Metab Disord 17:241–244

EMEA (2000) Note for guidance on the investigation of bioavailability and bioequivalence. http://www.ema.europa.eu/docs/en_GB/document_library/Scientific_guideline/2009/09/WC500003519.pdf

EMEA (2007) Scientific discussion. http://www.ema.europa.eu/docs/en_GB/document_library/EPAR_-_Scientific_Discussion/human/000726/WC500028287.pdf

EMEA (2009) Guideline on the requirements for clinical documentation for orally inhaled products (OIP) including the requirements for demonstration of therapeutic equivalence between two inhaled products for use in the treatment of asthma and chronic obstructive pulmonary disease (COPD) in adults and for use in the treatment of asthma in children and adolescents. http://www.ema.europa.eu/docs/en_GB/document_library/Scientific_guideline/2009/09/WC500003504.pdf

Evans C, Cipolla D, Chesworth T, Agurell E, Ahrens R, Conner D, Dissanayake S, Dolovich M, Doub W, Fuglsang A, Garcia Arieta A, Golden M, Hermann R, Hochhaus G, Holmes S, Lafferty P, Lyapustina S, Nair P, O'Connor D, Parkins D, Peterson I, Reisner C, Sandell D, Singh GJ, Weda M, Watson P (2012) Equivalence considerations for orally inhaled products for local action-ISAM/IPAC-RS European Workshop report. J Aerosol Med Pulm Drug Deliv 25:117–139

FDA (1995) Guidance for industry-topical dermatologic corticosteroids: in vivo bioequivalence. http://www.fda.gov/downloads/Drugs/GuidanceComplianceRegulatoryInformation/Guidances/ucm070234.pdf

FDA (2000) Waiver of in vivo bioavailability and bioequivalence studies for immediate-release solid oral dosage forms based on a biopharmaceutics classification system. http://www.fda.gov/downloads/Drugs/.../Guidances/ucm070246.pdf

FDA (2003a) Guidance for industry: bioavailability and bioequivalence studies for nasal aerosols and nasal sprays for local action. http://www.fda.gov/downloads/Drugs/GuidanceCompliance RegulatoryInformation/Guidances/ucm070111.pdf

FDA (2003b) Guidance for industry: exposure-response relationships—study design, data analysis, and regulatory applications. http://www.fda.gov/downloads/Drugs/GuidanceCompliance RegulatoryInformation/Guidances/ucm072109.pdf

FDA (2009) Draft guidance on acarbose. http://www.fda.gov/downloads/Drugs/Guidance ComplianceRegulatoryInformation/Guidances/UCM170242.pdf

FDA (2010a) Draft guidance on orlistat. http://www.fda.gov/downloads/Drugs/Guidance ComplianceRegulatoryInformation/Guidances/UCM201268.pdf

FDA (2010b) Guidance for industry-bioequivalence recommendations for specific products. http://www.fda.gov/downloads/Drugs/GuidanceComplianceRegulatoryInformation/Guidances/UCM072872.pdf

FDA (2011a) Draft guidance on enoxaparin sodium. http://www.fda.gov/downloads/Drugs/GuidanceComplianceRegulatoryInformation/Guidances/UCM277709.pdf

FDA (2011b) Draft guidance on fluticasone propionate. http://www.fda.gov/downloads/Drugs/GuidanceComplianceRegulatoryInformation/Guidances/UCM244386.pdf

FDA (2011c) Draft guidance on lanthanum carbonate. http://www.fda.gov/downloads/Drugs/GuidanceComplianceRegulatoryInformation/Guidances/UCM270541.pdf

FDA (2012) Draft guidance on dalteparin sodium. http://www.fda.gov/downloads/Drugs/GuidanceComplianceRegulatoryInformation/Guidances/UCM319988.pdf

FDA (2013a) Bioequivalence studies with pharmacokinetic endpoints for drugs submitted under an ANDA. http://www.fda.gov/downloads/Drugs/GuidanceComplianceRegulatory Information/Guidances/UCM377465.pdf

FDA (2013b) Draft guidance on albuterol sulfate. http://www.fda.gov/downloads/Drugs/GuidanceComplianceRegulatoryInformation/Guidances/UCM346985.pdf

FDA (2013c) Draft guidance on fluticasone propionate; salmeterol xinafoate. http://www.fda.gov/downloads/Drugs/GuidanceComplianceRegulatoryInformation/Guidances/UCM367643.pdf

Health-Canada (1999) Guidance to establish equivalence or relative potency of safety and efficacy of a second entry short-acting beta2-agonist metered dose inhaler. http://www.hc-sc.gc.ca/dhp-mps/alt_formats/hpfb-dgpsa/pdf/prodpharma/mdi_bad-eng.pdf

Health-Canada (2011) Data requirements for safety and effectiveness of subsequent market entry inhaled corticosteroid products for use in the treatment of asthma. http://www.hc-sc.gc.ca/dhp-mps/consultation/drug-medic/draft_inhal_ebauche_corticost-eng.php

Jatakanon A, Kharitonov S, Lim S, Barnes PJ (1999) Effect of differing doses of inhaled budesonide on markers of airway inflammation in patients with mild asthma. Thorax 54:108–114

Lee S, Chung JY, Hong KS, Yang SH, Byun SY, Lim HS, Shin SG, Jang IJ, Yu KS (2012) Pharmacodynamic comparison of two formulations of Acarbose 100-mg tablets. J Clin Pharm Ther 37:553–557

Lee S, Raw A, Yu L, Lionberger R, Ya N, Verthelyi D, Rosenberg A, Kozlowski S, Webber K, Woodcock J (2013) Scientific considerations in the review and approval of generic enoxaparin in the United States. Nat Biotechnol 31:220–226

Lehman PA, Franz TJ (2012) Assessing the bioequivalence of topical retinoid products by pharmacodynamic assay. Skin Pharmacol Physiol 25:269–280

Lionberger RA (2008) FDA critical path initiatives: opportunities for generic drug development. AAPS J 10:103–109

Lissy M, Ode M, Roth K (2011) Comparison of the pharmacokinetic and pharmacodynamic profiles of one US-marketed and two European-marketed epoetin alfas: a randomized prospective study. Drugs R D 11:61–75

MacDonald AJ, Parrott N, Jones H, Lave T (2004) Modelling and simulation of pharmacokinetic and pharmacodynamic systems-approaches in drug discovery. In: Beilstein-Institut Workshop, May 24–28, Bozen

Mastan S, Latha TB, Ajay S (2011) The basic regulatory considerations and prospects for conducting bioavailability/bioequivalence (BA/BE) studies—an overview. Comp Eff Res 1:1–25

N'Dri-Stempfer B, Navidi WC, Guy RH, Bunge AL (2009) Improved bioequivalence assessment of topical dermatological drug products using dermatopharmacokinetics. Pharm Res 26:316–328

Navidi W, Hutchinson A, N'Dri-Stempfer B, Bunge A (2008) Determining bioequivalence of topical dermatological drug products by tape-stripping. J Pharmacokinet Pharmacodyn 35:337–348

Saudi-FDA (2005) Bioequivalence requirements guidelines (draft). http://old.sfda.gov.sa/NR/rdonlyres/6A114B70-4201-46EF-B4C7-127FD66D3314/0/BioequivalenceRequirementGuidelines.pdf

Swainston Harrison T, Scott LJ (2004) Lanthanum carbonate. Drugs 64:985–996; discussion 997–998

Swystun VA, Bhagat R, Kalra S, Jennings B, Cockcroft DW (1998) Comparison of 3 different doses of budesonide and placebo on the early asthmatic response to inhaled allergen. J Allergy Clin Immunol 102:363–367

Taylor DA, Jensen MW, Kanabar V, Engelstatter R, Steinijans VW, Barnes PJ, O'Connor BJ (1999) A dose-dependent effect of the novel inhaled corticosteroid ciclesonide on airway responsiveness to adenosine-5'-monophosphate in asthmatic patients. Am J Respir Crit Care Med 160:237–243

Treffel P, Gabard B (1993) Feasibility of measuring the bioavailability of topical ibuprofen in commercial formulations using drug content in epidermis and a methyl nicotinate skin inflammation assay. Skin Pharmacol 6:268–275

WHO (2006) WHO expert committee on specifications for pharmaceutical preparation. Geneva. http://apps.who.int/prequal/info_general/documents/TRS937/WHO_TRS_937_eng.pdf#page = 359

Wiedersberg S, Leopold CS, Guy RH (2008) Bioavailability and bioequivalence of topical glucocorticoids. Eur J Pharm Biopharm 68:453–466

Zhang M, Yang J, Tao L, Li L, Ma P, Fawcett JP (2012) Acarbose bioequivalence: exploration of new pharmacodynamic parameters. AAPS J 14:345–351

Zhi J, Melia AT, Eggers H, Joly R, Patel IH (1995) Review of limited systemic absorption of orlistat, a lipase inhibitor, in healthy human volunteers. J Clin Pharmacol 35:1103–1108

Chapter 10
Clinical Endpoint Bioequivalence Study

John R. Peters

10.1 Introduction

Definition: A clinical endpoint bioequivalence study is a clinical study utilizing a patient population in which two products containing the same active moiety (chemically equivalent) in the same dosage form (pharmaceutically equivalent) are administered delivering the active moiety to the local site of action. Clinical effect is evaluated using a predetermined clinical endpoint to evaluate comparative clinical effect in the chosen population. From this analysis a determination of clinical equivalence is made from which an inference of bioequivalence of the two products is concluded. The general design of these studies is a blinded, randomized, balanced, parallel study. A placebo arm is usually included in these studies in order to demonstrate that the study is sufficiently sensitive to identify the clinical effect in the patient population enrolled in the study.

The clinical endpoint bioequivalence study is an expensive and time-consuming way to attempt to infer the bioequivalence of drug products. It also exposes a large number of volunteers to the risk of adverse reactions to a test product with no certain expectation of potential benefit except for possible financial reimbursement for use of their time and body. Of the in vivo methods for demonstration of bioequivalence it is the least accurate and reproducible methodology. Nonetheless, clinicians and the public generally misunderstand this fact, and demand clinical studies to approve generic products (Kesselheim et al. 2008). The clinical endpoint bioequivalence study is often confused with the randomized controlled study. Clinicians believe a clinical study in patients is the only way to prove that a generic product is as safe and efficacious as the reference innovator product. Unfortunately, that is not usually true when the goal of the study is to demonstrate that two

J.R. Peters (✉)
Center for Drug Evaluation and Research, U.S. Food and Drug Administration,
10903 New Hampshire Avenue, Silver Spring, MD 20993, USA
e-mail: Johnr.Peters@fda.hhs.gov

L.X. Yu and B.V. Li (eds.), *FDA Bioequivalence Standards*, AAPS Advances in the Pharmaceutical Sciences Series 13, DOI 10.1007/978-1-4939-1252-0_10, © The United States Government 2014

products are equivalent and can be substituted or interchanged with one another. In this chapter we shall endeavor to strip away some of the myths and misunderstandings, and also try to identify the situations when such a study is the only path forward for a generic approval as well as the benefits and limitations of the clinical endpoint bioequivalence study as used in the approval of generic drug products.

In 21 CFR Part 320 subpart B (Bioavailability and Bioequivalence Requirements) five approaches to determining bioequivalence are recommended. These approaches are listed in declining order of precision.

- In vivo measurement of active moiety(ies) in a biological fluid (**Pharmacokinetic Testing**)
- In vivo comparison of pharmacodynamic response(s) (**Pharmacodynamic Testing**)
- In vivo clinical comparison (**Clinical Endpoint Bioequivalence Testing**)
- In vitro comparison
- Any other approach deemed appropriate by the FDA

With the exception of the in vitro comparison, it is critical to note that the measured endpoint in each of these approaches is progressively less reliable, less reproducible, and more variable due to the nature of biological systems. While the Pharmacokinetic Testing measures the rate and extent of absorption of the active pharmaceutical moiety, Pharmacodynamic Testing measures a biological/physiological response to the presence of that pharmaceutical moiety. In the Clinical Endpoint Bioequivalence Testing approach there is only a measurement of a clinical endpoint that suggests that the active moiety is being delivered to the site of action to the same rate and extent in both test and reference products. Clinical endpoints are somewhat nebulous, nonspecific to the drug product, and subject to a number of conscious and unconscious biases. As we will discuss further, choice of endpoint or surrogate and the manner in which it is measured or quantified can have enormous impact on the determination of equivalence.

In vitro comparisons have the advantage of reproducible methods for comparing formulations quantitatively and qualitatively. Physicochemical characteristics can be identified and quantitatively compared. This is a fairly sensitive and reproducible way to compare formulations that are chemically equivalent. However, this method is only reliable if there is a reasonably well characterized mechanism of action and the specific contributions of the various physicochemical characteristics to the therapeutic effect of the active pharmaceutical ingredient (API) at the site of action are well characterized.

The clinical study is also not a very sensitive indicator of bioequivalence, but in some situations it is the only option given the currently available methodologies. Generally it must be applied for those drug products that have negligible systemic uptake, for which there is no identified pharmacodynamic measure, and for which the site of action is local. Local action is defined broadly as action at the surface to which the drug is applied rather than to which it is delivered in some body fluid (blood, bile, lymphatic fluid). These surfaces may include skin, mucosal surfaces of the eye, nose, lung, gut, bladder, rectum, and vagina. The indication for the drug and pathophysiology of the disease guides us to the known or proposed site of local action. The chosen clinical endpoint or surrogate provides us with the measure of clinical effect.

It is important to note that this is **not** a study of therapeutic efficacy. The efficacy of the active pharmaceutical moiety was established in the approval of the innovator product (reference) to which a proposed generic product is to be compared. The clinical endpoint bioequivalence study is a simple, quantitative comparison of the clinical (therapeutic) effect of a test and reference product. The successful quantitative comparison of therapeutic effect of two products that have already demonstrated both chemical and pharmaceutical equivalence allows for the inference that the two products are bioequivalent (*same active chemical moiety, same pharmacological dosage form [tablet, solution, cream, patch, etc.], and same clinical effect*). It should be clear that the high variability found in both patients and diseases combined with the subjective nature of many clinical responses and clinical evaluations make these studies more difficult to design than it is to design a randomized controlled clinical study designed to demonstrate therapeutic superiority over placebo or even to show non-inferiority in an active comparator study. Nonetheless, the clinical bioequivalence study is the least accurate, reproducible, and sensitive of all the in vivo methods used to demonstrate bioequivalence (Rahsid 2012; Davit and Conner 2010).

A frequent complaint about generic drug products is that approval without a randomized clinical study testing in the sort of patients targeted for the drug is inadequate. Physicians and the public believe that performance of a clinical endpoint bioequivalence study is the best way to establish the therapeutic equivalence of a generic drug. This concern arises from a general misunderstanding of the principles of the randomized controlled clinical study and how the clinical endpoint bioequivalence study differs. There is also a lack of understanding of the concepts of bioequivalence, therapeutic equivalence, and non-inferiority.

There are situations in which there is no alternative to performance of a clinical endpoint bioequivalence study. When a drug is topically acting in the broad sense of having a site of action to which it is directly administered a pharmacokinetic study is not possible, since there is no systemic absorption. This situation includes products applied to the skin as well as those applied or administered to mucous membrane sites such as the gut, oropharynx, nasal mucosa, tracheopulmonary mucosa, and vagina. Most of these products have little or no systemic absorption, and so the only way to evaluate them is by means of some biomarker that has been identified as indicating the desired therapeutic clinical effect.

In this chapter, we will examine these misconceptions and address the practical considerations for designing a clinical endpoint bioequivalence study. In addition, we will consider biomarkers, clinical endpoints, and surrogate endpoints and how they figure into the reliability of a study in demonstrating therapeutic equivalence. Finally, we will present some examples of what can and cannot be accomplished with a clinical endpoint bioequivalence study.

10.2   Background

Medical experimentation has a long and somewhat spotty history dating back into at least Neolithic times. But it was in the tenth century that Ibn Sina-Avicenna (980-1037 CE) published "The Canon of Medicine" in which he formalized the general approach of the randomized study. He described seven requirements for "the recognition of the strengths of. . .medicines through experimentation":

- Ensure use of pure drugs
- Test for only one disease
- Use control groups
- Use dose escalation
- Require long-term observation
- Require reproducible results
- Require human over animal testing

Over the ensuing centuries, clinical scientists incorporated gradually improving scientific methods into this basic framework (Gallin 2012). In the eighteenth century Lind performed the first modern comparative clinical study on board the HMS Salisbury. In this study he evaluated the clinical effect of oranges and lemons on scurvy. In the early twentieth century Fisher introduced the concepts of randomization and statistical analysis to clinical studies (Friedman et al. 2010). Principles of pathophysiology, chemistry, biochemistry, and ethical standards became formalized during the twentieth century and were incorporated in the planning of clinical studies for demonstrating the safety and efficacy of therapeutic interventions. Statistical methodologies were developed from which a quantitative assessment of beneficial effect might be made. The clinical judgment of risk:benefit ratio was, to some extent, given a statistical basis. For the most part the early history of drug development involved demonstration of a clinically significant and beneficial effect of a drug product in comparison to a placebo. In other words, proof that treatment with a particular drug was better than no treatment was demonstrated in a blinded, randomized, controlled clinical study.

The 1910 Flexner Report triggered a massive effort to enforce a scientifically based medical education system in the United States. This was intended to provide a high level of quality medical care by scientifically trained professionals. The practice of medicine was being molded into a science based profession, taking full advantage of the remarkable growth of truly life-saving medications (Cook et al. 2006; Maeshiro et al. 2010). Exciting new therapeutic agents were being discovered. Both physicians and the general public were kept well informed of the struggles made to ensure that high quality, effective, and safe drug products were available and scientifically demonstrated to be effective.

At the same time the American pharmaceutical industry was encouraged and later required to make use of new scientific methodology as it evolved. Along with these developments regulatory science grew as a means of ensuring maximal benefit to the public with a minimum of risk. As problems with medicinal products were discovered, regulations were updated to deal with them. Changes in regulations came quickly starting with the 1902 Biologics Control Act, mandating licensing of vaccine, antitoxin, and sera. In 1906 the Pure Food, Drug, and Cosmetic Act required the ingredients contained in medications be identified. By 1962 the Kefauver-Harris Amendment required demonstration of safety and efficacy. All of these regulatory changes were in response to growing physician demand for more quality drug products in the face of well publicized failures in the marketplace (http://www.fda.gov/centennial/history/history.html). Following passage of this amendment, there was a reassessment of 4,000 currently approved drugs. This was carried out by the National Academy of Sciences as the Drug Efficacy Study Implementation (DESI-1968). The concept of a generic drug, as we know it today, began to evolve over the decade of the 1970s.

Since the early part of the twentieth century, the pharmaceutical industry has been adhering to scientific methodology in the development of drug therapies, forever severing ties with the bleak history of the "Medicine Show," patent medicines, and the variable quality apothecary preparations that were common throughout the eighteenth and nineteenth centuries (Anderson 2005). At the center of this massive development of the industry was the randomized controlled study that rightly became the absolute gold standard for demonstration of drug efficacy. When tested in a random controlled, blinded fashion, in a population of patients with a condition for which the drug is thought to have a potentially therapeutic effect, the randomized controlled study is without a doubt the most scientific way in which to demonstrate therapeutic efficacy and safety. Marketing methods for these products firmly established the concept of the need of a clinically tested product in a patient population.

In the 1970s, as drugs in the same therapeutic class became more common than completely innovative new therapies, the concepts of bioequivalence and therapeutic equivalence began to develop. FDA regulations discussing bioequivalence did not appear until 1977. In 1980 the "Orange Book" (officially titled Approved Drug Products with Therapeutic Equivalence Evaluations) was first published listing products that were determined to be therapeutically equivalent. By 1984 the Hatch-Waxman Act (officially Drug Price Competition and Patent Term Restoration Act) authorized the FDA to approve generic drugs under an abbreviated process that did not require repetition of the research done by the innovator that established safety and efficacy (Peters et al. 2009). This development led to an explosion of generic drugs marketed to the public as well as an evolution of new methods by which products could be demonstrated to be equivalent. This new methodology developed largely within the realms of academia, pharmaceutical science and regulation. Unfortunately, the concepts of bioequivalence did not penetrate to the professional and public, and so remain poorly understood and not quite trusted.

The fact of the matter is that bioequivalence is integral to innovator products as well as to the development of generic products. The same bioequivalence methodologies are applied to innovator drugs to link study drugs to the final marketed product as well as to demonstrate equivalence of product produced at different manufacturing plants or of different lots of product. It is with this in mind that development of an accurate and reproducible methodology was not simply a regulatory issue. Demonstration of bioequivalence is a crucial step in both new and generic drug products. Industry and academia provided significant input, and the combined efforts of industry, academia, and regulatory authorities resulted in the current methods for demonstrating and monitoring product quality and equivalence to the initially approved product.

10.3   Clinical Endpoint Bioequivalence Study

The concept of bioequivalence has only been evolving since the late 1970s when Bioequivalence Regulations were issued in 1977. The "Orange Book" (officially titled Approved Drug Products with Therapeutic Equivalence Evaluations) was first published in 1977 and listed drug products that could be used as the reference listed drug (RLD) for bioequivalence studies. But it was with the passage of the Hatch-Waxman Amendment in 1984, that the development and approval of generic drugs really expanded. This act provided for three essential elements to the current generic drug approval process:

1. Generic drug approvals must be based on scientific considerations and minimize duplicative testing.
2. All generic and brand name drugs must meet the same quality standards.
3. Generic versions of drugs must be equivalent to a degree, calculated statistically, which ensures that therapeutically equivalent drugs have the same clinical effect and no greater chance of adverse effect.[1]

The clinical endpoint bioequivalence study developed in order to meet the third of these elements. Like the randomized controlled clinical study, it is also a study in a patient population, unlike the pharmacokinetic bioequivalence study that is generally performed in healthy volunteers. Use of healthy volunteers is a preferred method for pharmacokinetic testing since it eliminates much of the variability of the disease state from the evaluation and provides a more uniform (less variable) population is available for testing. Furthermore, in the pharmacokinetic study it is often possible to use the same subject in measuring PK for both test and reference products, further eliminating variability. Unfortunately, this is not a luxury available in any clinical endpoint study.

The clinical endpoint bioequivalence study provides a comparison of the clinical effect of a proposed generic product to that of an innovator product and of both to a placebo group. The major challenges faced when undertaking a clinical endpoint bioequivalence study include **precision of diagnosis** with which to enroll the most uniform and correct patients, **clear understanding of the drug mechanism of**

---

[1] Op. Cit. (2009) Generic drugs—safe, effective, and affordable. Dermatol Ther 22(3): 229–240.

action with which to eliminate concomitant conditions and drugs that could obscure the study, and **clear understanding of the pathophysiology of the disease** to be studied with which to identify the most accurate, valid, and quantifiable clinical endpoint or surrogate endpoint and when to best make this measurement.

The size of the population studied is generally smaller than that used to determine therapeutic efficacy of a new drug, except in the case of a placebo controlled study. But that is acceptable since the intent of this study is not to determine efficacy but only to compare the clinical effect of a known active drug to a proposed generic version of that same active drug. The generic and innovator products haves already been shown to have the same API and in it has the same dose form and route of administration (i.e., it is pharmaceutically equivalent). The next step is to provide evidence that it is also bioequivalent. For this there are the three previously mentioned in vivo methods (PK study, PD study, or Clinical Endpoint BE study) and an in vitro method. The choice of method is dependent on the nature of the product, the clinical indication for which it will be used, and the critical characteristics important to the delivery of the API to the appropriate site of action. For the purposes of this chapter we will concentrate on those products that are locally acting, do not have a defined pharmacodynamic marker, but have an acceptable clinical or surrogate endpoint.

Pharmaceutical equivalents contain the same active ingredient(s), in the same dosage form and route of administration, and are identical in strength or concentration, and meet the same compendial or other applicable standards (i.e., strength, quality, purity, and identity). If this is true AND we can demonstrate bioequivalence, then we can infer that the products will be therapeutically equivalent. This stems from the regulatory definition of bioequivalence: Bioequivalence refers to equivalent delivery of the same drug substance, in the same amount, to the intended site of action at an equivalent rate and extent as the reference product. It is reasonable then to conclude that a bioequivalent and pharmaceutical equivalent can be expected to have the same clinical effect and safety profile when administered to patients under the same conditions.[2]

The clinical endpoint bioequivalence study is not a study of efficacy or safety. It does not use the same kind of statistical methodology as that of bioequivalence studies, and it does not have the same goal. A clinical endpoint bioequivalence study is only a comparison of therapeutic effect not efficacy. The goal is to show an equivalent therapeutic effect of a test and reference product and to demonstrate that both of these are superior to the effect of a placebo. In the pharmacokinetic bioequivalence study two products are evaluated for equivalent bioavailability at the presumed site of action and must show the same rate and extent of availability. The C_{max} and AUC are used as surrogates for this rate and extent of bioavailability.[3] In this case demonstration of bioequivalence is used to infer therapeutic

[2] Op. Cit. (2010) Generic drug product development: international regulatory requirements for bioequivalence. In: Kanfer I, Shargel L (eds). Informa Healthcare, USA, Inc., New York, NY.
[3] Ibid.

equivalence. The clinical endpoint bioequivalence study does the reverse, inferring bioequivalence when equivalent therapeutic effect is demonstrated by a product containing the same chemical active ingredient as the comparator. This is a conclusion that requires careful consideration of the study design, choice of subjects, nature of the condition treated, concomitant medications, and clinical conditions. In addition, the choice of clinical or surrogate endpoint and choice of time of measurement of that endpoint can be pivotal to the sensitivity and accuracy of the result. Given the vagaries and variability of patients, disease states, biomarkers, and course of treatment, this can be a challenge.

The clinical endpoint bioequivalence study is unique in that it is not designed nor sufficiently powered to demonstrate non-inferior efficacy. The point of comparison is the anticipated therapeutic effect of two products demonstrated in a well-designed study. Whenever ethically permissible, a placebo arm will be included simply to demonstrate that the study was sufficiently sensitive to show some benefit of both test and reference drugs over placebo. In addition it is generally recommended that the study be conducted with the lowest approved strength of the reference product in order to improve the sensitivity of the study to detect differences in the test and reference product. This recommendation is only made if the standard of care for the clinical context of use and ethical concerns for the patient allow for it. In this way we can compare the relative effect of two active drugs that are known to be chemically equivalent.

10.3.1   Biomarkers, Clinical Endpoints, and Surrogate Endpoints

One of the most critical aspects of drug development over the past 30 years has been research into biomarkers. Biomarkers are used to estimate the delivery of drugs to the intended target, to better understand and predict pathophysiology, and how it is altered by various therapeutic interventions (Rolan 1997; Berns et al. 2007).

A **biological marker (Biomarker)** is a characteristic that is objectively measured and evaluated as an indicator of normal biologic processes, pathogenic processes, or pharmacologic responses to a therapeutic intervention (e.g., Lowering of Total Cholesterol in hypercholesterolemia). It is important to note here that the biomarker is also a measure used by clinicians to monitor the effect of their therapeutic intervention in an individual. The clinically preferred biomarker may not be the most scientifically accurate or reliable marker for purpose of a bioequivalence study. It is chosen in clinical practice because of its utility and cost. It is acceptable for the monitoring of an individual patient, but in a clinical bioequivalence study it may not be sufficiently sensitive. The most reliable and reproducible biomarker must be used in the context of a bioequivalence study. Biomarkers are presumed to be predictive of the more absolute goal of the therapy. For example, lowering cholesterol is a presumed indicator of reduction in cardiovascular events and death. This may or may not be true and may or may not be predictive that the drug is actually arriving at the site of action proposed for disease modification.

A **surrogate endpoint** is an indicator that is intended to substitute for a clinical endpoint. The surrogate endpoint has been defined by the FDA as "...a laboratory measurement or physical sign that is used in therapeutic studies as a substitute for a clinically meaningful endpoint that is a direct measure of how a patient feels, functions, or survives and is expected to predict the effect of the therapy." The surrogate endpoint is chosen using epidemiologic, therapeutic, pathophysiologic, or other scientific evidence indicating that the surrogate endpoint is expected to predict the long-term clinical benefit, harm, or lack of benefit or harm. Surrogate endpoints may include specific biomarkers, such as reduction in blood sugar or glycosylated hemoglobin (HgBA1C) in a diabetic, but could also be a clinical scale such as a visual analog pain scale or Beck Depression Inventory. In some cases the scales used in diagnosis or monitoring of a patient's progress are totally inadequate when used to compare therapeutic effects of two products. This will become obvious in one of the examples given later in this chapter of the evolution of a guidance for a clinical endpoint bioequivalence study for rifaximin.

The **clinical endpoint** is a characteristic or variable that reflects how a patient feels (e.g., Cessation of Diarrhea in Travelers' Diarrhea) or functions (e.g., Quality of Life measures), or how long a patient survives (e.g., 5 year or overall survival in cancer therapy). In some cases the clinical endpoint may be a biomarker of choice for a study. In those situations it is absolutely critical that the chosen biomarker be well validated, easily and reproducibly measurable, and quantifiable for comparison of therapies. In general it is a surrogate endpoint that is ultimately chosen as the clinical endpoint that defines the accuracy of the bioequivalence determination made in the evaluation of a clinical endpoint bioequivalence study.

The distinctions among these definitions are very important in the design of the clinical endpoint bioequivalence study as well as to the understanding of how such a study might be interpreted. Physicians will use biomarkers in order to monitor the impact of a therapy on the individual patient. This biomarker may or may not be suitable for quantitative comparison of a therapeutic effect between a test and reference drug product. Similarly, the surrogate endpoint that may be important to determination of efficacy in studies supporting an innovator drug, may not lend itself to quantitation suitable for bioequivalence comparison of a test and reference product (Temple 1999; Colburn 2000). DeGrutolla et al. discussed it in this way:

> For a biomarker to serve as a surrogate for the effect of an intervention on a clinical endpoint at the population level, more is required than just the ability of the marker measured on an individual to predict that individual's clinical endpoint. The extent to which a biomarker is appropriate for use as a surrogate endpoint in evaluating a new treatment depends on the degree to which the biomarker can reliably predict the clinical benefit of that therapy, as compared to a standard therapy. (De Gruttola et al. 2001)

This statement recognizes the intrinsic need for the quantifiability of a surrogate endpoint.

10.4   Value and Limitation of the Clinical Endpoint Bioequivalence Study

In pharmacokinetic (PK) bioequivalence studies, healthy volunteers are enrolled rather than patients for whom the drug is intended. For the practicing clinician this raises a question as to the "true" equivalence of generic products because the scientific methodological approach is misunderstood. From their perspective the role of the study is to show that the drug is effective in patients who have the condition meant to be treated. A clinical study in patients with the disease is considered a "gold standard." However, logically, it is more scientifically rigorous to conduct such studies in healthy individuals since this eliminates the variability produced by disease states. The method of testing in healthy subjects creates confusion in clinicians. Thus clinicians prefer the clinical endpoint bioequivalence study despite its poor sensitivity and reproducibility. But even the clinical endpoint bioequivalence study becomes suspect if it uses a different surrogate marker as clinical endpoint than that which is the standard of care used to monitor patient progress in therapy. In both of these situations the choices (to not test in patients [PK study] or not use a familiar but less appropriate surrogate endpoint) are made to improve the sensitivity and accuracy of the evaluation. Unfortunately this is not intuitively clear to clinicians or the public.

The medical literature is full of misleading articles on the failure of generic products to demonstrate equivalence. For the most part such reports or studies are not able to demonstrate that the innovator product would not also have failed in the same circumstance. Patients and diseases are dynamic and highly variable. Some clinical or surrogate endpoints are subject to considerable investigator judgment, and inter-investigator consistency is sometimes difficult to achieve. No therapy is 100 % effective in everyone to whom it is administered or at every stage of an illness. Furthermore, depending on the specific condition treated placebo or nocebo response can by highly significant leading to additional confusion regarding both therapeutic response rate and extent of adverse events (AEs).

To add to the confusion of the non-research oriented clinician, the clinical endpoint is not necessarily the ultimate clinical effect for which a drug is used. The Biomarkers Definitions Working Group presented a conceptual model that clearly shows the relationship of biomarkers to clinical endpoints (Frank and Hargreaves 2003). Using their recommendations and considering the ultimate goal of the study, Fig. 10.1 shows the conceptual model for a clinical endpoint bioequivalence study evaluating the relationship of biomarkers and surrogate endpoints to therapeutic interventions.

In the case of the clinical endpoint bioequivalence study there must be a robust linkage of the biomarker chosen and the clinical effect for which the reference product was approved. The biomarker chosen as surrogate clinical endpoint for the desired clinical effect must be quantifiable. Finally, there must be a clear understanding of when in the course of a study the biomarker can most sensitively predict

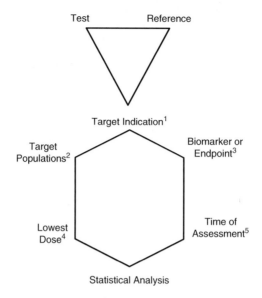

Test   Reference

Target Indication[1]

Target
Populations[2]

Biomarker or
Endpoint[3]

Lowest
Dose[4]

Time of
Assessment[5]

Statistical Analysis

[1]Indication selected from label of reference product
[2]Selected for both safety reasons and in order to limit variability
[3]Biomarker or Endpoint selected from evaluation of in use monitoring for therapeutic efficacy with validation of
methodology and evaluation of inter and intra patient variability and reproducibility.
[4]Selected as the lowest dose that reliably produces the desired effect in order to maximize sensitivity in the study
[5]Selected as the earliest time point at which the selected biomarker can be accurately and reproducibly measured

Fig. 10.1   Conceptual model for a clinical endpoint bioequivalence study evaluating the relationship of biomarkers and surrogate endpoints to therapeutic interventions

the ultimate clinical effect. Measurement of the *correct biomarker* at the *correct time* in the *correct patient population* is the only way a clinical effect can be used to infer bioequivalence (Ilyin et al. 2004).[4]

10.4.1   *Bioequivalence Perspective of a Clinical Endpoint Bioequivalence Study*

All of the considerations that are detailed in the ICH E6 Guidance "Good Clinical Practice: Consolidated Guidance" should be observed in the clinical endpoint bioequivalence study (ICH E6 1996). Like any study using human subjects this study must be ethically conducted, follow rigorous scientific principles, be fully documented and transparent, and have narrowly focused, well defined goals that are

---

[4]Op. Cit. (1999), JAMA 282:790–795.

statistically supportable. In order to design a study both the pathophysiology of the disease and the pharmacological characteristics of the therapy must be considered. Designing such a study is very challenging because both disease and clinical endpoints are highly variable. Hundreds of patients may be required for long study periods, making these quite expensive to conduct. Intersubject variability and inter-investigator variability can confound evaluation of even the most validated of clinical endpoints. For some products, such as acne products, it may be necessary to have multiple endpoints, further increasing complexity of analysis and expense of the study. Other products that are known to have only small therapeutic effects may require enormous studies to detect and quantitate the small effect found in each of the study arms.

While it is possible to review the pivotal studies that led to the approval of the innovator product, the endpoint used may now be obsolete or non-quantifiable. Diagnostic methodology may have changed, leading to differing criteria for inclusion or exclusion in a study. Newer therapies may have changed the incidence and prevalence of the target indication, such that patient enrollment may be problematic. Studies of safety and efficacy published subsequent to approval may have identified significant issues for certain population subgroups shifting the risk: benefit ratio and further complicating the task of patient enrollment. The decision to go forward with designing a clinical endpoint bioequivalence study is not to be made lightly and is definitely not for the faint of heart.

A final critical consideration is the comparability of drug product formulations. Formulations of generic drugs do not have to be identical to that of the reference drug. All drug products contain a number of "inactive" ingredients (excipients) meant to serve as preservatives, binders, or carriers of the active drug product. These ingredients are often different in the formulation of a generic compared to the innovator product. Inactive ingredients and their highest concentrations in approved products can be found in the Inactive Ingredient for Approved Drug Products listings (http://www.accessdata.fda.gov/scripts/cder/iig/index.cfm). The inactive ingredients in a generic formulation are generally ones that have previously been used in another approved drug with the same route of administration and in at least the same amount in maximum daily exposure. However, there must also be adequate evidence to demonstrate that the use of different excipients will not change the safety or effectiveness demonstrated by the innovator product (Code of Federal Regulations). Unfortunately, the acceptable amount listed in the Inactive Ingredients for Approved Drug Products listings does not mention the indication for the drug product that serves as the basis for the excipient amount recommended. What might be considered acceptable as a safe level in one indication or for one type of patient may not translate into safety for all indications or all patients. The clinical context of drug use must always be considered before a formulation containing differing excipients can be judged to be safe in the context of substitution for the reference drug. For generic products we must always consider the safety and risk in the context of both target population and reference drug risk profile.

A few of the considerations that should go into the planning of such a study include:

1. **What is known about the pathophysiology of the disease to be studied?**

   (a) Is there a proposed disease mechanism?
   (b) What are the criteria for diagnosis?
   (c) Are there various stages of the disease?
   (d) Are recommended therapies at different stages?
   (e) Are there known indicators for therapeutic responders?
   (f) Are there known indicators for increased risk AEs during therapy?
   (g) How is disease progression monitored in the clinician's office?

   - Are there validated disease scores?
   - Is there an easily measureable surrogate marker?
   - Can that marker be adequately quantified for comparison?

   (h) If it is not possible to include a placebo arm in the study, what is known of the background rate of placebo response or spontaneous resolution of the disease?
   (i) What is the current Standard of Care for this disease, and will your study provide acceptable care as required in clinical ethics?

2. **What is known about the drug to be studied?**

   (a) Is the site of action known?
   (b) Is the mechanism of action known or is there a proposed mechanism of action?
   (c) Are there multiple sites and mechanisms of action?
   (d) Is the active component the parent drug or a metabolite(s) or both?
   (e) Are there specific metabolites that are important to the safety profile?
   (f) For a locally acting drug will it be necessary to demonstrate comparability of local irritation or sensitization resulting from the drug product?
   (g) If the product is a transdermal patch, are overlays used and will adhesion studies be required?

3. **What is known about your proposed formulation?**

   (a) Are you planning a product that is qualitatively and quantitatively the same as the reference product?
   (b) Are there physicochemical characteristics that may be important to the formulation (e.g., particle size, particle distribution, melting point, viscosity)?
   (c) Are there other considerations about the formulation that may impact how the product reaches the site of action (such as pH for locally acting gastro-intestinal (GI) products) or distributes at the site of action (such as physico-chemical characteristics for ophthalmologic products or locally acting nasal sprays).

(d) Are there other considerations about the formulation that may impact the product safety (such as different excipients or novel excipients)?

(e) What is known about the toxicology of all excipients **in the clinical context of expected use** for this product?

This is a crucial consideration since the acceptable amount for different excipients as listed in the Inactive Ingredient Database does not take this into consideration. However, both patient population and disease pathophysiology may significantly impact the safe level of some excipients.

4. **What is known of the proposed clinical endpoint?**

(a) Is the proposed endpoint valid for all drugs treating the disease of interest?

(b) Is the proposed endpoint valid regardless of the specific mechanism of action for any drug treating this same condition?

(c) Is the proposed endpoint equally valid during asymptomatic phases of a chronic disease, particularly if the disease has relapsing and remitting phases?

(d) If the gold-standard clinical assessment surrogate cannot be quantified, is the proposed alternate endpoint at least as good as the standard followed in clinical practice?

(e) Are there genomic considerations for choice of endpoint or patient inclusion criteria?

5. **For drugs with multiple indications, which indication will be most sensitive to potential differences in therapeutic effect between the test and reference products?**

(a) Is there an indication that can be more accurately diagnosed?

(b) Is there an indication that has well defined stages and clear characteristics with improvement or cure?

(c) Is there a more obvious marker for therapeutic effect in one of the indications?

(d) Are there more specific or validated scoring systems for improvement or cure in one of the indications?

It will probably be much easier to understand these concepts when they are seen in action. First, we will describe a clinical endpoint bioequivalence study that was conducted comparing test and reference drug products for the most obvious of local uses, a topical cream for treating a skin fungal infection. Finally, we will look at the evolution of a guidance for a clinical endpoint bioequivalence study in a product that highlights the limits of clinical endpoint design in bioequivalence.

10.5   Examples of Clinical Endpoint Bioequivalence Studies

10.5.1   Ciclopirox Cream (http://www.accessdata.fda.gov/ Scripts/cder/DrugsatFDA/index.cfm)

Ciclopirox cream is a hydroxypyridone antifungal agent that is commonly used in the treatment of the following dermal infections: tinea pedis (Athlete's Foot), tinea cruris (Jock Itch), and tinea corporis due to *Trichophyton rubrum*, *Trichophyton mentagrophytes*, *Epidermophyton floccosum*, and *Microsporum canis;* candidiasis (moniliasis) due to *Candida albicans*; and tinea (pityriasis) versicolor due to *Malassezia furfur*. The innovator product was approved in 1982 under NDA 018748 and marketed under the brand name, Loprox® cream. According to the Orange Book (Approved Drug Products with Therapeutic Equivalence Evalua-tions), there are currently three approved generic formulations of the topical cream. All of these products have the Therapeutic Equivalence Code (TE Code) "AB." The "A" means that they are therapeutic equivalents and the "B" means actual or potential bioequivalence problems have been resolved with adequate in vivo and/or in vitro evidence supporting bioequivalence.

Methodology: This was a multicenter study of the topical application of ciclopirox cream in the treatment of tinea pedis. There were 462 patients in the intent-to-treat (ITT) analysis. Patients were randomized in a 2:2:1 ratio to receive one of the three treatments (ciclopirox cream, Loprox® cream, or the placebo vehicle used in the test product). The patients applied a thin layer of study medication topically twice daily for 28 days. Patients returned for clinical evaluations at Days 15 (Visit 2), 29 (Visit 3), and 43 (Visit 4). At all visits, the foot was examined for signs and symptoms, which were graded on a 5 point scale relative to both investigator and patient evaluation of the following characteristics of the lesion(s):

Skin grading scale		
Investigator evaluation	Patient evaluation	Scale
Erythema	Itching	0 = none
Scaling (desquamation)	Burning	1 = mild
Fissuring		2 = moderate
Bullae		3 = moderately severe
		4 = severe or extensive

There is a posted FDA draft guidance for this product that offers a detailed discussion of protocol design (Guidance Documents & Ciclopirox 2010). *The methodology of this particular study is consistent with the draft guidance.*

Patient Selection: Patients were required to have a definite clinical and mycological diagnosis of interdigital tinea pedis, with the typical signs and symptoms and a minimum score of 2 for scaling, in addition to a positive microscopic examination of skin scrapings with potassium hydroxide (KOH), showing the presence of fungal hyphae. The mean total severity score was approximately 7 for all three treatment groups. A fungal culture was taken at enrollment and, if fungal culture was negative at visit 3 (29 days), the patient was removed from the study and excluded from the analysis.

These specific instructions highlight the importance of ensuring the most accurate diagnosis possible. It is critical to the accuracy of the study that patients who clearly meet all diagnostic criteria be enrolled in a study evaluating comparative therapeutic effects. Fortunately, in this case the condition does not have significant morbidity and so it is ethically acceptable to include a placebo arm in the study. Both test and reference products will need to show superiority over placebo, thus ensuring sufficient sensitivity to detect differences in response in the study.

Clinical Endpoints: There were three endpoints planned for this study. Efficacy variables included the rates of therapeutic cure, mycological cures, and clinical cure. Clinical cure was defined as a total clinical score (sum of the severity scores on signs and symptoms) of 2 or less, with a score of no more than 1 for any of the six clinical parameters at visit 3. Mycological cure was defined as a negative KOH wet mount for pseudohyphae and a negative post treatment culture at visit 3. The final clinical endpoint was a composite of the above endpoints, therapeutic cure at week 6, defined as both clinical cure and mycological cure.

This is a good example of how an objective endpoint, mycological cure, can be combined with a subjective endpoint, signs and symptoms score, to form an endpoint that reflects a comparative, quantifiable therapeutic effect in a reproducible way.

Mycology: At all visits, the investigator obtained a scraping from the subject's target test site (designated at baseline as the most severely affected area) for microscopic examination using a KOH solution to determine the presence of characteristic fungal hyphae. If the baseline KOH examination was positive for hyphae, the subject was enrolled into the study and a culture plate was inoculated with material from the same scraping. The culture plate was sent to the mycology laboratory for speciation. To be included in efficacy analyses, the baseline culture was required to be positive for a qualified dermatophyte *(T. rubrum, T. mentagrophytes, E. floccosum*, or another causative dermatophyte). Patients whose baseline cultures failed to grow a dermatophyte were discontinued from the study at Day 29 and excluded from the analysis. Scrapings were taken at all post-baseline visits for KOH test and culture, even if the lesion had completely healed.

Safety Variables: Safety variables included AEs.

Although these studies are not designed to identify safety risks, it is possible to evaluate any major differences in safety between the test and reference should they occur. Such evaluation may suggest a reconsideration of formulation differences, study methodology, or flaws in patient selection.

Statistical Evaluation: Three patient populations were identified by the sponsor:

- ITT: received at least one dose of study medication.
- Modified intent-to-treat (MITT): met inclusion/exclusion criteria, including identification of a qualifying dermatophyte at baseline, and had at least one post-baseline efficacy evaluation.
- Per-protocol (PP): completed the entire study without any protocol violations and had data for all three major efficacy variables for Visit 3 and Visit 4 (i.e., KOH preparation, fungal culture, and investigator's clinical evaluation of Clinical Response to Treatment).

What the sponsor has identified as an ITT population is really a modified ITT (mITT) population. This minor misnaming will not affect the outcome of the analysis, since it is acceptable to use an mITT population in this situation. The test and reference must both show superiority to the placebo. Once that has been done, we switch to using the PP population to evaluate equivalence. In this way we have established study sensitivity and will then be using comparable groups of patients who received treatment according to the protocol. The ITT population is more sensitive for evaluating therapeutic efficacy, but the PP population is the more logical choice for simple comparison of therapeutic effect of test and reference.

10.5.1.1   Results

10.5.1.1.1   Mycology

The majority of patients had a baseline fungal culture positive for *Trichophyton rubrum* (84 %). The other dermatophytes isolated were *T. mentagrophytes* (11.5 %) and *Epidermophyton floccosum* (4.5 %). According to the sponsor's analysis, the percentage of patients with baseline fungal culture positive for *T. rubrum* was comparable in all treatment groups ($p = 0.62$).

10.5.1.1.2   Clinical

Therapeutic Cure = Clinical cure (the sum of the severity scores on signs and symptoms of 2 or less, with a score of no more than 1 for any of the six clinical parameters) and mycological cure (negative KOH and culture) at visit 3. Results for the per protocol population are shown in the following table:

Per protocol population results				
	Generic	RLD	Mean difference (90 % confidence interval)	Is confidence interval within ± 20 %?
Therapeutic cure	50 %	46 %	+4 % (−5.4, 14.1)	Yes
Clinical cure	64 %	63 %	+1 % (−8.7, 10.2)	Yes
Mycologic cure	78 %	73 %	+5 % (−2.1, 15.1)	Yes

10.5.1.1.3   Safety

Safety results			
	Generic	RLD	Placebo
All AEs	11.1 %	10.9 %	14.7 %
AEs probably or possibly related to study medication	5.4 %	6.6 %	4.9 %

Reviewer's Conclusion

This generic formulation was determined to be bioequivalent to the RLD (Loprox®). Both generic and RLD products were superior to placebo in this study, confirming that the study was sufficiently sensitive to identify a difference between products.

10.5.2   Evolution of the Rifaximin Draft Guidance

Probably the best way to understand the clinical and scientific considerations that are important is to review an example of how the development of recommendations for a clinical endpoint study design came about. The following is a good example of the sorts of considerations that must go into the development of a clinical endpoint bioequivalence study. Rifaximin is a locally acting gastrointestinal antibiotic product with extremely limited systemic absorption. It has two very disparate indications; in one case for a simple condition that has a short timespan for therapy and expected clinical effect, the other for a complex condition requiring careful subject selection and long-term monitoring to observe the clinical effect. Looking at the way draft guidances evolved for this drug illustrates how pathophysiology, knowledge of drug characteristics, and thorough evaluation of potential clinical endpoints are all essential to designing a clinical endpoint bioequivalence study that can reasonably relate the observed comparative therapeutic effect to an inference of bioequivalence.

10.5.2.1   Why Do We Need a Clinical Endpoint Bioequivalence Study for This Drug?

Rifaximin is an antibiotic, non-aminoglycoside, semi-synthetic derivative of rifamycin that is poorly absorbed from the gastrointestinal (GI) tract. It appears to function locally in the GI tract. It is also practically insoluble in water. Currently there are two approved indications for this drug: treatment of travelers' diarrhea (TD) and reduction in the risk of recurrence of hepatic encephalopathy (HE). The innovator product (RLD) is marketed in 200 and 550 mg tablet strengths. Clinical studies of the reference drug failed to demonstrate clinical correlation of pathogen eradication with clinical response to treatment in either approved indication. Consequently, an in vitro pathogen eradication assay is also not a relevant parameter for establishing bioequivalence (BE) of the test and reference products. Furthermore, lack of significant systemic absorption suggests that systemic effects are not the primary mode of action for the drug. This is true even for the much higher doses used for HE, in which there is a modest measureable blood level.

Rifaximin is very poorly absorbed into the blood following an oral dose. According to Su et al.,

> ...rifaximin is virtually not absorbed from the gastrointestinal tract in healthy volunteers. The pharmacokinetics of radiolabeled rifaximin (^{14}C-rifaximin) administered as a single oral dose (400 mg) to four healthy men aged 30–41 years were assessed via radioactivity measurement for the 168 hours after dosing and via validated liquid chromatography/tandem mass spectrometry assay of plasma and urine samples. No radioactivity was measurable in blood, and rifaximin concentrations were undetectable in most plasma samples. (Su et al. 2006)

Similar to other rifamycins, rifaximin is bound to the β subunit of bacterial DNA dependent RNA polymerase. This causes an inhibition of the initiation of chain formation in RNA synthesis. Additional actions include alteration of bacterial pathogenicity and alteration in attachment and tissue toxicity of bacteria (Adachi and DuPont 2006).

It must be noted, however, that it is clearly stated in the product label: "...this drug does not appear to perform as a conventional antimicrobial in eradicating possible pathogenic organisms." Rifaximin does not demonstrate superior microbiologic activity to that of placebo versus any pathogens. Thus, it appears that a clinical endpoint bioequivalence study will be needed for approval of a generic product for this indication and that microbiology will only play a limited role in the evaluation.

In the case of hepatic encephalopathy, the presumed mechanism of action for rifaximin is inhibition of the metabolism of urea by deaminating bacteria resulting in reduction in production of ammonia and other toxins. Similar to the other broad spectrum antibiotics used in this condition (e.g., Neomycin, Vancomycin), rifaximin inhibits the division of urea deaminating bacteria, thereby reducing the production of ammonia and other compounds thought to be important in the pathogenesis of hepatic encephalopathy.

However, Adachi and DuPont note "... rifaximin has little effect on the normal gastrointestinal flora. After 3 cycles of receiving 1800 mg per day for 10 days, followed by 25 days free of medication, there was an initial decrease in the gastrointestinal flora (enterococcus, *Escherichia coli*, Lactobacillus species, Bifidobacterium species, Bacteroides species, and *Clostridium perfringens*), followed by normalization after 1 month...Our group has also shown that rifaximin has minimal effects on *E. coli* and enterococcus flora after 3–14 days of therapy."[5]

---

[5] Ibid.

Thus, the exact mechanism of action is not clearly defined, and determination of the site of action cannot be made with any certainty. Both plasma concentration and local GI concentration may be critical to function for the HE indication.

Therefore, in both of these indications (TD and HE) the initial approval of rifaximin was based on clinical endpoints, not microbiologic efficacy. Furthermore, PK measurements were at or below the minimal level of detection in the studies for TD using the 200 mg strength. At multiples of the 200 mg tablet or at the higher (550 mg) dose used in the HE indication, minimally adequate plasma levels were obtained, particularly in fed studies (see below). After a high fat meal, absorption can increase as much as twofold. This would suggest that it might be possible to design a bioequivalence PK study for the higher dose product used to treat HE. Unfortunately, it is not at all clear that the therapeutic effect in this indication is produced by the systemic absorption of rifaximin. Uncertainty regarding the mechanism of action and site of action in HE makes the PK option less viable. Thus, in this indication too it appears that a clinical endpoint bioequivalence study will be needed for approval of a generic product.

10.5.2.2   Indication #1: Travelers' Diarrhea

This indication is relatively straightforward for both initial diagnosis and methodology of monitoring the course of the disease. A reasonably reproducible and quantifiable clinical endpoint can be established and used in the design of a clinical endpoint bioequivalence study.

For travelers' diarrhea (TD), the expected clinical effect is a cessation of diarrhea. The course of the disease is short and the expectation of a cure (clinical endpoint) can be quickly identified by monitoring the patient for a time to last unformed stool (TLUS). We know that the clinical endpoint of TLUS was reduced from approximately 65 h in the untreated patients to approximately 32 h in those treated with rifaximin. The microbiologic eradication of pathogenic organisms did not differ greatly between the rifaximin and placebo treated populations. Stools at patient screening (Day 0) and end of study (Day 5) should be cultured for pathogenic organisms, but microbiological information is useful only to support the similarity of enrolled populations, and to screen out patients with more serious pathogenic organisms that cause diarrhea but are not classified as TD. Measurements made in the development of the reference product show that the clinical endpoint can be measured at 3 days to identify a likely cure. This is consistent with the recommended treatment course with rifaximin. Some patients relapse when drug administration is completed at 3 days, but this is identifiable within 2 more days. Consequently, a 5 day study would likely be needed with monitoring of patients for TLUS. The condition is not life threatening or involving significant morbidity, and so inclusion of a placebo arm in the study is ethically acceptable. Now we can look in more detail at the evidence supporting the above statements.

In the United States, there are approximately 200–375 million cases of acute diarrheal illness each year. This is definitely an adequate pool from which a suitable study population might be enrolled. Many of these cases are mild, but there are still over 900,000 hospitalizations and approximately 6,000 deaths each year (Cottreau et al. 2010). Travelers' diarrhea is defined as passage of at least three loose stools in a 24 h period during time of travel, accompanied by at least one gastrointestinal symptom including:

- Abdominal Pain
- Cramping
- Nausea
- Vomiting
- Fever
- Tenesmus

Such diarrhea can occur at any time during travel and for up to 10 days following return, but is most common on the third day following arrival in a developing country. The condition is self-limited although rarely it can be fatal. A limited number of patients require hospitalization (1 %) or are confined to bed (20 %), and approximately 40 % require a change in their travel itinerary.

The primary cause of this clinical syndrome is ingestion of fecally contaminated foods or water. The ingested pathogens result in a watery diarrhea due to secretion of fluid and electrolytes into the gastrointestinal tract without damage to the intestinal mucosa (Robins and Wellington 2005). Most commonly, noninvasive bacterial agents, account for 80 % of cases. The remaining causes are parasites and viruses. The most commonly identified bacterial causes of all travelers' diarrheas include Enterotoxigenic *Escherichia coli* (ETEC), enteroaggregative *E. coli* (EAEC), *Shigella* spp., *Salmonella* spp., *Campylobacter jejuni*, *Aeromonas* spp., *Plesiomonas* spp., and non-cholerae *Vibrio*. However, not all travelers' diarrhea is bacterial in origin (Boggess 2007). The following table summarizes the common causes of travelers' diarrhea (Palmgren et al. 1997; Theilman and Guerrant 1998).

Travelers' diarrhea		
Pathogenic bacteria	Parasitic agents	Enteric viruses
Escherichia coli (Enterotoxigenic and enteroaggregative)	*Giardia lamblia*	Rotavirus
Campylobacter jejuni	*Entamoeba histolytica*	Noroviruses
Salmonella species	*Cryptosporidium parvum*	Enteric adenoviruses
Shigella species	*Cyclospora cayetanensis*	Astrovirus
Vibrio species	*Dientomoeba fragilis*	
Yersina enterocolitica	*Isospora belli*	
Clostridium difficile	*Microsporidia*	
	Strongyloides	
	Schistosoma	
	Trichiuris	

Travelers' diarrhea is typically a self-limited illness, but may result in persistent diarrhea in at least 2 % of cases, irritable bowel syndrome with post-infectious diarrhea in 4–10 % of travelers, and other chronic complications, such as reactive arthritis with *Salmonella*, *Shigella*, and *Campylobacter*, and Guillain–Barre syndrome from *Campylobacter* infection (Koo et al. 2009). Antibacterial treatment is considered one of the primary treatment options for travelers' diarrhea. Bacterial enteropathogens can cause up to 85 % of the cases of diarrhea among persons visiting high-risk tropical and semitropical areas (DuPont and Ericsson 1993).

Empirical evidence demonstrates that antibiotic therapy can reduce the duration of diarrhea by 1–2 days.

Therapies related to non-bloody "Watery Diarrhea Syndrome" encountered worldwide are the TD syndromes that would be treated with rifaximin. Rifaximin is generally preferred to systemic acting antibiotics for TD due to its broad safety profile and the apparent local mechanism of action. Little antimicrobial resistance has been encountered with rifaximin used in such a short course.

10.5.2.2.1   Case Summary

From this short overview, it is clear that TD is sufficiently common and easy to diagnose, thus making it a reasonably simple task to enroll a suitable population of patients for a parallel designed study. The condition can be simply distinguished from the more severe diarrheal syndromes, and so inclusion of a placebo study arm is acceptable. The clinical endpoint coincides with the expected endpoint: cessation of diarrhea, defined as no unformed stool. Thus, a plan to monitor the time to the last unformed stool (TLUS) would capture all the information needed to allow for a comparison of test, reference, and placebo. The condition is self-limited and of a relatively short duration and the administration of drug is limited to only 3 days of therapy. Therefore, a 5 day study would allow for identification of both true cures and partial cures (recurrence between Day 3 and Day 5). These data are quantifiable such that it is possible to demonstrate both superiority to placebo and comparative cure rates for both test and reference drugs. A clinical endpoint bioequivalence study is feasible.

10.5.2.3   Indication #2: Hepatic Encephalopathy

This is an extremely difficult condition to identify and study. As will be seen in the overview, nowhere in the currently known pathophysiology of this condition is there consensus on how to monitor, grade, or quantify the highly variable relapsing and remitting signs and symptoms. Even diagnosis is mired in some controversy and variability.

10.5.2.3.1 Overview of Hepatic Encephalopathy

There are approximately 5.5 million cases of chronic liver disease and cirrhosis in the United States and 30–50 % are estimated to demonstrate overt hepatic encephalopathy. This would seem to provide us with an adequate source of patients for a study. Unfortunately, this may not be helpful if we cannot make a precise diagnosis. The specific cause(s) and mechanism of disease are not known. Minimal Hepatic Encephalopathy (MHE) is defined as impaired performance on psychometric or neurophysiologic testing in the presence of apparently normal mental status (Bajaj et al. 2010). It is characterized by subtle motor and cognitive deficits that are clinically very difficult to identify. MHE affects approximately 20–60 % of patients with liver disease. This should not be considered as a definitive diagnosis, but rather as a continuum that fluctuates unpredictably up and down the scale. It is part of the spectrum of neurocognitive impairments in cirrhosis (SONIC) as depicted graphically below:

Progression of hepatic encephalopathy		
Patient status	Clarity of diagnosis	Cognitive status
Normal	Inconsistent, variable among rating physicians, arbitrary	Normal cognitive functioning
Minimal encephalopathy		Variable cognitive functioning ranging from normal to mild impairment to mild dysfunction
Overt encephalopathy stages		Near imperceptible movement from normal to mild dysfunction returning to normal
I		
II	Relatively clear and consistent diagnostic placement among raters	Clear cognitive dysfunction
III		
IV		

In 1998, the World Congress of Gastroenterology met in Vienna to recommend a standard definition and nomenclature for hepatic encephalopathy. Hepatic encephalopathy was defined as, "...a spectrum of neuropsychiatric abnormalities seen in patients with liver dysfunction after exclusion of other known brain disease. A multiaxial definition of HE is required that defines both the type of hepatic abnormality and the duration/characteristics of neurologic manifestations in chronic liver disease." The diagnosis is based on a careful neuropsychiatric evaluation (Ferenci et al. 2002).

Generally, HE is recognized as a metabolic-neurophysiologic disorder related to a reduction of hepatic functional mass and the presence of portosystemic shunts resulting in circulating toxins originating from the intestinal metabolism of nitrogenous compounds and from alteration of neurotransmission (Festi et al. 1993). Unfortunately, clinical scales used in HE classification, such as the West Haven (Conn) scale, do not take into account the continuous nature of impairments that characterize chronic liver disease. Similarly, the variability in symptoms over short time periods is not easily measurable. In addition, the more detailed psychometric tests, while sensitive, lack specificity and are fairly complicated and time consuming to administer. Division of patients from grade 0 (normal or MHE) and grade 1 is controversial and the shift from grade 1 to grade 2 is not necessarily clear cut (Bajaj et al. 2009a).

Psychometric testing is complex and not easy to quantify between subjects undergoing testing. The following table illustrates some of the tests used in trying to identify cognitive dysfunction: (Kiernan et al. 1987; Kircheis et al. 2002; Bajaj et al. 2009b).

Psychometric testing			
Sample tests	Cognitive domain tested	Complexity of test	Degree of cognitive dysfunction identifiable
Inhibitory control test	Executive functions[a]	Complex, time consuming, and requiring highly trained tester with adequate equipment	Mild changes in cognition or function
Clock drawing		Relatively objective	
Sternberg test			
Continuous performance tests	Conscious vigilance and attention		
Choice reaction times			
Hopkins verbal learning test			Moderate cognitive dysfunction
Digit symbol test	Psychomotor responsiveness and reaction times		
Critical flicker fusion test			
Clinical tests	General alertness, orientation, and motor function	Simple, requiring considerable tester judgment for interpretation	Gross cognitive dysfunction
Conn score			
Asterixis score			
Epworth sleepiness scale			

[a]Executive functions generally include working memory, judgment, response inhibition, and ability to plan and problem solve.

Clearly a categorical approach to HE is not reliable, "...due to (1) an inherent subjectivity of the clinical assessments in the earlier and middle stages of the disease; (2) no consensus regarding the multiplicity of tools to diagnose cognitive dysfunction; (3) a lack of population norms for most tools in the United States, which invalidates data interpretation; and (4) imperfect prediction of important clinical and psycho-social outcomes based on the test performances." (Poordad 2007)

In HE, treatment with rifaximin is intended to modify fluctuations in mental status and neurocognitive functioning, such that patients are not cured of their condition, but rather are limited in the degree to which they manifest cognitive and motor dysfunction. As can be seen in the graphic above, the difference between grade 1 and grade 2 HE on the Conn scale would rest in the upper 1/2 of the domains, including the executive functions (judgment, response inhibition, problem solving) and those of vigilance and attention.

When rifaximin was approved for this indication, a shift in function from grade 0 or 1 to grade 2 or greater on the Conn Scale was considered to be the indicator for recurrence of overt HE. This is a rather coarse distinction, but one that is, at least, reasonably within clinical judgment. It is not quantitative, however, and is subject to a clinical observer bias that would be very difficult to quantitate in a clinical endpoint bioequivalence study. This can be seen when the specifics of the West Haven (Conn) Criteria are seen:

Conn score:	
Grade	Characteristic
1	Minimal lack of awareness
	Short attention span
	Anxious or euphoric state
2	Apathetic or lethargic
	Mildly disoriented to time and/or place
	Subtle personality changes
	Inappropriate behavioral responses
3	Somnolent or semistuporous
	Arousable and responsive with verbal stimulation
4	Comatose, unresponsive to verbal or noxious stimuli

Inclusion of a placebo arm in the study may help, but the difficulty in how to compare the two active treatment arms would remain, and it is clinically unethical to withhold treatment in a condition for which there is significant morbidity when the clinical standard of care provides for potentially supportive therapy. A placebo arm is not acceptable in a study of HE.

So it is clear that the tests used to diagnose and monitor HE are themselves qualitative assessments. By their nature, it would be impossible to use them in a quantitative, comparative way. This can be seen clearly in the chart below that summarizes the type of clinical assessments used to determine efficacy of rifaximin for this indication:

- Conn Score
- Asterixis Score
- Critical Flicker Frequency Score
- Epworth Sleepiness Scale
- Short Form 36 (SF-36)
- Evaluation of mental control, based on counting numbers and listing the alphabet
- Assessment of vision (using a picture of a cross held 12 in. from the subject
- Hopkins Verbal Learning Test (delayed recall and recognition components)
- Simple and complex computations
- Depression rating
- Anxiety and nervousness rating

- Digit span
- Figure copying

Thus, as stated by Bajaj et al., "The continuous nature of this neurocognitive impairment is missed by the current system of arbitrary cutoffs, most often 2 standard deviations impaired beyond comparable controls. In therapeutic HE study, the decision for HE reversal versus persistence can therefore often boil down to a few seconds or limited change in raw scores on the individual tests, which can result in important changes in patient classification."[6]

Despite the difficulty in diagnosing and staging this condition, it can significantly interfere with the patient's functioning, social interaction and work activities (Poordad 2007). Clearly, there is a breakpoint in terms of psychosocial function for the patient somewhere between the grade 1 and grade 2 categories of HE.

The Model for End Stage Liver Disease (MELD) score is a somewhat more objective indicator of the severity of liver disease and is entirely based on laboratory data, specifically the serum creatinine, total bilirubin, and international normalized ratio (INR). It was developed in order to evaluate patients as candidates for liver transplant, not as a means of monitoring severity of HE. It is also a predictor of survival for patients with end stage liver disease. There were no patients with MELD scores > 25 in the clinical studies of the reference drug and only 8.6 % of the enrolled patients had scores over 19 according to the product label. An additional scale for hepatic impairment, the Child-Turcotte-Pugh (CTP) score is based on the total bilirubin, albumin, prothrombin time and ascites. CTP score is not referenced in the product label, but the two scoring systems can be combined in such a way as to stratify HE stages to some extent.

A Conn Score of 0–1 and a MELD score of ≤25 theoretically indicate that the patient is in remission from HE. This could serve as a starting point for designing a study comparing a generic product to the approved rifaximin. Unfortunately, as can be seen in the chart above, the median MELD scores for Grades 0 and 1 are essentially indistinguishable, as are those for Grades 2, 3, and 4. Thus the more objective measure of disease in the liver, the MELD score, does not correlate well with the primary measure of HE, the Conn score.

Hepatic encephalopathy is a serious consequence of liver failure and is associated with brain edema, intracranial hypertension, and neurologic death. Nonetheless, the term has not been clearly defined clinically. It is a condition difficult to treat and even more difficult to study. Clinical tools for the diagnosis of HE are often subjective while psychometric tools are not widely accepted. As a consequence, it is even more complex to perform studies that are meant to compare a test and reference product in this condition than would be the case in TD.

---

[6] Op. Cit. (2009) Hepatology 50(6):2014–2021.

10.5.2.3.2   A Brief Consideration of Mechanism of Action

HE is also referred to as Portal Systemic Encephalopathy (PSE). It can have a number of precipitating causes, ultimately resulting in severe impairment of hepatic function. We might consider further investigation into this aspect of the condition to consider possible biomarkers that may be more quantitative and therefore useful in a clinical endpoint bioequivalence study. Greenberger suggested the following precipitating causes: (Greenberger 2009)

- Hypoperfusion and N2 load due to dietary protein, surgery, anesthesia, or gastrointestinal bleeding
- Cytokines produced in protein catabolism due to sepsis
- Production of ammonia and ureases due to azotemia, fecal bacteria, dehydration, diuretics, or hypokalemia
- Activation of inhibitory neurotransmission due to medications (narcotics, benzodiazepines)
- Liver injury, other drugs, acute hepatitis

Fortunately, correction of the precipitating cause usually results in regression of encephalopathy without permanent neurologic sequelae in many of the above conditions. However, in those patients with chronic liver disease, especially those with portacaval shunts or TIPS, continuous therapy will be required and correction of the underlying problem may only be by means of liver transplant. Occurrences of HE during the course of chronic liver disease will be variable, triggered by the above factors primarily related to ammonia production or some other toxin such as gamma-aminobutyric acid (GABA), tryptophan, or other false neurotransmitters resulting from disturbed amino acid metabolism (Fischer and Baldessarini 1971). There are a number of hypotheses as to the pathogenesis of HE including (Vaquero et al. 2003).

- Direct ammonia neurotoxicity
- Synergistic neurotoxins
- Plasma amino acid imbalance
- GABA/benzodiazepine
- Cerebral edema/astrocytic swelling
- Tryptophan and its metabolites
- Histamine
- Opioids
- Brain edema

A complete discussion of the presumed pathogenesis of HE is unnecessary for our purpose here. It is clear that there are many factors that both independently contribute to HE and interact with arterial ammonia levels to trigger HE (Vaquero et al. 2003). Lack of a clear mechanism of pathogenesis means that no clear and validated biomarker can be derived from what is currently known about this condition. A specific site of action for a therapeutic agent cannot be defined. The complexity of HE cannot be reduced to a simple assumption that antimicrobial effect in the gut alone accounts for the efficacy of rifaximin in prevention of recurrence of HE in patients with chronic liver failure. The beneficial effect of rifaximin in HE may not be only due to antimicrobial action in the gut (locally acting). The much higher doses used in the various regimens for rifaximin in HE, coupled with the generally limited alteration in the gut flora suggest a systemic effect may be of importance. At the doses appropriate to this condition, there is systemic absorption.

Diagnosis and monitoring of a patient with this condition is by means of careful and repetitive neuropsychiatric testing. It is apparent that the evaluation of the mental status and neurological changes are not generally quantitative. Psychometric testing is required and the most appropriate tests for this purpose have not been definitely determined or validated. As stated by Bajaj et al. the categorical approach to HE is not reliable, "...due to (1) an inherent subjectivity of the clinical assessments in the earlier and middle stages of the disease; (2) no consensus regarding the multiplicity of tools to diagnose cognitive dysfunction; (3) a lack of population norms for most tools in the United States, which invalidates data interpretation; and (4) imperfect prediction of important clinical and psycho-social outcomes based on the test performances."[7]

Scoring is imprecise and subject to clinician observation bias and inter-evaluator variability. This poses an insurmountable problem for quantitative comparison between a proposed generic product and the reference drug in a clinical endpoint bioequivalence study.

Furthermore, any clinical endpoint study in this condition would need to be between 3 and 6 months in duration since it is known that the efficacy of rifaximin in HE does not become evident for a minimum of 2 months of therapy. Thus, it appears that, unlike the case of TD where use of TLUS is an easily identifiable, reasonable, and quantifiable surrogate for the therapeutic effect of rifaximin, no such quantifiable biomarker has been found for the indication of HE. Therefore, no study can be designed for the hepatic encephalopathy indication due to the subjective and non-quantifiable nature of the biomarkers used in diagnosis and monitoring of this condition, and due to the highly variable nature of expression, severity, and recurrence of signs and symptoms of this condition.

10.5.2.3.3   Case Summary

Hepatic encephalopathy is a very complicated condition, and clearly many more factors must be considered in contemplating a clinical endpoint bioequivalence test for HE than were needed for TD. The condition not only has multiple, poorly distinguishable stages, but likely has a presymptomatic phase in which there are some soft neurologic findings, but not overt HE. This alone would make enrollment of patients to a parallel design study extremely challenging. The pathogenesis of the

---

[7] Op. Cit. (2009) Hepatology 50(6):2019.

condition is not well defined and is most likely multifactorial. This means that a site of action for any therapeutic agent will be difficult if not impossible to define. The significant morbidity inherent in this condition makes it clinically unethical to design a study using a placebo arm. The overall impact of therapeutic intervention in HE is itself ill defined, and so lack of a placebo arm would raise the issue that comparison of two drugs demonstrating no difference in effect may simply be due to the fact that neither drug had an effect and not because the drugs were bioequivalent or non-inferior.

Surrogate markers or clinical endpoints are nebulous and not easily subject to quantification. Finally, the study would need to be prolonged, lasting at least 6 months, due to the known fact that observable clinical effect cannot be identified before 2–3 months of therapy and the variability of recurrence and remission needs to be taken into account. A clinical endpoint bioequivalence study is not feasible for this indication.

10.5.2.3.4   Reviewer's Conclusion

Based on these considerations two draft guidance documents for rifaximin were devised. In the first, a clinical endpoint bioequivalence study is detailed for the indication of travelers' diarrhea and specifying the 200 mg tablet strength. The second represents a best compromise for sponsors hoping to develop a generic version of the 550 mg strength. A pharmacokinetic bioequivalence study is recommended to establish systemic equivalence using the 550 mg strength, and a clinical endpoint bioequivalence is also required for the 200 mg strength. This recommendation also suggests that the 550 mg strength should be dose proportional to the 200 mg strength. In this way, it is believed that evidence of both local and systemic equivalence can be achieved.

10.6   Final Summary and Conclusion

The clinical endpoint bioequivalence study is, unfortunately, often confused with the well-designed and well-controlled, randomized, double blind clinical study by both physicians and the general public. These groups have a deep seated belief that the only way to demonstrate equivalence is by means of a clinical study in patients for whom the drug is intended. This is unfortunate because it is in actuality the least accurate, specific, and reproducible of the three in vivo methods of bioequivalence determination. It is the only acceptable way to evaluate bioequivalence in those products that are locally acting and have little or no systemic absorption. Acceptability is based on the prior demonstration that the products are both chemically and pharmaceutically equivalent. Therefore a demonstration of equivalence of therapeutic effect allows for an inference that the products are bioequivalent.

Unlike the qualitative studies used to demonstrate efficacy of a drug product for a clinical indication, the clinical endpoint bioequivalence study attempts to quantify the therapeutic effect obtained in the study in order to demonstrate that both test and reference products produce the same therapeutic effect within a clinically acceptable range relative to the variability of a biologic system. This range has been established historically and remains controversial. We have not dealt with this aspect of the clinical endpoint bioequivalence study because it requires extensive statistical discussion and support. For the purpose of this chapter it suffices to say that the intent of analysis of data from a clinical endpoint study is exactly the same as for analysis of a pharmacokinetic bioequivalence study, to ensure that the acceptable range of values in the study population is no different from that which would be expected if the reference product was tested against itself under the same conditions.

References

Adachi JA, DuPont HL (2006) Rifaximin: a novel nonabsorbed rifamycin for gastrointestinal disorders. Clin Infect Dis 42:541–547

Anderson A (2005) Snake oil, hustlers and hambones: the American medicine show. McFarland & Company, Inc., Jefferson, NC

Approved Drug Products with Therapeutic Equivalence Evaluations (Orange Book), on line edition. http://www.accessdata.fda.gov/scripts/cder/ob/default.cfm

Bajaj JS, Wade JB, Sanyal AJ (2009a) Spectrum of neurocognitive impairment in cirrhosis: implications for the assessment of hepatic encephalopathy. Hepatology 50(6):2014–2021

Bajaj JS, Wade JB, Sanyal AJ (2009b) Spectrum of neurocognitive impairment in cirrhosis: implications for the assessment of hepatic encephalopathy. Hepatology 50(6):2014–2021

Bajaj JS, Schubert CM, Heuman DM, Wade JB et al (2010) Persistence of cognitive impairment after resolution of overt hepatic encephalopathy. Gastroenterology 138:2332–2340

Berns B, De'molis P, Scheulen ME (2007) How can biomarkers become surrogate endpoints. EJC Suppl 5(9):37–40

Boggess BR (2007) Gastrointestinal infections in the traveling athlete. Curr Sports Med Rep 6:125–129

Code of Federal Regulations, 21, food and drugs, part 314.94(a)(9)(ii), content and format of an abbreviated application, inactive ingredients, p 128

Colburn WA (2000) Optimizing the use of biomarkers, surrogate endpoints, and clinical endpoints for more efficient drug development. J Clin Pharmacol 40:1419–1427

Cook M, Irby DM, Sullivan W, Ludmerer KM (2006) American medical education 100 years after the flexner report. N Engl J Med 355:1339–1344

Cottreau J, Baker SF, DuPont HL, Garey KW (2010) Rifaximin: a nonsystemic rifamycin antibiotic for gastrointestinal infections. Expert Rev Anti Infect Ther 8(7):747–760

Davit BM, Conner DP (2010) United States of America (Chapter 12). In: Kanfer I, Shargel L (eds) Generic drug product development: international regulatory requirements for bioequivalence. Informa Healthcare USA, Inc., New York, NY

De Gruttola VG, Clax P, DeMets DL, Downing GJ et al (2001) Considerations in the evaluation of surrogate endpoints in clinical trials: summary of a National Institutes of Health Workshop. Control Clin Trials 22:485–502

Drugs @ FDA searchable database. http://www.accessdata.fda.gov/Scripts/cder/DrugsatFDA/index.cfm

DuPont H, Ericsson C (1993) Prevention and treatment of travelers' diarrhea. N Engl J Med 328:1821–1827

FDA, CDER, Inactive ingredient search for approved drug products. http://www.accessdata.fda. gov/scripts/cder/iig/index.cfm

FDA A history of the FDA and drug regulation in the United States. http://www.fda.gov/centen nial/history/history.html

Ferenci P, Lockwood A, Mullen K, Tarter R et al (2002) Hepatic encephalopathy—definition, nomenclature, diagnosis, and quantification: final report of the working party at the 11th World Congresses of Gastroenterology, Vienna, 1998. Hepatology 35:716–721

Festi D, Mazzella G, Orsini M, Sottili S et al (1993) Rifaximin in the treatment of chronic hepatic encephalopathy; results of a multicenter study of efficacy and safety. Curr Ther Res 54 (5):598–609

Fischer JE, Baldessarini RJ (1971) False neurotransmitters and hepatic failure. Lancet 298 (7715):75–79

Frank R, Hargreaves R (2003) Clinical biomarkers in drug discovery and development. Nat Rev Drug Discov 2:566–580

Friedman LM, Furberg CD, DeMets DL (2010) Introduction to clinical trials (Chapter 1). In: Fundamentals of clinical trials, 4th edn. Springer, New York, NY

Gallin JI (2012) A historical perspective on clinical research (Chapter 1). In: Gallin JI, Ognibene FP (eds) Principles and practice of clinical research, 3rd edn. Elsevier, Academic, Waltham, MA

Greenberger NJ (2009) Portal systemic encephalopathy & hepatic encephalopathy (Chapter 44). In: Current diagnosis & treatment gastroenterology, hepatoloty, & endoscopy, 3rd edn. McGraw Hill Medical, New York

FDA Guidance Documents, Ciclopirox, Cream/Topical, 2010. http://www.fda.gov/downloads/ Drugs/GuidanceComplianceRegulatoryInformation/Guidances/UCM199627.pdf

ICH E6 (1996) Guidance for industry, good clinical practice: consolidated guidance. www.fda. gov/downloads/Drugs/GuidanceComplianceRegulatoryInformation/Guidances/ucm073122. pdf

Ilyin SE, Belkowski SM, Plata-Salaman CR (2004) Biomarker discovery and validation: technologies and integrative approaches. Trends Biotechnol 22(8):411–416

Kesselheim AS, Misono AS, Lee JL, Stedman MR et al (2008) Clinical equivalence of generic and brand-name drugs used in cardiovascular disease. JAMA 300(21):2514–2526

Kiernan RJ, Mueller J, Llangston JW, Van Dyke C (1987) The neurobehavioral cognitive status examination: a brief but differentiated approach to cognitive assessment. Ann Intern Med 107:481–485

Kircheis G, Wettstein M, Timmermann L, Schnitzler A, Haussingern D (2002) Critical flicker frequency for quantification of low-grade hepatic encephalopathy. Hepatology 35:357–366

Koo HL, DuPont HL, Huang DB (2009) The role of rifaximin in the treatment and chemoprophylaxis of travelers' diarrhea. Ther Clin Risk Manag 5:841–848

Maeshiro R, Johnson I, Koo D, Parboosingh J et al (2010) Reflections on the flexner report from a public health perspective. Acad Med 85(2):211–219

Mullens KD (2003) Pathogenesis of hepatic encephalopathy (Chapter 20). In: Jones EA, Meijer AJ, Chamuleau RAFM (eds) Encephalopathy and nitrogen metabolism in liver failure. Kluwer Academic, Norwell, MA, pp 177–183

Palmgren H, Sellin M, Olsen B (1997) Enteropathogenic bacteria in migrating birds arriving in Sweden. Scand J Infect Dis 29:565–568

Peters JR, Hixon D, Davit BM, Conner D, Catterson D, Parise C (2009) Generic drugs—safe, effective, and affordable. Dermatol Ther 22(3):229–240

Poordad FF (2007) Review article: the burden of hepatic encephalopathy. Alim Pharmacol Ther 25 (Suppl 1):3–9

Rahsid A (2012) Handbook of clinical endpoint bioequivalence studies. Harris Grier, Appian International Research, Inc., Charlotte, NC

Robins GW, Wellington K (2005) Rifaximin: a review of its use in the management of traveller's diarrhoea. Drugs 65(12):1697–1713

Rolan P (1997) The contribution of clinical pharmacology surrogates and models to drug development—a critical appraisal. Br J Clin Pharmacol 44:219–225

Su CG, Aberra F, Lichtenstein GR (2006) Utility of the nonabsorbed (<0.4%) antibiotic rifaximin in gastroenterology and hepatology. Gastroenterol Hepatol 2(3):186–197

Temple R (1999) Are surrogate markers adequate to assess cardiovascular disease drugs. JAMA 282:790–795

Theilman NM, Guerrant RL (1998) Persistent diarrhea in the returned traveler. Infect Dis Clin North Am 12(2):489–501

Vaquero JV, Chung C, Cahill ME, Blei AT (2003) Pathogenesis of hepatic encephalopathy in acute liver failure. Semin Liver Dis 23(3):259–269

Chapter 11
Bioequivalence for Liposomal Drug Products

Nan Zheng, Wenlei Jiang, Robert Lionberger, and Lawrence X. Yu

11.1 Introduction

Liposomal drug products are defined as drug products containing drug substances encapsulated in liposomes (FDA 2002). Liposomes are formed by amphiphilic molecules such as phospholipids, upon dispersion in an aqueous environment. Structurally, liposomes are composed of a bilayer that encloses a central aqueous core or multiple bilayers separated by aqueous compartments. With the advance of novel lipid characterization and modification techniques, a variety of liposomes are being developed as drug delivery system to improve the pharmacokinetics (PK), biodistribution, safety, and efficacy profiles of specific therapeutic agents. Currently nine innovator liposomal drug products are available on the US and European markets for human use (Table 11.1). All of them are injections.

Compared with traditional injectable formulations for the same indication, liposomal drug products have shown greater success. For instance, Doxil (doxorubicin hydrochloride (HCl) liposome injection) prolongs doxorubicin circulation by effectively protecting the active drug substance from systemic elimination and attenuates side effects by preferentially accumulating drug at the target tumor sites. In 2011, doxorubicin HCl liposome injection has a worldwide revenue of 400 million US dollars and US revenue of 140 million US dollars; in contrast, its small molecular weight drug counterpart, doxorubicin HCl injection, has an anticipated US revenue of 14 million US dollars (EvaluatePharma 2013). As patents and exclusivities on the early liposomal drug products expire, many pharmaceutical companies initiate endeavor to make generic copies of these products. It becomes more and more imperative for regulatory bodies to develop scientifically sound bioequivalence (BE) criteria for liposomal drug products.

N. Zheng (✉) • W. Jiang • R. Lionberger • L.X. Yu
Center for Drug Evaluation and Research, U.S. Food and Drug Administration,
10903 New Hampshire Avenue, Silver Spring, MD 20993, USA
e-mail: nan.zheng@fda.hhs.gov; wenlei.jiang@fda.hhs.gov; robert.lionberger@fda.hhs.gov;
lawrence.yu@fda.hhs.gov

L.X. Yu and B.V. Li (eds.), *FDA Bioequivalence Standards*, AAPS Advances
in the Pharmaceutical Sciences Series 13, DOI 10.1007/978-1-4939-1252-0_11,
© The United States Government 2014

Table 11.1 Innovator liposomal drug products that are on the US and European markets for human use

NDA/MAN	Trade name	Generic name	Route of administration	Approved date	Patent expiration date
050718	Doxil	Doxorubicin hydrochloride	Intravenous	17 November 1995	Expired
050704	DaunoXome	Daunorubicin citrate	Intravenous	8 April 1996	Expired
050740	AmBisome	Amphotericin B	Intravenous	11 August 1997	12 October 2016
021041	DepoCyt	Cytarabine	Intrathecal	1 April 1999	3 March 2015
021119	Visudyne	Verteporfin	Intravenous	12 April 2000	11 March 2016
EU/1/00/141/001-002	Myocet	Doxorubicin hydrochloride	Intravenous	13 July 2000	N.A.
EU/1/08/502/001	Mepact	Mifamurtide	Intravenous	3 June 2009	N.A.
022496	Exparel	Bupivacaine	Infiltration into soft tissue	28 October 11	18 September 2018
202497	Marqibo	Vincristine sulfate	Intravenous	9 August 2012	25 September 2020

A product-specific bioequivalence recommendation is available at FDA webpage for doxorubicin hydrochloride liposome injection

The challenge of demonstrating BE between a generic liposomal drug product and its reference listed drug (RLD) is usually twofold (Zhang et al. 2013). Firstly, after administration, drug substance often exists in multiple forms both in systemic circulation and at the target site; hence, it is critical to identify the most therapeutically relevant moiety for establishing BE. For example, after intravenous injection of Doxil, both encapsulated doxorubicin and free doxorubicin can be detected in systemic circulations. Secondly, drug level in systemic circulation may not always reflect drug concentration at the target site; as a result, in most cases, PK endpoint-based BE study alone may not be sufficient to ensure equivalent safety and efficacy. Additional measures such as comparative physicochemical testing (including in vitro release studies) must be undertaken to correlate BE surrogate to its availability at the site of action. These physicochemical characterizations are often listed in BE guidelines as an indispensable component of BE requirements. Under these general rules, individual BE recommendation should be based on the mechanism of action, biodistribution of therapeutically relevant drug species, and the metabolism and elimination of specific drug substance.

This chapter will summarize the application of liposome technology in drug development, review the physicochemical and biological properties of available liposomal drug products, and discuss the challenges, approaches, and opportunities in establishing BE recommendations for liposomal drug products. Specifically we use PEGylated liposomal doxorubicin injection as an example to illustrate FDA's common considerations in developing product-specific BE guidance for liposomal drug products.

11.2 The Application of Liposome in Drug Development

Liposomes were discovered in the 1960s by electron microscopic observations of phospholipids that were negatively stained with surface-active agents (Bangham and Horne 1964). Initially used as a model for biological membranes (Sessa and Weissman.G 1968), liposome has been tailored for use in drug development for a diversity of purposes. Today liposomes are available in different sizes, shapes, and functions. Based on the internal structure, liposomes can be classified as unilamellar, multilamellar, or multivesicular liposomes (Fig. 11.1). Unilamellar liposomes are composed of a single lipid bilayer. The sizes of unilamellar liposomes vary between 25 and 400 nm in diameter (Hofheinz et al. 2005). Multilamellar liposomes are composed of several concentric lipid bilayers and are between 100 nm to several microns in diameter (Hofheinz et al. 2005). More recently a multivesicular liposomal platform, the DepoFoam particle, is introduced in the drug development (Kim and Sankaram 2000; Sankaram and Kim 1996). DepoFoam is an aqueous suspension of multivesicular liposomes (MVL) where the MVLs are formed by a double emulsion process forming a water-in-oil-in-water emulsion and differ in sizes from 1 to 100 μm.

The liposome carriers are often used to stabilize the encapsulated drugs in vivo and to avoid rapid systemic clearance (e.g., hepatic and/or renal elimination). For example, liposomal doxorubicin formulations, Doxil and Myocet, protect doxorubicin from hepatic clearance by stable encapsulation of the drug substance.

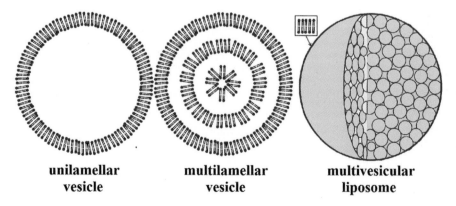

unilamellar vesicle **multilamellar vesicle** **multivesicular liposome**

Fig. 11.1 Schematic presentation of the internal structure of a unilamellar liposome, a multilamellar liposome, and a DepoFoam particle (a multivesicular liposome)

After intravenous administration, unilamellar and multilamellar liposome vesicles share one major route of elimination: the reticuloendothelial system (RES) clearance. RES uptake is mediated by opsinization after liposome binds to plasma proteins (Yan et al. 2005). Liposomes may also interact with plasma protein via lipid exchange, which in turn can compromise the integrity of the liposome membrane and result in drug release. Surface modification, especially PEGylation, has been shown to effectively prevent liposome opsinization. The design of liposome should be adjusted based on the intended use of drug product. When prolonged systemic exposure is desirable, PEGylation is likely a favorable option in the liposome design. Nevertheless, with PEGylated liposome, formulation scientists need to carefully evaluate the accelerated blood clearance phenomenon, defined as a loss of long-circulating characteristic upon repeated dosing (Ishida and Kiwada 2008). Ishida and colleagues show that the administration of PEGylated liposome induces the formation of anti-PEG IgM, which leads to unexpected immune-response and faster clearance of subsequent doses. The accelerated blood clearance phenomenon can be controlled by the physicochemical properties (e.g., particle size, surface modification) of injected liposomes and dosing regimen of liposomal drug product (Ishida and Kiwada 2008; Koide et al. 2008).

Liposomes with controlled sizes exhibit preferential accumulation in the tumor tissues via the enhanced permeability and retention effect. This is achieved because circulating liposome is not able to penetrate normal blood vessel due to its large size, but can leak out in areas of discontinuous capillaries, usually in the tumor tissues (Maeda 2001). Additionally, tumor tissues generally lack effective lymphatic drainage. Hence, after extravasation into the tumor tissue liposomes may accumulate at the site for a prolonged period. This enhanced permeability and retention effect can be used to reduce drug exposure at unwanted sites and increase therapeutic effect at the target sites.

In addition to improved pharmacokinetics and biodistribution profiles, the application of liposomes may be used to circumvent the multiple drug resistance phenomenon (Ma and Mumper 2013). By optimizing liposome composition and fine tuning its physicochemical properties, properly designed liposomal formulations are proposed to reverse the multiple drug resistance phenomenon by interacting with efflux transporter or by regulating MDR-related gene expression (Krishna and Mayer 1997; Pakunlu et al. 2006; Chen et al. 2010).

11.3   Properties of Liposomal Drug Products

Physicochemical and biological properties of the approved liposomal drug products vary largely depending on drug substance and liposome properties. Based on drug loading mechanism, the nine liposomal drug products currently on the market can be divided into three categories: active loading of weakly basic drug substances such as Doxil; passive loading of lipophilic drug substances such as AmBisome;

Table 11.2 Classification of approved liposomal drug products based on drug substance properties

Properties		Lipophilic molecules	Weakly basic amphiphilic molecules	Hydrophilic molecules
Morphology	Unilamellar	AmBisome, Visudyne	Doxil, Myocet, DaunoXome, Marqibo	–
	Multilamellar	Mepact	–	–
	Multivesicular	–	Exparel	DepoCyt
Drug loading mechanisms		Passive loading	Active loading or DepoFoam technique	DepoFoam technique
Drug location		Lipid bilayer	Aqueous interior	Aqueous interior
In vivo drug release		Rapid to intermediate	Intermediate to slow	Slow

and the DepoFoam technique for hydrophilic or weakly basic drug substances such as DepoCyt (Table 11.2). This section reviews the unique properties of different types of liposomal drug products based on drug loading mechanism.

11.3.1 Liposomal Drug Products by Active Loading Mechanism

Significant advances have been made in achieving desirable pharmacokinetics and drug release kinetics of weakly basic amphiphilic drugs (doxorubicin, daunorubicin, and vincristine) that are actively loaded into liposomal interior by a transmembrane pH gradient or ion gradient. The general manufacturing process of actively loaded liposomal drug products has been extensively studied (Lasic et al. 1992; Haran et al. 1993; Barenholz 2001) (Fig. 11.2). In the first step, liposomes are prepared in the presence of loading buffer (e.g., ammonium sulfate in the case of Doxil) and extruded to control liposome size. Then, the exterior loading buffer is exchanged with a second buffer which contains drug substance. The cation entrapped in the liposome interior plays a role in establishing a pH gradient across the membrane. The pH gradient drives the diffusion of deprotonated drug into the liposome interior and the drug immediately forms stable complex or precipitate with the counter ion present in the liposome interior. This further promotes the diffusion of deprotonated drug until almost all the drug has been transferred inside the liposomes. The formation of stable complex or the precipitation stabilizes the formulation and achieves sustained drug release. However, the success of active loading method is limited to weakly basic or acidic amphiphilic drugs. Highly lipophilic molecules are entrapped in lipid bilayers while highly hydrophilic molecules are unable to diffuse membrane. The controlled particle size, prolonged

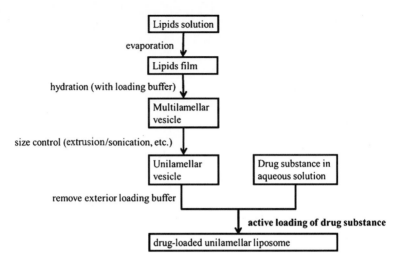

Fig. 11.2 Diagram of the typical manufacturing process of actively loaded liposomal drug products

Table 11.3 Approved liposomal drug products by active loading mechanism

Product	Drug substance	Indication	Particle size (nm)
Doxil	Doxorubicin hydrochloride	Ovarian cancer and Kaposi's sarcoma	~100
Myocet	Doxorubicin hydrochloride	Breast cancer	~180
DaunoXome	Daunorubicin	HIV-associated Kaposi's sarcoma	35–65
Marqibo	Vincristine sulfate	Leukemia	100

liposome circulation lifetimes, and sustained release of entrapped drugs results in drug accumulation in tumors and decreased systemic toxicity.

Among the approved liposomal drug products that were prepared by the active loading method, Doxil is most successful in the market. The drug substance, doxorubicin, belongs to an anthracycline family of compounds that are effective anticancer agents (Table 11.3). Based on the proposed mechanism, doxorubicin exerts its pharmacological effects by intercalating into DNA and disrupting of DNA repair, and by generating free radicals which damage cellular membranes (Drummond et al. 1999; Barenholz 2001). Cardiac toxicity is the most common and serious adverse effect observed in treatments with conventional doxorubicin HCl formulations. The stable encapsulation of doxorubicin and the tumor targeting effect of Doxil improve safety profile by reducing doxorubicin exposure in normal tissues.

The formulation composition of Doxil and physicochemical testing methods are available to the public. The lipid bilayer of Doxil is composed of cholesterol

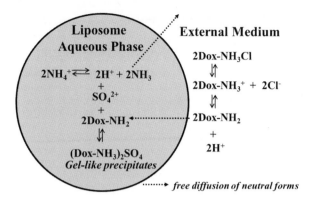

Fig. 11.3 The active loading mechanism in the manufacturing of Doxil. Adapted from *Bioanalysis* (2011) 3(3), 333–344 with permission of Future Medicine Ltd: Jiang et al. In vitro and in vivo characterizations of PEGylated liposomal doxorubicin (Jiang et al. 2011)

(3.19 mg/mL), fully hydrogenated soy phosphatidylcholine (HSPC, 9.58 mg/mL), and *N*-(carbonyl-methoxypolyethylene glycol 2000)-1,2-distearoyl-sn-glycero-3-phospho-ethanolamine (MPEG-DSPE, 3.19 mg/mL) (Janssen Products 2012). The loading buffer contains 250 mM ammonium sulfate to establish a pH gradient (Martin 2001a) (Fig. 11.3). The final product contains doxorubicin HCl at a concentration of 2 mg/mL (greater than 90 % as encapsulated), roughly 2 mg/mL ammonium sulfate, histidine as buffer, hydrochloric acid and/or sodium hydroxide for pH adjustment, and sucrose for maintaining isotonicity (Janssen Products 2012).

Plasma PK of Doxil and its relation to formulation variations have been described in many reports (Gabizon et al. 1994; Janssen Products 2012). Most circulating doxorubicin exists as encapsulated drug. Compared with conventional doxorubicin formulation, the AUC of doxorubicin, representing systemic exposure to the drug substance, increases by two to three orders of magnitude after Doxil injection. The plasma clearance of doxorubicin is slow after Doxil administration, with a mean clearance value of 0.041 L/h/m^2 at a dose of 20 mg/m^2. The volume of distribution of doxorubicin is small (2.7 L/m^2) at the dose of 20 mg/m^2. This is in contrast to conventionally dosed doxorubicin, which displays a plasma clearance value ranging from 24 to 35 L/h/m^2 and a large volume of distribution ranging from 700 to 1,100 L/m^2. The small steady state volume of distribution of Doxil shows that Doxil is confined to the vascular fluid volume. Compared with conventional doxorubicin formulation, Doxil also exhibits special PK behavior including the accelerated blood clearance phenomenon (Laverman et al. 2001) in preclinical models and preferential accumulation in the tumor tissues via the enhanced permeability and retention effect.

Myocet is a non-PEGylated liposomal doxorubicin injection product, which was approved in Europe and Canada for treatment of metastatic breast cancer in combination with cyclophosphamide. Myocet has not been approved by the FDA for sale on US market. Myocet liposome is a large unilamellar vesicle composed of

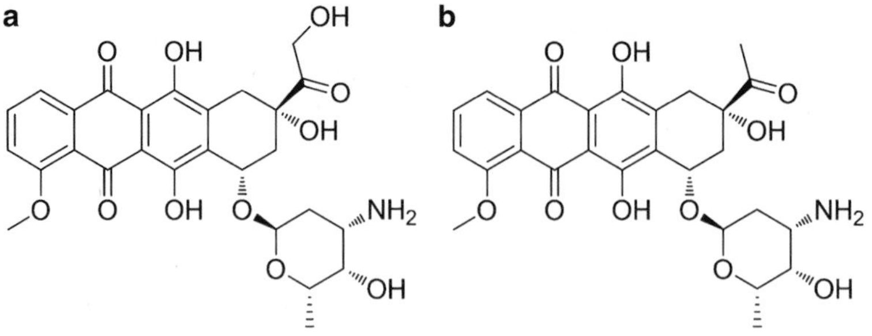

Fig. 11.4 Structure of (**a**) doxorubicin and (**b**) daunorubicin

egg phosphatidylcholine and cholesterol at molar ratio of 55:45 (Swenson et al. 2001). Doxorubicin is actively loaded into aqueous interior by a pH gradient as established with citric acid loading buffer. Inside liposome, doxorubicin forms organized fiber buddle. Overall the drug to lipid ratio is 0.27 (w/w) and over 95 % of drug substance is encapsulated (Batist et al. 2002). Due to the different lipid composition, surface properties, particle size, and liposomal internal environment, the pharmacokinetics, tissue distribution, and in vivo drug release are different between Doxil and Myocet. In dogs receiving a single slow bolus injection of 1.5 mg/kg Myocet or conventional doxorubicin formulation, the AUC for doxorubicin after Myocet injection is roughly 100 times that of conventional doxorubicin formulation (Kanter et al. 1993, 1994). Compared with Doxil, the clearance of Myocet is much more rapid and the half-life of Myocet is much shorter (i.e., less than 0.2 h in small rodents) (Martin 2001a; Swenson et al. 2001). Eighty-five to 93 % of circulating doxorubicin remains encapsulated after Myocet administration. Due to the shorter liposome circulation lifetime, it is unlikely that Myocet could effectively penetrate into skin tissues (Alberts and Garcia 1997). Consequently, Myocet has a lower incidence of palmar-plantar erythrodysaesthesia than Doxil.

DaunoXome is a liposomal preparation of daunorubicin and indicated as a first-line cytotoxic therapy for advanced HIV-associated Kaposi's sarcoma (DailyMed 2013). Daunorubicin is an analogue of doxorubicin and has similar physicochemical properties (Fig. 11.4). DaunoXome liposome is unilamellar and composed of cholesterol and distearoylphosphatidylcholine at a molar ratio of 1:2 (Forssen 1997). Similar to Myocet, daunorubicin was loaded into DaunoXome liposome via a pH gradient established by low pH citric acid in the liposome interior and relatively higher pH in the outside. Compared with conventional, non-liposomal daunorubicin formulation, DaunoXome has a slower clearance (17.3 vs. 236 mL/min), smaller volume of distribution (6.4 vs. 1,006 L), and longer distribution half-life (4.41 vs. 0.77 h) in human. It also demonstrates reduced toxicity and comparable antitumor efficacy compared with conventional formulation.

Marqibo is a vincristine sulfate liposomal injection for the treatment of adult patients with Philadelphia chromosome-negative acute lymphoblastic leukemia. Similar to doxorubicin, the conventional vincristine formulation results in diffusive

Table 11.4 Properties of liposomal drug products by passive loading mechanism

Product	Drug substance	Indication	Particle size
AmBisome	Amphotericin B	Fungal infections	45–80 nm
Mepact	Mifamurtide	Osteosarcoma	1–5 μm
Visudyne	Verteporfin	Blood vessel disorders in the eye	18–104 nm

distribution and extensive tissue binding, thereby limiting optimal drug exposure and tumor targeting (Silverman and Deitcher 2013). Marqibo was designed to increase the circulation time and preferential drug delivery into tumor tissues. The liposome is unilamellar and its size ranges from 90 to 140 nm in diameter. The lipid bilayer of Marqibo is comprised of sphingomyelin and cholesterol in a molar ratio of 58:42 (FDA 2012). Different from other liposomal drug products in this category, Marqibo is presented to end users as a 3-vial kit (Talon Therapeutics 2012): a vial containing vincristine sulfate injection, 5 mg/5 mL; a vial containing sphingomyelin/cholesterol liposome injection, 103 mg/mL; and a vial containing sodium phosphate injection, 355 mg/25 mL. Drug encapsulation is performed at the medical facilities under instructions provided by the manufacturer.

11.3.2 Liposomal Drug Products by Passive Loading Mechanism

At present, three intravenously administrated liposomal drug products containing a poorly water-soluble active ingredient are available in the market for human use (Table 11.4). These include AmBisome (amphotericin B), Mepact (mifamurtide), and Visudyne (verteporfin). In a passive loading process, the active substances are co-dispersed with the lipids and are entrapped in the lipid bilayer during liposome formation (Cullis et al. 1988; Akbarzadeh et al. 2013). Three types of methods are often used in the passive loading process: the mechanical dispersion method (e.g., sonication, membrane extrusion), the solvent dispersion method (e.g., ethanol injection, solvent vaporization), and the detergent removal method (e.g., dialysis, gel-permeation). The specific passive loading process should be designed based on properties of the active ingredients such as lipophilicity and active ingredient–lipid interaction.

The in vivo release rates of drug substances passively loaded are usually much faster than the drug substance actively loaded in the aqueous interior. The in vivo release rates of poorly water-soluble compounds depend on their binding affinity to liposomal bilayer. Some lipophilic drugs do not bind to liposomal lipids with high affinity, but are entrapped in liposomal bilayer due to hydrophobic interaction (Fahr et al. 2005; Shabbits et al. 2002). Upon intravenous injection, these lipophilic drugs will be immediately released and bind to serum lipoproteins, blood cell membranes, and endothelial cell membranes. For these lipophilic drugs, liposomes serve as a solubilizer.

In contrast, compounds with high lipophilicity and a lipid-like structure are likely to exhibit extended in vivo release when entrapped in intravenously administered liposomes. These lipid-like compounds tightly bind to the liposomal membrane as a part of the bilayer (Awiszus and Stark 1988). The binding affinity to liposomal membrane depends on the lipid compatibility and lipophilicity of entrapped drug substances. For example, short-chain ceramide with a six-carbon side chain showed rapid in vivo release from liposomes after intravenous administration while long-chain ceramide with a 16-carbon side chain exhibited higher binding affinity to liposomal membrane (Shabbits et al. 2002; Zolnik et al. 2008). The release rate of a lipid-like molecule also depends on the physical state of liposomal bilayer. For example, amphotericin B was almost completely transferred from liposomes composed of dimyristoyl phosphatidylcholine (DMPC) and dimyristoyl phosphatidylglycerol (DMPG) to serum lipoproteins at 37 °C within 5 min (Wasan et al. 1993). With a phase transition temperature (T_m) of 23 °C, the DMPC/DMPG bilayer is in liquid crystalline state at 37 °C, which explains quick release of amphotericin B in serum. Amphotericin B and DMPG may be co-transferred to serum lipoproteins. In contrast, AmBisome liposomes are composed of hydrogenated soy phosphatidylcholine, distearoyl-phosphatidylglycerol, cholesterol, and alpha tocopherol and exhibit a T_m of approximately 55 °C (Iman et al. 2011). Only 5 % of amphotericin B was released when AmBisome was incubated in 50 % human plasma for 72 h at 37 °C (Adler-moore and Proffitt 1993), indicating the slow release of amphotericin B from gel state lipid bilayer.

AmBisome is a unilamellar liposome loaded with amphotericin B for intravenous infusion and treatment of fungi infection. As described in the drug label (Gilead Sciences 2012), the success of AmBisome is inherently correlated to the fact that amphotericin B is a membrane-bound drug by nature, which can easily be seen by its lipid-like polyene structure. Amphotericin B is intercalated within the liposomal bilayer and the lipophilic moiety in the amphotericin B molecule is an integral part of the liposomal bilayer. Amphotericin B interacts more strongly with fungal ergosterol than with liposomal phospholipids. The interaction of AmBisome with fungal membrane will stimulate drug release from liposomes. For AmBisome, the liposome vesicle is not only a drug solubilizer but also a targeted delivery carrier. Human pharmacokinetic study of AmBisome showed that the central compartment volume of amphotericin B is 50 mL/kg, which is similar to human plasma volume. This finding suggests that initial burst release of amphotericin B is limited (Bekersky et al. 2002). However, it is worthy to mention that after intravenous administration of ^{14}C-cholesterol-labeled AmBisome, plasma concentration ratio of amphotericin B and liposomal ^{14}C-cholesterol rapidly decreased by tenfold during the first 36 h (Bekersky et al. 2001), suggesting a faster clearance of amphotericin B than the liposomes and an intermediate release rate of amphotericin B with a complete release within 36 h.

Mepact is a multilamellar liposomal formulation with active substance mifamurtide entrapped in liposomal bilayers (Meyers 2009). Mepact was approved by the EMA in 2009 for the treatment of osteosarcoma. Mifamurtide (muramyl tripeptide phosphatidyl ethanolamine, MTP-PE) is a synthetic derivative of muramyl dipeptide (MDP), the smallest naturally occurring immune stimulatory component in bacterial cell walls (Meyers 2009). MTP-PE has similar immunostimulatory effects as natural MDP. MTP-PE is a potent activator of monocytes and macrophages. Activation of human macrophages by MTP-PE induces the production of cytokines, which can selectively kill tumor cells (EMA 2009). Large multilamellar liposomes (1–5 µm) are used as a carrier to deliver MTP-PE to macrophages and monocytes due to the rapid uptake by RES. MTP-PE is synthesized by coupling hydrophilic MDP and lipophilic dipalmitoyl phosphatidyl ethanolamine. Due to their high lipophilicity and lipid compatibility, the palmitoyl chains of MTP-PE can anchor MTP-PE into liposomal bilayers with high affinity. Hence, preclinical and clinical pharmacokinetic studies showed that both liposomes and MTP-PE were rapidly cleared from circulation after intravenous administration and distributed to liver, spleen, nasopharynx, thyroid, and lung (a RES uptake dependent distribution), suggesting that a majority of MTP-PE is associated with liposomes in vivo. However, free MTP-PE was detected in plasma samples indicating leakage of MTP-PE from liposomes even during their very short residence time in the blood circulation.

Unlike lipid-like amphotericin B and mifamurtide, verteporfin doesn't bind to liposomal bilayer with high affinity. In vitro release studies revealed immediate and complete transfer of verteporfin from Visudyne liposomes to serum proteins (Chowdhary et al. 2003). In addition, human pharmacokinetic study showed a relatively large volume of distribution of verteporfin with the liposomal formulation (0.6 L/kg), indicating an extensive extravascular distribution of released verteporfin (Novartis 2012). Furthermore, biodistribution study showed that the liposome formulation did not cause accumulation of verteporfin in mouse liver, lung, and spleen compared with DMSO solubilized verteporfin (FDA 1999). All the results revealed no verteporfin retention in liposomes after intravenous administration. Liposomes only serve as a verteporfin solubilizer and Visudyne can be regarded as a parenteral solution.

11.3.3 Liposomal Drug Products by the DepoFoam Technique

To achieve extended drug release, hydrophilic drug substances can be passively loaded into liposomal aqueous interior. Typically, liposomes are formed upon hydration of a dry lipid film and a hydrophilic drug substance dissolved in hydration medium is loaded into liposomal interior. However, drug loading efficiency is usually very low and it is mainly dependent on lipid concentration and liposome size (Xu et al. 2012). Increasing liposome size can improve drug loading efficiency but may cause instability of liposomes. DepoFoam is a technology for preparation of MVLs (Mantripragada 2002) (Fig. 11.5). To prepare hydrophilic drug loaded MVLs, the drug substance is dissolved in an acidic aqueous solution. Meanwhile,

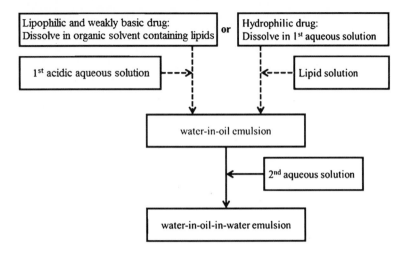

Fig. 11.5 Diagram of the typical manufacturing process of liposomal drug products by the DepoFoam technique. *Dashed lines* means the process may be elected based on drug substance property (i.e., lipophilic or hydrophilic)

Table 11.5 Properties of liposomal drug products by DepoFoam technique

Product	Drug substance	Indication	Particle size (μm)
DepoCyt	Cytarabine	Lymphomatous meningitis	17–23
Exparel	Bupivacaine	Postsurgical analgesia	24–31

lipid solution containing cholesterol, neutral triglyceride, and lipids is formulated in chloroform. The two solutions are combined and mixed at high speeds to obtain a water-in-oil (w/o) emulsion. Next, the w/o emulsion is mixed with a second aqueous phase to form a water-in-oil-in-water (w/o/w) double emulsion. Chloroform is removed under nitrogen gas flow, causing the lipid layer to collapse around the first (drug-containing) aqueous droplets, forming the MVLs. MVLs contain numerous nonconcentric chambers filled with drug solution. The large size (10–30 μm) and the non-lamellar honeycomb structure of MVLs result in a high drug loading efficiency and improved liposomal stability. Meanwhile, the large size of MVLs precludes their intravenous administration. The routes of administration most viable for MVLs include intrathecal, epidural, subcutaneous, intramuscular, intra-articular, and intraocular injection (Mantripragada 2002). Three approved MVL drug products prepared by the DepoForm technique are DepoCyt (cytarabine), DepoDur (morphine sulfate), and Exparel (bupivacaine). DepoDur was discontinued in the US market (Table 11.5).

A key ingredient that governs the formation of MVLs over other liposomes is the inclusion of neutral triglycerides in the formulation, which facilitates the formation of membrane corners between vesicles through reduction of internal membrane surface tension (Mantripragada 2002). This provides increased stability of the honeycomb structure. In addition, the triglycerides can affect the volume and encapsulation efficiency of the aqueous components. Drug release rate depends on erosion and/or reorganization of the lipid membranes which is further influenced by the length of the fatty acid chain of the triglycerides. Neutral triglycerides with long fatty acid chains have been shown to reduce drug release rate, in comparison to short-chain triglycerides (Willis 1999). The drug release profiles of MVLs can be tailored by the use of mixtures of long-chain and short-chain triglycerides. In addition to triglycerides, the osmolarity of interior aqueous phase has also been shown to influence drug release. Increasing osmolarity in the first aqueous phase can decrease drug release rate from MVLs (Sankaram and Kim 1996). Similarly important, the selection of acid (counter ion) and the concentration of the acid in the first aqueous phase of MVL manufacturing can affect drug encapsulation efficiency and drug release rate (Mantripragada 2002).

DepoCyt is a sustained-release MVL formulation of the active ingredient cytarabine indicated for intrathecal treatment of lymphomatous meningitis (Sigma-Tau Pharmaceutical 2011). Cytarabine is a cell cycle phase-specific anti-neoplastic agent, affecting cells only during the S-phase of cell division. DepoCyt MVLs are composed of dioleoylphosphatidylcholine (DOPC, 5.7 mg/mL), dipalmitoylphosphatidylglycerol (DPPG, 1.0 mg/mL), cholesterol (4.4 mg/mL), and triolein (1.2 mg/mL) (Sigma-Tau Pharmaceutical 2011). Cytarabine is freely soluble in aqueous solution and can be passively encapsulated in nonconcentric chambers of MVLs at a concentration of 10 mg/mL. The half-life of intrathecally administered, conventional cytarabine injection in human cerebrospinal fluid (CSF) is 3.4 h. With the aid of triolein, a triglyceride, DepoCyt extends the terminal half-life of cytarabine in human CSF to 5.9–82.4 h (Sigma-Tau Pharmaceutical 2011). The cytotoxic CSF concentrations are maintained for prolonged periods and thus inhibit proliferation of lymphoma cells. Systemic exposure to cytarabine is negligible following intrathecal administration of 50 mg of DepoCyt. Due to the abnormal CSF flow in lymphomatous meningitis patients and the differences in patient posture, there are large inter- and intra-patient variations of cytarabine exposure in CSF after intrathecal administration of DepoCyt (Phuphanich et al. 2007). The patient variations pose a challenge for demonstrating in vivo bioequivalence between two cytarabine liposome formulations.

Exparel is a MVL injection of bupivacaine, an amide local anesthetic, indicated for single-dose infiltration into the surgical site to produce postsurgical analgesia (Pacira Pharmaceuticals 2011). Ingredients of Exparel and their nominal concentrations are: bupivacaine (13.3 mg/mL), cholesterol (4.7 mg/mL), DPPG (0.9 mg/mL) tricaprylin (2.0 mg/mL), and 1, 2-dierucoylphosphatidylcholine (DEPC, 8.2 mg/mL) (Pacira Pharmaceuticals 2011). DEPC is used to replace DOPC in DepoCyt formulation to facilitate the formation of MVL and to enhance the stability of MVL structure. To improve encapsulation efficiency, the loading method of bupivacaine is modified (FDA 2011). Lipophilic bupivacaine in free base form is dissolved in methylene chloride solution containing lipids instead of aqueous solution (Fig. 11.5). An aqueous solution of phosphoric acid is introduced into the lipid solution to obtain water-in-oil (w/o) emulsion. The phosphoric acid protonates bupivacaine, facilitating its partition into the aqueous phase. Then, a second aqueous phase (containing dextrose and lysine, pH 9.8) is added to form a water-in-oil-in-water (w/o/w) emulsion. The phosphoric acid in the first aqueous solution is neutralized. After removing organic solvent, MVLs loaded with bupivacaine are obtained. Local infiltration of Exparel results in significant systemic plasma levels of bupivacaine which can persist for 96 h.

11.4 Bioequivalence Recommendation for Doxorubicin Hydrochloride Liposome Injection

As discussed above, current liposomal drug products cover a diversity of drug substances, physicochemical properties, and in vitro/in vivo release behaviors. Individual bioequivalence recommendations should be developed based on the unique properties of each product. Using PEGylated liposomal doxorubicin hydrochloride injection as an example, this section reviews the critical considerations in developing bioequivalence guidance for liposomal drug products in the US market.

11.4.1 Physicochemical Characterization

Similar to BE requirements for conventional injection products, generic PEGylated liposomal doxorubicin HCl injection products must be qualitatively (Q1) and quantitatively (Q2) the same as the RLD, except differences in buffers, preservatives, and antioxidants provided that the applicant identifies and characterizes these differences and demonstrates that the differences do not impact the safety/efficacy profile of the drug product. Lipid excipients are critical in the liposome formulation. FDA recommends generic applicant to use lipids from the same category of synthesis route (natural or synthetic) and have similar specifications to lipids used in the RLD (FDA 2010).

In addition to the Q1/Q2 requirement, FDA recommends the demonstration of equivalent critical liposome properties, such as surface modification, surface charge, lipid content, and lipid bilayer transition profiles, particle size distribution, and the number of lamellae. Studies have shown that changes in liposome properties can profoundly alter the in vivo fate of doxorubicin-liposome complex by influencing the stability of lipid bilayer, the location of in vivo release, and the release rate of encapsulated drug. The following liposome characteristics are explicitly stated in the FDA draft BE guidance for PEGylated liposomal doxorubicin HCl injection products (FDA 2010):

- Liposome composition, including lipid content, free and encapsulated drug, internal and total sulfate and ammonium concentration, histidine concentration, and sucrose concentration.

- State of encapsulated drug. The formation of stable precipitation inside liposomes is critical for achieving stable formulation and extended drug release.
- Internal environment, including volume, pH, sulfate, and ammonium ion concentration. The internal environment maintains the state of encapsulated drug.
- Liposome morphology and the number of lamellae. Liposome morphology and lamellarity should be determined as drug loading, drug retention, and liposome interaction with RES cells are likely influenced by the morphology and lamellarity.
- Lipid bilayer phase transitions. Equivalence in lipid bilayer phase transitions contributes to equivalence in bilayer fluidity and uniformity as well as equivalence in the stability of encapsulation and the rate of drug release from liposomes. For example, doxorubicin encapsulated in liposomes with different lipid composition resulted in different membrane fluidity and thus demonstrated variations in vitro release rates (Charrois and Allen 2004). In vivo study showed that these liposomes were cleared from the body at different rates.
- Liposome size distribution. Particle size distribution of the liposome is a critical property that predicts in vivo biodistribution pattern. Compared with free doxorubicin, Doxil demonstrates preferential distribution into target site and reduced systemic toxicity (Gabizon 2001; Judson et al. 2001; O'Brien et al. 2004). The mechanism for the tumor accumulation is that leaky microvasculatures and impaired lymphatics at tumor sites allow long-circulating liposomes in this size range (~100 nm) to extravasate into and accumulate within tumors tissues (Martin 2001b; Reynolds et al. 2012). Because of this particle-size related tissue distribution mechanism, equivalent mean particle size and size distribution should be always required so that plasma PK will reflect efficacy/toxicity in targeted tissues. Cui and colleagues have demonstrated in mice that liposomes with different size and internal ammonium sulfate concentration could have similar plasma concentration but different tissue distribution and clinical effectiveness (Cui et al. 2007). Charrois and Allen examined the tissue distribution of liposomes of different sizes in tumor-bearing rats. It was observed that the large liposomes (241 nm in diameter) demonstrated different and remarkably lower accumulation in tumor than the smaller liposomes (82, 101, and 154 nm in diameter) (Charrois and Allen 2003). To demonstrate equivalent particle size distribution, generic applicants are recommended to perform in vitro BE study during which at least three lots of both test and reference products will be evaluated using proper analytical methods. Bioequivalence will be evaluated using the population bioequivalence criterion and will be determined by the 95 % upper confidence interval on D50 and SPAN (D90–D10)/D50 or polydispersity index.
- Grafted PEG at the liposome surface. Surface modification, such as PEGylation and coating, may substantially change plasma clearance of encapsulated drugs. As described in previous sections, the plasma clearance of doxorubicin varies based on the formulation in the following order: Doxil (PEGylated liposome) < Myocet (non-PEGylated liposome) < freely dosed doxorubicin HCl. The slow plasma elimination of Doxil is partially attributed to the protective effect of

surface-bound polymer coating. MPEG-DSPE provides the steric stabilization of doxorubicin-liposome injection. MPEG polymer coating also protects liposomes from clearance by RES, reduces liposome adhesion to cells and blood vessel walls, and thereby increases blood circulation time (Senior et al. 1991; Ishida et al. 2002).

• Electrical surface potential or charge. Surface charge is believed to be responsible for mediating RES uptake, tumor targeting, and cellular uptake of liposome and its encapsulated drugs. After reaching the target tissue, encapsulated drug needs to be picked up by tumor cells to exert its effect. Although the exact mechanism of drug release is not fully understood, three theories with regard to drug release from the liposomes are widely endorsed (Drummond et al. 1999; EMA 2005; Minko et al. 2006): (1) the drug leaks very slowly and thus the liposome provides a slow infusion of the drug specifically near the cells; (2) fusion of the liposomal membrane with the cell membrane; and (3) cellular uptake of intact liposomes by tumor cells themselves or by tumor macrophages. Whichever the mechanism, modification to surface charge may affect liposome stability as well as liposome–cell interaction, thereby altering cellular uptake of encapsulated drugs. For example, studies have shown that neutral PEGylated liposomes were more efficiently internalized by proliferating tumor cells than negatively charged liposomes (Miller et al. 1998). Krasnici et al. revealed that cationic liposomes was preferred for delivering drugs into angiogenic tumor vessels whereas anionic and neutral liposomes may be preferred to targeting the extravascular compartment of tumor (Krasnici et al. 2003).

In order to meet the comprehensive equivalence tests as outlined above, FDA expects generic applicant to use the same active loading process with ammonium sulfate gradient. The four major steps are described in the draft BE guidance: (1) formation of liposomes containing ammonium sulfate, (2) liposome size reduction, (3) creation of ammonium sulfate gradient, and (4) active drug loading. An active loading process uses an ammonium sulfate concentration gradient between the liposome interior and the exterior environment to drive the diffusion of doxorubicin into the liposomes.

11.4.2   In Vivo PK study

Although the sameness in the comprehensive physicochemical characterization should be able to guarantee the sameness in the drug products, in vivo BE studies are also recommended as a conservative measure for most liposomal drug products. The analytes to be measured and the proper sampling of biofluid should be determined on a case-by-case basis, depending on the nature of drug substance, the mechanism of action, and the unique PK behavior of individual product.

After intravenous injection of Doxil, the PK profile of encapsulated doxorubicin is indicative of systemic clearance and availability of encapsulated drugs for delivery to the target tissues. The level of free doxorubicin is indicative of in vivo stability as well as the availability of toxic species at the nontarget sites. Because both free and encapsulated doxorubicin have physiological meaning, the FDA draft BE guidance on PEGylated liposomal doxorubicin HCl injection requires measurement of both free and encapsulated doxorubicin in a two-way crossover study. BE should be established based on the 90 % confidence interval of Cmax and AUC of both species. Because doxorubicin is a cytotoxic compound, a bio-IND is required before the in vivo PK study. Special cautions should be taken in determining the study population for the in vivo PK study.

11.4.3 In Vitro Release Studies

In vivo drug release kinetics, in circulation or at the site of action, is critical for correlating plasma pharmacokinetics with the therapeutic effects of PEGylated liposomal doxorubicin HCl injection. As a result, in vitro drug release studies under biorelevant conditions are recommended as an indispensable part of bioequivalence guidance.

When designing an in vitro release assay for drug products, it is ideal to select appropriate conditions and method that can distinguish formulation differences, demonstrate method robustness, and allow for good in vivo predictability. The conventional in vitro release method relies on dialyzing the liposomal formulation against a large volume of buffer, serum protein solution, or serum at physiological temperatures. Even though the dialysis method has certain advantages and provides important information, it is often not a good predictor of in vivo drug release (Shabbits et al. 2002). The poor in vivo predictability of current dialysis method is probably caused by its inability to mimic the following in vivo conditions: (1) the large membrane pool present in blood cells and endothelial cells; (2) drug release in macrophages after RES uptake; and (3) elimination of released drug through hepatic metabolism and urinary excretion.

To circumvent these limitations, modified in vitro release assays have been proposed. One such method is to conduct in vitro release study in the presence of excess acceptor multilamellar (Shabbits et al. 2002) or unilamellar (Fahr et al. 2005; Hefesha et al. 2011) vesicles. After incubation, the donor and acceptor liposomes are separated by centrifugation or an ion exchange column. The rate of drug transfer from donor to acceptor liposomes depends on incubation temperature, ratio of donor and acceptor liposomes, and liposomal properties (Hefesha et al. 2011). The in vitro release kinetics can be optimized by changing the ratio of donor and acceptor liposomes. Using the modified method, an improved in vitro and in vivo correlation of release kinetics has been observed for liposomal encapsulated doxorubicin, verapamil, cyclosporine, and paclitaxel (Shabbits et al. 2002; Fahr et al. 2005; Hefesha et al. 2011). Additionally, serum proteins such as albumin was found to facilitate drug release from liposomes (Makriyannis et al. 2005).

Table 11.6 Examples of in vitro leakage conditions to compare generic PEGylated liposomal doxorubicin injection to its reference listed drug

In vitro drug leakage condition	Purpose	Rationale
At 37 °C in 50 % human plasma for 24 h	Evaluate liposome stability in blood circulation	Plasma mostly mimics blood conditions
At 37 °C with pH values 5.5, 6.5, and 7.5 for 24 h in buffer	Mimic drug release in normal tissues, around cancer cells, or inside cancer cells	Normal tissues: pH 7.3; Cancer tissues: pH 6.6; Inside cancer cells (endosomes and lysosomes): pH 5–6
At a range of temperatures (43, 47, 52, 57 °C) in pH 6.5 buffer for up to 12 h or until complete release	Evaluate the lipid bilayer integrity	The phase transition temperature (T_m) of lipids is determined by lipid bilayer properties such as rigidity, stiffness, and chemical composition. Differences in release as a function of temperature (below or above T_m) will reflect small differences in lipid properties
At 37 °C under low-frequency (20 kHz) ultrasound for 2 h or until complete release	Evaluate the state of encapsulated drug in the liposome	Low-frequency ultrasound (20 kHz) disrupts the lipid bilayer via a transient introduction of pore-like defects and will render the release of doxorubicin controlled by the dissolution of the gel inside the liposome

Therefore, in vitro drug release in the presence of both acceptor liposomes and serum proteins might mimic in vivo conditions.

To be actively loaded into aqueous interior, amphipathic drugs must have high partition coefficient between aqueous phase and membrane bilayer. They can readily cross liposomal membranes and cell membranes (Drummond et al. 1999). In addition, release rates of amphipathic drugs are sensitive to the pH gradient across liposomal bilayer. Therefore, generic applicants are recommended to investigate the release of doxorubicin under pH 6.5 (to mimic pH in solid tumor environment) and pH 5.5 (to mimic pH in lysosomes of cancer cells).

Based on the in vivo release mechanisms and physicochemical properties, FDA provides the following in vitro leakage conditions as an example in the draft BE guidance for the PEGylated liposomal doxorubicin HCl injection (Table 11.6):

11.5 Summary

In February 2010, FDA published a draft BE guidance on generic doxorubicin hydrochloride liposome injection (FDA 2010). This is the first BE guidance on a generic liposomal drug product. On Feb 4 of 2013, FDA approved Sun Pharma's Lipodox as the first generic version of Doxil based on its pharmaceutical equivalence and bioequivalence to Doxil.

Given the diversity and complexity of liposomal drug products, we recognize tremendous opportunities for regulatory agencies in developing BE regulations for these products. A significant portion of on market products still lacks product-specific BE guidance. Product-specific BE guidance is important because products in the same category may have subtle different properties that require different in vivo BE study design, in vitro release/dissolution test, and/or physical and chemical characterization methods. For example, for liposomal products prepared by a different loading mechanism from Doxil, changes in the in vitro testing conditions are expected depending on lipophilicity of the encapsulated drugs and their sites of action.

References

Adler-Moore JP, Proffitt RT (1993) Development, characterization, efficacy and mode of action of Ambisome, a unilamellar liposomal formulation of amphotericin B. J Liposome Res 3:429–450

Akbarzadeh A, Rezaei-Sadabady R, Davaran S, Joo SW, Zarghami N, Hanifehpour Y, Samiei M, Kouhi M, Nejati-Koshki K (2013) Liposome: classification, preparation, and applications. Nanoscale Res Lett 8:102

Alberts DS, Garcia DJ (1997) Safety aspects of pegylated liposomal doxorubicin in patients with cancer. Drugs 54(suppl 4):30–35

Awiszus R, Stark G (1988) A laser-T-jump study of the adsorption of dipolar molecules to planar lipid membranes. II. Phloretin and phloretin analogues. Eur Biophys J 15:321–328

Bangham AD, Horne RW (1964) Negative staining of phospholipids and their structural modification by surface-active agents as observed in electron microscope. J Mol Biol 8:660–668

Barenholz Y (2001) Liposome application: problems and prospects. Curr Opin Colloid Interface Sci 6:66–77

Batist G, Barton J, Chaikin P, Swenson C, Welles L (2002) Myocet (liposome-encapsulated doxorubicin citrate): a new approach in breast cancer therapy. Expert Opin Pharmacother 3:1739–1751

Bekersky I, Fielding RM, Dressler DE, Kline S, Buell DN, Walsh TJ (2001) Pharmacokinetics, excretion, and mass balance of 14C after administration of 14C-cholesterol-labeled Am Bisome to healthy volunteers. J Clin Pharmacol 41:963–971

Bekersky I, Fielding RM, Dressler DE, Lee JW, Buell DN, Walsh TJ (2002) Pharmacokinetics, excretion, and mass balance of liposomal amphotericin B (AmBisome) and amphotericin B deoxycholate in humans. Antimicrob Agents Chemother 46:828–833

Charrois GJ, Allen TM (2003) Rate of biodistribution of STEALTH liposomes to tumor and skin: influence of liposome diameter and implications for toxicity and therapeutic activity. Biochim Biophys Acta 1609:102–108

Charrois GJ, Allen TM (2004) Drug release rate influences the pharmacokinetics, biodistribution, therapeutic activity, and toxicity of pegylated liposomal doxorubicin formulations in murine breast cancer. Biochim Biophys Acta 1663:167–177

Chen Y, Bathula SR, Li J, Huang L (2010) Multifunctional nanoparticles delivering small interfering RNA and doxorubicin overcome drug resistance in cancer. J Biol Chem 285:22639–22650

Chowdhary RK, Shariff I, Dolphin D (2003) Drug release characteristics of lipid based benzoporphyrin derivative. J Pharm Pharm Sci 6:13–19

Cui J, Li C, Guo W, Li Y, Wang C, Zhang L, Hao Y, Wang Y (2007) Direct comparison of two pegylated liposomal doxorubicin formulations: is AUC predictive for toxicity and efficacy? J Control Release 118:204–215

Cullis PR, Mayer LD, Bally MB, Madden TD, Hope MJ (1988) Generating and loading of liposomal systems for drug-delivery applications. Adv Drug Deliv Rev 3:267–282

Dailymed (2013) http://dailymed.nlm.nih.gov/dailymed/lookup.cfm?setid=bf2b4eaf-5227-42d9-af78-88d921e565f6 [Online]. Accessed 2013

Drummond DC, Meyer O, Hong K, Kirpotin DB, Papahadjopoulos D (1999) Optimizing liposomes for delivery of chemotherapeutic agents to solid tumors. Pharmacol Rev 51:691–743

EMA (2005) Review of Caelyx. http://www.ema.europa.eu/docs/en_GB/document_library/EPAR_-_Scientific_Discussion/human/000089/WC500020175.pdf [Online]. Accessed Aug 2013

EMA (2009) Assessment report for Mepact. http://www.ema.europa.eu/docs/en_GB/document_library/EPAR_-_Public_assessment_report/human/000802/WC500026564.pdf [Online]. Accessed 2013

Evaluatepharma (2013) http://www.evaluategroup.com/Public/EvaluatePharma-Overview.aspx [Online]. Accessed 2013

Fahr A, van Hoogevest P, May S, Bergstrand N, S Leigh ML (2005) Transfer of lipophilic drugs between liposomal membranes and biological interfaces: consequences for drug delivery. Eur J Pharm Sci 26:251–265

FDA (1999) Nonclinical pharmacology and toxicology review for verteporfin for injection. [Online]. Accessed 2013

FDA (2002) Guidance for industry: liposome drug products (Draft). http://www.fda.gov/downloads/Drugs/GuidanceComplianceRegulatoryInformation/Guidances/ucm070570.pdf [Online]. Accessed 2013

FDA (2010) Bioequivalence recommendation for doxorubicin hydrochloride (Draft). http://www.fda.gov/downloads/Drugs/GuidanceComplianceRegulatoryInformation/Guidances/UCM199635.pdf [Online]. Accessed 2013

FDA (2011) Exparel (bupivacaine liposomal injection) chemistry review #1. http://www.accessdata.fda.gov/drugsatfda_docs/nda/2011/022496Orig1s000ChemR.pdf [Online]. Accessed 2013

FDA (2012) Marqibo® (vincristine sulfate liposomes injection) for the treatment of advanced relapsed and/or refractory Philadelphia chromosome negative (PH-) adult acute lymphoblastics leukemia: sponsor briefing documents. http://www.fda.gov/downloads/AdvisoryCommittees/CommitteesMeetingMaterials/Drugs/OncologicDrugsAdvisoryCommittee/UCM296424.pdf [Online]. Accessed 2013

Forssen EA (1997) The design and development of DaunoXome(R) for solid tumor targeting in vivo. Adv Drug Deliv Rev 24:133–150

Gabizon AA (2001) Pegylated liposomal doxorubicin: metamorphosis of an old drug into a new form of chemotherapy. Cancer Invest 19:424–436

Gabizon A, Catane R, Uziely B, Kaufman B, Safra T, Cohen R, Martin F, Huang A, Barenholz Y (1994) Prolonged circulation time and enhanced accumulation in malignant exudates of doxorubicin encapsulated in polyethylene-glycol coated liposomes. Cancer Res 54:987–992

Gilead Sciences, Inc. (2012) AmBisome label. http://www.accessdata.fda.gov/drugsatfda_docs/label/2012/050740s021lbl.pdf [Online]. Accessed 2013

Haran G, Cohen R, Bar LK, Barenholz Y (1993) Transmembrane ammonium sulfate gradients in liposomes produce efficient and stable entrapment of amphipathic weak bases. Biochim Biophys Acta 1151:201–215

Hefesha H, Loew S, Liu X, May S, Fahr A (2011) Transfer mechanism of temoporfin between liposomal membranes. J Control Release 150:279–286

Hofheinz RD, Gnad-Vogt SU, Beyer U, Hochhaus A (2005) Liposomal encapsulated anti-cancer drugs. Anticancer Drugs 16:691–707

Iman M, Huang Z, Szoka FC Jr, Jaafari MR (2011) Characterization of the colloidal properties, in vitro antifungal activity, antileishmanial activity and toxicity in mice of a di-stigma-steryl-hemi-succinoyl-glycero-phosphocholine liposome-intercalated amphotericin B. Int J Pharm 408:163–172

Ishida T, Kiwada H (2008) Accelerated blood clearance (ABC) phenomenon upon repeated injection of PEGylated liposomes. Int J Pharm 354:56–62
Ishida T, Harashima H, Kiwada H (2002) Liposome clearance. Biosci Rep 22:197–224
Janssen Products, Ltd. (2012) Doxil label. http://www.accessdata.fda.gov/drugsatfda_docs/label/2012/050718s043lbl.pdf [Online]. Accessed 2013
Jiang W, Lionberger R, Yu LX (2011) In vitro and in vivo characterizations of PEGylated liposomal doxorubicin. Bioanalysis 3:333–344
Judson I, Radford JA, Harris M, Blay JY, van Hoesel Q, le Cesne A, van Oosterom AT, Clemons MJ, Kamby C, Hermans C, Whittaker J, Donato di Paola E, Verweij J, Nielsen S (2001) Randomised phase II trial of pegylated liposomal doxorubicin (DOXIL/CAELYX) versus doxorubicin in the treatment of advanced or metastatic soft tissue sarcoma: a study by the EORTC Soft Tissue and Bone Sarcoma Group. Eur J Cancer 37:870–877
Kanter PM, Bullard GA, Pilkiewicz FG, Mayer LD, Cullis PR, Pavelic ZP (1993) Preclinical toxicology study of liposome encapsulated doxorubicin (TLC D-99): comparison with doxorubicin and empty liposomes in mice and dogs. In Vivo 7:85–95
Kanter PM, Klaich G, Bullard GA, King JM, Pavelic ZP (1994) Preclinical toxicology study of liposome encapsulated doxorubicin (TLC D-99) given intraperitoneally to dogs. In Vivo 8:975–982
Kim S, Sankaram MB (2000) Multivesicular liposomes with controlled release of encapsulated biologically active substances. http://www.google.co.in/patents/US6132766 [Online]. Accessed 2013
Koide H, Asai T, Hatanaka K, Urakami T, Ishii T, Kenjo E, Nishihara M, Yokoyama M, Ishida T, Kiwada H, Oku N (2008) Particle size-dependent triggering of accelerated blood clearance phenomenon. Int J Pharm 362:197–200
Krasnici S, Werner A, Eichhorn ME, Schmitt-Sody M, Pahernik SA, Sauer B, Schulze B, Teifel M, Michaelis U, Naujoks K, Dellian M (2003) Effect of the surface charge of liposomes on their uptake by angiogenic tumor vessels. Int J Cancer 105:561–567
Krishna R, Mayer LD (1997) Liposomal doxorubicin circumvents PSC 833-free drug interactions, resulting in effective therapy of multidrug-resistant solid tumors. Cancer Res 57:5246–5253
Lasic DD, Frederik PM, Stuart MC, Barenholz Y, Mcintosh TJ (1992) Gelation of liposome interior. A novel method for drug encapsulation. FEBS Lett 312:255–258
Laverman P, Carstens MG, Boerman OC, Dams ETM, Oyen WJG, van Rooijen N, Corstens FHM, Storm G (2001) Factors affecting the accelerated blood clearance of polyethylene glycol-liposomes upon repeated injection. J Pharmacol Exp Ther 298:607–612
Ma P, Mumper RJ (2013) Anthracycline nano-delivery systems to overcome multiple drug resistance: a comprehensive review. Nano Today 8:313–331
Maeda H (2001) The enhanced permeability and retention (EPR) effect in tumor vasculature: the key role of tumor-selective macromolecular drug targeting. Adv Enzyme Regul 41:189–207
Makriyannis A, Guo J, Tian X (2005) Albumin enhances the diffusion of lipophilic drugs into the membrane bilayer. Life Sci 77:1605–1611
Mantripragada S (2002) A lipid based depot (DepoFoam technology) for sustained release drug delivery. Prog Lipid Res 41:392–406
Martin F (2001a) Liposome drug products. http://www.fda.gov/ohrms/dockets/ac/01/slides/3763s2_08_martin.ppt [Online]. Accessed 2013
Martin F (2001b) Product evolution and influence of formulation on pharmaceutical properties and pharmacology. http://www.fda.gov/ohrms/dockets/AC/01/slides/3763s2_08_martin.ppt [Online]. Accessed 2013
Meyers PA (2009) Muramyl tripeptide (mifamurtide) for the treatment of osteosarcoma. Expert Rev Anticancer Ther 9:1035–1049
Miller CR, Bondurant B, Mclean SD, Mcgovern KA, O'Brien DF (1998) Liposome-cell interactions in vitro: effect of liposome surface charge on the binding and endocytosis of conventional and sterically stabilized liposomes. Biochemistry 37:12875–12883
Minko T, Pakunlu RI, Wang Y, Khandare JJ, Saad M (2006) New generation of liposomal drugs for cancer. Anticancer Agents Med Chem 6:537–552

Novartis (2012) Visudyne label. http://dailymed.nlm.nih.gov/dailymed/lookup.cfm? setid=31512723-9ff0-4e18-aa3a-55ab833038c6 [Online]. Accessed 2 June 2013

O'Brien ME, Wigler N, Inbar M, Rosso R, Grischke E, Santoro A, Catane R, Kieback DG, Tomczak P, Ackland SP, Orlandi F, Mellars L, Alland L, Tendler C (2004) Reduced cardiotoxicity and comparable efficacy in a phase III trial of pegylated liposomal doxorubicin HCl (CAELYX/Doxil) versus conventional doxorubicin for first-line treatment of metastatic breast cancer. Ann Oncol 15:440–449

Pacira Pharmaceuticals, Inc. (2011) Exparel label. http://www.accessdata.fda.gov/drugsatfda_ docs/label/2011/022496s000lbl.pdf [Online]. Accessed 2013

Pakunlu RI, Wang Y, Saad M, Khandare JJ, Starovoytov V, Minko T (2006) In vitro and in vivo intracellular liposomal delivery of antisense oligonucleotides and anticancer drug. J Control Release 114:153–162

Phuphanich S, Maria B, Braeckman R, Chamberlain M (2007) A pharmacokinetic study of intra-CSF administered encapsulated cytarabine (DepoCyt) for the treatment of neoplastic meningitis in patients with leukemia, lymphoma, or solid tumors as part of a phase III study. J Neurooncol 81:201–208

Reynolds JG, Geretti E, Hendriks BS, Lee H, Leonard SC, Klinz SG, Noble CO, Lucker PB, Zandstra PW, Drummond DC, Olivier KJ, Nielsen UB, Niyikiza C, Agresta SV, Wickham TJ (2012) HER2-targeted liposomal doxorubicin displays enhanced anti-tumorigenic effects without associated cardiotoxicity. Toxicol Appl Pharmacol 262:1–10

Sankaram MB, Kim S (1996) Preparation of multivesicular liposomes for controlled release of active agents. http://www.google.com/patents/CA2199004A1 [Online]. Accessed 2013

Senior J, Delgado C, Fisher D, Tilcock C, Gregoriadis G (1991) Influence of surface hydrophilicity of liposomes on their interaction with plasma protein and clearance from the circulation: studies with poly(ethylene glycol)-coated vesicles. Biochim Biophys Acta 1062:77–82

Sessa G, Weissman G (1968) Phospholipid spherules (liposomes) as a model for biological membranes. J Lipid Res 9:310–318

Shabbits JA, Chiu GN, Mayer LD (2002) Development of an in vitro drug release assay that accurately predicts in vivo drug retention for liposome-based delivery systems. J Control Release 84:161–170

Sigma-Tau Pharmaceutical, Inc. (2011) DepoCyt label. http://www.accessdata.fda.gov/ drugsatfda_docs/label/2011/021041s023lbl.pdf [Online]. Accessed 2013

Silverman JA, Deitcher SR (2013) Marqibo(R) (vincristine sulfate liposome injection) improves the pharmacokinetics and pharmacodynamics of vincristine. Cancer Chemother Pharmacol 71:555–564

Swenson CE, Perkins WR, Roberts P, Janoff AS (2001) Liposome technology and the development of Myocet (TM) (liposomal doxorubicin citrate). Breast 10:1–7

Talon Therapeutics, Inc. (2012) Marqibo label. http://www.accessdata.fda.gov/drugsatfda_docs/ label/2012/202497s000lbl.pdf [Online]. Accessed 2013

Wasan KM, Brazeau GA, Keyhani A, Hayman AC, Lopez-Berestein G (1993) Roles of liposome composition and temperature in distribution of amphotericin B in serum lipoproteins. Antimicrob Agents Chemother 37:246–250

Willis RC (1999) Method for utilizing neutral lipids to modify in vivo release from multivesicular liposomes. US patent US_5891467_A. Available at: http://www.lens.org/images/patent/US/ 5891467/A/US_5891467_A.pdf

Xu X, Khan MA, Burgess DJ (2012) Predicting hydrophilic drug encapsulation inside unilamellar liposomes. Int J Pharm 423:410–418

Yan X, Scherphof GL, Kamps JA (2005) Liposome opsonization. J Liposome Res 15:109–139

Zhang X, Zheng N, Lionberger RA, Yu LX (2013) Innovative approaches for demonstration of bioequivalence: the US FDA perspective. Ther Deliv 4:725–740

Zolnik BS, Stern ST, Kaiser JM, Heakal Y, Clogston JD, Kester M, Mcneil SE (2008) Rapid distribution of liposomal short-chain ceramide in vitro and in vivo. Drug Metab Dispos 36:1709–1715

Chapter 12
Bioequivalence for Drug Products Acting Locally Within Gastrointestinal Tract

Xiaojian Jiang, Yongsheng Yang, and Ethan Stier

12.1 Background

In an abbreviated new drug application (ANDA) submitted to US Food and Drug Administration (US-FDA), the proposed generic drug products must be both pharmaceutically equivalent and bioequivalent (BE) to the corresponding reference product to establish that the two products are therapeutically equivalent (TE). To be considered pharmaceutically equivalent, two products must contain the same amount of the same drug substance and be of the same dosage form with the same indications and uses. Under the Code of Federal Regulations Title 21 Part 320 (21 CFR. 320.1), BE is defined as "the absence of a significant difference in the rate and extent to which the active ingredient or active moiety in pharmaceutical equivalents or pharmaceutical alternatives becomes available at the *site of drug action* when administered at the same molar dose under similar conditions in an appropriately designed study."

Therapeutically equivalent products are expected to have the same clinical effect and safety profiles and can be substituted. The most efficient method of assuring therapeutic equivalence is to assure that the formulations of two pharmaceutically equivalent drug products perform in an equivalent manner. Formulation performance is defined as the drug substance releasing from the drug product and becoming available at the site of action to eventually exert its pharmacologic effects. If formulation performance from two products is equivalent, then the safety and clinical effects are likely to be equivalent (Conner and Davit 2008).

The views expressed in this chapter are those of the authors and do not reflect the official policy of the FDA. No official support or endorsement by the FDA is intended or should be inferred.

X. Jiang (✉) • Y. Yang • E. Stier
Center for Drug Evaluation and Research, U.S. Food and Drug Administration,
10903 New Hampshire Avenue, Silver Spring, MD 20993, USA
e-mail: Xiaojian.Jiang@fda.hhs.gov

L.X. Yu and B.V. Li (eds.), *FDA Bioequivalence Standards*, AAPS Advances
in the Pharmaceutical Sciences Series 13, DOI 10.1007/978-1-4939-1252-0_12,
© The United States Government 2014

Section 320.24 (b) of FDA's regulations describes preferred bioequivalence methods in what, for systemically absorbed products, is the descending order of accuracy, sensitivity, and reproducibility. They include: (1) in vivo pharmacokinetic studies, (2) in vivo pharmacodynamics effect studies, (3) clinical endpoint studies, and (4) in vitro studies. In addition, consistent with section 505 (j) (8)(C) of the FD&C Act, section 320.24 (b) (6) of the regulation states that FDA has the authority to use "[a]ny other approach deemed adequate by FDA to establish bioequivalence." The selection of the method used to meet an in vivo *or* in vitro testing requirement depends upon the purpose of the study, the analytical methods available, and the nature of the drug product (21 CFR 320.24 (a)). Applicants shall conduct the bioequivalence testing using the most accurate, sensitive, and reproducible approach.

For drugs intended to be absorbed and systemically delivered to the site(s) of activity, bioequivalence is generally achieved by measuring drug concentrations in an accessible biological fluid, typically plasma. However, for drugs that are not intended to be absorbed in the blood stream and whose site of action is the GI tract, determination of bioequivalence is more complicated. This is because systemic exposure may not be directly correlated to the local drug concentration in the GI tract (FDA 2008a, b).

In the past, the FDA has recommended clinical endpoint study to demonstrate BE for many locally acting GI products (e.g., orally administered Mesalamine extended or delayed-release products, Vancomycin capsules and Orlistat capsules). However, the regulation 21 CFR 320.24 (b) (4) states that clinical endpoint studies are "the least accurate, sensitive, and reproducible of the general approaches for demonstrating BE." The comparative clinical trial is also expensive and time consuming, which often limits the development of generic products (Lionberger 2008). The FDA General BA and BE guidance further states that comparative clinical studies are generally less favored, when other in vivo PK study, in vivo PD study or in vitro tests are feasible (FDA 2003).

The FDA has been actively investigating alternative BE methods, including in vitro studies, in vivo PK studies, in vivo PD studies, and other approaches that may be more efficient and more sensitive at detecting product differences (Zhang et al. 2013). As a result, the product-specific BE recommendations for Vancomycin capsules, Orlistat capsules, and Mesalamine modified-release (MR) products and rectally administered products were recently developed. The drug-specific BE recommendations for these products were developed by the FDA on a case-by-case basis. The recommendation of the most appropriate BE method depends upon the ability, sensitivity, and reproducibility of the method to compare drug delivered by the two products at the particular site of action, taking into account the site and mechanism of drug action, systemic absorption of drug, drug-specific physicochemical properties, product design, and drug product safety and efficacy profiles (Davit 2013).

12.2 General Considerations in Selecting the BE Approaches for Local Acting GI Products

12.2.1 Role of Pharmacokinetic Studies

For systemically acting drug products, the most sensitive approach in evaluating BE of two formulations is to measure drug concentration in biological fluids. Following oral dosing, the active ingredient is released from the drug product, dissolves in the GI tract, and is absorbed through the gut wall, then appears in the systemic circulation. Since the two products being compared contain the same active ingredient, the differences in the systemic absorption result from differences in formulation performance occurring earlier in the absorption process. By comparing PK profiles we make a conclusion as to whether there is a significant difference in formulation performance (FDA 2004).

For the locally acting GI products, the active ingredient is released from drug product, dissolves in the GI tract, and becomes available in the GI tract for therapeutic effect. The drug dissolved locally is also potentially available for the systemic absorption. Although plasma drug levels may not be related to the therapeutic effect, a PK study may still provide a comparison of the relative performance of the generic and reference products (FDA 2004). Comparison of PK profiles can distinguish two products that release drug in different regions of the GI tract.

Some locally acting GI products may not produce measurable concentrations of drug or metabolites in an accessible biologic fluid. For these products the Agency can review data from PD effect studies to assess bioequivalence if applicable. For others, when no PD endpoints can be readily measured, the Agency relies, if it can, on "appropriately designed comparative clinical trials" (21 CFR 320.24(b)(4)) or, in appropriate cases, in vitro studies, to assess bioequivalence (FDA 2010a).

It should be noted that if there is a safety concern related to systemic exposure for locally acting GI products, then the FDA will recommend a PK study intended to demonstrate equivalent or less systemic exposure, in addition to any other study requested to demonstrate equivalent local delivery.

12.2.2 Role of In Vitro Dissolution Testing

The regulation 21 CFR 320.24 (b) (5) states that under certain circumstances, product quality BA and BE can be documented using an in vitro test acceptable to FDA (usually a dissolution test) that ensures human in vivo BE.

The transit and dissolution of the locally acting GI products in vivo determine the presentation of drug to the site of action (Amidon 2004). In general, for locally acting GI drugs, in vivo dissolution testing is related to the rate and extent to which the active ingredient becomes available at the site of action and detects the most significant potential formulation difference between generic and reference product (FDA 2009b). In contrast, for systemically acting drug products, the rate and extent of drug delivery to the site of action involves both in vivo dissolution and subject-determined processes, such as absorption, metabolism, distribution, and excretion, which may not be related to product performance (Polli 2008).

In vitro dissolution testing alone may suffice as a BE approach for locally acting immediate release (IR) oral products which contain drug substances with high aqueous solubility and are formulated to be qualitatively and quantitatively ($Q1$ and $Q2$) the same as the reference products. The high solubility over pH covering the GI tract ensures that drug is largely in solution by the time when the drug enters the site of action in the lower GI tract; thus in vitro dissolution is highly predictive of in vivo dissolution of the drug product. If the formulation of the generic and reference products is $Q1$ and $Q2$ the same, the effect of inactive ingredients on the transport of drug through the GI tract will be equivalent. If the generic drug product has a different formulation than the reference drug product, additional studies may be recommended to demonstrate that any formulation differences between generic and reference drug products will not affect the safety and effectiveness of the drug product (FDA 2008a). Figure 12.1 summarizes the decision tree for selecting in vitro dissolution testing as a BE approach for IR locally acting generic products (Davit 2010). For a GI product that has low solubility, the traditional dissolution media may not predict in vivo performance, thus dissolution testing has not been recommended as a BE approach. The FDA generally recommends a clinical endpoint study in addition to in vivo PK study to demonstrate BE for locally acting GI products with low aqueous solubility.

As one of the FDA's regulatory science initiatives for generic drugs, the FDA is working on developing appropriate in vitro testing for low solubility IR and MR products and assessing biorelevant dissolution media and its utility in predicting in vivo dissolution. In the future, a more extensive dissolution testing or biorelevant media may be used to demonstrate BE for locally acting GI products with low aqueous solubility.

Some locally acting MR oral products are formulated to target different regions of the GI tract, often via coatings that lead to pH-dependent dissolution. Comparative dissolution testing at different pH could demonstrate that generic and reference products have comparable drug release in each region of the GI tract (FDA 2004). Therefore, the selection of dissolution testing conditions representative of conditions in the target GI tract should be the focus of in vitro dissolution method development (Yang et al. 2013a, b). However, due to the complexity of the in vivo GI transit and environment, limitations exist on how well the in vitro dissolution testing conditions can mimic the in vivo dissolution, especially for complicated dosage forms, e.g., MR products. The FDA does not recommend the use of only in vitro dissolution testing to establish BE for MR oral products. An in vivo PK study or a PD study will be necessary, in addition to dissolution testing, to establish BE.

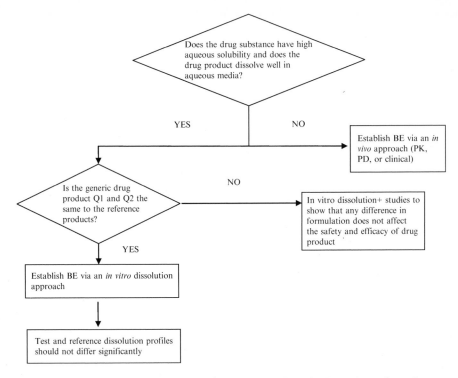

Fig. 12.1 Selection of an appropriate BE approach for locally acting IR oral generic products

In vitro dissolution testing together with an in vivo PK study can be used to establish BE for locally acting MR oral products if the drug substance dissolves in aqueous media, in vitro dissolution is predictive of in vivo performance and systemic exposure is measurable. Figure 12.2 summarizes the decision tree for selecting in vitro dissolution testing as a BE approach for a locally acting MR generic drug product (Davit 2010).

12.2.3   Role of In Vitro Assay

For some locally acting GI drug products (e.g., Calcium Acetate, Sevelamer, and Cholestyramine (FDA 2011e, g, e)), the FDA recommends that BE may be assessed, with suitable justification, by in vitro BE studies that reflect the drug mechanism of action, since it is recognized that in vitro methods are less variable, easier to control, and are more likely to detect differences between products. However, the clinical relevance and sensitivity of the in vitro test should be clearly established (Chow and Liu 2009).

Binding agents including insoluble resins such as Cholestyramine and Colstipol, and phosphate binders, such as Calcium, Sevelamer, and Lanthanum, exert their

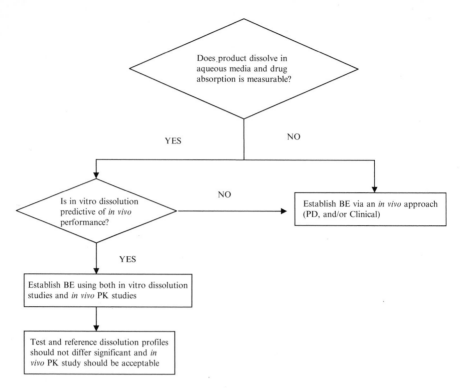

Fig. 12.2 Selection of an appropriate BE approach for locally acting MR oral generic products

therapeutic effectiveness by binding bile acid or phosphate in the upper GI tract to form an insoluble complex which is excreted in the feces. In vitro binding assays is considered to be directly related to the pharmacological action of the products and can ensure equivalent local availability to the site of action between the generic and reference products. Consequently, the FDA recommended in vitro equilibrium and kinetic binding assays for demonstrating bioequivalence for these locally acting binding agents (FDA 2008a). The kinetic binding study assesses the rate of binding and the time to reach the binding equilibrium, and the equilibrium binding studies determine the binding affinity and capacity (Zhang et al. 2013). The in vitro binding assay may be combined with disintegration/dissolution testing to demonstrate BE for phosphate binders.

12.2.4 Role of Pharmacodynamic Studies

Pharmacodynamic or pharmacological effects studies are useful in cases when a PK study or an in vitro method cannot be used to determine BE. The pharmacodynamics is defined as relationship of drug concentration at the site of action to the therapeutic effect. As illustrated in Figs. 12.1 and 12.2, after the drug or active metabolite is delivered to the site of action, it elicits a PD response. In general, measurement of PD endpoint has more variability than PK and in vitro studies (Conner and Davit 2008). As such, a PD or clinical approach is generally considered not as accurate, sensitive, and reproducible for determining difference in performance of generic and reference products in comparison to a PK measurement or an in vitro testing.

Generally, the PD response plotted against the logarithm of dose appears as a sigmoidal curve. It is critical to select a dose on the linear portion of the dose–response curve so as that the change of the PD responses is sensitive to small changes in dose. For a PD endpoint BE study, the FDA recommends a pilot study in additional to a pivotal study to determine the appropriate dose that is on the linear portion of the PD dose–response curve as well as the number of study subjects to provide adequate statistical power. This approach has been recommended to demonstrate BE for Acarbose tablets that are not $Q1$ and $Q2$ the same formulation as the reference product (FDA 2009c).

12.2.5 Role of Clinical Endpoint Studies

Even though, the clinical endpoint BE study has limitation in sensitivity, in some cases, the FDA recommends clinical endpoint study when there is no other feasible methods available (Lionberger 2008).

BE studies with clinical endpoints generally employ a randomized, blinded, balanced, and parallel design. Studies compare the efficacy of the generic product, reference product, and the placebo to determine if the two products containing the same active ingredient are bioequivalent. The study uses a product-specific clinical indication in a patient population according to the labeled dosing regimen of the reference product. Both the generic and reference products must be statistically superior to placebo ($p < 0.05$) in order to ensure that the study is sensitive enough to show a difference between products. If the reference product is labeled for multiple indications, the indication that is most sensitive to formulation difference is usually preferred (Lionberger 2008).

Clinical endpoint studies are frequently conducted at a dose that is on the plateau phase of the dose–response curve. The large difference in dose ($D2$) produces only a small difference in response ($R2$) as illustrated in Fig. 12.5. Thus clinical endpoint studies are the least sensitive method to detect differences in performance between generic and RLD products, and therefore least preferred. The clinical endpoint study often requires a large number of subjects and patient recruitment maybe an issue.

12.2.6 Summary of Various BE Approaches

For locally acting GI products, such as Mesalamine MR products, that have measurable systemic absorption, in vivo BE studies with PK endpoints as well as in vitro dissolution studies are suitable methods to demonstration BE. For highly soluble, IR oral products that have little or undetectable systemic absorption, e.g., Vancomycin capsule, in vitro dissolution testing in multimedia with profile comparison is adequate to ensure BE for the generic product that is $Q1$ and $Q2$ the same formulation to the reference product. In cases where the systemic absorption is undetectable and in vitro dissolution is not predictive of in vivo dissolution (e.g., Orlistat capsule), a PD endpoint or clinical endpoint study can be used. For locally acting GI products that have low aqueous solubility (e.g., Rifaximin and Lubiprostone), the FDA recommended in vivo BE studies with clinical endpoints as well as in vivo BE studies with PK endpoints.

For locally acting MR oral products, the FDA generally recommends establishing BE by "weight-of-evidence" approach or combined approaches to overcome the limitation of individual endpoint and provide adequate assurance of equivalence. Table 12.1 shows examples of products, the BE approaches, and the factors that were considered in selecting these approaches.

12.3 BE Recommendation Case Study

12.3.1 Mesalamine

Ulcerative colitis (UC) is a chronic inflammatory disease characterized by mucosal inflammation in the colon. The clinical course involves recurrent episodes of active disease separated by periods of remission, with symptoms of bloody diarrhea, rectal urgency, and abdominal pain (Kornbluth and Sachar 2010). UC most commonly affects teenagers and young adults, but can occur in any age group. It has a prevalence of 238 per 100,000 in the US adult population and an incidence rate of 2.2–14.3 cases per 100,000 person-years in North America (Loftus 2004).

Sulfasalazine (SSZ) was the first medication used to successfully treat UC for 40 years but less than 20 years ago it was recognized that the active moiety is mesalamine. SSZ contains mesalamine bound to sulfapyridine via an azo-bond. The azo-bonded sulfapyridine, an inactive moiety, protects the drug from absorption until the mesalamine is released by azoreductases present in high concentration in the bacteria-rich colon (Lichtenstein and Kamm 2008).

Mesalamine is 5-aminosalicylic acid (5-ASA) compound (Fig. 12.3) that is the first-line treatment for the induction of remission and maintenance therapy of mild-to-moderate UC. The mechanism of action of mesalamine is not fully understood, but appears to have a topical anti-inflammatory effect on the colonic epithelial cells. Mucosal production of arachidonic acid metabolites, both through the

Table 12.1 Summary of BE approaches and factors considered

Bioequivalence methods	Example drug/drug product	Characteristics	Product category
In vitro disintegration and binding assay	Cholestyramine, Colestipol, Lanthanum carbonate, Calcium acetate and Sevelamer	Little or no systemic absorption; binding can be quantitatively measured; adverse events are local	Binding agents
In vitro dissolution	Vancomycin HCl oral capsules Acarbose tablets (*Q1 and Q2 the same generic formulation*)	High solubility, little or no systemic absorption; dissolution is highly predictive of in vivo release	High solubility immediate release dosage forms
In vivo PK studies and in vitro dissolution	Mesalamine ER and DR products (pAUC); mesalamine prodrugs; mesalamine enema	Plasma concentration is measurable; product dissolve in aqueous media and dissolution is predictive of in vivo release	Modified release dosage forms and immediate release products
In vivo PK studies and in vitro physical–chemical characterization	Mesalamine suppositories	Plasma concentration is measurable; dissolution is not predictive of in vivo release	Rectal products
In vivo PD studies	Acarbose tablets (*Q1 and Q2 not the same*); Orlistat capsules	PD endpoint is readily measureable and sensitive; little or no systemic absorption; no systemic safety concern	High solubility and low solubility immediate release products
Clinical endpoint study and/or PK studies	Rifaximin capsules, lubiprostone capsules	Low solubility drug; dissolution is not predictive of in vivo; systemic absorption may be measurable; clinical study is feasible	Low solubility immediate release dosage forms

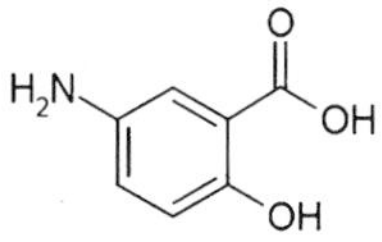

Fig. 12.3 Chemical Structure of mesalamine

cyclooxygenase and lipoxygenase pathways, is increased in patients with chronic inflammatory bowel disease, and it is possible that mesalamine diminishes inflammation by blocking cyclooxygenase and inhibiting prostaglandin production in the colon (FDA 2011a).

Mesalamine is rapidly and completely absorbed from the upper GI tract when administered orally, but poorly absorbed from the colon (Schroder and Campbell 1972). To prevent proximal small intestinal absorption and allow mesalamine to reach the inflamed small bowel and/or colon, a variety of mesalamine delivery systems have been developed (Qureshi and Cohen 2005). These include (Table 12.2):

- Orally Administrated Pro-drugs: creating a larger unabsorbed molecule by binding mesalamine to a carrier or another 5-ASA via an azo-bond, which subsequently is cleaved by bacterial azoreduction in the colon to release equimolar quantities of the active mesalamine moiety. Products include SSZ (Azulfidine®, Azulfidine EN-tabs®), olsalazine sodium (Dipentum®), and balsalazide disodium (Colazal® and Giazo®).
- Orally Administrated Delayed-Release Products: formulating the product with outer-layer pH-sensitive enteric coating, which dissolves in the basic environment of the distal ileum and colon. Products include Asacol®, Asacol® HD and Liada® (mesalamine delayed-release tablets) and Delzicol® (mesalmine delayed-release capsules).
- Orally Administrated Extended Release Products: formulating the product with controlled-release excipients to provide releasing of mesalamine throughout the GI in a time delayed fashion. Products include Pentasa® and Apriso® (mesalamine extended release capsules).
- Topical Products: administering mesalamine as an enema (Rowasa® (mesalamine) rectal enema) or suppository (Canasa™ (mesalamine) rectal suppository) directly into the body at the site of rectum and distal colon, effectively bypassing the threat of small bowel absorption.

While all of these products deliver the same active moiety (mesalamine) to the inflamed bowel, the differences in formulation, as well as colonic condition and transit time, result in varying release profiles, PK profiles, safety profiles, and differences in drug availability at the level of the colonic mucosa (Qureshi and Cohen 2005).

The plasma concentrations of these mesalamine products listed in Table 12.2 can be reliably measured but have high variability. Since mesalamine has a short half-life (about 42 min after i.v. administration of 500 mg) (Myers et al. 1987), the systemic exposure of mesalamine is presumably due to the absorption of mesalamine after it is released from the dosage form at the target site. The PK profiles for these products are distinct from each other with varying T_{lag}, T_{max}, but some overlap in AUC and C_{max}. For example, the mesalamine PK profile of Colazal® (Balsalazide disodium capsules, mesalamine released in the colon) is different from the PK profiles of Asacol®, Pentasa®, Lialda®, and Apriso® (mesalamine MR products) (FDA 2010a). As discussed before, systemic absorption at the target site of action is expected to be proportional to the local concentration. Therefore, PK profile can detect the formulation differences and provides information of the local availability of mesalamine.

Due to the differences in the formulation design, oral mesalamine products exhibit different dissolution profiles in multiple pH media. For example, for a two-stage dissolution test, acid stage (0.1 N HCl) followed by a basic stage (pH 6.0–7.5), the dissolution profiles of Asacol®, Asacol® HD, and Liada®

Table 12.2 Mesalamine formulation currently marketed in the US

Brand name	Generic name	Strength	Dosing regimen	Site of 5-SAS released	Formulation
Oral pro-drugs					
Azulfidine	Sulfasalazine (SSZ) tablet and suspension	500 mg	3–4 g daily in evenly divided dose for initial therapy; 2 g daily for maintenance therapy	Colon	Sulfa moiety linked to 5-ASA; azo-bond cleaved by colonic bacteria. DR tablets can retard disintegration of the tablet in the stomach and reduce potential irritation
Azulfidine EN-tabs	SSZ delayed-release tablet	500 mg			
Dipentum	Olsalazine sodium capsule	250 mg	2 × 0.25 g Bid (1 g/day) for maintenance of remission	Colon	Two 5-ASA molecules linked; azo-bond cleaved by colonic bacteria
Colazal	Balsalazide disodium capsule	750 mg	3 × 0.75 g TID (6.75 g/day) for treatment of mildly to moderately active UC	Colon	Inert carrier linked to 5-ASA; azo-bond cleaved in colon
Giazo	Balsalazide disodium tablet	1.1 g	3 × 1.1 g BID (6.6 g/day) for treatment of mildly to moderately active UC	Colon	
Oral mesalamine MR products					
Asacol	Mesalamine delayed-release tablet	400 mg	2 × 400 mg TID (2.4 g daily) to treat mildly to moderately active UC. For the maintenance of remission of UC, 1.6 g a day in divided doses	Terminal ileum and colon	Tablets are coated with Eudragit S resin; releases 5-ASA at pH = 7
Asacol HD	Mesalamine delayed-release tablet	800 mg	2 × 800 mg TID (4.8 g daily) for treatment of moderately active UC	Terminal ileum and colon	Same as Asacol but contains an additional coating layer of Eudragit L
Lialda	Mesalamine delayed-release tablet	1.2 g	2.4 or 4.8 g once daily for induction of remission; 2.4 g once daily for maintenance of remission	Terminal ileum and colon	Tablets are coated with Eudragit S resin; releases 5-ASA ≥ 6.8; hydrophilic and lipophilic excipients contained in the core to provide extended release

(continued)

Table 12.2 (continued)

Brand name	Generic name	Strength	Dosing regimen	Site of 5-SAS released	Formulation
Delzicol	Mesalamine delayed-release capsule	400 mg	2×400 mg TID (2.4 g daily) to treat mildly to moderately active UC. For the maintenance of remission of UC, 1.6 g a day in divided doses	Terminal ileum and colon	Tablets are coated with Eudragit S resin; releases 5-ASA at pH = 7
Pentasa	Mesalamine extended release capsule	250 and 500 mg	4×250 or 2×500 mg QID for a total daily dosage of 4 g for induction of remission	Duodenum, Jejunum, ileum, colon	granules are coated with Ethylcellose to provide extended release
Apriso	Mesalamine extended release capsule	375 mg	4×0.375 g (1.5 g) once daily for maintenance of remission	Jejunum, ileum and colon	Granules are composed of 5-ASA in a polymer matrix with a Eudragit L coating dissolves at pH 6
Rectal mesalamine products					
Canasa	Mesalamine rectal suppository	1 g	1 g once daily for treatment of ulcerative proctitis	Rectum and distal colon	5-ASA suppository
Rowasa	Mesalamine rectal enema	4 g/60 mL	4 g once daily for treatment of distal ulcerative colitis and proctitis	Rectum and distal colon	5-ASA suspension

(mesalamine DR tablets), begin to release at pH 6.8–7.0. However at pH greater than 7, the dissolution profiles of Asacol® and Asacol® HD (mesalamine DR tablets) are similar to an immediate release profile whereas Lialda® (mesalamine DR tablets) has extended release dissolution profile. Different dissolution profiles for different mesalamine products in simulated gastric fluid and phosphate buffer pH 6.8, 7.2, and 7.8 are also reported in the literature (Rudolph et al. 2001). Thus, in vitro dissolution testing is a sensitive method to detect formulation differences. Therefore, to compare the dissolution profiles in multimedia representative of the expected conditions in the GI tract will evaluate the effect of pH and transit time on drug release and to demonstrate BE for products that have different release mechanisms. When two products that have an equivalent in vitro release (under different pH conditions) and PK profiles, the FDA can conclude that drug availability at those sites of action is the same (FDA 2010a).

Clinical endpoint studies are generally considered not sensitive to discriminate the drug release pattern between products. It was reported that clinical studies failed to demonstrate a significant difference among existing mesalamine formulations listed in Table 12.1 or different doses for the same formulation, even though these products have different PK profiles and in vitro release profiles (Sandborn 2002).

In 2005, the FDA recommended to establish BE for mesalamine pro-drugs (balsalzide, olsalazine, and SSZ) by in vivo PK study and in vitro dissolution testing. Since 2005, the FDA has implemented a reference-scaled BE approach for highly variable drugs that the sponsors can use the variability of the reference product to set appropriate limits on the generic-reference difference. As a result, the reference-scaled BE approach makes the PK studies on the mesalamine feasible. In addition, it was recognized that a comparison of AUC and C_{max} may not distinguish two mesalamine products that release drug in different regions of the GI tract if the total amount absorbed is similar. Therefore partial AUC has been recommended to evaluate PK profile similarity to ensure therapeutic equivalence. Furthermore, partial AUC could distinguish the drug release "exclusively" in the colon versus mesalamine released partially outside the colon. For oral modified-release mesalamine products, the FDA's current position is that PK profile equivalence using partial AUC under fasting and fed conditions and dissolution profile equivalence in multiple media using f_2 similarity comparison are the best demonstration of BE.

12.3.1.1 Pro-drug Products

SSZ is the first-line treatment of inflammatory bowel disease for decades. This drug consists of 5-ASA linked by an azo-bond to sulfapyridine (SP). The 5-ASA is liberated when the azo-bond is cleaved by bacterial azoreductases in the colon, with the SP functioning as a carrier molecule. There are currently three SSZ reference products on the market: Azulfidine® (SSZ tablets and suspension) and Azulfidine EN-tabs® (SSZ delayed-release tablets). Azulfidine EN-tabs® (SSZ) DR tablets are film coated with cellulose acetate phthalate to retard disintegration of the tablet in the stomach and reduce potential irritation of the gastric mucosa (FDA 2013a). SSZ is poorly absorbed, with less than 15 % of the dose being absorbed as parent drug. Sulfapyridine is well absorbed from the colon, with an estimated bioavailability of 60 %, while 5-ASA is much less absorbed from the GI tract. Peak plasma levels of both SP and 5-ASA occur approximately 10 h after dosing, which is indicative of the gastrointestinal transit time to the lower intestine where bacteria-mediated metabolism occurs (FDA 2013a).

Olsalazine is composed of two molecules of 5-ASA radicals joined by an azo-bond, which is subsequently split in the colon by bacterial azoreductase, releasing two 5-ASA molecules. After oral administration of olsalazine sodium capsule (Dipentum®), the systemic bioavailability is low. Based on oral and intravenous dosing studies, approximately 2.4 % of a single 1.0 g oral dose is absorbed (FDA 2006). The remaining fraction of the dose is delivered to the colon, where each molecule is rapidly converted into two molecules of 5-aminosalicylic acid (5-ASA) by colonic bacteria. The liberated 5-ASA is absorbed slowly, resulting in very high local concentrations in the colon.

Balsalazide is a pro-drug of 5-aminosalicylic acid (5-ASA) linked via a diazo-bond to the inert carrier 4-aminobenzoyl-B-alanine (4-ABA). Balsalazide capsules (Colazal®) and tablets (Glazo®) contain a powder of balsalazide disodium that is insoluble in acid and designed to be delivered to the colon as intact pro-drug. Animal and human pharmacokinetics studies have confirmed the suitability of the balsalazide molecule as an efficient pro-drug source of luminal 5-ASA in the colon, with minimal systemic absorption of both the parent compound and its metabolites (Qureshi and Cohen 2005).

For all three products, the FDA recommended that the generic applicants compare the generic formulation and the reference products using in vivo PK studies and in vitro dissolution in testing in dissolution media covering the pH range encountered in the GI tract to demonstrate BE. Healthy subjects should be used in the in vivo PK studies as the low variability associated with healthy subjects provides greater sensitivity for the detection of differences in the bioavailability of pharmaceutically equivalent products (FDA 2007). Dissolution profiles of the generic and reference products in the multiple media should be compared using a similarity factor (f_2).

As summarized in Table 12.3, in the PK studies, parent drugs (sulfasalzine, olsalazine, and balsalazide) should be measured in plasma because as per FDA's general BA/BE guidance (FDA 2003), plasma parent drug concentrations are the most sensitive to detect changes in formulation performance. Since mesalamine absorption from the colon is relevant to the availability of the active moiety at the site of action, mesalamine concentrations should also be determined. With the advance in scientific technology, plasma concentration of parent drugs and 5-ASA can be accurately and reproducibly measured by current assays.

For the pro-drugs, AUC and C_{max} are a good proxy for the amount of mesalamine available at the site of colon, as the pro-drugs are only converted to mesalamine by bacterial action at or near the sites of action, followed by its absorption primarily at the sites of action (FDA 2010a). The 90 % confidence

Table 12.3 BE recommendation for pro-drugs (some of the posted guidance maybe outdated and require revision)

Brand name/ generic name	BE recommendation				
	In vivo PK studies using healthy subjects	Analytes to be measured	BE based on 90 % of AUC and C_{max}	In vitro Dissolution (BE)	Guidance posting date (FDA 2008d, e, 2010c, d, 2013b)
Azulfidine (SSZ tablets and capsules)	Fasting and Fed	SSZ, Sulfapyridine and mesalamine	SSZ and mesalamine	0.1 N HCl; pH 4.5 buffer;	2/2010
Azulfidine EN-tabs (SSZ DR tablets)				pH 6.8 buffer; pH 7.4	2/2010
Dipentum (Olsalazine sodium capsules)	Fasting and Fed	Olsalazine and mesalamine	Olsalazine and mesalamine	0.1 N HCl; pH 4.5 buffer; pH 6.8 buffer	5/2008
Colazal (balsalazide disodium capsules)	Fasting and Fed	Balsalazide and Mesalamine	Balsalazide and Mesalamine	0.1 N HCl; pH 4.5 buffer; pH 6.8 buffer; pH 7.4	1/2008
Glazo (balsalazide disodium tablets)	Fasting and Fed	Balsalazide and Mesalamine	Balsalazide and Mesalamine	0.1 N HCl; pH 4.5 buffer; pH 6.8 buffer; pH 7.4	6/2013

intervals of the generic/reference geometric mean ratios for AUC and C_{max} of both parent drugs and mesalamine should fall within the range of 0.8–1.25. The FDA's request that the presystemically formed active metabolite, mesalamine, meet bioequivalence limits is an exception to FDA's bioequivalence guidance, because generally the FDA asks that plasma metabolite data be used as supportive information only (FDA 2003). The FDA determined that evaluating the mesalamine data using the confidence internal approach is important to ensure pro-drugs reach the colon and be converted to the active moiety, 5-ASA (FDA 2007). It is not necessary to measure plasma concentrations of N-Ac-5-ASA since it is pharmacologically inactive and not likely associated with a safety issue. For SSZ products, sulfapyridine, the carrier moiety is requested to be measured due to its association with a considerable number of adverse events.

Although each of dissolution and pharmacokinetics studies, only partially reflect drug appearance at the local site(s) of action, these parameters together provide adequate assurance of formulation performance to support a determination of bioequivalence.

12.3.1.2 Mesalamine Delayed-Release Products

The oral mesalamine delayed-release products are formulated with a pH-sensitive enteric coating method to target mesalamine deliver drugs to the lower GI tract. Currently marketed oral mesalamine delayed-release tablets include Asacol® (400 mg), Asacol® HD (800 mg), and Liada® (1.2 g).

Asacol® (mesalamine) delayed-release tablets are coated with acrylic-based resin, Eudragit S (methacrylic acid copolymer B, NF), which dissolves at pH 7 or greater, releasing mesalamine in the terminal ileum and beyond for topical anti-inflammatory action in the colon (FDA 2011a). After oral administration, approximately 28 % of the mesalamine in Asacol® (mesalamine) DR tablets is absorbed, leaving the remainder available for topical action and excretion in the feces. Absorption of mesalamine is similar in fasted and fed subjects. The T_{max} for mesalamine ranges from 4 to 12 h, reflecting the delayed release.

Asacol® HD (mesalamine) delayed-release tablets have an outer protective coat consisting of a combination of acrylic-based resins, Eudragit S (methacrylic acid copolymer B, NF) and Eudragit L (methacrylic acid copolymer A, NF). The Eudragit S dissolves at pH 7 or greater, releasing mesalamine in the terminal ileum and beyond for topical anti-inflammatory action in the colon (FDA 2010e). The Asacol® HD (mesalamine) delayed-release tablet was shown not bioequivalent to Asacol® (mesalamine) delayed-release tablet administered at the same dose. The T_{max} for mesalamine (10–16 h) is also delayed for Asacol® HD (mesalamine) DR tablets compared to Asacol® (mesalamine) DR tablets. Based on cumulative urinary recovery of mesalamine and N-Ac-5-ASA from single dose studies in healthy volunteers, approximately 20 % of the orally administered mesalamine in Asacol® (mesalamine) HD tablets, is systemically absorbed. A high-fat meal does not affect AUC, but mesalamine C_{max} decreases by 47 % and T_{max} is delayed by 14 h under fed conditions.

Liada® (mesalamine) delayed-release tablets are coated with a pH-dependent polymer film, which breaks down at or above pH 6.8, normally in the terminal ileum where mesalamine then begins to be released from the tablet core. The tablet core contains mesalamine with hydrophilic and lipophilic excipients and provides for extended release of mesalamine from ileum to rectum (FDA 2011b). The total absorption of mesalamine from Liada® (mesalamine) DR tablets was found to be approximately 21–22 % of the administered dose. T_{max} ranges from 9 to 12 h. A high-fat meal increased systemic exposure of mesalamine (mean C_{max}: ↑ 91 %; mean AUC: ↑ 16 %) compared to results in the fasted state (label).

In August 2010, the FDA decided that applicants for generic version of mesalamine modified-release product must demonstrate bioequivalence through a combination of pharmacokinetic studies and in vitro dissolution testing over a range of pH expected in the GI tract. As per the product-specific BE guidance for mesalamine DR products posted in September 2012 (Table 12.4), the FDA recommended fasting and fed studies using healthy subjects and in vitro dissolution study to demonstrate BE. A fed study is requested as per the general FDA policy for

Table 12.4 BE recommendations for masalamine delayed-release products

RLD	In vivo fed/fast PK studies, BE matrics	In vitro dissolution						Guidance posting date (FDA 2012a, b, c)
Asacol® (masalamine DR tablets)	AUC_{8-48}, AUC_{0-t} and C_{max}	II (paddle)	100 rpm for pretreatment stage; 50 rpm for evaluation stage	2 h in 0.1 N HCl (500 mL)	Each of (1) pH 4.5 Acetate buffer (2) pH 6.0 PB (3) pH 6.5 PB (4) pH 6.8 PB (5) pH 7.2 PB (6) pH 7.5 PB	900	0, 10, 20, 30, 45, 60, 75, 90, 120, 150 min or as needed for profile comparison	9/2012
Asacol® HD (mesalamine DR tablets)								9/2012
Lialda® (mesalamine DR tablets)		II (paddle)	100 rpm for pretreatment stage and buffer stages	2 h in 0.1 N HCl (750 mL)	Buffer stage 1: (950 mL) pH 6.4 PB for 1 h Buffer stage 2: (960 mL) Each of (1) pH 6.5 PB (2) pH 6.8 PB (3) pH 7.2 PB (4) pH 7.5 PB		1, 2, 4, 6 and 8 h or as needed for profile comparison	9/2012

Table 12.5 Comparison of different parts of the gastrointestinal tract (Chuong et al. 2008; Wilson 2010 and Gao et al. 2010)

GI compartment	pH	Transit time (h) under fasting condition
Stomach	1–2	1–5
Duodenum	5–6.5	>5 min
Jejunum	6.0–7.0	1–2
Ileum	6.6–7.5	2–3
Ascending colon	5–8	3–5
Transverse colon	6–8	0.2–4
Descending colon	6–8	5–72
Sigmoid colon/rectum		

modified-release dosage forms (FDA 2002). In addition, food appears to affect the AUC_t, C_{max} and T_{max} for all three products; thus a fed BE study is important to demonstrate generic mesalamine products interact with food in a similar manner to the reference product. The generic applicants may consider using a reference-scaled average bioequivalence approach to address the high variability of plasma mesalamine concentration associated with these products (Haidar et al. 2008). The FDA does not request measurement of plasma concentrations of N-Ac-5-ASA because this metabolite does not contribute significantly to the safety or efficacy.

Since mesalamine delayed-release products release mesalamine in the small and large intestine (colon), plasma concentrations reflect total mesalamine absorption, not just absorption at the site of action. The FDA recommended that partial $AUC_{8\text{-}48}$, reflecting drug absorption in the colon, should be applied to evaluate PK profile similarity in addition to AUC_t and C_{max} (FDA 2013c).

In the case of the mesalamine delayed-release products, the multistage dissolution testing reflects the pH values in each segment of the GI tract. The selected pH levels include pH levels of stomach (0.1 N HCl), upper GI tract (5–7.0), and lower GI tract (6–8), where the mesalamine starts to release as illustrated in Table 12.5. Dissolution profiles of the generic and reference products in the multiple media should be compared using f_2 similarity factor. The equivalent in vitro dissolution testing over a range of pH serves as a surrogate of equivalent in vivo drug release in the GI tract and a confirmation of in vivo BE study with PK endpoints

Because of the complexity of the GI tract, some differences between the generic and reference products can be masked by relying explicitly only on a PK study or only on a dissolution study. PK profile similarity and in vitro BE dissolution similarity comparison will ensure similar local delivery of mesalamine.

12.3.1.3 Mesalamine Rectal Administrated Products

There are two marketed mesalamine products for rectal administration, which are effective for topical therapy for ulcerative proctitis and left-sided colitis. Rowasa® (mesalamine) rectal suspension enema contains 4 g of mesalamine in a 60 mL bottle with an applicator tip. Mesalamine administered as enema is poorly absorbed from the colon, approximately 10–30 % of the dose. The extent of absorption is dependent upon the retention time of the drug product and is variable among individual subjects (FDA 2008f). Canasa® (mesalamine) rectal suppository contains 1 g of mesalamine in a base of hard fat. Systemic absorption from Mesalamine administered as a rectal suppository is low (about 12 %) but measurable (Aumais et al. 2003).

For mesalamine enema, the FDA recommended a fasting BE study with PK endpoint and comparative dissolution testing in 0.1 N HCl, and buffers at pH 4.5, 6.8, and 7.2 using apparatus 2 (paddle) at 25 and 50 rpm, provided that the generic product is qualitatively ($Q1$) and quantitatively ($Q2$) the same as the RLD (FDA 2008g). For mesalamine suppository, the FDA previously recommended an in vivo bioequivalence study with clinical endpoints; and an in vivo bioequivalence study with PK endpoints under fasting condition. Based on the recent development of the BE recommendation for oral mesalamine products, in March 2013, the FDA revised the BE recommendation to ask an in vivo PK study under fasting condition and in vitro physicochemical characterization testing for demonstration BE, given that the generic product is qualitatively ($Q1$) and quantitatively ($Q2$) the same as the RLD (FDA 2013d). Table 12.6 summarized the BE recommendations for rectal suspension and suppository. Similar to oral mesalamine products, the BE study with clinical endpoint is no longer requested because of its insensitivity in discriminating formulations. However, unlike oral mesalamine products, the partial AUC metric is not needed because mesalamine appears to be absorbed primarily at the site of action following administration of enema and suppository (Brown et al. 1997). Since the enema and suppositories are confined in the rectum and distal colon, food is less likely to interact with the products. Thus, the fed study is not requested.

The reasons for the use of in vitro physicochemical characterization testing for demonstrating BE for the suppository instead of dissolution testing for the suspension enema are: (1) release of drug from suppositories is a complex process and not as well understood as release from a suspension; (2) for the suppository the dissolution test is not relevant to drug release considering the very low fluid volume in the rectum; (3) dosage form physicochemical properties other than in vitro dissolution contribute substantially to in vivo drug release from suppositories.

12.3.2   Orlistat Capsules

Orlistat, a chemically synthesized hydrogenated derivative of lipstatin, is a reversible and selective inhibitor of gastric and pancreatic lipase enzymes, as shown in Fig. 12.4. It is indicated for obesity management when used along with a reduced-calorie diet. Orlistat exerts its therapeutic activity in the lumen of the stomach and small intestine by forming a covalent bond with the active serine residue site of gastric and pancreatic lipases (FDA 2012d). The dietary fat in the form of triglycerides thus cannot be hydrolyzed by lipases into absorbable free fatty acids and monoglycerides, resulting

Table 12.6 BE Recommendations for rectal suspension and suppository

Brand name/ generic name	BE recommendation			In vitro dissolution (BE)	In vitro physicochemical characterization	Guidance posting date
	In vivo PK studies using healthy subjects	Analytes to be measured	BE based on 90 % of AUC and C_{max}			
Rowasa (mesalamine rectal enema)	Fasting	mesalamine	Mesalamine	0.1 N HCl; pH 4.5 buffer; pH 6.8 buffer; pH 7.2 using paddle at 25 and 50 rpm	NA	1/2008
Canasa (mesalamine rectal suppository)	Fasting	mesalamine	Mesalamine	NA	Differential scanning calorimetry: viscosity; melting point; and density	3/2013

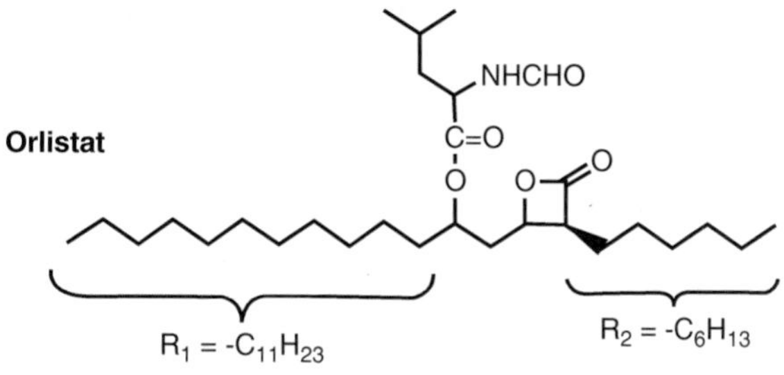

Fig. 12.4 Chemical structure of orlistat

in increased excretion of fat in the feces. Orlistat capsule, 120 mg is marketed as a prescription (Rx) product under the trade name Xenical®. Orlistat capsule, 60 mg is also marketed as Over-the-Counter (OTC) product as Alli®. Alli® (orlistat) capsule is the only FDA-approved weight loss medication as OTC.

As per Xenical® (orlistat) capsule Label, systemic absorption of orlistat is minimal (<2 % of the dose); approximately 97 % of the dose is excreted in feces. Plasma concentrations of orlistat are low (below 10 ng/mL), sporadic, and inadequate for pharmacokinetic analysis (Zhi et al. 1995). The low systemic exposure of orlistat at a dose of 120 mg three times daily is not expected to produce significant systemic lipase inhibition; thus, it is unlikely that systemic side effects occur.

Several publications have demonstrated that there is a dose–response relationship between the daily dose of orlistat and the percent of fecal fat excreted relative to daily fat intake (Zhi et al. 1994). A simple maximum effect (E_{max}) model was used to define the dose–response curve between Xenical® (orlistat) capsule daily dose and fecal fat excretion as representative of gastrointestinal lipase inhibition:

$$E = E_0 + E_{max} \cdot D/(ED_{50} + D)$$

where E is the intensity of the effect produced by orlistat treatment expressed as the percent of ingested fat excreted, E_0 is the intensity of the basal effect (no drug), E_{max} is the maximum attainable intensity of effect produced by orlistat alone, D is the orlistat daily dose (mg/day) administered as three divided doses, and ED_{50} is the orlistat daily dose which produces 50 % of the maximum effect. The dose–response curve exhibits a linear portion for doses up to approximately 400 mg daily, followed by a plateau for higher doses. The calculated ED_{50} is 98.1 mg/day. At doses greater than 120 mg three times a day, the percentage increase in effect is minimal.

Orlistat has low aqueous solubility; thus dissolution testing is not likely predictive of in vivo release in GI. The PD endpoint of fecal fat excretion (% FFE) after drug administration is a quantitative pharmacodynamic marker for measurement of orlistat bioavailability at the site of its local action. The FDA currently recommended the generic applicant conduct in vivo bioequivalence study with PD endpoint (%FFE) to assess BE for orlistat capsule (FDA 2010b).

The PD response generally shows nonlinear behavior with a linear portion only around the EC_{50} (O'Connor et al. 2011). The use of the PD response for BE evaluation to detect different delivered doses between generic and reference products is only possible at the linear portion of the dose–response curve. Therefore, a pilot study must be conducted in addition to a pivotal study to determine the appropriate dose that is on the linear portion of the PD dose–response curve. This approach has been used to demonstrate BE for topical corticosteroid products.

For orlistat capsules, it is apparent that a 60 mg dose given TID (above ED_{50}) is already in the nonlinear range of the curve. A given difference in response between two products in this dose range may be indicative of a considerably large difference in dose delivered to the site of action. Due to the nonlinear relationship at the labeled dosing regimen, the response scale assessments cannot accurately reflect difference in relative bioavailability of generic and reference products on the dose scale. Therefore, the FDA recommended a dose-scale method to establish BE of generic orlistat to the reference product. In this approach as presented in Fig. 12.5, we estimate the parameters of the dose–response relationship (assumed to be an E_{max} model) based on the response data for the reference product at different doses, and then use this fitted dose–response curve to determine what dose of the reference product (60 mg Orlistat capsule TID) would give the same mean response as observed in the study for the generic product (Gillespie et al. 1997). The assessment of BE is made in terms of relative bioavailability, F, the ratio of that hypothetical dose to the administered generic product dose. In this way, we are mapping the mean responses seen on the "response scale" back to the "dose scale."

As described in the FDA's product-specific BE recommendation for Orlistat capsules (FDA 2012f), a randomized three-way crossover study consisting of two doses of the reference product (60 mg and 2×60 mg) and one dose of generic product (60 mg) should be conducted. The baseline response is determined by including a run-in period in the beginning of the study. The mean or pooled, dose–response data of the reference product (baseline, 60 mg dose and 120 mg dose) are used to estimate the model parameters, E_{0R}, E_{maxR}, and ED_{50R} for the E_{max} model:

$$E_R = \varphi_R(D_R) = E_{0R} + \frac{E_{max\,R} \times D_R}{ED_{50\,R} \times D_R}$$

$E_{0R} =$ Baseline response
$E_{maxR} =$ Fitted maximum drug effect
$ED_{50R} =$ Dose required to produce 50 % of the fitted maximum effect

The relative bioavailability F of a dose of 60 mg generic product to that of the reference product can be calculated by applying the inverse of φ_R to the mean of response data of the test product, E_{Test}: $F = \varphi_R - 1(E_{Test})/60$ mg. A 90 % confidence interval for "F" of the test product is estimated by a bootstrap procedure. Each bootstrap estimation includes calculation of "F" by fitting the above model to a "sample dose–response dataset," which is generated by repetitive sampling with replacement. The FDA recommended that the 90 % confidence interval for the relative bioavailability, F, must fall within 80–125 % in order to establish bioequivalence. The dose-scale method using a single dose of the test product is acceptable. However, it is acknowledged that the use of multiple doses of both test and reference products may enrich the study data and enhance precision of the estimated values.

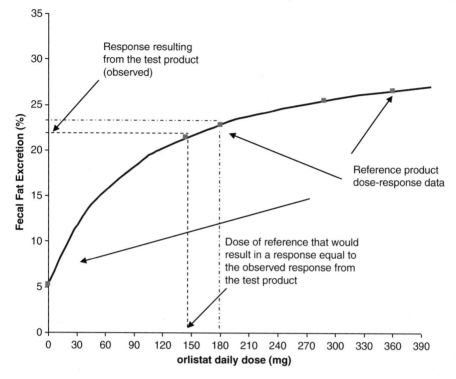

Fig. 12.5 Comparison of difference in PD response and delivered dose for orlistat. PD response ratio: test/reference = 21.41/22.83 = 0.94. Delivered dose ratio: test/reference = 144 mg/ 180 mg = 0.8

12.3.3   Vancomycin Capsules

Vancomycin is a tricyclic glycopeptide antibiotic derived from *Amycolatopasis orientalis* (formerly *Nocardia orientalis*). Orally administered Vancomycin capsule is acting locally for the treatment of enterocolitis caused by *Staphylococcus aureus* (including methicillin-resistant strains) and antibiotic-associated pseudomembranous colitis caused by *C. difficile* (*label*). Specifically, vancomycin acts by inhibition of cell-wall biosynthesis of *Staphylococcus aureus* and the vegetative cells of *Clostridium difficile*. In addition, vancomycin alters bacterial-cell-membrane permeability and RNA synthesis. Vancomycin capsules are marketed in strengths of 125 and 250 mg. The usually daily dosage for adults is 500 mg to 2 g administered orally in three or four divided doses for 7–10 days (FDA 2011c).

Vancomycin capsule is poorly absorbed after oral administration. As per the label, after multiple dosing of 250 mg every 8 h for seven doses, no measurable blood concentrations were attained in normal volunteers and urinary recovery did not exceed 0.76 %. However, the fecal vancomycin concentrations exceeded 100 μg/g in the majority of samples, which is 10 to 100-fold higher than the highest minimum inhibitory concentration (MIC), about 1–8 μg/mL (Bartlett 2008). Therefore, plasma and urine concentrations of vancomycin are generally undetectable following oral administration (FDA 2008c).

Based on the FDA's publicly available solubility study, vancomycin is considered a highly soluble drug substance over the pH range of 1–7.5, according to the BCS guidance (FDA 2000, 2008h). The FDA's laboratory also conducted studies to determine the dissolution characteristics of the reference product, Vancocin[®] (vancomycin HCl) capsules. The dissolution data showed that Vancocin[®] (vancomycin HCl) capsules met the FDA's definition of a rapidly dissolving drug product at pH 1.2, where greater than 85 % was dissolved at 30 min (FDA 2008i). At pH 4.5, Vancocin[®] (vacomycin HCl) capsules generally dissolve more than 85 % in 45 min. It requires 60 min for Vancocin[®] (vancomycin HCl) capsules to dissolve more than 85 % at pH 6.8 (FDA 2009a, b).

As per the product-specific guidance on Vancomycin HCl Capsules issued in December 2008, the FDA recommended two options: in vitro bioequivalence studies or in vivo BE studies with a clinical endpoint. In order for an in vitro BE study approach to be acceptable, the test product should be $Q1$ and $Q2$ the same as the reference product with respect to inactive ingredients. The test product must also contain the same amount of active pharmaceutical ingredient (API) as the reference product. The in vitro BE study is comprised of dissolution testing in three media, representative of pH conditions throughout the GI tract: 0.1 N HCl (or 0.1 N HCl with NaCl at pH 1.2), pH 4.5 Acetate buffer, and pH 6.8 Phosphate buffer. The dissolution profile similarity factor metric f_2 comparing test (T) versus reference (R) product in each medium should meet criteria established by the FDA. If the test product formulations are not $Q1$ and $Q2$ the same as the reference product with respective to inactive ingredients, then an in vivo study with clinical endpoints in patients with C. difficile-Associated Diarrhea (CDAD) should be conducted.

The FDA determined that in vitro dissolution testing in different media (pH 1.2, 4.5, and 6.8) alone are appropriate to demonstrate equivalent release of vancomycin for two formulations that are $Q1$ and $Q2$ the same for the following reasons:

- In vitro dissolution is the most sensitive method to detect differences in manufacturing processes and related to rate and extent to which vancomycin becomes available at the local site of action (Davit 2009).
- The high solubility over pH range from 1 to 7.5 and relatively rapid dissolution in stomach and upper GI tract (>85 % in 60 min) ensures that vancomycin is largely in solution by the time (about 3–4 h transit times) the drug enters the site of action in the lower GI tract (allowing it to attain concentration well in excess of the MIC of the target organism) (FDA 2008c). In light of this fact, in vitro dissolution is highly predictive of in vivo dissolution for vancomycin capsules.
- Equivalent dissolution profiles using similarity factor (f_2) comparison across the pH ranges ensure that there are no significant differences between the generic

and reference vancomycin dissolution profile, i.e., the dissolution of a generic Vancomycin HCl Capsule product is neither faster nor slower than that of the reference product (FDA 2009b). As such, equivalent local availability will be ensured in patients that have variable fluid volume, transition time, and pH.

- The $Q1$ and $Q2$ the same ensure the equivalent effect of inactive ingredients on the systemic absorption, GI transit, and drug-excipient interaction at the site of action between the generic and reference formulations (FDA 2008c).
- Minimal systemic exposure indicates there is no systemic safety concern. Furthermore, formulation factors that could change systemic exposure are ensured to be equivalent by dissolution tests and use of the same inactive ingredients (Davit 2009).

Since microbial assay as listed in the USP monograph appears to be less sensitive, specific, and reproducible than the HPLC method, the FDA asks the applicants to use a validated HPLC method to assay for vancomycin in the three dissolution media (pH 1.2, 4.5, and 6.8) (FDA 2012f). The FDA uses the f_2 metric to compare dissolution profiles of the generic and reference products in the vancomycin in intro bioequivalence study. The f_2 metric is widely used by the FDA to compare dissolution profiles. It is calculated according to the following algorithm (FDA 1995):

$$ f_2 = 50 \log \left\{ \left[1 + \left(\frac{1}{n} \right) \sum_{t=1}^{n} (R_t - T_t)^2 \right]^{-0.5} \times 100 \right\} $$

where f_2 is the similarity factor, n is the number of time points, R_t is the dissolution value of the reference batch at time t, and T_t is the dissolution value of the test batch at time t. f_2 values not less than 50 indicates the equivalence of the two profiles.

The dissolution guidance describes several prerequisites to use the f_2 test: (1) the f_2 calculation should include three or four or more time points but only one should be more than 85 % of drug release; (2) the dissolution measurements of the two compared products should be made under the same conditions and using the same sampling intervals; (3) to allow use of mean percentage dissolution data for the f_2 test, the percent coefficient of variation at the earlier time points should not be more than 20 %, and at other time points should not be more than 10 %. A review of the vancomycin dissolution data available to the FDA demonstrates that dissolution data of vancomycin exhibit high variability. Such high variability precludes the use of f_2 calculation using the mean values. In this case, the FDA recommended the applicants to use a statistical approach (bootstrapping method) to evaluate the f_2 confidence interval (Shah et al. 1998). The FDA also recommended the firm to conduct dissolution testing on at least 24 capsules to provide a better estimate of the mean difference.

If a generic vancomycin capsule is not $Q1$ and $Q2$ the same to the reference product, it is possible that different excipients or different amounts of an excipient may interact with the ability of the drug to inhibit bacterial-cell-wall synthesis, change the GI transit times of the drug, or alter the systemic exposure of vancomycin. Under this circumstance, the FDA recommended an in vivo bioequivalence study with clinical endpoints in patient with *C. difficile*-associated diarrhea to demonstrate BE (FDA 2009b).

12.3.4 Lanthanum Carbonate Chewable Tablets

Lanthanum carbonate was approved by the FDA on October 26, 2004 for treatment of hyperphosphatemia in the management of patients with end-stage renal diseases (ESRDs). Lanthanum carbonate is marketed as a chewable tablet in three dosage strengths for oral administration: 500, 750, and 1,000 mg. Each tablet contains lanthanum carbonate hydrate equivalent to 500, 750, or 1,000 mg of elemental lanthanum and the following inactive ingredients: dextrates (hydrate), colloidal silicon dioxide, magnesium stearate, and talc. Currently there are no generic lanthanum carbonate tablets on the market.

Hyperphosphatemia is an electrolyte disbalance in which there is an abnormally elevated level of phosphate in serum. It develops in the majority of patients with ESRD and is associated with secondary hyperparathyroidism, metabolic bone disease, soft tissue calcification, and possibly cardiovascular calcification resulting in significant morbidity and mortality (Berner and Shike 1988; Block et al. 1998, 2004; Goodman et al. 2000). Adequate control of serum phosphate remains a cornerstone in the clinical management of patients with ESRD. These measures include dietary phosphorus restriction, dialysis, and oral phosphate binders (Coladonato 2005). Dietary phosphate restriction and dialysis often are insufficient to control serum phosphorus to the levels of 3.5–5.5 mg/dL recommended by the National Kidney Foundation's Kidney Disease Outcomes Quality Initiative (KDOQI) (Kidney Disease: Improving Global Outcomes KDIGO CKD-MBD Work Group 2009). The use of phosphate binders is therefore a major therapeutic consideration to maintain phosphate balance and to prevent hyperphosphatemia (Loghman-Adham 1999; McIntyre 2007). Because of the toxicity, aluminum-containing phosphate binders are no longer used (Alfrey et al. 1976; Bellasi et al. 2006). For calcium-based phosphate binders, such as calcium carbonate and calcium acetate, large doses are often required; thus, hypercalcemia is a potential complication that maybe linked to adynamic bone disease and cardiovascular calcification (Slatopolsky et al. 1986; Braun et al. 2004). Recently, the aluminum-free and calcium-free phosphate binding agents including sevelamer hydrochloride and lanthanum carbonate have been approved for the prevention and treatment of hyperphosphatemia in patients with ESRD (Albaaj and Hutchison 2005; Pai et al. 2009; Wrong and Harland 2007; Tonelli et al. 2007; Sprague 2007; Martin et al. 2011).

Lanthanum carbonate is acting locally in the GI tract. When given orally, lanthanum carbonate dissociates in the acidic environment of the upper gastrointestinal tract to release lanthanum ions which are trivalent cations with a high affinity for oxygen-donor atoms, especially phosphates. These cations bind to ingested phosphates to form insoluble, nonabsorbable lanthanum-phosphate complexes, which are excreted in the feces (Behets et al. 2004). The oral absorption of lanthanum carbonate is very low with a bioavailability of less than 0.002 % (Damment and Pennick 2008; Pennick et al. 2006). Therefore, a conventional in vivo bioequivalence study with a PK endpoint is not feasible. Several approaches including theoretical calculations, in vitro binding test and in vivo study were used to compare and predict the binding effectiveness of a variety of phosphate binders. According to the published data, the in vivo performance of phosphate binders could be predicted through theoretical and in vitro methods (Sheikh et al. 1989). The FDA product-specific guidance on lanthanum carbonate recommended that the generic applicants can conduct either (1) in vitro dissolution testing and phosphate binding studies in varying pH expected throughout the GI tract comparing the generic and the reference products, or (2) a study using pharmacodynamic endpoints in healthy subjects, to demonstrate BE (FDA 2011d).

The FDA determined that in vitro dissolution testing and phosphate binding studies are appropriate to demonstrate BE for the following reasons: (1) In vitro dissolution testing is a sensitive method to detect differences in manufacturing processes and related to rate and extent to which lanthanum carbonate becomes available at the local site of action. Equivalent dissolution profiles using similarity factor (f_2) comparison across the pH ranges ensure that there are no significant differences between the generic and reference lanthanum carbonate in dissolution profiles; (2) In vitro binding test is used to compare the extent and rate of phosphate binding between the generic and reference products. Equivalent binding extent using the ratio of k_1 and the 90 % confidence interval of k_2 between the generic and reference products ensure that the effect of the inactive ingredients in the generic or reference products is equivalent on phosphate binding extent. The 90 % confidence interval of k_2 between the generic and reference products should be within 80–125 %. Equivalent kinetic binding profiles using similarity factor (f_2) comparison across the pH ranges ensure that there are no significant differences between the generic and reference lanthanum carbonate in binding rate and profiles. Therefore, the $Q1$ and $Q2$ the same is not required.

12.3.4.1 Dissolution Testing

Lanthanum carbonate is practically insoluble in water. It has poor aqueous solubility at an alkaline pH with increasing solubility in acid environment (Product Monograph 2010). The formulation, being a chewable tablet, does not contain any disintegrating agent. The chewing process of patients could help to release the drug, which then can become available at the site of action. Occasionally, some patients might swallow the whole lanthanum carbonate tablet instead of chewing it as needed. Hence, it is useful to determine drug release from both whole and crushed tablets. In an effort to simulate the chewing condition, tablets were crushed using a mortar and pestle to very small uniform pieces, but not ground.

It is important to determine the dissolution profile at various pH conditions which simulate GI tract conditions. Dissolution media with pH 1.2, 3.0, 4.5, and 6.8 were selected for conducting drug release studies of lanthanum carbonate chewable tablets. Selection of the dissolution apparatus and test conditions is important in determining dissolution of chewable tablets. Maintenance of sink conditions for an extremely low aqueous soluble drug such as lanthanum carbonate should be considered. For lanthanum carbonate chewable tablets, the goal was to develop a comparative dissolution test for the reference as well as the generic products. For the study, USP apparatus II (paddle) was selected with 900 mL of dissolution media (0.1 N HCl, pH 3.0, 4.5, and 6.8 buffers). Phosphate containing dissolution media were not used as the drug binds to phosphate to form an insoluble complex, thus retarding dissolution. Therefore, acetate and borate buffers should be used. The paddle was stirred at 50 rpm, and the temperature of the dissolution media was maintained at 37 °C. All standards and samples should be analyzed by a validated method. An f_2 test should be performed using mean profiles to compare test (T) and reference (R) product drug release under a range of pH conditions.

12.3.4.2   Phosphate Binding Studies (Yang et al. 2013a, b)

The FDA recommended both, (1) kinetic and (2) equilibrium binding studies for the in vitro binding studies as per the guidance documents (FDA 2011d).

12.3.4.2.1   Kinetic Phosphate Binding Study

The main purpose of conducting the kinetic binding study is to assess the rate of binding as well as the time to reach the binding equilibrium when the concentration of phosphate binding solution is fixed. The highest strength (1,000 mg tablet) should be tested for both kinetic and equilibrium binding study. Since the formulation of lanthanum carbonate is a chewable tablet and it needs to be chewed before swallowing, crushed tablets were also tested. For normal subjects, average gastric emptying time (80 % of content) is approximately 30 min for a low-calorie bland meal, and 3.5 h for a high-fat liquid meal (Houghton et al. 1990). The intestinal residence time in normal subjects varies from 20 to 30 h due to the influence by many factors (Read et al. 1980). To simulate the physiologically relevant circumstance, the kinetic binding study should be performed for at least eight time points up to 24 h.

The rate at which lanthanum carbonate dissolves in a given medium depends mainly upon solubility, pH, amount of lanthanum carbonate, the stirring speed, and temperature. The pH of the medium affects solubility of the phosphate binder. Lanthanum carbonate, for example, is more soluble at low pH (Product Monograph 2010). Thus in 0.1 N HCl more lanthanum was released from lanthanum carbonate tablets, whereas only a small amount was released at higher pH (pH 6.8). To be effective, lanthanum carbonate must be dissociated to release lanthanum ions which in turn bind to phosphates to form highly insoluble lanthanum-phosphate complexes. This dissociation process is a pH-dependent process with the more dissociation occurring in the acid environment than in the basic condition (Behets et al. 2004; Damment and Pennick 2008). Therefore, it can be speculated that pH should have significant impact on the lanthanum carbonate binding to phosphate.

To reduce dietary phosphate absorption, a phosphate binder must mix with food and precipitate or adsorb meal phosphate before it is absorbed by the small intestine. Therefore, a phosphate binder should be administered with or immediately after food to ensure adequate contact and mixing with dietary phosphate. The mixing of food phosphate and the binder can occur in the stomach and upper small intestine as food phosphate is readily solubilized in the upper gastrointestinal tract. Because most phosphate is believed to be absorbed by the small intestine (Davis et al. 1983), and since most of ingested food passes through the stomach and small intestine in 4–6 h (Read et al. 1980), the binding reaction needs to occur within this time period if phosphate absorption is to be prevented.

To take these factors into consideration and in order to mimic the physiologically relevant pH conditions encountered by lanthanum carbonate when traveling through the GI tract after oral administration, the phosphate binding to lanthanum carbonate should be conducted under the pH conditions of 1.2–5.0 with up to 24 h of incubation time. Both whole and crushed tablets of reference and test products should be tested.

In general 1,000–1,200 mg of phosphate is ingested from a regular meal per day (Hruska et al. 2008). Assuming that phosphate exists as a mixture of PO_4^{3-} (MW $= 95$ g/mol), HPO_4^{2-} (MW $= 96$ g/mol), $H_2PO_4^-$ (MW $= 97$ g/mol), or H_3PO_4 (MW $= 98$ g/mol), 345–1,000 mg of phosphate in 250 mL equals to 14–41 mM. Therefore, the concentration of phosphate should fully cover the range of 14–41 mM in the kinetic binding study.

12.3.4.2.2 Equilibrium Binding Study

The equilibrium study is used to determine the affinity and capacity binding constants. It should be conducted under conditions of constant time and varying concentrations of phosphate. The constant time represents the time at which the binding equilibrium is reached that is determined by a kinetic binding study. A sufficient number of phosphate concentrations need to be studied to provide accurate estimates of k_1 and k_2. Thus, concentrations studied should be spaced along the spectrum from the linear binding range until maximum binding is clearly established. Typically, this would include at least two concentrations that vary linearly with concentrations, and at least two concentrations resulting in maximum binding. In addition, two concentrations falling below k_d and two concentrations falling above k_d on the convex portion of the curve should be included.

The binding of phosphate molecules from a solution to the drug product at constant temperature can be described by the following Langmuir-type equation (Gessner and Hasan 1987).

$$\frac{x}{m} = \frac{k_1 k_2 C_{eq}}{1 + k_1 C_{eq}} \qquad (12.1)$$

Upon rearranging, Eq. 12.2 is obtained:

$$\frac{C_{eq}}{x/m} = \frac{1}{k_1 k_2} + \frac{1}{k_2} C_{eq} \qquad (12.2)$$

where:
C_{eq} = phosphate concentration remaining in the solution at equilibrium.
x = amount of phosphate bound to the drug product at equilibrium.
m = the amount of drug product used.
k_1 = affinity binding constant which is related to the magnitude of the forces involved in the binding process.
k_2 = the Langmuir-capacity constant which indicates the apparent maximum amount of phosphate that can be bound per unit weight of drug product.

From equilibrium binding experiment, C_{eq} expressed in micromoles can be obtained; m expressed in g is known factor; x can be obtained by subtraction of the C_{eq} from total amount of added phosphate.

A plot of $C_{eq}/(x/m)$ versus C_{eq} on rectilinear coordinates may yield a straight line. Application of regression analysis will yield a slope (a) and intercept (b) of the line. The affinity constant k_1, and capacity constant k_2 can be calculated from the slope and intercept as follows: $k_1 = a/b$; $k_2 = 1/a$; $k_d = b/a$.

12.3.5 Colesevelam Hydrochloride

Colesevelam hydrochloride is a nonabsorbed lipid-lowering agent approved for use alone or in combination with hydroxymethylglutaryl–coenzyme A (HMGCoA) reductase inhibitors for the reduction of low-density-lipoprotein (LDL) cholesterol in patients with primary hypercholesterolemia (Steinmetz 2002). Colesevelam hydrochloride is marketed as film-coated solid tablet containing 625 mg of drug. The clinical efficacy studies have revealed significant lipid- and glucose-lowering effects of colesevelam as add-on treatment to existing metformin, sulfonylurea, or insulin therapy in type 2 diabetes (Bays 2011; Bays et al. 2008; Fonseca et al. 2008; Goldberg et al. 2008). Therefore colesevelam has been approved as adjunctive therapy to diet and exercise to improve glycemia in patients with type 2 diabetes in USA (Younk and Davis 2012; Aggarwal et al. 2012). After oral administration, it undergoes very limited systemic absorption with minimal or trace tissue concentrations (Heller et al. 2002). This locally acting polymeric gel contains both cationic and hydrophobic sites, making it apt to bind to anionic, hydrophobic bile acids in the intestinal tract with higher affinity than conventional bile acid sequestrants (Braunlin et al. 2000). Conventional in vivo bioequivalence study with pharmacokinetic (PK) endpoints such as C_{max} and AUC is neither appropriate nor feasible for this locally acting drug. Therefore, the FDA draft guidance on colesevelam hydrochloride (FDA 2011f) recommends that In vitro bile salt binding study be used to demonstrate bioequivalence of colesevelam hydrochloride, since in vitro binding studies are less variable, easier to control, and more likely to detect differences between drug products if they exist (Chow et al. 2003).

Equivalent binding extent using the ratio of k_1 and the 90 % confidence interval of k_2 between the generic and reference products ensure that effect of the inactive ingredients in the generic or reference products is not significant on phosphate binding extent. Equivalent kinetic binding profiles using similarity factor (f_2) comparison across the pH ranges ensure that there are no significant differences between the generic and reference lanthanum carbonate in binding rate and profiles. Therefore, the $Q1$ and $Q2$ the same is not required.

The in vitro bile salt binding study is based on the pharmacological action of colesevelam hydrochloride (Steinmetz 2002). Colesevelam forms nonabsorbable complexes with bile acids in the GI tract and subsequently eliminated. Impairment of bile acid return leads to up-regulation of hepatic bile acid synthesis through breakdown of cholesterol which results in an increased clearance of LDL-C. Thus, the clinical efficacy of colesevelam depends on its binding capacity to intestinal bile acids. The in vitro binding capacity of colesevelam hydrochloride with bile acid sodium salts of glycocholic acid (GC), glycochenodeoxycholic acid (GCDA), and taurodeoxycholic acid (TDCA) is an alternative approach to evaluate the efficacy of the locally acting colesevelam drug products. The in vitro equilibrium binding study is considered as a pivotal bioequivalence study for colesevelam hydrochloride tablets. The kinetic binding study is used to support the pivotal equilibrium binding study. Kinetic binding study was carried out with constant initial bile salt concentrations as a function of time. Equilibrium binding studies were conducted under conditions of constant incubation time and varying initial concentrations of bile acid sodium salts. The unbound concentration of bile salts was determined in the samples of these studies. Langmuir equation was utilized to calculate the binding constants k_1 and k_2.

Recently published data (Krishnaiah et al. 2013) showed that the bile salt binding to both test and reference colesevelam hydrochloride tablets reached equilibrium at about 3 h. The similarity factor (f_2) was 99.5 based on the binding profile of total bile salts to the test and reference colesevelam tablets as a function of time. This suggests that both the test and reference tablet products have similar in vitro bile acid binding profiles indicating no difference in the binding rate. The equilibrium binding studies were conducted under conditions of constant time and varying initial concentrations of bile acid sodium salts in simulated intestinal fluid (SIF). The mean values of capacity constants (k_2) of GC, GCDA, and TDCA for test and reference tablets exhibited low binding capacity to GC and high binding capacity to GCDA and TDCA. The 90 % confidence intervals for the test to reference ratio of k_2 values were 96.06–112.07 which is within the acceptance criteria of 80–120 % to demonstrate BE.

12.4 Conclusion

As discussed above, the selection of the bioequivalence method for a local drug is based on product-specific factors and a scientific understanding of the product's mechanism of action (Lionberger 2008). Though a PK study is not a good surrogate for pharmacological effect but it is sensitive to detect formulation difference and reflect local drug available. The recent development in the utility of the pAUC as a profile comparison tool further enhanced the role of PK profiles in comparing formulation difference. Dissolution has been recognized as the most sensitive and direct method to measure the local drug availability for locally acting GI drugs. Therefore, PK studies together with dissolution testing in different media are recommended for demonstration of BE for mesalamine products, as mesalamine can be rapidly absorbed through the GI tract. Dissolution testing in multimedia with profile comparison alone is adequate to ensure BE for vancomycin capsule that has $Q1$ and $Q2$ the same formulation to the reference product since vancomycin is highly soluble and poorly absorbed. For orlistat capsule, where the systemic absorption is undetectable and dissolution is not predictive of in vivo dissolution, the PD endpoint or clinical endpoint study can be used.

With the latest scientific advances and new data available, the FDA developed in vivo or in vitro alternative method to replace clinical endpoint study in establishing BE for locally acting GI products. The improved BE methodology greatly reduce the unnecessary human studies and industries' regulatory burden, and accelerate drug development and approval process. Recent approval of generic vancomycin capsules and guidance issuing for mesalamine product is the success in this area. For those locally acting GI products, the current BE recommendation still rests on clinical endpoint study (e.g., low solubility drugs), the FDA is actively working on identifying alternative methods.

References

Aggarwal S, Loomba RS, Arora RR (2012) Efficacy of colesevelam on lowering glycemia and lipids. J Cardiovasc Pharmacol 59:198–205

Albaaj F, Hutchison AJ (2005) Lanthanum carbonate for the treatment of hyperphosphataemia in renal failure and dialysis patients. Expert Opin Pharmacother 6(2):319–328

Alfrey AC, LeGendre GR, Kaehny WD (1976) The dialysis encephalopathy syndrome. Possible aluminum intoxication. N Engl J Med 294(4):184–188

Amidon GL (2004) Bioequivalence testing for locally acting gastrointestinal drugs: scientific principles. Presented at FDA meeting of the Advisory Committee for Pharmaceutical Science and Clinical Pharmacology, Oct 20. Available from http://www.fda.gov/ohrms/dockets/ac/04/slides/2004-4078S2_10_Amidon_files/frame.htm. Accessed Oct 2013

Aumais G, Lefebvre M, Tremblay C, Bitton A, Martin F, Giard A, Madi M, Spénard J (2003) Rectal tissue, plasma and urine concentrations of mesalamine after single and multiple administrations of 500 mg suppositories to healthy volunteers and ulcertive proctitis patients. Aliment Pharmacol Ther 17(1):93–97

Bartlett JG (2008) Historical perspectives on studies of Clostridium difficile and C. difficile infection. Clin Infect Dis 46(suppl 1):S4–S11

Bays HE (2011) Colesevelam hydrochloride added to background metformin therapy in patients with type 2 diabetes mellitus: a pooled analysis from 3 clinical studies. Endocr Pract 17:933–938

Bays HE, Goldberg RB, Truitt KE, Jones MR (2008) Colesevelam hydrochloride therapy in patients with type 2 diabetes mellitus treated with metformin—glucose and lipid effects. Arch Intern Med 168:1975–1983

Behets GJ, Verberckmoes SC, Haese PC, De Broe ME (2004) Lanthanum carbonate: a new phosphate binder. Curr Opin Nephrol Hypertens 13(4):403–409

Bellasi A, Kooienga L, Block GA (2006) Phosphate binders: new products and challenges. Hemodial Int 10(3):225–234

Berner YN, Shike M (1988) Consequences of phosphate imbalance. Annu Rev Nutr 8(1):121–148

Block GA, Hulbert-Shearon TE, Levin NW, Port FK (1998) Association of serum phosphorus and calcium x phosphate product with mortality risk in chronic hemodialysis patients: a national study. Am J Kidney Dis 31(4):607–617

Block GA, Klassen PS, Lazarus JM, Ofsthun N, Lowrie EG, Chertow GM (2004) Mineral metabolism, mortality, and morbidity in maintenance hemodialysis. J Am Soc Nephrol 15(8):2208–2218

Braun J, Asmus HG, Holzer H, Brunkhorst R, Krause R, Schulz W, Neumayer HH, Raggi P, Bommer J (2004) Long-term comparison of a calcium-free phosphate binder and calcium carbonate–phosphorus metabolism and cardiovascular calcification. Clin Nephrol 62(2):104–115

Braunlin WH, Holmes-Farley SR, Smisek D, Guo A, Appruzese W, Xu QW, Hook P, Zhorov E, Mandeville H (2000) In vitro comparison of bile acid-binding to colesevelam hydrochloride and other bile acid sequestrants. Abstr Pap Am Chem Soc 219:U409–U410

Brown J, Haines S, Wilding IR (1997) Colonic spread of three rectally administered mesalazine (Pentasa) dosage forms in healthy volunteers as assessed by gamma scintigraphy. Aliment Pharmacol Ther 11(4):685–691

Chow SC, Liu JP (2009) In vitro bioequivalence testing. In: Design and analysis of bioavailability and bioequivalence studies, 3rd edn. Taylor & Francis Group, Boca Raton, FL, pp 451–452

Chow SC, Shao J, Wang HS (2003) In vitro bioequivalence testing. Stat Med 22:55–68

Chuong MC, Christensen JM, Ayres JW (2008) New dissolution method for mesalamine tablets and capsules. Dissolution Technol 15:7–14

Coladonato JA (2005) Control of Hypophosphatemia among Patients with ESRD. J Am Soc Nephrol 16(suppl 2):107–114

Conner DP, Davit BM (2008) Bioequivalence and drug product assessment, in vivo. In: Shargel L, Kanfer I (eds) Generic drug development: solid oral dosage forms. Marcel Dekker, New York, pp 275–277

Damment SJ, Pennick M (2008) Clinical pharmacokinetics of the phosphate binder lanthanum carbonate. J Clin Pharmacokinet 47(9):553–563

Davis GR, Zerwekh JE, Parker TF, Krejs GJ, Pak CTC, Fordtran JS (1983) Absorption of phosphate in the jejunum of patients with chronic renal failure before and after correction of Vitamin D deficiency. Gastroenterology 85(4):908–916

Davit BM (2009) FDA recommendation for vancomycin HCl Capsule BE studies. Presented at FDA Meeting of the Advisory Committee for Pharmaceutical Science and Clinical Pharmacology, 4 Aug 2009. Available from http://www.fda.gov/downloads/AdvisoryCommittees/CommitteesMeetingMaterials/Drugs/AdvisoryCommitteeforPharmaceuticalScienceand ClinicalPharmacology/UCM179419.pdf. Accessed Oct 2013

Davit BM (2010) Regulatory perspective on developing dissolution method and specifications for generic drug products in the US. Presented at Informa Life Sciences second annual dissolution conference, Barcelona, Spain, Oct 2010

Davit BM (2013) Regulatory approaches for generic drugs: BE of topical drug product. Presented at PQRI workshop, evaluation of new and generic topical drug product, Bethesda, MD, 11–13 March 2013

FDA (1995) Guidance for industry: immediate release solid oral dosage forms: scale-up and post-approval changes. Available from http://www.fda.gov/downloads/Drugs/Guidances/UCM070636.pdf. Accessed Oct 2013

FDA (2000) Guidance for industry: Waiver of in vivo bioavailability and bioequivalence studies for immediate release solid oral dosage forms based on a biopharmaceutics classification system. Available from http://www.fda.gov/downloads/Drugs/GuidanceComplianceRegulatory Information/Guidances/ucm070246.pdf. Accessed Oct 2013

FDA (2002) Guidance for industry: food-effect bioavailability and fed bioequivalence studies. Available from http://www.fda.gov/downloads/Drugs/GuidanceComplianceRegulatory Information/Guidances/UCM070241.pdf. Accessed Dec 2013

FDA (2003) Guidance for industry: bioavailability and bioequivalence studies for orally administered drug products—general considerations. Available from http://www.fda.gov/down loads/Drugs/GuidanceComplianceRegulatoryInformation/Guidances/UCM070124.pdf. Accessed Oct 2013

FDA (2004) FDA meeting of the Advisory Committee for Pharmaceutical Science and Clinical Pharmacology, Oct 2004 briefing information. Available from http://www.fda.gov/ohrms/dockets/ac/04/briefing/2004-4078B1_07_Bioequivalence-Testing.pdf. Accessed Oct 2013

FDA (2006) FDA approved label: Dipentum® (Olsalazine sodium) capsule. UCB, Inc., Smyrna, GA

FDA (2007) FDA response to Citizen's Petition for Colazal®, Doc#FDA-2005P-0314, Dec. Available from http://www.regulations.gov/#!documentDetail;D=FDA-2005-P-0314-0004. Accessed Oct 2013

FDA (2008a) FDA meeting of the Advisory Committee for Pharmaceutical Science and Clinical Pharmacology, July 2008 briefing information. http://www.fda.gov/ohrms/dockets/ac/08/brief ing/2008-4370b1-01-FDA.pdf. Accessed Oct 2013

FDA (2008b) FDA meeting of the Advisory Committee for Pharmaceutical Science and Clinical Pharmacology, July 2008 summary minutes. http://www.fda.gov/ohrms/dockets/ac/08/minutes/2008-4370m2-Final%20Minutes.pdf. Accessed Oct 2013

FDA (2008c) Draft guidance on vancomycin capsules. Available from http://www.fda.gov/down loads/Drugs/GuidanceComplianceRegulatoryInformation/Guidances/UCM082278.pdf. Accessed Oct 2013

FDA (2008d) Draft guidance on balsalazide disodium capsule. Available from http://www.fda. gov/downloads/Drugs/GuidanceComplianceRegulatoryInformation/Guidances/ucm082854. pdf. Accessed Oct 2013

FDA (2008e) Draft guidance on olsalazine sodium capsule. Available from http://www.fda.gov/downloads/Drugs/GuidanceComplianceRegulatoryInformation/Guidances/ucm089229.pdf. Accessed Oct 2013

FDA (2008f) FDA approved label: Rowasa® (mesalamine) enema. Alaven Pharmaceutical LLC, Marietta, GA

FDA (2008g) Draft guidance on Rowasa. Available from http://www.fda.gov/downloads/Drugs/GuidanceComplianceRegulatoryInformation/Guidances/UCM088662.pdf. Accessed Oct 2013

FDA (2008h) Vancomycin solubility study report. Available from http://www.fda.gov/downloads/Drugs/GuidanceComplianceRegulatoryInformation/Guidances/ucm082291.pdf. Accessed Oct 2013

FDA (2008i) Vancomycin dissolution study report. Available from http://www.fda.gov/down loads/Drugs/GuidanceComplianceRegulatoryInformation/Guidances/ucm082295.pdf. Accessed Oct 2013

FDA (2009a) FDA meeting of the Advisory Committee for Pharmaceutical Science and Clinical Pharmacology, Aug 2009 summary minutes. Available from http://www.fda.gov/downloads/

AdvisoryCommittees/CommitteesMeetingMaterials/Drugs/
AdvisoryCommitteeforPharmaceuticalScienceandClinicalPharmacology/UCM237493.pdf.
Accessed Oct 2013

FDA (2009b) FDA meeting of the Advisory Committee for Pharmaceutical Science and Clinical
Pharmacology, Aug, 2009 Briefing information. Available from http://www.fda.gov/down
loads/AdvisoryCommittees/CommitteesMeetingMaterials/Drugs/
AdvisoryCommitteeforPharmaceuticalScienceandClinicalPharmacology/UCM173220.pdf.
Accessed Oct 2013

FDA (2009c) Draft guidance on Acarbose tablets. Available from http://www.fda.gov/downloads/
Drugs/GuidanceComplianceRegulatoryInformation/Guidances/UCM170242.pdf.   Accessed
Oct 2013

FDA (2010a) FDA response to Citizen's Petition for Asacol® and Pentasa®, Doc#FDA-2010-P-
0111 and FDA-2008-P-0507, Aug 2010. Available from http://www.regulations.gov/#!
documentDetail;D=FDA-2010-P-0111-0011. Accessed Oct 2013

FDA (2010b) Draft guidance on Orlistat capsules. Available from http://www.fda.gov/downloads/
Drugs/GuidanceComplianceRegulatoryInformation/Guidances/UCM082278.pdf.   Accessed
Oct 2013

FDA (2010c) Draft guidance on Sulfasalazine DR tablets. Available from http://www.fda.gov/
downloads/Drugs/GuidanceComplianceRegulatoryInformation/Guidances/UCM199673.pdf.
Accessed Oct 2013

FDA (2010d) Draft guidance on Sulfasalazine tablets. Available from http://www.fda.gov/down
loads/Drugs/GuidanceComplianceRegulatoryInformation/Guidances/UCM199674.pdf.
Accessed Oct 2013

FDA (2010e) FDA approved label: Asacol® HD (mesalamine) DR tablets. Warner Chilcott
(US) LLC, New Jersey

FDA (2011a) FDA approved label: Asacol® (mesalamine) DR tablet. Warner Chilcott (US) LLC,
Rockaway, NJ

FDA (2011b) FDA approved label: Liada® (mesalamine) DR tablet. Shire US manufacturing Inc.,
Owings Mills, MD

FDA (2011c) FDA approved label: Vancocin® (vancomycin) capsule. ViroPharma Inc, Exton, PA

FDA (2011d) Draft guidance on lanthanum carbonate. Available from http://www.fda.gov/down
loads/Drugs/GuidanceComplianceRegulatoryInformation/Guidances/UCM270541.pdf.
Accessed Oct 2013

FDA (2011e) Draft guidance on calcium acetate capsules. Available from http://www.fda.gov/
downloads/Drugs/GuidanceComplianceRegulatoryInformation/Guidances/UCM179176.pdf.
Accessed Oct 2013

FDA (2011f) Draft guidance on colesevelam hydrochloride. Available at http://www.fda.gov/
downloads/Drugs/GuidanceComplianceRegulatoryInformation/Guidances/ucm083337.pdf.
Accessed Oct 2013

FDA (2011g) Draft guidance on Sevelamer carbonate tablets. Available from http://www.fda.gov/
downloads/Drugs/GuidanceComplianceRegulatoryInformation/Guidances/UCM089620.pdf.
Accessed Oct 2013

FDA (2012a) Draft guidance on Asacol. Available from http://www.fda.gov/downloads/Drugs/
GuidanceComplianceRegulatoryInformation/Guidances/UCM320002.pdf. Accessed Oct 2013

FDA (2012b) Draft guidance on Asacol HD. Available from http://www.fda.gov/downloads/
Drugs/GuidanceComplianceRegulatoryInformation/Guidances/UCM320003.pdf.   Accessed
Oct 2013

FDA (2012c) Draft guidance on Liada. Available from http://www.fda.gov/downloads/Drugs/
GuidanceComplianceRegulatoryInformation/Guidances/UCM320004.pdf. Accessed Oct 2013

FDA (2012d) FDA approved label: Xenical® (orlistat) capsule. Roche Laboratories Inc., Basel

FDA (2012e) Draft guidance on Cholestyramine powder. Available from http://www.fda.gov/
downloads/Drugs/GuidanceComplianceRegulatoryInformation/Guidances/UCM273910.pdf.
Accessed Oct 2013

FDA (2012f) FDA response to Citizen's Petition for Vancocin®, Doc#FDA-2006P-0007, Dec. Available from http://www.regulations.gov/#!documentDetail;D=FDA-2006-P-0007-0051. Accessed Oct 2013

FDA (2013a) FDA approved label: Azufidine EN-Tabs® (mesalamine) DR tablet. Pharmacia and Upjohn Company, Peapack, NJ

FDA (2013b) Draft guidance on Balsalazide Disodium tablets. Available from http://www.fda. gov/downloads/Drugs/GuidanceComplianceRegulatoryInformation/Guidances/ucm082854. pdf. Accessed Oct 2013

FDA (2013c) FDA response to Citizen's Petition for Asacol® and Pentasa®, Doc#FDA-2012-P-1087, March 2013. Available from http://www.regulations.gov/#!documentDetail;D=FDA-2012-P-1087-0005. Accessed Oct 2013

FDA (2013d) Draft guidance on Canasa. Available from http://www.fda.gov/downloads/Drugs/ GuidanceComplianceRegulatoryInformation/Guidances/UCM088666.pdf. Accessed Oct 2013

Fonseca VA, Rosenstock J, Wang AC, Truitt KE, Jones MR (2008) Colesevelam HCl improves glycemic control and reduces LDL cholesterol in patients with inadequately controlled type 2 diabetes on sulfonylurea-based therapy. Diabetes Care 31:1479–1484

Gao W, Chan J, Farokhzad OC (2010) pH-responsive nanoparticles for drug delivery. Mol Pharm 7(6):1913–1920

Gessner PK, Hasan MM (1987) Freundlich and Langmuir isotherms as models for the adsorption of toxicants on activated charcoal. J Pharm Sci 76(4):319–327

Gillespie WR et al (1997) Bioequivalence assessment based on pharmacodynamics response: application to albuterol metered dose inhalers. Pharm Res 14(11):S139–S140

Goldberg RB, Fonseca VA, Truitt KE, Jones MR (2008) Efficacy and safety of colesevelam in patients with type 2 diabetes mellitus and inadequate glycemic control receiving insulin-based therapy. Arch Intern Med 168:1531–1540

Goodman WG, Goldin J, Kuizon BD, Yoon C, Gales B, Sider D, Wang Y, Chung J, Emerick A, Greaser L, Elashoff RM, Salusky IB (2000) Coronary-artery calcification in young adults with end-stage renal disease who are undergoing dialysis. N Engl J Med 342(20):1478–1483

Haidar SH, Davit B, Chen ML, Conner D, Lee L, Li QH, Lionberger R, Makhlouf F, Patel D, Schuirmann DJ, Yu LX (2008) Bioequivalence approaches for highly variable drugs and drug products. Pharm Res 25(1):237–241

Heller DP, Burke SK, Davidson DM, Donovan JM (2002) Absorption of colesevelam hydrochloride in healthy volunteers. Ann Pharmacother 36:398–403

Houghton LA, Mangnall YF, Read NW (1990) Effect of incorporating fat into a liquid test meal on the relation between intragastric distribution and gastric emptying in human volunteers. Gut 31(11):1226–1229

Hruska KA, Mathew S, Lund R, Qiu P, Pratt P (2008) Hyperphosphatemia of chronic kidney disease. Kidney Int 74(2):148–157

Kidney Disease: Improving Global Outcomes (KDIGO) CKD-MBD Work Group (2009) KDIGO clinical practice guideline for the diagnosis, evaluation, prevention, and treatment of chronic kidney disease-mineral and bone disorder (CKD-MBD). Kidney Int 76(suppl 113):1–130

Kornbluth A, Sachar DB (2010) Ulcerative colitis practice guidelines in adults: American College of gastroenterology, Practice Parameters Committee. Am J Gastroenterol 105(3):501–523

Krishnaiah YS, Yang Y, Bykadi S, Sayeed VA, Khan MA (2013) Comparative evaluation of in vitro efficacy of colesevelam hydrochloride tablets. Drug Dev Ind Pharm. Early Online: 1–7

Lichtenstein GR, Kamm MA (2008) Review article: 5-aminosalicylate formulations for the treatment of ulcerative colitis-methods of comparing release rates and delivery of 5-aminosalicylate to the colonic mucosa. Aliment Pharmacol Ther 28(6):663–673

Lionberger RA (2008) FDA critical path initiatives: opportunities for generic drug development. AAPS J 10(1):103–109

Loftus EV Jr (2004) Clinical epidemiology of inflammatory bowel disease: incidence, prevalence and environmental influences. Gastroenterology 126(6):1504–1517

Loghman-Adham M (1999) Phosphate binders for control of phosphate retention in chronic renal failure. Pediatr Nephrol 13(8):701–708

Martin P, Wang P, Robinson A, Poole L, Dragone J, Smyth M, Pratt R (2011) Comparison of dietary phosphate absorption after single doses of lanthanum carbonate and sevelamer carbonate in healthy volunteers: a balance study. Am J Kidney Dis 57(5):700–706

McIntyre CW (2007) New developments in the management of hyperphosphatemia in chronic kidney disease. Semin Dial 20(4):337–341

Myers B, Evans DN, Rhodes J, Evans BK, Hughes BR, Lee MG, Richens A, Richards D (1987) Metabolism and urinary excretion of 5-amino salicylic acid in healthy volunteers when given intravenously or released for absorption at different sites in the gastrointestinal tract. Gut 28:196–200

O'Connor D, Adams WP, Chen ML, Daley-Yates P, Davis J, Derendorf H, Ducharme MP, Fuglsang A, Herrle M, Hochhaus G, Holmes SM, Lee SL, Li BV, Lyapustina S, Newman S, Oliver M, Patterson B, Peart J, Poochikian G, Roy P, Shah T, Singh GJ, Sharp SS (2011) Role of pharmacokinetics in establishing bioequivalence for orally inhaled drug products: workshop summary report. J Aerosol Med Pulm Drug Deliv 24(3):119–135

Pai AB, Conner TA, McQuade CR (2009) Therapeutic use of the phosphate binder lanthanum carbonate. Expert Opin Drug Metab 5(1):71–81

Pennick M, Dennis K, Damment SJ (2006) Absolute bioavailability and disposition of lanthanum in healthy human subjects administered lanthanum carbonate. J Clin Pharmacokinet 46(7):738–746

Polli JE (2008) In vitro studies are sometimes better than conventional human pharmacokinetic in vivo studies in assessing bioequivalence of immediate-release solid oral dosage forms. AAPS J 10(2):289–299

Product Monograph (2010) Fosrenol (Lanthanum carbonate hydrate). Available from http://www.shirecanada.com/en/shire-canada/Fosrenol_M_100203_En.pdf. Accessed Oct 2013

Qureshi AI, Cohen RD (2005) Mesalamine delivery system: do they really make much difference? Adv Drug Deliv Rev 57(2):281–302

Read NW, Miles CA, Fischer D, Halgate AM, Konie ND, Mitchell MA, Reeve AM, Roche TB, Walker M (1980) Transit of a meal through the stomach, small intestine and colon in normal subjects and its role in the pathogenesis of diarrhea. Gastroenterology 79(6):1276–1282

Rudolph MW, Klein S, Beckert TE, Petereit H, Dressman JB (2001) A new 5-aminosalicylic acid multi-unit dosage form for the therapy of ulcerative colitis. Eur J Pharm Biopharm 51(3):183–190

Sandborn WJ (2002) Rational selection of oral 5-aminosalicylate formulations and prodrugs for the treatment of ulcerative colitis. Am J Gastroenterol 97(12):2939–2941

Schroder H, Campbell DE (1972) Absorption, metabolism and excretion of salicylazosulfapyridine in man. Clin Pharmacol Ther 13(4):539–551

Shah VP et al (1998) In vitro dissolution profile comparison-statistics and analysis of the similarity factor, f2. Pharm Res 15(6):1998

Sheikh MS, Maguire JA, Emmett M, Santa Ana CA, Nicar MJ, Schiller LR, Fordtran JS (1989) Reduction of dietary phosphorus absorption by phosphorus binders. A theoretical, in vitro, and in vivo study. J Clin Invest 83(1):66–73

Slatopolsky E, Weerts C, Lopez-Hiker S, Norwood K, Zink M, Windus D, Delmez J (1986) Calcium carbonate as a phosphate binder in patients with chronic renal failure undergoing dialysis. N Engl J Med 315(3):157–161

Sprague SM (2007) A comparative review of the efficacy and safety of established phosphate binders: calcium, sevelamer, and lanthanum carbonate. Curr Med Res Opin 23(12):3167–3175

Steinmetz KL (2002) Colesevelam hydrochloride. Am J Health Syst Pharm 59:932–939

Tonelli M, Wiebe N, Culleton B, Lee H, Klarenbach S, Shrive F, Manns B (2007) Systematic review of the clinical efficacy and safety of sevelamer in dialysis patients. Nephrol Dial Transplant 22(10):2856–2866

Wilson CG (2010) The transit of dosage forms through the colon. Int J Pharm 395(1–2):17–25

Wrong O, Harland C (2007) Sevelamer and other anion-exchange resins in the prevention and treatment of hyperphosphataemia in chronic renal failure. Nephron Physiol 107(1):17–33

Yang Y, Shah RB, Yu LX, Khan MA (2013a) In vitro bioequivalence approach for a locally acting gastrointestinal drug: lanthanum carbonate. Mol Pharm 10(2):544–550

Yang Y, Bykadi S, Carlin AS, Shah RB, Yu LX, Khan MA (2013b) Comparative evaluation of the in vitro efficacy of lanthanum carbonate chewable tablets. J Pharm Sci 102(4):1370–1381

Younk LM, Davis SN (2012) Evaluation of colesevelam hydrochloride for the treatment of type 2 diabetes. Expert Opin Drug Met 8:515–525

Zhang X, Zheng N, Lionberger RA, Yu LX (2013) Innovative approaches for demonstration of bioequivalence: the US FDA perspective. Ther Deliv 4(6):725–740

Zhi J, Melia AT, Guerciolini R, Chung J, Kinberg J, Hauptman JB, Patel IH (1994) Retrospective population-based analysis of the dose–response (Fecal fat excretion) relationship of orlistat in Normal and obese volunteers. Clin Pharmacol Ther 56(1):82–85

Zhi J, Melia AT, Eggers H, Joly R, Patel IH (1995) Review of limited systemic absorption of orlistat, a lipase inhibitor, in healthy human volunteers. J Clin Pharmacol 35(11):1103–1108

Chapter 13
Bioequivalence for Topical Drug Products

April C. Braddy and Dale P. Conner

13.1 Background

A topical drug product is designed to deliver drug to the targeted site of action, via the skin or mucous membranes for the mitigation, treatment, prevention, or cure of certain diseases and/or disorders. As the largest organ of the integumentary system, the skin covers the entire body surface and combines with the mucosal lining of the digestive, excretory, nervous, reproductive, and respiratory systems of the body (Buxton 2006). Thus a topically applied drug can be considered a dermatologic (skin), ophthalmic (eyes), or otic (ears) drug product. In addition, topical drug products can also be used for rectal and vaginal administration (Electronic Orange Book 2013). Under most circumstances topical drug products are intended to act locally. However, there are some instances such as gastrointestinal drug products which are not locally applied but like topical drug products due to the site of activity drug concentrations may not be measured directly. Although, topical drug products can be applied to multiple areas of the body either locally, regionally or in some cases for systemic absorption, it is most commonly applied to the skin for direct therapeutic effects on the surface of the skin or within the underlying subcutaneous tissue, Fig. 13.1. These types of products are considered as topical dermatologic drug products and indicated for skin diseases and/or disorders.

A.C. Braddy (✉) • D.P. Conner
Center for Drug Evaluation and Research, U.S. Food and Drug Administration,
10903 New Hampshire Avenue, Silver Spring, MD 20993, USA
e-mail: April.Braddy@fda.hhs.gov

L.X. Yu and B.V. Li (eds.), *FDA Bioequivalence Standards*, AAPS Advances
in the Pharmaceutical Sciences Series 13, DOI 10.1007/978-1-4939-1252-0_13,
© The United States Government 2014

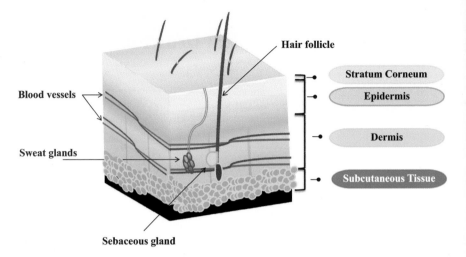

Fig. 13.1 A schematic of the human skin. There are three (3) major layers of the human skin: stratum corneum, epidermis, and dermis. The stratum corneum is the outermost layer of the epidermis. The epidermis consists of a series of cells and is the body's major barrier. The dermis is the layer between the epidermis and subcutaneous tissue. The dermis contains blood vessels, the sweat glands (produce sweat), sebaceous glands (create oily/waxy matter, called sebum to lubricate and waterproof the skin) along with hair follicles

13.1.1 Most Common Class of Topical Drug Products: Dermatologic

Topical dermatologic drug products consist of multiple therapeutic drug classes and are available in array of different dosage forms ranging from simple to complex. The therapeutic drug classes mainly include analgesic, anesthetic, antibacterial, antifungal, anti-inflammatory (non-steroidal), anti-mitotic, antiviral, glucocorticoid (corticosteroid), oncologic, and retinoid. The multiple dosage forms range from solutions to semi-solids, such as creams, foams, gels, lotions, ointments, pastes, solutions (aqueous or oily), and sprays. These differences in dosage form are also referred to as vehicles and can have an impact on drug absorption and thus efficacy of the drug product. In general, the therapeutic response elicited by these drug products is based on a sequential process (Shah 2001), Fig. 13.2. The process begins with the release of the drug from the vehicle and ends with the activation of the desired therapeutic response.

Most topical dermatologic drug products are generally not intended for systemic absorption. Therefore, the utilization of in vivo pharmacokinetic (PK) studies which are an established approach for BE assessment of solid oral dosage forms is often not feasible for topical dosage forms. Oftentimes even when a topical dermatologic drug product is systemically absorbed there is no established link between the concentration of the drug in the systemic circulation and the therapeutic efficacy. The only link that can generally be made is to the undesired therapeutic

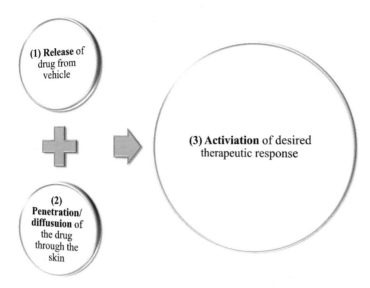

Fig. 13.2 A schematic representation of the process for inducing a therapeutic response after application of a topical dermatologic drug product

effect that may occur due to systemic absorption. Thus, other approaches for BE assessment must usually be employed in order to approve generic topical dermatologic drug products. In some rare cases, systemic treatment may be needed if the skin disease is severe, recalcitrant, or fails to respond to topical drug treatment (Long 2007). As a caveat there are some topical drug products in which the amount of systemically absorbed drug has a correlation to clinical safety and/or efficacy for such drug products as Lidocaine (ointment and patch), a local anesthetic drug, along with anti-inflammatory/analgesic drugs, such as Diclofenac Gel. It should be noted, that although the drug delivery system for transdermal products is applied to the skin, they are not considered topical dermatologic drug products.

13.1.2 Guidelines, Policies, and Regulatory Requirements

The US FDA regulatory requirements for approval of a generic topical drug product are in part based on the approval pathway and date of the reference product. Legislative changes were made to the Federal Food, Drug, and Cosmetic Act (FFD&C Act) of 1938 (ensuring drug safety) due to the passage of the Drug Price Competition and Patent Restoration Act (Waxman-Hatch) in 1984. The Waxman-Hatch Act allowed for the submission of abbreviated new drug applications (ANDAs) referencing the safety and effectiveness of the already approved innovator drug product (United States Code 1984). In general, in order for an ANDA to be approved, the generic must demonstrate BE to the reference product. It also must contain the same active ingredient, conditions of use, route of administration, dosage form, strength and labeling (with some exceptions) as the reference product. Overall, the generic should be pharmaceutically equivalent, bioequivalent and thus therapeutically equivalent to the reference product. In most cases, all ANDA submissions for topical drug products whose reference product was approved after 1962 require some form of BE assessment. This is due to the passage of the Kefauver-Harris amendments (Also known as the Drug Efficacy Amendment) which required that all drugs be proven both safe and effective based on the labeling indications (US Food and Drug Administration 2012).

At the behest of the US FDA in 1966, the National Academy of Sciences/ National Research Council evaluated the efficacy of all drug products approved between 1938 and 1962. Based on the results of this extensive project, many of products were deemed effective and thus classified as drug efficacy study implementation (DESI) drugs (Federal Register 2012; National Academy of Sciences 1974). The US FDA currently has a complied listing of those drug products. Most of the topical drug products on that list that are considered as "DESI" drugs have a therapeutic equivalence code of "AT" in the FDA's Drug Listing Database (Electronic Orange Book 2013). The defined criteria for a topical drug product to have a therapeutic rating of "AT" is a follows:

> There are a variety of topical dosage forms available for dermatologic, ophthalmic, otic, rectal, and vaginal administration, including creams, gels, lotions, oils, ointments, pastes, solutions, sprays and suppositories. Even though different topical dosage forms may contain the same active ingredient and potency, these dosage forms are not considered pharmaceutically equivalent. Therefore, they are not considered therapeutically equivalent. All solutions and DESI drug products containing the same active ingredient in the same topical dosage form for which a waiver of in vivo bioequivalence has been granted and for which chemistry and manufacturing processes are adequate to demonstrate bioequivalence, are considered therapeutically equivalent and may be coded **AT**. Pharmaceutically equivalent topical products that raise questions of bioequivalence, including all post-1962 non-solution topical drug products, are coded **AB** when supported by adequate bioequivalence data, and **BT** in the absence of such data.

Thus, based on the current Code of Federal Regulations—Title 21 (21 CFR), Chap. 1, Part 320.22, a waiver of evidence of in vivo bioavailability (BA) or BE studies can be granted for these products, as long as they do not contain an inactive ingredient or other change in formulation that may affect the absorption of the drug product. The defined waiver criteria for "DESI" drugs as outlined in 21 CFR § 320.22 (c) is as follows:

> FDA shall waive the requirement for the submission of evidence measuring the in vivo bioavailability or demonstrating the in vivo bioequivalence of a solid oral dosage form (other than a delayed release or extended release dosage form) of a drug product determined to be effective for at least one indication in a Drug Efficacy Study Implementation notice or which is identical, related, or similar to such a drug product under § 310.6 of this chapter unless FDA has evaluated the drug product under the criteria set forth in § 320.33, included the drug product in the Approved Drug Products with Therapeutic Equivalence Evaluations List, and rated the drug product as having a known or potential bioequivalence problem. A drug product so rated reflects a determination by FDA that an in vivo bioequivalence study is required.

Also, as per 21 CFR § 320.22 (b) (3), waivers from in vivo BA/BE studies may be granted for topical solutions, as well. The defined waiver criteria for solutions as outlined in 21 CFR § 320.22 (b) (3) is as follows:

> (3) The drug product: (i) Is a solution for application to the skin, an oral solution, elixir, syrup, tincture, a solution for aerosolization or nebulization, a nasal solution, or similar other solubilized form; and (ii) Contains an active drug ingredient in the same concentration and dosage form as a drug product that is the subject of an approved full new drug application or abbreviated new drug application; and (iii) Contains no inactive ingredient or other change in formulation from the drug product that is the subject of the approved full new drug application or abbreviated new drug application that may significantly affect absorption of the active drug ingredient or active moiety for products that are systemically absorbed, or that may significantly affect systemic or local availability for products intended to act locally.

In general for all topical products, the proposed generic and reference product should be qualitatively Q1 and Q2 the similar for the active pharmaceutical ingredients (API) and excipients. In particular for topical solutions, there should be no more than ±5 % difference between the concentration of the excipients. The overall formulation of a topical solution should be Q1/Q2 the same between the generic and reference product. Any changes to the formulation of topical solutions that will impact penetration, such as addition of penetration enhancers then in most cases a clinical endpoint will be required for submission (US Food and Drug Administration 2009).

In addition, in order to further address the limitations in the assessment of BA for drug products that are not intended for absorption into the bloodstream, in 2003 an addition to the FFD&C Act at Section 505(j)(8)(A)(ii) was made to indicate the following:

> For a drug that is not intended to be absorbed into the bloodstream, the Secretary may assess bioavailability by scientifically valid measurements intended to reflect the rate and extent to which the active pharmaceutical ingredient or therapeutic ingredient becomes available at the site of drug action.

An historical regulatory timeline for the approval of generic and reference products, along with approval milestones, guidelines, and policies for topical drug products is provided in Fig. 13.3.

13.2   Characteristics of Topical Drug Products

In order for a topical, in particular a locally acting dermatologic drug product, to be clinically effective, there must be efficient drug delivery to the target site of action in the skin. This is largely a function of the dosage form, which will be referred to as vehicle henceforth, presenting the API to the surface of the skin. As outlined in Fig. 13.2, the primary objective is for the drug to be released from the vehicle and reach the target site of action to elicit the desired therapeutic response. A more detailed outline of the significance of the selection of the vehicle in the overall

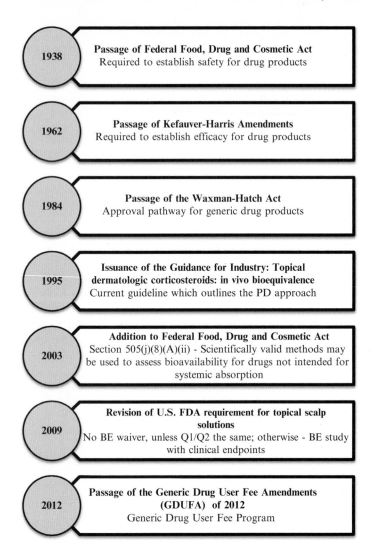

Fig. 13.3 An historical timeline of the US FDA's regulation of generic and reference drug products, with specific emphasis on approval milestones, guidelines, and policies impacting topical drug products

characteristics of the topical drug product is as follows: (1) the vehicle of the topical drug product should efficiently deposit the drug on the skin with even distribution, (2) release of the drug from the vehicle into the skin so it can migrate to the site of action, (3) delivery of the drug to the target site of action, and (4) sustaining a therapeutic level in the target tissue for a sufficient duration of time to elicit the desired therapeutic response (Shah 2001, Kircik et al. 2010; Weiss 2011). The extent of drug absorption will depend on the interaction between the drug, vehicle, and the skin. This interaction controls partitioning into and diffusion through the stratum corneum, the outer barrier layer of the skin, along with the site of activity and disease/state. This section will discuss all of these critical characteristics of a topical drug product, along with the disease state/disorder and absorption process through the skin in order to achieve clinically relevant drug delivery:

- Skin disease state/disorder (*only taken into consideration for BE studies with clinical endpoints*)
- Route(s) of topical drug absorption through the skin
- Drug-specific characteristics of the API
- Different vehicles
- Excipients
- Overall formulation characteristics

13.2.1   Skin Disease/Disorder

13.2.2   Type of Disease/Disorder

The type and site of the disease state/disorder are important in selecting the appropriate vehicle for topical drug delivery. Table 13.1 provides a listing of the numerous skin diseases/disorders, along with common sites of action on or in the skin. The most common dermatologic conditions include acne (over 80 % of the population), dermatitis, eczema, and psoriasis. Acne normally develops due to the blockage of follicles. There are several causes of acne: genetic, hormonal, psychological, and infectious or even possibly diet. For the other skin conditions, such as dermatitis, eczema, and psoriasis it may be due to a dysfunction in the stratum corneum based on environmental, genetic, or changes within the body itself. One of the vital skin functions is as a barrier to water and maintenance of hydration. Oftentimes, for these particular conditions, the skin and/or scalp, become dry, itchy and may even become inflamed due to excessive water loss from skin (Harding 2004). Therefore, the topical drug products which are used to treat these conditions often can be used to increase hydration of the skin as well. Since, the disease/disorder can result in different manifestations, it is critical to have an understanding of the skin disease/disorder to properly alleviate the condition.

13.2.3   Site of Action

The site of action is also critical. There has been previous discussion that the rate of penetration between the skin of the sole of the foot, face, or scalp, may differ (Stoughton 1989). The hair-bearing areas like the scalp are more suitable for

Table 13.1 Common skin diseases/disorders and topical drug treatments

Disease/disorder state	Clinical features	Common sites on body	Example of some topical dermatologic therapeutic drug products for treatment
Acne (Acne vulgaris)	Increased sebum (sebaceous glands) secretion, comedone (a plug blocking the hair follicle), papules (a small raised lesion, ≤ 0.5 diameter), pustules (a small collection of pus) *Common in adolescents	Face, chest, and upper back	Antibacterial(s): Erythromycin, Clindamycin, and Tetracycline Retinoid(s): Adapalene, Tretinoin and Isotretinoin Antibacterial and Antiproliferative: Azelaic acid
Dermatitis (Atopic Eczema) Contact Dermatitis	Inflammation of skin and itchy rash Allergic: Resulting from interaction with allergen Irritant: Resulting from direct interaction with a detergent	Face, wrist, and flexural aspects of elbows and knees Usually at the site of contact but may spread	Corticosteroid(s), e.g., Betamethasone Valerate, Clobetasol Propionate, Dexamethasone, Fluocinolone acetonide, Hydrocortisone (acetate), *not all inclusive list*
Seborrheic Eczema	Greasy adherent scale on the scalp (cradle cap)	Scalp, eyebrows, eyelids, nasal-labial, and chin	Antifungal(s): Miconazole, Clotrimazole or Ketoconazole (in combination with low-potency steroids); Selenium Sulfide
Psoriasis	Overactive immune system which leads to flacking, inflammation, and thick white, silvery or red patches of skin	Elbows, knees, scalp, and lower back	Anti-psoriatic: Anthralin, Calcipotriene—Vitamin D analogue or corticosteroids
Rosacea	Redness and pimples	Face, nose, cheeks, chin, and forehead	Retinoid(s): Isotretinoin
Actinic (Solar) Keratoses	Crusty or scaly bump formed on skin exposed to light Can sometimes progress to cancer *Common in elderly	Skin	Anti-mitotic: Fluorouracil and Imiquimod NSAID: Diclofenac gel
Yeast infections Pityriasis Versicolor	Brown scaly area (of child)	Trunk and sometimes limbs	Selenium sulfide
Dermatophyte Infection (Ring worm)	Red ring of scaly skin	Skin (stratum corneum), nails, and hair	Antifungal(s): Clotrimazole, Econazole, or Imidazole

(continued)

Table 13.1 (continued)

Disease/disorder state	Clinical features	Common sites on body	Example of some topical dermatologic therapeutic drug products for treatment
Bacterial Infections	Various bacteria can yield different infections of the skin		Antibacterial(s) and Antiseptic(s)
Viral infections Herpes Simplex Type I, II; Herpes Zoster Viral Warts & Molluscum Contagiosum (poxvirus)	Periodic blisters or rash (Zoster-similar to chicken pox-elderly); Cause skin to grow excessively creating a wart. A papule on the skin (children)	Skin: around lips (Type I) or genitals (Type II); no specific region; trunk or limbs	Antiviral: Acyclovir Salicylic acid, Lactic acid (for warts) Mollusum contagiosum: normally resolve itself without treatment
Scabies	Tiny mites that burrow into the skin. Intense itching	Fingers, wrists, elbows, and behind	Benzyl Benzoate, Permethrin, Lindane, or Malathion
Alopecia	Hair loss	Scalp	Minoxidil, Cyclosporine

solutions, while the palms and soles of the feet are more responsive to semi-solids, such as ointments due to the thickness of the skin (Weiss 2011).

In addition, topical drug delivery to the scalp involves a greater increase in skin and appendage surface area which generally requires the application of more formulation per square centimeter of skin. Also, the scalp is covered by a thin layer of sebum (oily, waxy, secretion from the sebaceous gland). This secretion can sometimes be incorporated into the formulation after application to the scalp, and potentially alter the kinetics of intracellular transport. For this reason effective topical drug product formulations must take into consideration the in vivo environment from which the API is delivered to the site of action within the skin (Kircik et al. 2010).

13.2.4 Route(s) of Topical Drug Absorption in the Skin

As stated previously, there are several layers of the skin, see Fig. 13.1. Percutaneous absorption is the process of absorption through the skin from topical drug application. It is a passive process in which these products can permeate through the skin of the body. This means the movement of the drug across the membranes is without the input of energy. The major barrier to percutaneous drug absorption is the stratum corneum. In particular, the transport of hydrophilic or charged molecules is especially difficult and attributable to the lipid-rich nature of the stratum corneum and its low water content. The layer is composed of about 40 % lipids, 40 % protein,

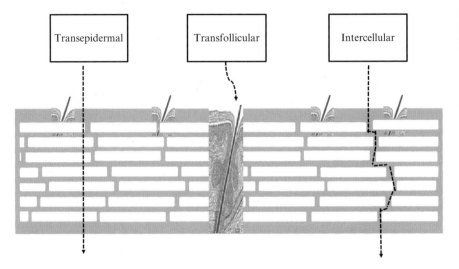

Fig. 13.4 The three (3) routes of absorption of topical drug products across the stratum corneum. The transepidermal route is passive diffusion through the stratum corneum. The transfollicular route is through the hair shaft in the skin for transport to the different layers of the skin. The intercellular route is between the cell junctions of the stratum corneum and other layers of the skin

and only 20 % water. The three principle routes of absorption through the skin are transepidermal, transfollicular, and intercellular, Fig. 13.4 (Hueber et al. 1994; Topical Drug Delivery Systems, PharmaInfo Net. 2008).

Transepidermal: Permeation by this route involves partitioning into the stratum corneum. Diffusion takes place across the stratum corneum. In the case of systemic absorption via this route, when a permeating drug exits at the stratum corneum, it enters the wet cell mass of the epidermis and since the epidermis has no direct blood supply, the drug is forced to diffuse across it to reach the vasculature immediately beneath.

The viable epidermis is considered a single field in diffusion models. The epidermal cell membranes are tightly joined and there is little to no intercellular space for ions and polar nonelectrolytes molecules to diffusionally squeeze through. Passage through the dermis represents the final hurdle, for systemic entry. Permeation through the dermis is through the interlocking channels of the ground substance. Diffusion through the dermis is facile and without molecular selectivity since gaps between the collagen fibers are far too wide to filter large molecules. Since, the viable epidermis and dermis lack measureable physicochemical distinction, they are generally considered as a single field of diffusion, except when penetrants of extreme polarity are involved, as the epidermis polarity is involved. This is due to the fact that the epidermis offers measureable resistance. This route is preferred for topical drug products that are intended to act directly on the stratum corneum.

Transfollicular: Permeation is through the skin appendages as a "secondary avenue". The sweat glands in the dermis are considered as a shunt bypassing the stratum corneum. These stunts are over the entire body. Though the glands are numerous, their orifices are tiny. Also, further diffusion into the epidermis may occur due to the follicular pores, where the hair shaft exits the skin. This area is relatively large and sebum aids in diffusion and subsequent passage into the dermis for systemic absorption. However, these openings are barely 1 % of the surface area. Thus, the small surface of these alternate pathways can limit the amount of drug absorption by these routes. Nonetheless, it is a favorable route for hydrophilic drug molecules. It is the preferred delivery route for topical dermatologic drug products intended for the scalp, acne or folliculitis, since it acts as a reservoir for the drug as well (Wosicka and Cal 2010).

Intercellular: Permeation by this route involves transport of the drug between the cells in the skin. The current belief is that most drugs diffuse across the stratum corneum via this route. It is a favorable route for lipophilic drug molecules.

13.2.5 Drug-Specific Characteristics

The specific characteristics of the drug itself should be considered in the selection of the vehicle. The stability of the API and its BA are primary considerations. Based on the physicochemical properties of the drug, a strategy can be developed. Key factors include degree of solubility or insolubility in various excipients, compatibility, or incompatibility with potential excipients and sensitivities of the molecule resulting in degradation and instability. In the case of topical corticosteroids, the drugs in this class vary in potency; therefore, the selection of a vehicle based on the severity of the disease state can also be important.

13.2.6 Different Vehicles

The US FDA currently recognizes multiple topical dosage forms. The recognized dosage forms are consistent with the current United States Pharmacopeia (US Pharmacopeia 2013). Table 13.2 lists the current descriptors/definitions for topical dosage forms.

13.2.7 Excipients

Excipients are used in virtually all drug products and are essential to product performance. Thus, for successful design and manufacture of a robust product

Table 13.2 The US pharmacopeia definitions of the topical dosage forms

Descriptors/Dosage form	Definition of terms
Topical	A route of administration characterized by application of the body
Dermal	A topical route of administration where the article is intended to reach or be applied to the dermis
Liquid[a,b]	A dosage form consisting of a pure chemical in its liquid state
Aerosols	A dosage form consisting of a liquid or solid preparation packaged under pressure and intended for administration as a fine mist.[c] An aerosol contains the therapeutic agent(s) and propellant that are released upon actuation of an appropriate valve system
Emulsion[d]	A dosage form consisting of a two-phase system composed of at least two immiscible liquids, one of which is dispersed as droplets (internal or dispersed phase) within the other liquid (external or continuous phase), generally stabilized with one or more emulsifying agents
Cream[e]	An emulsion dosage form often containing more than 20 % water and volatiles and/or containing less than 50 % hydrocarbons, waxes, or polyols as the vehicle for the API. Creams are generally intended for external application to the skin or mucous membranes
Ointment	A semisolid dosage form, usually containing less than 20 % water and volatiles and more than 50 % hydrocarbons, waxes, or polyols as the vehicle. This dosage form is generally for external application to the skin or mucous membranes
Emollient[a]	Attribute of a cream or ointment indicating an increase in the moisture content of the skin following application of bland, fatty or oleaginous substances
Semisolid[a]	Attribute of a material characterized by a reduced ability to flow or conform to its container at room temperature. A semisolid does not flow at low shear stress and generally exhibits plastic flow behavior
Foam	An emulsion dosage form containing dispersed phase of gas bubbles in a liquid continuous phase containing the API. Foams are packaged in pressurized containers or special dispensing devices and are intended for application to the skin or mucous. The foam is formed at the time of application. Surfactants are used to ensure the dispersion of the gas and the two phases. When dispensed it has a fluffy, semisolid consistency. It can also be formulated to break to a liquid quickly or to remain as foam to ensure prolonged contact
Gel[f,g]	A dosage form that is a semisolid dispersion of small inorganic particles or a solution of large organic molecules containing a gelling agent to provide stiffness. A gel may contain suspended particles
Lotion[h]	An emulsion dosage form applied to the outer surface of the body. Historically, this term has also been applied to suspensions and solutions
Paste	A semisolid dosage form containing a high percentage (e.g., 20–50 %) of finely dispersed solids with a stiff consistency. This dosage form is intended for application to the skin, oral cavity, or mucous membranes
Patch (Transdermal system)	Dosage forms designed to deliver the API(s) through the skin into the systemic circulation. Transdermal systems are typically composed of an outer covering (barriers), a drug reservoir (that may incorporate rate a rate-controlling membrane), a contact adhesive to affix the transdermal system to the administration site, and a protective covering that is removed immediately prior to the application of the transdermal system

(continued)

Table 13.2 (continued)

Descriptors/Dosage form	Definition of terms
Powder[i]	A dosage form composed of a solid or mixture of solid reduced to an finely divided state and intended for internal or external use
Shampoo	A solution or suspension dosage form used to clean the hair and scalp. May contain an API intended for topical application to the scalp
Soap	The alkali salt(s) of a fatty acid or mixture of fatty acids used to cleanse the skin. Soaps used as dosage form may contain an API intended for topical application to the skin. Soaps have been used as liniment and enemas
Spray	Attribute that describes the generation of droplets of a liquid or solution to facilitate application to the intended area
	By definition and in accordance with the USP drug product monographs, a spray dosage form drug product delivers an accurately metered spray through the delivery system, i.e., device. A spray drug product is a preparation that contains an API(s) in either solution or suspension form
Suspension	A liquid dosage form that consists of solid particles dispersed throughout a liquid phase
Solution	A clear, homogenous liquid dosage form that contains one or more chemical substances dissolved in a solvent or mixture of mutually miscible solvents
Tape, medicated	A dosage or device composed of a woven fabric or synthetic material onto which an API is placed, usually with an adhesive on one or both sired to facilitate topical application

[a]This term should be used in the article names (official name of the medicinal drug products)
[b]This dosage form should not be applied to solutions. When it is used as a descriptive term, it indicates a material that is pourable and conforms to its container at room temperature
[c]The descriptive term aerosol also refers to the fine mist of small droplets or solid particles that are emitted from the product
[d]Emulsion is not used as a dosage form term if a more specific term is applicable (e.g., cream, lotion or ointment)
[e]Creams have a relatively soft, spreadable consistency and can be formulated as either a water-in-oil emulsion (e.g., Cold Cream or Fatty Cream as in the European Pharmacopeia) or as an oil-in water emulsion (e.g., Betamethasone Valerate Cream)
[f]Gels can be classified either as a single-phase or two-phase systems. A single-phase consists of organic macromolecules uniformly distributed throughout a liquid in such a manner that no apparent boundaries exist between the dispersed macromolecules and the liquid. A two-phase system consists of a network of small discrete particles
[g]Jellies are a type of gel that typically will have a higher water content
[h]Lotions share many characteristics with creams. The distinguishing factor is that they are more fluid than semisolid and thus pourable
[i]Powders are often used topically as dusting powders

requires the use of well-defined excipients that, when combined together yield a consistent and effective product (Chang et al. 2013a, 2013b). All of the excipients should comply with compendial standards (US Pharmacopeia 2013). For topical drug products, based on the function, there are several different categories of excipients. The selection of the excipient has an impact on the physicochemical

Table 13.3 Categorization of excipients based on function

Function (Agent)	Description	Examples (not all inclusive listing)
Antioxidant	It is used to inhibit oxidation of other molecules	Butylated hydroxyanisole, Butylated hydroxytoluene, Glyceryl monostearate, Tocopherol (α or, dl)
Chelating/ Complexing	It is used to form a soluble complex molecules with certain metal ions and essentially removes the ions from solution to minimize or eliminate the ability to react with other elements and/or precipitate	Calcium acetate, Citric acid, Disodium edetate, Sodium phosphate (Monobasic, Dibasic)
Emollient	It is used as a lubricant, imparts spreading ease, texture, and softening of the skin. It also counters the potential for drying/irritating impact of surfactants on the skin	Butyl stearate, Glycol distearate, Isopropyl myristate, Mineral oil and Lanolin alcohols
Emulsifier	It is used to serve as a protective barrier and also stabilizes the emulsion by reducing interfacial tension of the system	Cetostearyl alcohol, Glyceryl monostearate, Hypromellose, Lecithins
Humectants	It is used to increase the solution of the API, elevate the skins penetration, and increase its activity time. It can also be used to elevate the hydration of the skin	Propylene glycol, Sodium lactate, Linoleic acid, Glycerin
Odor modifier	It is used to enhance the fragrance	–
Ointment Base	It is used as major components of ointments, which control its physical properties	White petrolatum, Lanolin, Polyethylene glycol mineral oil (medium chain Triglycerides (fatty acids))
Penetration Enhancer	It is used to promote absorption of drug	Isopropyl palmitate—Propylene glycol, Ethanol, Oleic acid, Isopropyl myristate
pH modifier (acidifying/ alkalizing/ buffer)	It is used to regulate the pH	Maleic acid, Lactic acid, Undecylenic acid, Diisopropanolamine sodium acetate or amines
Preservative	It is used to prevent microbial growth	Methylparaben, Propylparaben, Boric acid, Sodium lactate
Solvent	It is used to dissolve another substance	Isopropyl alcohol, Acetone, Water, Hydroxy(ethyl/methyl) Cellulose
Stiffening	It is used alone or as a mixture of agents to increase the viscosity or harness of a preparation, especially ointments and creams	Stearyl alcohol, Cetostearyl alcohol, Paraffin, White or Yellow wax
Surfactant	It is used to reduce the surface tension between two liquids or between a liquid and solid	Myristyl alcohol, Sodium lauryl sulfate, Sorbitan esters, Tocopherol (α)
Vehicle	It is used to deliver the API for administration	Isopropyl myristate, Peanut oil, Mineral oil, Isopropyl palmitate

(continued)

Table 13.3 (continued)

Function (Agent)	Description	Examples (not all inclusive listing)
Viscosity Enhancer	It is used to stabilize disperse systems, to reduce the rate of solute or particulate transport, or to decrease the fluidity of liquid formulations	Carbomer, Methylcellulose, Ethylcellulose, Cetyl palmitate

Note: The information in the table is complied mainly from information in the Handbook of Pharmaceutical Excipients (Rowe et al. 2012, updated 2013). (Other useful sources are FDA's Inactive Ingredient Search for Approved Products 2013, along with Chang (Chang et al. 2013a, 2013b) and USP-Excipient Performance 2013)

properties of the API, release of the API from the vehicle, overall stability and formulation characteristics. Table 13.3 provides a listing of the different functional categories for excipients used in the formulation of these particular drug products.

In general, in order for a generic to be bioequivalent to the reference product there should be no major alternations to the formulation, especially changes that can impact the overall formulation performance.

13.2.8 Overall Formulation Characteristics

As outlined in this section, there are several contributing factors to the overall characteristics which impact the final formulation of a topical drug product. The generic and reference topical drug products formulation should be similar if not identical in all aspects of its formulation. The overall formulation has a significant impact on the clinical efficacy of the topical drug product with several key factors playing a role in the drugs' absorption into the skin. Furthermore, in an effort to facilitate a better understanding of the generic development process for a topical, the US FDA has begun recent discussion in the utilization of the quality-by-design (QbD) for pharmaceutical development and manufacturing of topical semi-solids drug products (Chang et al. 2013a, 2013b).

13.3 BE Approaches

In all cases, in order for a BE study to be acceptable, no significant difference should be seen between the rate and extent of the availability of the generic and reference product at the site of action. According, to the current US FDA regulations, 21 CFR § 320.24 (US Code of Federal Regulations 2013), there are several approaches which have been deemed acceptable for demonstrating BA and/or BE, Fig. 13.5. In some cases, as stated previously in Sect. 13.1.2, a waiver may be

Fig. 13.5 A listing of the current BA/BE approaches acceptable to the US FDA. The approaches are listed in order of accuracy, sensitivity, and reproducibility (*left to right*). The most common approach currently recommended for topical drug products is well-controlled clinical trials

appropriate in which no in vivo studies are required when BE is "self-evident". This section will discuss all of the listed approaches and their feasibility for topical drug products.

13.3.1   In Vivo PK Studies

As stated previously throughout this chapter, most topical drug products, specifically topical dermatologic drug products, are generally not intended for systemic absorption. And if the drug product is systemically absorbed there tends to be a lack of obvious correlation between the systemic levels and clinical efficacy. Furthermore, drug concentrations in biological fluids do not necessarily represent drug concentrations at the site of action. It only represents the drug concentrations after passage through the target compartment. In addition, the amounts in blood may not be measured consistently and accurately. Therefore, the measurement of the drug concentration in biological fluids as a function of time is often not feasible or appropriate for these drug products. However, as previously stated, some topical drug products, especially those which are applied locally and may be intended for some systemic absorption, then this type of study may be feasible. It should be noted that currently this approach is limited to very few topical drug products.

13.3.2   In Vivo PD Studies

The in vivo PD approach is based on the measurement of a pharmacological effect of a drug product being assessed as a function time. Currently the McKenzie-Stoughton vasoconstrictor assay (VCA) is the only PD approach accepted by the US FDA. However, this approach is limited only to topical dermatologic corticosteroid drug products. It is based on the premise that a corticosteroid product will produce a visible blanching response as a result of vasoconstriction of the skin microvasculature over time (McKenzie and Stoughton 1962; Stoughton 1987, 1992). In 1995, the US FDA issued a Guidance for Industry: Guidance Topical

Dermatologic Corticosteroids: In Vivo Bioequivalence (issued June 2, 1995), which describes the VCA approach. Both a pilot study and pivotal BE study are recommended and the data analysis is based on E_{max} model—population modeling, (Guidances 1995; Singh et al. 1999).

The E_{max} model:

$$E = E_o + \frac{E_{max} \times D}{D + ED_{50}}$$

The E_{max} model for dermatological products data:

$$AUEC = \frac{AUEC_{max} \times Dose\ Duration}{Dose\ Duration + ED_{50}},$$

where, AUEC = Pharmacodynamic effect metric, $AUEC_{max}$ = Maximum fitted value of "AUEC", D = Dose, E = Pharmacodynamic effect, E_0 = Baseline effect, E_{max} = Maximum fitted value "E", ED_{50} = Dose duration required to achieve 50 % of the E_{max} value.

The study is conducted using healthy volunteers. Duration of exposure of the skin to the products is used to control the dose. The initial pilot dose duration-response study is recommended in order to determine the appropriate dose duration for use in the subsequent pivotal BE study. In the pilot study, the initial estimation of the E_{max} is conducted using the reference drug product. The appropriate dose duration times (ED_{50}, dose duration that results in half-maximal response, D_1-ED_{50}/2-shorter dose duration time and D_2-longer dose duration time, $ED_{50} \times 2$) are selected based on the outcome of the pilot study. These dose duration times are used to perform the pivotal BE study. The selection of the dose is critical in being able to distinguish between two products that may be quite different but appear bioequivalent. Figure 13.6 is an illustration of the relationship between the PD/clinical response and dose. For BE assessment, the pivotal BE study incorporates a replicate design and documentation of acceptable individual subject dose duration responses by comparing the generic and reference products. Based on this model, the generic and reference drug products should meet the acceptable BE criteria in which the 90 % confidence interval (CI) is between 80.00 and 125.00 % for the observed vasoconstrictor response. For almost 20 years, this PD approach has proven to be a direct and efficient indicator of local BE. It is relatively inexpensive and requires a considerably less number of subjects to obtain a sufficient level of sensitivity than BE studies with clinical endpoints performed in patients.

13.3.3 BE Studies with Clinical Endpoints

Bioequivalence studies with clinical endpoints allow for the establishment of BE between two drug products based on the demonstration of equivalent safety and

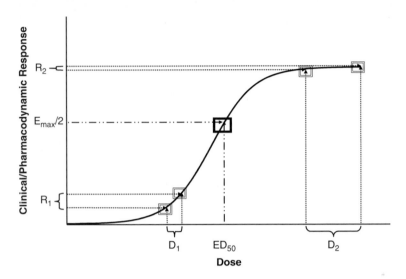

Fig. 13.6 A schematic representation of the relationship between the PD response and dose. Depending on the dose given may or may not detect the pharmacological response signifying an important difference in drug BA between the comparison products. A low dose correlates to a low response, designated, respectively, as D_1 and R_1. This low response is often not sufficiently sensitive to detect differences in the two products. Whereas, a high dose correlates to an oversaturation of the response in which you hit a plateau and cannot achieve any more response even if you continue to increase the dose, designated as D_2 and R_2. In those two areas, one may be unable to differentiate between two products, since the PD response to either product is insensitive. However, when a dose is given which elicits 50 % of the maximum response, ED_{50}, and correlates to 50 % of the E_{max}, which is the maximum effect, one is able to differentiate between two products based on the most sensitive range of the PD response

efficacy in patients. Currently, this is the most common approach for BE assessment of topical drug products. A BE study with clinical endpoints will use a product-specific indication recommended by the US FDA. If the reference product is labeled for multiple indications, the indication that is most sensitive to difference in local delivery of the drug is usually preferred. This is due to the fact that in some instances the selected clinical endpoint may not be sensitive to formulation differences. Therefore, the selection of the clinical endpoint is critical, since the detection of differences in formulation performance between the products depends on the exposure-response relationship for the particular drug and indication.

BE studies with clinical endpoints are often designed as randomized, blinded, parallel studies using a placebo. The placebo arm ensures that the study is conducted at a sufficiently sensitive dose to assure that an effect was achieved (in other words the both treatments were active) and that the differences between treatments can be detected. The studies are usually conducted over several weeks and the endpoints are mostly visually assessed based on scoring scales or dichotomous endpoints where it is based on success (completely resolved) or failure (not resolved) scales. Also, for some topical drug products, the establishment of BE is based on a continuous variable, in which the statistical analysis is performed based on the mean change from baseline.

Under these conditions in order for a generic to be deemed bioequivalent to the reference product, then the 90 % CI of the generic-to-reference ratio of means must be within [0.80, 1.25], see Table 13.4. For dichotomous types of endpoints, statistical analysis is conducted on the data and in order to establish BE, the 90 % CI of the difference between the products must be within [−0.20, +0.20]. Although BE studies with clinical endpoints are currently the "gold standard" for all topical drug products, these studies are still very expensive, time-consuming, labor intensive, lacking in sensitivity, and require a large study population (averaging from 200 to 300 patients and often times more).

13.3.4 BE Studies with In Vitro Endpoints

13.3.4.1 Basis of Approval

Bioequivalence studies with in vitro endpoints rely on in vitro characterization of the topical drug product with a simple formulation (Advisory Committee 2004). Recently, the US FDA has begun accepting of in vitro studies for simple topical drug formulations (non-solution). These tests include in vitro rheological tests (physiochemical properties of the formulation) and in vitro drug release testing (IVRT) using diffusion cells, such as Franz diffusion cell system, fitted usually with a synthetic membrane, Fig. 13.7. The IVRT method is used to estimate the rate of drug release from its formulation. A difference in drug release should reflect changes in the characteristics of the drug product formulation or the thermodynamic properties of the drug. Overall, the threshold requirements for this approach to be considered is the generic must be Q1 and Q2 the same based on formulation. In addition, the generic and reference product should not differ significantly in physicochemical properties (Q3), which includes in vitro testing (Guidances 2012).

13.3.4.2 Post-approval Changes

The US FDA Guidance for Industry: Nonsterile Semisolid Dosage Forms, Scale-Up and Post-approval Changes: Chemistry, Manufacturing, and Control; In Vitro Release Testing and In Vivo Bioequivalence Documentations (issued May 1997) addresses nonsterile topical preparations such as creams, gels, lotions, and ointments. It provides recommendations for IVRT approach and/or in vivo BE testing to support changes to (1) components or composition, (2) the manufacturing (process and equipment), (3) scale-up/scale-down of manufacture, and/or (4) the

Table 13.4 Some notable drug-specific BE recommendations for topical drug products

Drug name, strength	Dosage form	Innovator[a]	Drug class	BE assessment	Draft recommended
Acyclovir, 5 %	Ointment	Zovirax® Approved: 29 March 1982 (Valeant Bermuda)	Antiviral	2 Options: (1) In vitro studies (Q1, Q2, and Q3 the same) or (2) BE study w/clinical endpoint BE criteria, 90 % CI—T/R ratio of means [0.80, 1.25]	March 2012
Lidocaine, 5 %	Patch	Lidoderm® Approved: 19 March 1999 (Teikoku Pharm, USA)	Anesthetic	2 Studies: (both recommended) (1) PK study: Lidocaine in plasma BE criteria—90 % CI—AUCT, AUCI, Cmax,—80.00-125.00 % (2) Skin irritation/sensitization study BE criteria—scoring scale	May 2007
Butenafine Hydrochloride, 1 %	Cream	Mentax® & Mentax-TC® Approved: 18 October 1996 and 17 October 2002, respectively (Mylan)	Antifungal	BE study w/clinical endpoint BE criteria: Dichotomous variable, success versus failure [−0.20, +0.20]	March 2012
Naftifine, 1 %	Cream	Naftin® Approved: 29 February 1988 (Merz)	Antifungal	BE study w/clinical endpoint BE criteria: Dichotomous variable, success versus failure [−0.20, +0.20]	March 2012
	Gel	Naftin® Approved: 18 June 1990			
Tacrolimus, 0.1 % and 0.03 %	Ointment	Protopic® Approved: 8 December 2000 (Astellas)	Dermatologic	BE study w/clinical endpoint BE criteria: Dichotomous variable, success versus failure [−0.20, +0.20]	March 2012

| Diclofenac Sodium, 1% | Gel | Voltaren® Approved: 17 October 2007 (Novartis) Solaraze® Approved: 16 October 2000 (Fougera) | NSAID | 2 studies: (both recommended) (1) PK study: Diclofenac in plasma BE criteria—90 % CI—AUCT, AUCI, Cmax,—80.00-125.00 % (2) BE study w/clinical endpoint BE criteria, 90 % CI T/R ratio of means [0.80, 1.25] | February 2011 January 2011 |

[a]As of August 2013, all of these innovator drug products were designated as the reference product (Electronic Orange Book 2013)

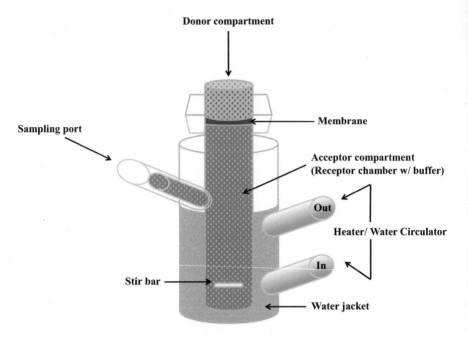

Fig. 13.7 A schematic of Franz diffusion cell system. The IVRT method for a topical drug product is based on an open chamber diffusion cell. The drug of interest (generic/test/reference) is placed in the donor compartment of the diffusion cell and a sampling fluid is placed on the side of the membrane in the acceptable compartment (receptor chamber). Diffusion of the drug from the topical product across the membrane is monitored by assay of sequentially collected samples of the receptor fluid

site of manufacture of a semisolid formulation during the post-approval period. The US FDA does not currently require IVRT testing to be submitted in order to document the development of its drug product or validation of the method in order to gain approval of drug applications (NDAs or ANDAs). Also, it is not required for routine batch-to-batch quality control testing. The IVRT method will be discussed in greater detail in Sect. 13.4.2.

13.3.5 Any Another Acceptable Approach by the US FDA

Currently, the US FDA does not deem other approaches acceptable for BE assessment of topical drug products. Section 13.4 will discuss the previous and developing approaches that have been evaluated and/or are currently under consideration by the US FDA.

13.4 Case Studies

The US FDA continues to encourage and focus on scientific research for the development of new in vivo and/or in vitro approaches for BE assessment of generic topical drug products, along with the improvement of previous and current approaches such as in vivo PK and BE studies with clinical endpoints. In 2004, the US FDA introduced its Critical Path Initiatives in order to identify and address some critical scientific challenges in the development of new and generic drug products through collaborative solutions with different governmental agencies, pharmaceutical industry, and academia.

Subsequently, in 2007, the US FDA released a document which identified some specific challenges involved in the development of generic drug products, in particular BE assessment of topical dermatological drug products (Critical Path Opportunities 2007; Lionberger 2008). A variety of potential BE approaches were listed: PK studies, design of new BE studies with clinical endpoints, in vitro characterization, such as rheological test methods and diffusion cells, along with dermatopharmacokinetics (DPK or "skin stripping"), dermal microdialysis (DMD), and near infrared (NIR) spectroscopy.

13.4.1 Prior to 2007

13.4.1.1 DPK Approach—Skin Stripping

In the 1990s, the US FDA strongly considered the DPK approach for BE assessment of topical drug products (Braddy and Conner 2011). This approach employs the tape stripping of successive layers of the stratum corneum after topical administration over a specified time period. In fact, in July 1992, the US FDA issued the interim guidance, Topical Corticosteroids: In Vivo Bioequivalence and In Vitro Release Methods, which included the skin stripping technique, along with PK and IVRT. However, it was later concluded that there was insufficient data to recommend this approach. Therefore, it was removed from the current guidance issued in 1995 for topical corticosteroids.

Three years later in 1998, the US FDA issued the draft guidance: Topical Dermatological Drug Product NDAs and ANDAs-In Vivo Bioavailability, Bioequivalence, In Vitro Release, and Associated Studies. In this guidance the DPK approach was once again recommended as an approach for BE assessment of topical dermatological drug products. The approach was deemed comparable to the PK approach used in systemically available drug products by being able to determine drug concentration as a function of time (Shah et al. 1998). The guidance also discussed stratum corneum and follicular penetration. In the draft guidance, a pilot study and a pivotal BE study were proposed. It also included the recommendations for performance and validation of the technique. The metrics of the pivotal

BE study were time to reach the maximum concentration (T_{max}), the maximum concentration (C_{max}), and the area under the concentration t-time curve (AUC). The BE criteria was 80.00–125.00 % and 70.00–143.00 % for AUC and C_{max}, respectively. However, the guidance was withdrawn in 2002 after years of scientific research, along with public comments and Advisory Committee meetings in 1998, 2000, and 2001 (Federal Register 2002). The major scientific concerns were the doubt of adequacy to assess BE of topical dermatological drug products since they are used to treat a variety of diseases in different parts of the skin, not just the stratum corneum and the reproducibility of the method across laboratories. There were collaborative research studies conducted prior to and during the issuance of this guidance.

One study was conducted in order to establish a possible correlation between clinical safety/efficacy and the DPK method in BE determination for Tretinoin Gel, 0.025 % (Pershing et al. 2003). This drug product is a retinoid and indicated for the treatment of acne. The study was conducted in forty-nine patients using three approved Tretinoin Gel, 0.25 % products: Retin-A® (tretinoin) Gel (innovator and reference, Ortho Pharmaceutical, US), Avita® (tretinoin) Gel (Mylan Bertex, US) and Tretinoin Gel (Spear Pharmaceuticals, US). Based on these studies, it was determined that the drug products with similar composition and therapeutic equivalence (**AB** in the US FDA's Electronic Orange Book) met the 90 % criteria to establish BE set forth by the US FDA, 80.00–125.00 %; whereas Avita® and Retin-A® did not meet the BE criteria, Avita® has a therapeutic equivalence rating of **BT** and is therefore not designated as demonstrating BE to the other pharmaceutically equivalent products. The results of this study further confirmed the US FDA rating of this particular product.

Another study was conducted in order to assess BE of Triamcinolone Acetonide Cream products using IVRT, DPK, and VCA approaches (Pershing et al. 2002). Triamcinolone Acetonide is a synthetic corticosteroid and intended for the treatment of multiple skin diseases/disorders such as dermatitis, psoriasis, and eczema. The study was conducted in healthy volunteer subjects (no more than ten subjects per in vivo study) using two products: Kenalog® (triamcinolone acetonide) Cream, 0.025 %, 0.1 %, and 0.5 % (Bristol-Myers Squibb, US) and Triamcinolone Acetonide Cream, USP, 0.1 % and 0.5 % (Fougera, US). From these studies, the data demonstrated that with an increase in concentration there was an increase in the rate and extent of drug uptake and skin blanching. The authors stated they did not find a statistically significant ($p < 0.05$ %) difference between the two sources of the 0.1 % and 0.5 % of the creams based on using the DPK and VCA approaches. Based on the overall results of study, the authors concluded that DPK could be used for assessment of BA/BE of topical dermatological drug products.

Although, the DPK approach is currently not accepted by the US FDA for BE assessment of topical drug products it is still being evaluated as a possible method in that will be potentially accepted in the future. As such, there has been recent proposed refinements to the method for BE assessment of topical products (N'Dri-Stempfer et al. 2008) and further method evaluation (Boix-Montanes 2011).

13.4.1.2 Pharmaceutical Equivalence

In 2004, a topical bioequivalence update was given at the US FDA Advisory Committee for Pharmaceutical Science (Lionberger 2004). During this presentation one of the discussion points was pharmaceutically equivalent products and the application of in vitro characterization through the utilization of rheological tests and diffusion cells, such as the Franz cell method. As outlined in Sect. 13.3.4, this involves the generic product being Q1, Q2, and Q3 the same as the reference product. Furthermore, as previously stated in 2012, the US FDA for the first time issued a guidance recommending this method for Acyclovir Ointment (antiviral). Therefore, it is evident that the US FDA is continually progressing towards finding alternative approaches for BE assessment of topical drug products.

13.4.2 Post 2007

13.4.2.1 New Approaches and Improvements to Current BE Approaches

In 2007, the US FDA began to publish for the public guidances for industry describing product-specific BE recommendations (Federal Register 2008). Currently, approximately 10 % of the 1,100+ drug-specific guidances posted are for topical drug products, in particular recommendations for the design and conduct of BE studies with clinical endpoints and/or PK studies (Guidances 2013). In the last few years, some of the notable guidances for topical drug products are listed in Table 13.4. It should be noted that this list is not all inclusive of the guidances that have been issued over this time period. The guidances for BE studies for clinical endpoints include a detailed outline of the study design, clinical endpoint, and statistical methods and BE criteria to be used in order to establish BE. Whereas, for guidances that include PK studies, the outline of the study design, analyte to be measured, statistical BE criteria and any other pertinent information to performing an acceptable BE study are included.

As we continue to progress from a regulatory and scientific perspective, the US FDA will continue to provide additional guidances for BE assessment of topical products as well as possible revisions to some of the current recommendations in the future.

Moreover, as a follow-up to these initiatives, the US FDA has continued to reiterate the importance of addressing these scientific challenges by incorporating this into the agreed upon regulatory initiatives in 2013 as a part of GDUFA implementation (Generic Drug User Fee Act Program Performance Goals and Procedure 2012). In addition, a recent public hearing was held by the US FDA to discuss the current successes, ongoing and new projects for addressing these challenges (Regulatory Science Initiatives Part 15 Public Meeting 2013). Some of

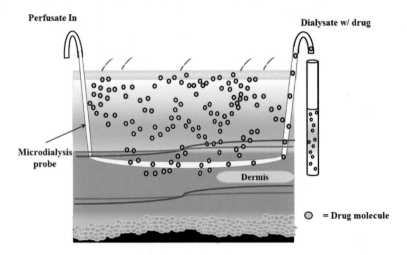

Fig. 13.8 A schematic of microdialysis probe inserted into the dermis. The microdialysis probe can be inserted into a specific region in the skin. The probe is composed of a semipermeable membrane, partly covered by an impermeable coating. After the drug begins to penetrate and/or diffuse across the skin it will transverse the semipermeable membrane of the microdialysis probe and be collected within the dialysate and subsequently analyzed in order to determine the drug concentration levels. Some of the drug will be bound to protein. The microdialysis probe is most often inserted into the volar aspect of the forearm

the successes are outlined in Table 13.4, such as Lidocaine and Acyclovir guidances. While, the ongoing and new projects involve the DMD and IVRT approaches. These research projects involve collaborative efforts with academia and the National Institutes of Health, respectively.

13.4.2.2   DMD Approach

The in vivo method of DMD is currently being studied as a promising surrogate to BE studies with clinical endpoints for potential assessment of BA/BE of topical drug products (Chaurasia et al. 2007; Holmgaard et al. 2010). It is used for sampling of free drug concentrations in extracellular tissue or organs. This approach consists of placing an ultrathin semipermeable hollow fiber structure called a probe into the dermis and perfusing the probe with a tissue compatible sterile buffer at a very low rate by means of a microdialysis pump (a very precise syringe driver). The probe functions as an "artificial blood vessel" in the dermis and thus exchanges small diffusible molecules from the probe to the tissue and vice versa, Fig. 13.8. This is driven by a concentration gradient. As the molecules diffuse across the membrane, aliquots (dialysate) of sample are collected over time and then analyzed in order to determine drug concentrations. This makes this approach suitable for PK studies. DMD allows for sampling at multiple sites in the same healthy volunteer.

The first reported human study was in 1991, in which the percutaneous absorption of the solvent, ethanol into the dermis was measured using microdialysis (Anderson et al. 1991). There have been several domestic and international reports on the application of DMD in the literature spanning the last 20 years (Groth et al. 2006; Holmgaard et al. 2010). However, the approach has been limited especially for lipophilic or highly protein-bound drugs due to low recovery. In addition, further development and validation of this approach has also limited its acceptability.

Recently a study was conducted to investigate the DMD approach for assessment of BA/BE of a topical drug product, Ketoprofen Gel (Tettey-Amlalo et al. 2009). This drug product is an NSAID and is indicated for the relief of localized pain and inflammation associated with acute musculo-skeletal injuries. The study was conducted with eighteen healthy volunteer subjects using Fastum® (ketoprofen) Gel (Adcock Ingram, Bryanston, South Africa). The designation of the site of application was based on three (3) different sequences: A (TTRR/RRTT), B (TRTR/RTRT), and C (TRRT/RTTR). The authors reported that for all three sequences, the 90 % CI of 80.00–125.00 % were met for AUC0-5 h. However, the inter-subject variability was 68 % in the study. Based on the results of the study, the authors concluded that DMD can be used as a tool for performing comparative BA/BE studies with further optimization of the study design and application of the topical drug products to sites on the skin which have been appropriately identified. Another recent study, involved the evaluation of BE between three marketed Metronidazole Creams using DMD and DPK (Ortiz et al. 2011). This drug product is indicated for the treatment of rosacea. The study was conducted with fourteen healthy volunteer subjects using Metronidazol® (metronidazole) Cream, 1 % (Alpharma ApS, Denmark), Flagyl® (metronidazole) Cream, 1 % (Aventis Pharma A/S, Denmark), and Rozex® (metronidazole) Cream, 0.75 % (Galderma Nordic AB, Sweden). The authors reported there was no statistical difference between the penetration of the topical drug products based on the DMD approach. However, BE could not be determined due to high inter-subject variability which was greater than >100 %, while the intra-subject variability was >30 % for all of the products. Based on the results of the study, the authors still concluded that DMD provides more relevant information on drug BA than DPK.

In addition, another recent study involved the application of DMD for determination of BA for Clobetasol Propionate (Au et al. 2012). This study also investigated the impact of changing the perfusate to conduct the study. The basis of this proposed change was due to the fact that many topical drug products are lipophilic (such as Clobetasol Propionate), thus their poor aqueous solubility coupled with binding/adherence of these drugs to the membrane and other components of the microdialysis system as in the past have interfered with the feasibility of this approach. In an effort to overcome some of those limitations, an alternative perfusate, Intralipid® was investigated as compared to the usual aqueous isotonic saline and buffered electrolyte solutions. Clobetasol Propionate is a corticosteroid and is indicated for treatment of various skin diseases which include dermatitis, eczema, and psoriasis. The study was conducted with ten healthy volunteer subjects. The authors reported that by using a perfusate, Intralipid® lipid emulsion, one can greatly increase the recovery of the drug compared to a saline perfusate, allowing for the recovery and BA assessment.

13.4.2.3 In Vitro Release Tests

As discussed in Sect. 13.3.4, the US FDA currently accepts IVRT in some forms for the approval of generic drug products and post-approval changes for both generic and reference products. The IVRT can reflect the combined effect of several physical and chemical parameters, including solubility and particle size of the API, along with rheological properties of the dosage form. The IVRT is commonly referred to as the Franz diffusion cell system, has been used by drug companies for years during the development and screening of topical drug products. For IVRT, the US FDA currently accepts the usage of a synthetic membrane. In recent years possible consideration has been given to the usage of human skin obtained from surgical procedures or excised (cadaver) human skin instead of synthetic membranes. Based on studies referenced in the recent past, the results of the studies using human skin as the membrane have at times been limited or yielded widely variable data (Russell and Guy 2009). Despite these problems, there have been some results in literature which may support the usage of human cadaver skin.

A recent study was conducted for BE assessment of topical drug products using excised human skin (Franz et al. 2009). Generic drug products for the following reference products were studied: Tretinoin, 0.01 % and 0.25 % [Retin-A® (tretinoin) Gel, Valent Intl, US], Alclometasone Dipropionate, 0.05 % [Aclovate® (alclometasone dipropionate) Cream and Ointment, Fourgera, US], Halobetasol Propionate, 0.05 % [Ultravate® (halobetasol propionate) Cream and Ointment, Ranbaxy, US], and Mometasone Furoate, 0.1 % [Elocon® (mometasone) Ointment, Merck Sharp Dohme, US]. All of these drug products are corticosteroids and are indicated for various skin diseases/disorders. The US approval of the generics for these drug products were all based on the VCA approach. The authors reported that for all except one of the corticosteroids (Mometasone Furoate Ointment), the in vitro test-to-reference ratio were within 0.8–1.25. In addition, the excised human skin showed discriminatory evidence across the different vehicles. Based on the results, the authors concluded that this in vitro skin model study could be possible used as a surrogate for in vivo BE studies.

Also, in 2011 an article was published that examined the existing literature of using excised human skin model to match those of a living man (Lehman et al. 2011). A total of 92 datasets were collected from 30 published studies. For harmonized datasets (11 from 2 studies), the average IVIC correlation was 0.96. There was less than a twofold difference between the in vitro and in vivo results for any one compound. The dominant factors for exclusion of certain data is the use from different anatomical sites and vehicles of differing compositions. With the comparison of all datasets the average in vitro–in vivo correlation ratios across all values was 1.6, though for a single dataset there could nearly be a 20-fold difference between the in vitro and in vivo values. In 85 % of cases, however, the difference was less than threefold. From this data collection, it was determined that excised human skin can possibly be used to determine percutaneous absorption.

Another option that has been proposed is the usage of cultured cell lines. Although, this method has proven to be valuable in research on skin irritation assessment, it is still in development for quantitative predictions of percutaneous absorption (Netzlaff et al. 2007; Russell and Guy 2009).

13.4.2.4 Other Alternative Approaches

NIR Spectroscopy. NIR Spectroscopy is a relatively new noninvasive approach for quantitation of the drug concentration in the skin (Narkar 2010; Lademann et al. 2012). This spectroscopic (imaging) approach allows for the determination of the diffusion of drugs and chemicals into the human skin. In 2006, a study was conducted in order to quantify the dermal absorption in human skin and guinea pig animal skin of Econazole Nitrate and Estradiol using this spectroscopy approach (Medendorp et al. 2007). Econazole Nitrate is an antifungal that is indicated for the treatment of infections caused by susceptible dermatophyte and candida species. While, Estradiol is typically indicated for hormone replacement in post-menopausal women. The drugs of interest were saturated in solution prepared in propylene glycol and 1 % cream for Econazole Nitrate and a solution prepared in ethanol for Estradiol. After equilibration of the skin samples and drugs for 2 h, then the skin samples were washed to remove any residual drug product and immediately analyzed by NIR on both the epidermal and dermal sides. The NIR results were validated against known skin concentrations measured by high-performance liquid chromatography analysis of solvents. The authors reported that there was a strong correlation between the results. The r^2 ranged from 0.967 to 0.996, with a standard error of estimate ranging from 1.98 to 5.53 % and a standard error of performance ranging from 2.12 to 6.83 %. Based on the results, the authors concluded that it may be possible to develop an all-optical method for measurement of dermal drug absorption.

Other spectroscopic approaches have also been proposed such as Raman spectroscopy (Zhao et al. 2008), Simulated Raman scattered spectroscopy (Saar et al. 2011), Photoacoustic Fourier Transformed infrared spectroscopy, Photothermal Deflection spectroscopy, PDS (Gotter et al. 2008), and Confocal Raman spectroscopy (Mateus et al. 2013). However, the practicality of spectroscopy approaches is still in the exploratory phase. In the past, some spectroscopic approaches have been used to quantify the drug using the DPK approach (Reddy et al. 2002; Stamatas 2011).

13.5 Conclusion

The current US FDA guidelines, policies, and regulations for the submission and subsequent approval of topical drug products have led to the approval of safe and effective drugs for the American public. Despite the current limitations in the number of feasible approaches for BE assessment, the US FDA is accepting of alternative approaches to BE studies with clinical endpoints, such as in vitro endpoint studies and pharmacokinetic studies, along with the VCA approach for corticosteroids. As the US FDA moves forward it continues to make great efforts to expand the selection of approaches through regulatory scientific initiatives and collaborative research efforts. The US FDA will continue to explore the development and improvement of current approaches for BE assessment of topical drug products.

References

Anderson C, Andersson T, Molander M (1991) Ethanol absorption across human skin measured by in vivo microdialysis technique. Acta Derm Venereol 71(5):389–393

Au WL, Skinner MF, Benfeldt E, Verbeeck RK, Kanfer I (2012) Application of dermal microdialysis for the determination of bioavailability of clobetasol propionate applied to the skin of human subjects. Skin Pharmacol Physiol 25(1):17–24

Boix-Montanes A (2011) Relevance of equivalence assessment of topical products based on the dermatopharmacokinetics approach. Eur J Pharm Sci 42(3):173–1796

Braddy AC, Conner DP (2011) Regulatory Perspective of Dermatokinetic Studies. In: Murthy SN (ed) Dermatokinetic of therapeutic agents. CRC, Boca Raton, pp 193–201

Buxton ILO (2006) Pharmacokinetic and pharmacodynamics: the dynamics of drug absorption, distribution, action and elimination. In: Brunton LL, Lazo JS, Parker KL (eds) Goodman & Gilman's The pharmacological basis of therapeutics, 11th edn. McGraw-Hill, New York, pp 1–39

Chang RK, Raw A, Lionberger R, Yu L (2013a) Generic development of topical dermatologic products: formulation development, process development, and testing of topical dermatologic products. AAPS J 15(1):41–52

Chang RK, Raw A, Lionberger R, Yu L (2013b) Generic development of topical dermatologic products, part II: quality by design for topical semisolid products. AAPS J 15(3):674–682

Chaurasia CS, Muller M, Bashaw ED, Benfeldt E, Bolinder J et al (2007) AAPS-FDA workshop white paper: microdialysis principles, application and regulatory perspectives. Pharm Res 24 (5):1014–1025

Critical Path Opportunities for Generic Drugs (2007) U.S. Department of Health and Human Services, Food and Drug Administration, Silver Spring, MD. http://www.fda.gov/Science Research/SpecialTopics/CriticalPathInitiative/CriticalPathOpportunitiesReports/ucm077250.htm. Accessed 18 Nov 2013

Electronic Orange Book (2013) U.S. Department of Health and Human Services, Food and Drug Administration, Silver Spring, MD. http://www.accessdata.fda.gov/scripts/cder/ob/default. cfm. Accessed 18 Nov 2013

Federal Register (2002) Draft guidance for industry on topical dermatological drug product NDAs and ANDAs—In vivo bioavailability, bioequivalence, in vitro release and associated studies; withdrawal [Docket No. 98D-0388]. U.S. Department of Health and Human Services, Food and Drug Administration, Silver Spring, MD. https://www.federalregister.gov/articles/1998/06/18/98-16141/draft-guidance-for-industry-on-topical-dermatological-drug-product-ndas-and-andasin-vivo. Accessed 18 Nov 2013

Federal Register (2008) Publication of guidances for industry describing product-specific bioequivalence recommendations [Docket No. FDA 2007-D-0369]. U.S. Department of Health and Human Services, Food and Drug Administration, Silver Spring, MD. https://www.federalregister.gov/articles/2008/09/05/E8-20580/publication-of-guidances-for-industry-describing-product-specific-bioequivalence-recommendations. Accessed 18 Nov 2013

Federal Register (2012) Drugs for human use; drug efficacy study implementation; certain prescription drugs offered for various indications; opportunity to affirm outstanding hearing request. U.S. Department of Health and Human Services, U.S. Food and Drug Administration, Silver Spring, MD. https://www.federalregister.gov/articles/2012/07/24/2012-18015/drugs-for-human-use-drug-efficacy-study-implementation-certain-prescription-drugs-offered-for. Accessed 18 Nov 2013

Franz TJ, Lehman PA, Raney SG (2009) Use of excised human skin to assess the bioequivalence of topical products. Skin Pharmacol Physiol 22(5):276–286

General Chapter <1151> Pharmaceutical Dosage Forms (2013) U.S. Pharmacopeia, Rockville, MD. http://www.uspnf.com/uspnf/pub/index?usp=36&nf=31&s=1&officialOn=August 1, 2013. Accessed 18 Nov 2013

General Chapter <1059> Excipient Performance (2013). U.S. Pharmacopeia, Rockville. http://www.uspnf.com/uspnf/pub/index?usp=36&nf=31&s=1&officialOn=August%201,%202013. Accessed 18 Nov 2013

Generic Drug User Fee Act Program Performance Goals and Procedure (2012) U.S. Department of Health and Human Services, Food and Drug Administration, Silver Spring, MD. http://www.fda.gov/downloads/ForIndustry/UserFees/GenericDrugUserFees/UCM282505.pdf. Accessed 18 Nov 2013

Generic Drug User Fee Amendments (2012) U.S. Food and Drug Administration Safety and Innovation Act, S.3187. 3187. U.S. Government Printing Office, Washington. http://www.gpo.gov/fdsys/pkg/BILLS-112s3187enr. Accessed 18 Nov 2013

Gotter B, Faubel W, Neubert RHH (2008) Optical methods for measurements of skin penetration. Skin Pharmacol Physiol 21(3):151–156

Groth L, Garcia OP, Benfeldt E (2006) Microdialysis methodology for sampling in the skin. In: Serup J, Jemec GBE, Grove GL (eds) Handbook of non-invasive methods and the skin, 2nd edn. CRC, Boca Raton, pp 443–454

Guidance for Industry (1995) topical dermatologic corticosteroids: In vivo bioequivalence (1995) US Department of Health and Human Services, Food and Drug Administration, Silver Spring, MD. http://www.fda.gov/downloads/Drugs/GuidanceComplianceRegulatoryInformation/Guidances/UCM070234.pdf. Accessed 18 Nov 2013

Guidance for Industry (1997) Nonsterile semisolid dosage forms, scale-up and post-approval changes: Chemistry, manufacturing, and control; In vitro release resting and in vivo bioequivalence documentation. US Department of Health and Human Services, Food and Drug Administration, Silver Spring, MD. http://www.fda.gov/downloads/Drugs/GuidanceComplianceRegulatoryInformation/Guidances/UCM070930.pdf. Accessed 18 Nov 2013

Guidances (2012) Bioequivalence recommendations for Acyclovir Ointment. U.S. Department of Health and Human Services, U.S. Food and Drug Administration, Silver Spring, MD. http://www.fda.gov/downloads/Drugs/GuidanceComplianceRegulatoryInformation/Guidances/UCM296733.pdf. Accessed 18 Nov 2013

Guidances (2013) Bioequivalence recommendations for specific drug products. U.S. Department of Health and Human Services, U.S. Food and Drug Administration, Silver Spring, MD. http://www.fda.gov/Drugs/GuidanceComplianceRegulatoryInformation/Guidances/ucm075207.htm. Accessed 18 Nov 2013

Handbook of Pharmaceutical Excipients Association (2012) Rowe RC, Sheskey P, Cook WG, Fenton ME (2012) 7th edn. Pharmaceutical Press and American Pharmacists (Online edition-Medicines Complete, http://www.medicinescomplete.com/mc/)

Harding CR (2004) The stratum corneum: structure and function in health and disease. Dermatol Ther 17(Suppl 1):6–15

Holmgaard R, Nielsen JB, Benfeldt E (2010) Microdialysis sampling for investigations of bio-availability and bioequivalence of topically administered drugs: current state and future perspectives. Skin Pharmacol Physiol 23(5):225–243

Hueber F, Schaefer H, Weipierre J (1994) Role of transepidermal and transfollicular routes in percutaneous absorption of steroids: in vitro studies on human skin. Skin Pharmacol 75 (5):237–244

Interim Guidance of Industry (1998) Topical dermatological drug product NDAs and ANDAs—in vivo bioavailability, bioequivalence, in vitro release, and associated studies. U.S. Department of Health and Human Services, Food and Drug Administration, Silver Spring, MD. http://www.fda.gov/ohrms/dockets/ac/00/backgrd/3661b1c.pdf. Accessed 18 Nov 2013

Kircik LH, Bikowski JB, Cohen DE, Draelos ZD, Hebert A (2010) Supplement to practical dermatology—vehicles matter. Formulation development, testing, and approval. Practical Dermatology http://bmctoday.net/vehiclesmatter/pdfs/0310.pdf. Accessed 8 Aug 2013

Lademann J, Meinke MC, Schanzer S, Richter H, Darvin ME et al (2012) In vivo methods for the analysis of the penetration of topically applied substances in and through the skin barrier. Int J Cosmet Sci 34(6):551–559

Lehman PA, Raney SG, Franz TJ (2011) Percutaneous absorption in man: in vitro-in vivo correlation. Skin Pharmacol Physiol 24(4):224–230

Lionberger R (2004) Topical bioequivalence update at: Advisory Committee for Pharmaceutical Science. U.S. Department of Health and Human Services, Food and Drug Administration, Silver Spring, MD. http://www.fda.gov/ohrms/dockets/ac/04/slides/4034s2.htm. Accessed 18 Nov 2013

Lionberger R (2008) FDA critical path initiatives: opportunities for generic drug development. AAPS J 10(1):103–109

Long CC (2007) Common skin disorders and their topical treatment. In: Walter KA (ed) Dermatological and transdermal formulations. Informa Health Care, New York, pp 41–59

Mateus R, Abdalghafor H, Oliveira G, Hadgraft J, Lane ME (2013) A new paradigm in dermatopharmacokinetics—confocal raman spectroscopy. Int J Pharm 444(1–2):106–108

McKenzie AW, Stoughton RB (1962) Method for comparing percutaneous absorption of steroids. Arch Dermatol 86(5):608–610

Medendorp JP, Paudel KP, Lodder RA, Stinchcomb AL (2007) Near infrared spectrometry for the quantification of human dermal absorption of econazole nitrate and estradiol. Pharm Res 24 (1):186–193

N'Dri-Stempfer B, Navidi WC, Guy RH, Bunge AL (2008) Improved bioequivalence assessment of topical dermatological drug products using dermatopharmacokinetics. Pharm Res 26 (2):316–328

Narkar Y (2010) Bioequivalence for topical products—an update. Pharm Res 27(12):2590–2601

National Academy of Sciences (1974) Drug efficacy study of the national research council's division of medical sciences, 1966–1969. National Academy of Sciences, Washington, DC. http://www.nasonline.org/about-nas/history/archives/collections/des-1966-1969-1.html. Accessed 18 Nov 2013

Netzlaff F, Kaca M, Bock U, Haltner-Ukomadu E, Meiers P et al (2007) Permeability of the reconstructed human epidermis model Episkin® in comparison to various human skin preparations. Eur J Pharm Biopharm 66(1):127–134

Ortiz PG, Hansen SH, Shah VP, Sonne J, Benfeldt E (2011) Are marketed topical metronidazole creams bioequivalent? evaluation by in vivo microdialysis sampling and tape stripping methodology. Skin Pharmacol Physiol 24(1):44–53

Pershing LK, Nelson JL, Corlett JL, Shrivastava S, Hare DB, Shah VP (2003) Assessment of dermatopharmacokinetic approach in the bioequivalence determination of topical tretinoin gel products. J Am Acad Dermatol 48(5):740–751

Pershing LK, Bakhitan S, Poncelet CE, Corlett JL, Shah VP (2002) Comparison of skin stripping, in vitro release, and skin blanching response methods to measure does response and similarity of triamcinolone acetonide cream strengths from two manufactured sources. J Pharm Sci 91 (5):1312–1323

Reddy MB, Stinchcomb AL, Guy RH, Bunge AL (2002) Determining dermal absorption parameters in vivo from tape strip data. Pharm Res 19(2):292–298

Russell LM, Guy RH (2009) Measurement and prediction of the rate and extent of drug delivery into and through the skin. Expert Opin Drug Deliv 6(4):355–369

Saar BG, Contreras-Rojas LR, Xie XS, Guy RH (2011) Imaging drug delivery to skin with stimulated raman scattering microscopy. Mol Pharm 8(3):969–975

Shah VP, Flynn GL, Yacobi A, Maibach HI, Bon C et al (1998) Bioequivalence of topical dermatological dosage forms—methods of evaluation of bioequivalence. Pharm Res 15 (2):167–171

Shah VP (2001) Progress in the methodologies for evaluating bioequivalence of topical formulations. Am J Clin Dermatol 2(5):275–280

Singh GJP, Adams WP, Lesko LJ, Shah VP, Molzon JA, Williams RL, Pershing LK (1999) Development of in vivo bioequivalence methodology for dermatologic corticosteroids based on pharmacodynamic modeling. Clin Pharmacol Ther 66(4):346–357

Stamatas GN (2011) Spectroscopic techniques in dermatokinetic studies. In: Murthy SN (ed) Dermatokinetic of therapeutic agents. CRC, Boca Raton, pp 175–191

Stoughton RB (1987) Are generic formulations equivalent to trade name topical glucocorticoids? Arch Dermatol 123(10):1312–1314

Stoughton RB (1989) Percutaneous absorption of drugs. Annu Rev Pharmacol Toxicol 29:55–69

Stoughton RB (1992) Vasoconstrictor assay: specific application. In: Maibach HI, Surber C (eds) Topical corticosteroids. Karger, Switzerland, pp 42–53

Tettey-Amlalo RN, Kanfer I, Skinner MF, Benfeldt E, Verbeeck RK (2009) Application of dermal microdialysis for the evaluation of bioequivalence of a ketoprofen topical gel. Eur J Pharm Sci 36(2–3):219–225

Topical drug delivery systems. A review (2008) Pharmainfo.net. http://www.pharmainfo.net/reviews/topical-drug-delivery-systems-review. Accessed 18 Nov 2013

U.S. Code (1984) The Waxman-Hatch Act. The drug prove competition and patent term restoration act of 1984 (popularly known as the Waxman-Hatch or Hatch-Waxman Act) is codified as 21 U.S.C. 355 of the food, drug and cosmetic act and 35 U.S.C. 271(e) and 35 U.S.C. 156 of the patent act

U.S. Code of Federal Regulations. Title 21 Part 320 bioavailability and bioequivalence. (Revised as 01 January 2012) U.S. Government Printing Office, Washington http://www.accessdata.fda.gov/scripts/cdrh/cfdocs/cfcfr/CFRSearch.cfm?CFRPart=320&showFR=1. Accessed 18 Nov 2013

U.S. Food and Drug Administration (2009) Requests the commissioner of food & drugs not approve any generic equivalent version of the petitioner's proprietary drug product derma-smoothe/FS (Fluocinolone acetonide 0.01 % topical oil) unless & until applicants comply with statutory requirements [Docket No. FDA-2004-P-0215]. Regulations.gov, http://www.regulations.gov/#!home. Accessed 18 Nov 2013

U.S. Food and Drug Administration (2012) 50 Years: The Kefauver-Harris Amendments. U.S. Department of Health and Human Services, U.S. Food and Drug Administration, Silver Spring, MD. http://www.fda.gov/Drugs/NewsEvents/ucm320924.htm. Accessed 18 Nov 2013

Weiss SC (2011) Conventional topical delivery systems. Dermatol Ther 24(5):471–476

Wosicka H, Cal K (2010) Targeting to the hair follicles: current status and potential. J Dermatol Sci 57(2):83–89

Zhao J, Lui H, McLean DI, Zang H (2008) Integrated real-time raman system for clinical in vivo skin analysis. Skin Res Technol 14(4):484–492

Chapter 14
Bioequivalence for Orally Inhaled and Nasal Drug Products

Bhawana Saluja, Bing V. Li, and Sau L. Lee

14.1 Introduction

Locally acting orally inhaled and nasal drug products offer effective and rapid remedies for treating pulmonary and nasal diseases including, but not limited to, asthma, chronic obstructive pulmonary disease, emphysema, and allergic rhinitis. Out of the several drug classes available for treatment of these disease conditions, the most common therapeutic moieties include β-agonists, anticholinergics, and corticosteroids for the inhalation route, as well as antihistamines, anticholinergics, and corticosteroids for the nasal route. In addition, the drug delivery systems most commonly used to deliver these drugs to their site(s) of action include metered dose inhalers, dry powder inhalers, nebulizers (which will not be covered by this chapter), and nasal sprays. Figure 14.1 provides examples of these drug delivery devices.

Bioequivalence (BE)[1] is defined as the absence of a significant difference in the rate and extent of availability of the active ingredient or active moiety in a pharmaceutical equivalent[2] or pharmaceutical alternative[3] at the site(s) of action

---

[1][21 CFR 320.1(e)].

[2]Pharmaceutical equivalents mean drug products in identical dosage forms that contain identical amounts of the identical active drug ingredient, i.e., the same salt or ester of the same therapeutic moiety, or, in the case of modified release dosage forms that require a reservoir or overage or such forms as prefilled syringes where residual volume may vary, that deliver identical amounts of the active drug ingredient over the identical dosing period; do not necessarily contain the same inactive ingredients; and meet the identical compendial or other applicable standard of identity, strength, quality, and purity, including potency and, where applicable, content uniformity, disintegration times, and/or dissolution rates [21 CFR 320.1(c)].

[3]Pharmaceutical alternatives mean drug products that contain the identical therapeutic moiety, or its precursor, but not necessarily in the same amount or dosage form or as the same salt or ester.

B. Saluja • B.V. Li • S.L. Lee (✉)
Center for Drug Evaluation and Research, U.S. Food and Drug Administration,
10903 New Hampshire Avenue, Silver Spring, MD 20993, USA
e-mail: sau.lee@fda.hhs.gov

L.X. Yu and B.V. Li (eds.), *FDA Bioequivalence Standards*, AAPS Advances
in the Pharmaceutical Sciences Series 13, DOI 10.1007/978-1-4939-1252-0_14,
© The United States Government 2014

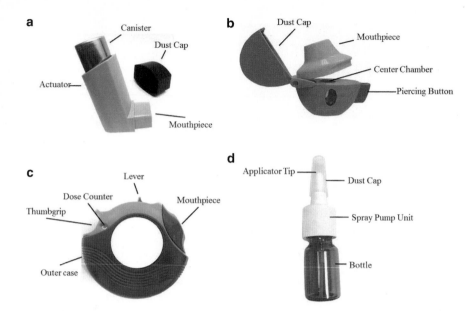

Fig. 14.1 (**a**) Schematic diagram of a metered dose inhaler device—It consists of a metal canister which holds the formulation, an actuator with a mouthpiece and a dust cap. (**b**) Schematic diagram of a pre-metered, single-unit dose dry powder inhaler device—It consists of a dust cap, a mouthpiece, a central chamber that holds the capsule containing the formulation and a piercing button. (**c**) Schematic diagram of a pre-metered, multi-unit dose dry powder inhaler device—It consists of an outer case, a mouthpiece, a thumbgrip, a lever and a dose counter. (**d**) Schematic diagram of a nasal spray device—It consists of a dust cap, an applicator tip, a spray pump, and a bottle that holds the formulation

when administered at the same molar dose under similar conditions. For orally administered drug products that act systemically, pharmacokinetic studies are generally considered adequate for the assessment of BE, based on a premise that equivalence of drug concentration in plasma or blood implies equivalence of drug delivery to the site(s) of action through systemic circulation. However, the use of pharmacokinetic studies alone is currently considered insufficient for establishing equivalence in local drug delivery for orally inhaled and nasal drug products, because these drugs do not rely on the systemic circulation for delivery to the local site(s) of action and entry of drugs into the systemic circulation may depend on multiple sites (e.g., gastrointestinal tract versus respiratory tract), as illustrated in Fig. 14.2.

Demonstration of BE for orally inhaled and nasal drug products presents unique challenges, primarily due to the lack of relevance of pharmacokinetics to

---

Each such drug product individually meets either the identical or its own respective compendial or other applicable standard of identity, strength, quality, and purity, including potency and, where applicable, content uniformity, disintegration times and/or dissolution rates [21 CFR 320.1(d)].

Fig. 14.2 Disposition of
dose emitted from an orally
inhaled drug product
(adopted from the figure
originally published in
Singh and Adams (2005))

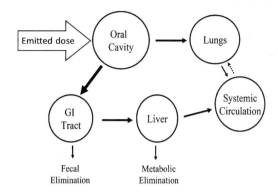

equivalence in the local drug delivery. Furthermore, the efficacy or performance of these drug-device combination products depends on a number of product-related factors such as formulation and device design. For instance, albuterol administered via a dry powder inhaler (Turbuhaler®) requires only half the dose, as compared to when administered via a metered dose inhaler, to achieve similar bronchodilatation effect in patients with asthma (Löfdahl et al. 1997). Similar effect is observed with terbutaline when administrated using Turbuhaler® and metered dose inhalers in patients with asthma (Borgström et al. 1996). Therefore, the design of BE studies for orally inhaled and nasal drug products should take into consideration the above factors. To overcome these challenges, the US Food and Drug Administration (FDA) proposed an aggregate weight-of-evidence approach, which emphasizes three elements: (1) in vitro studies, (2) pharmacokinetic studies, and (3) pharmaco-dynamic or clinical endpoint studies, to establish equivalent systemic exposure and local drug delivery (Lee et al. 2009; Adams et al. 2010). The weight-of-evidence approach was used to approve four Abbreviated New Drug Applications of chlorofluorocarbon-based albuterol metered dose inhalers in mid-1990s and several locally acting nasal suspension sprays (Li et al. 2013) in the United States (Table 14.1).

This chapter provides a scientific discussion of each key element of the weight-of-evidence approach, with primary focus on metered dose inhalers and dry powder inhalers containing β-agonists and/or corticosteroids as well as nasal sprays containing corticosteroids. For metered dose inhalers and nasal sprays, the BE discussion will focus on suspension formulations. The chapter also pro-vides a general discussion on the critical aspects of formulation and device design, particularly with respect to the development of test orally inhaled and nasal drug products that are expected to be interchangeable with their reference products.

Table 14.1 List of abbreviated new drug applications for metered dose inhalers and nasal sprays for local action approved by the FDA

Drug product	Applicant	Approval date
Albuterol Metered Dose Inhaler	Ivax/Teva Pharms	December 1995[a]
Albuterol Metered Dose Inhaler	Pliva	August 1996[a]
Albuterol Metered Dose Inhaler	Armstrong Pharms	August 1996[a]
Albuterol Metered Dose Inhaler	Genpharm	August 1997[a]
Azelastine Nasal Spray	Apotex	April 2009
Azelastine Nasal Spray	Sun Pharma Global	May 2012
Azelastine Nasal Spray	Apotex	August 2012
Flunisolide Nasal Spray	Bausch and Lomb Pharms	February 2002
Flunisolide Nasal Spray	Apotex	August 2007
Flunisolide Nasal Spray	HH and P	August 2006
Fluticasone Propionate Nasal Spray	Roxane	February 2006
Fluticasone Propionate Nasal Spray	Apotex	September 2007
Fluticasone Propionate Nasal Spray	Hi Tech Pharma	January 2008
Fluticasone Propionate Nasal Spray	Wockhardt	January 2012
Ipratropium Bromide Nasal Spray	Dey LP	March 2003
Ipratropium Bromide Nasal Spray	Dey LP	March 2003
Ipratropium Bromide Nasal Spray	Novex Pharma	April 2003
Ipratropium Bromide Nasal Spray	Novex Pharma	April 2003
Ipratropium Bromide Nasal Spray	Roxane	November 2003
Ipratropium Bromide Nasal Spray	Roxane	November 2003
Ipratropium Bromide Nasal Spray	Bausch and Lomb Pharms	March 2003
Ipratropium Bromide Nasal Spray	Bausch and Lomb Pharms	March 2003
Tetrahydrozoline Nasal Spray	Fougera Pharms	November 1979
Triamcinolone Hydrochloride Nasal Spray	Teva Pharms	July 2009

Data collected in June, 2013

[a]These metered dose inhaler products have been discontinued due to the environment concern associated with the use of chlorofluorocarbon propellants in these products

14.2 Bioequivalence Studies for Orally Inhaled and Nasal Drug Products

14.2.1 In Vitro Bioequivalence Studies

Present thinking on this topic spans several major aspects of in vitro performance of dry powder inhalers, metered dose inhalers, and nasal sprays.

14.2.1.1 Dry Powder Inhalers

The key in vitro tests for dry powder inhalers include demonstration of equivalence in single actuation content and aerodynamic particle size distribution, as indicated in Table 14.2. These two in vitro performance attributes can affect the total and

Table 14.2 In vitro bioequivalence tests for dry powder inhalers

Study type	Study design	In vitro equivalence based on	Lifestage(s)
Single Actuation Content (SAC)	SAC is performed with a single actuation to determine the delivered dose drug mass. It is performed at three flow rates of reference labeled flow rate, and ±50 % of the labeled flow rate *Equipment*: USP ⟨601⟩ Apparatus B or another appropriate apparatus	Delivered drug mass per single actuation using population BE (PBE) analysis (FDA 2012)	Beginning (B), middle (M), and end (E) lifestages
Aerodynamic Particle Size Distribution (APSD)	APSD determination is performed with a minimum number of inhalations justified by the sensitivity of the validated assay. It is performed at three flow rates of reference labeled flow rate, and ±50 % of the labeled flow rate *Equipment*: USP ⟨601⟩ Apparatus 3 or Apparatus 5, or another appropriate apparatus	Impactor-sized mass using PBE analysis. Cascade Impactor profiles representing drug deposition on the individual stages of the Cascade Impactor along with the mass median aerodynamic diameter (MMAD), geometric standard deviation (GSD), and fine particle mass (FPM) as supportive evidence for equivalent APSD	B and E lifestages

regional deposition of drug(s) in the lung, thereby affecting the safety and efficacy of dry powder inhalers.

Dry powder inhalers are generally breath-actuated or passive, meaning that the devices utilize the patient's inspiratory effort to generate drug fluidization, deaggregation, and release. Dry powder inhalers are used over a range of inspiratory flow rates due to variations in the inspiratory effort across the patient population; single actuation content and aerodynamic particle size distribution may vary with change in inspiratory flow rate. Therefore, it is important that equivalence in single actuation content and aerodynamic particle size distribution be established at a range of flow rates (i.e., minimum of three flow rates). More importantly, the flow rates selected for in vitro testing of single actuation content and aerodynamic particle size distribution are expected to reasonably cover flow rates generated by the relevant patient population. For example, the flow rates for in vitro testing of a test dry powder inhaler referencing Advair® Diskus® may include 30, 60 (reference labeled flow rate), and 90 L/min.

Dry powder inhalers are generally multi-dose products. The device material and formulation may have a subsequent effect on the in vitro performance of DPIs, in part due to their influence on the accumulation of electrostatic charge over time, and they may differ between the test and reference products (Lee et al. 2009). Thus, it is essential that equivalence of single actuation content and aerodynamic particle size distribution be demonstrated at multiple stages of product life, including the beginning, middle (e.g., for single actuation content only), and end lifestages. For instance, based on the labeled number of inhalations, beginning lifestage may represent the first inhalation(s), middle lifestage may represent the inhalation (s) corresponding to 50 % of the labeled number of inhalation(s), and end lifestage may represent the inhalation(s) corresponding to the labeled number of inhalations.

The above in vitro tests are sensitive enough to assess the effect of formulation and device on the product performance. The elongation ratio (defined as the ratio of length of a powder particle to its width) of dry powder inhaler carriers has been reported to impact the aerosolization properties of dry powder inhaler formulation of albuterol sulfate, leading to an increase in the emitted dose with increase in elongation ratio; although this increase is restricted to a certain point (Kalaly et al. 2011). In addition, the interaction of the dry powder inhaler formulation with its device also impacts product performance by affecting the powder fluidization and deaggregation, and therefore the aerodynamic particle size distribution of an emitted dose (Newman and Busse 2002).

To ensure that the targeted patients are able to operate the test device effectively and receive proper medication without any significant change in their inspiratory effort relative to use of the reference dry powder inhaler, the design of a test dry powder inhaler warrants consideration of device resistance comparability to a reference dry powder inhaler. More importantly, since single actuation content and aerodynamic particle size distribution may depend on flow rate, and such flow rate dependence may, in part, depend on device resistance (Lee 2012), the use of a test dry powder inhaler with a comparable air flow resistance to the reference dry powder inhaler is also expected to increase the likelihood of establishing single actuation content and aerodynamic particle size distribution equivalence at each of the three selected flow rates (Shur et al. 2012).

14.2.1.2 Metered Dose Inhalers

Like dry powder inhalers, demonstration of equivalence in the single actuation content and aerodynamic particle size distribution constitutes the key in vitro components used to support BE for metered dose inhalers. Other important in vitro performance attributes for equivalence evaluation of metered dose inhalers are spray pattern and plume geometry (Table 14.3). They can affect the drug deposition in the mouth and thus the total dose to the lung. Priming and repriming test is used to ensure that the number of actuations to be wasted for priming (the initial use) and repriming (following one or more periods of nonuse, if applicable) for a test metered dose inhaler either equal to or not greater than the reference metered dose inhaler.

In vitro testing of these attributes also provides a sensitive measure of detecting product differences between metered dose inhalers, because the single actuation content and aerodynamic particle size distribution are known to be affected by

Table 14.3 In vitro bioequivalence tests for metered dose inhalers

Study type	Study design	In vitro equivalence based on	Lifestage(s)
Single Actuation Content (SAC)	SAC is performed with a single actuation to determine the delivered dose drug mass. It is performed at flow rate of 28.3 L/min. *Equipment*: USP ⟨601⟩ Apparatus A or another appropriate apparatus	Delivered drug mass per single actuation using PBE analysis (FDA 2012)	Beginning (B), middle (M), and end (E) lifestages
Aerodynamic Particle Size Distribution (APSD)	APSD determination is performed with a minimum number of inhalations justified by the sensitivity of the validated assay. It is performed at flow rate of 28.3 or 30 L/min. *Equipment*: USP ⟨601⟩ Apparatus 1 or Apparatus 6, or another appropriate apparatus	Impactor-sized mass using PBE analysis. Cascade Impactor profiles representing drug deposition on the individual stages of the Cascade Impactor along with the mass median aerodynamic diameter (MMAD), geometric standard deviation (GSD), and fine particle mass (FPM) as supportive evidence for equivalent APSD	B and E lifestages
Spray Pattern	Spray pattern test is performed with a single actuation. It is determined at two different distances from the actuator orifice, with selected distances at least 3 cm apart. *Equipment*: Non-impaction (laser light sheet and high-speed digital camera), impaction (thin-layer chromatography plate impaction), or other suitable method	Ovality ratio (ratio of longest diameter to shortest diameter, D_{max}/D_{min}) and area within the perimeter of the true shape (not within the fitted geometric shape) for automated analysis, or ovality ratio and D_{max} for manual analysis using PBE analysis. Qualitative comparison of spray shape	B lifestage
Plume Geometry	Plume geometry is performed with a single actuation, and reported at a single delay time while the fully developed phase of the plume is still in contact with the actuator tip/nose piece tip. *Equipment*: High-speed photography, a laser light sheet and high speed digital camera, or other suitable method	Plume angle and plume width. Ratio of the geometric mean of three batches of a test product to that of three batches of a reference product (based on log-transformed data) for both plume angle and width, within 90–111 %	B lifestage

(continued)

Table 14.3 (continued)

Study type	Study design	In vitro equivalence based on	Lifestage(s)
Priming and Repriming	Priming and repriming tests are performed with a single actuation, immediately following the specified number of priming or repriming actuations specified in the reference product labeling. The repriming test is performed following storage for the specified period of nonuse after initial use and/or other conditions (e.g., dropping), if the reference product labeling provides such repriming information *Equipment*: USP ⟨601⟩ Apparatus A or another appropriate apparatus	PBE analysis of the emitted dose of a single actuation immediately following the specified number of priming or repriming actuations specified in the reference product labeling	

many product-related factors, such as physicochemical properties of the drug(s) and inactive ingredients (e.g., surfactant, cosolvent) as well as device geometries (Smyth 2003). For instance, use of ethanol as a cosolvent in a fluticasone propionate metered dose inhaler led to a decrease in the emitted dose (Murthy et al. 2010). In terms of device geometries, it has been reported that for hydrofluoroalkane-based metered dose inhalers, a reduction in orifice diameter led to a significant increase in fine particle mass (Lewis et al. 1998).

Since metered dose inhalers are also multi-dose products, it is important that equivalence of some key in vitro tests (single actuation content and aerodynamic particle size distribution) be demonstrated at multiple stages of product life, including the beginning, middle (e.g., for single actuation content only), and end lifestages, because differences in device (e.g., material of construction and design of valve stem) and formulation properties (e.g., surface properties of an inactive ingredient) between two products may lead to differences in aerosol charge profiles, thereby affecting their performance over the product's lifetime (Kwok et al. 2005).

14.2.1.3 Nasal Sprays

As indicated in Table 14.4, the key in vitro tests for nasal sprays include single actuation content, droplet size distribution, spray pattern, plume geometry, as well as priming and repriming. It is generally considered that these in vitro tests are expected to ensure equivalence of drug deposition pattern in the nasal cavity.

Table 14.4 In vitro bioequivalence tests for nasal sprays

Study type	Study design	In vitro equivalence based on	Lifestage(s)
Single Actuation Content (SAC)	SAC is performed with a single actuation to determine the delivered dose drug mass. *Equipment*: Dosage unit sampling apparatus described in USP ⟨601⟩ or another appropriate apparatus	Delivered drug mass per single actuation using PBE analysis (FDA 2012)	Beginning (B) and end (E) lifestages
Particle/Droplet Size Distribution by Cascade Impactor	APSD determination is performed with a minimum number of inhalations justified by the sensitivity of the validated assay. It is performed at 28.3 L/min. *Equipment*: USP ⟨601⟩ Apparatus 1 or Apparatus 6, or another appropriate apparatus	Drug in small particles/droplets using PBE analysis	B lifestage
Spray Pattern	Spray pattern test is performed with a single actuation. It is determined at two different distances from the actuator orifice, with selected distances at least 3 cm apart. *Equipment*: Non-impaction (laser light sheet and high-speed digital camera), impaction (thin-layer chromatography plate impaction), or other suitable method	Ovality ratio (ratio of longest diameter to shortest diameter, D_{max}/D_{min}) and area within the perimeter of the true shape (not within the fitted geometric shape) for automated analysis, or ovality ratio and D_{max} for manual analysis using PBE analysis. Qualitative comparison of spray shape.	B lifestage
Plume Geometry	Plume geometry is performed with a single actuation, and reported at a single delay time while the fully developed phase of the plume is still in contact with the actuator tip/ nose piece tip. *Equipment*: High-speed photography, a laser light sheet and high speed digital camera, or other suitable method	Plume angle and plume width. Ratio of the geometric mean of three batches of T to that of three batches of R (based on log-transformed data) for both plume angle and width, within 90–111 %	B lifestage
Priming and Repriming	Priming and repriming tests are performed with a	PBE analysis of the emitted dose of a single actuation	

(continued)

Table 14.4 (continued)

Study type	Study design	In vitro equivalence based on	Lifestage(s)
	single actuation, immediately following the specified number of priming or repriming actuations specified in the reference product labeling. The repriming test is performed following storage for the specified period of nonuse after initial use and/or other conditions (e.g., dropping), if the reference product labeling provides such repriming information *Equipment*: USP $\langle 601 \rangle$ Apparatus A or another appropriate apparatus	immediately following the specified number of priming or repriming actuations specified in the reference product labeling	
Droplet Size Distribution by Laser Diffraction	Droplet size distribution is performed with a single actuation. It is characterized at the fully developed phase at two distances from the nose piece tip. Mean D_{10}, D_{50}, and D_{90} values for a given bottle or canisters are computed from the mean of up to three consecutive sprays from that unit at each lifestage. Span $((D_{90} - D_{10})/D_{50})$ can be computed from these data *Equipment*: Laser diffraction or other appropriately validated methodology	D_{50} and span using PBE analysis	B and E lifestages

Particle/droplet size distribution by cascade impactor is also utilized to demonstrate equivalence of drug in small particles and droplets. This in vitro test addresses a potential safety concern, which is related to an excess of small droplets from a test product, relative to a reference product, that may deliver particles/droplets (with possible adverse effects) to regions beyond the nose. Like metered dose inhalers and dry powder inhalers, nasal sprays are generally multi-dose products. Therefore, it is important to conduct some critical in vitro tests (single actuation content and droplet size distribution) at different product lifestages to assess and compare the effect of lifestage on product performance of test and reference products.

The in vitro tests for nasal spray are sensitive to potential changes in formulation and device design. The interaction between the nasal spray formulation and pump device has been reported to influence the droplet size distribution as well as the spray pattern, which may affect intranasal drug deposition (Kublik and Vidgren 1998; FDA 2003a). Moreover, an increase in droplet particle size has been observed with increase in methylcellulose, a viscosity building agent, thereby affecting intranasal deposition pattern of the drug product (Harris et al. 1988). For solution nasal spray drug products, BE can be established based on the above in vitro tests only, provided the test formulation is (1) qualitatively $(Q_1)^4$ and quantitatively $(Q_2)^5$ the same as the reference product, (2) container and closure systems are comparable.

However, if a suspension formulation contains more than one suspending particles, the current particle sizing technologies are unable to provide accurate and precise measurements of the particle size distribution of the active ingredient for the purpose of demonstrating BE. Meanwhile, drug particle size distribution in suspension formulations has the potential to influence the rate and extent of drug availability to nasal site(s) of action and to the systemic circulation. The lack of this key information is one of the reasons why additional in vivo studies, i.e., PK and clinical BE studies, are needed to support BE of suspension nasal spray drug products.

14.2.2 Pharmacokinetic Bioequivalence Study

Orally inhaled and nasal drug products are intended for delivery to the site(s) of action in the lung/nose, but drug deposited in the target regions of the lung/nose may also enter the systemic circulation. For instance, in the case of orally inhaled drug products, a portion of the emitted drug may be deposited in the nontarget regions (i.e., oropharyngeal region), swallowed and subsequently become available for absorption from the gastrointestinal tract. Figure 14.2 provides an example illustrating how the orally inhaled drug reaches the systemic circulation. Determination of drug concentration in plasma is essential for establishing BE of test orally inhaled and nasal suspension drug products to the reference products, due to possible systemic side effects of these drug products (FDA 2003a; Fardon et al. 2004). Therefore, one of the components of the weight-of-evidence approach is demonstration of equivalent systemic exposure following administration of test and reference orally inhaled and nasal drug products.

The study design for a pharmacokinetic BE study for orally inhaled and nasal drug products is similar to that used for solid oral dosage forms. This study is

---

[4]$Q1$ (Qualitative sameness) means that the test product contains the same inactive ingredients as the reference listed drug.

[5]$Q2$ (Quantitative sameness) means that the test product contains all inactive ingredients at concentrations within ±5 % of the concentrations in the reference listed drug.

generally based on administration of a single dose in healthy human volunteers. The use of healthy volunteers in a pharmacokinetic BE study is based upon that they are generally less variable than patients, who may be associated with various sources of variability related to their disease condition. Therefore, healthy subjects are considered to provide a reliable and more sensitive measure for detecting differences in drug product characteristics, which may affect the systemic exposure of test and reference orally inhaled and nasal drug products. In addition, the dose for the pharmacokinetic study is typically selected based on minimizing the number of actuations/inhalations (preferably no more than the single maximum label adult dose), but justified by assay sensitivity. Administration of high doses may be considered necessary if the plasma drug concentration level is undetectable at the labeled dose of the reference product provided that an Investigational New Drug is supplemented to justify the safety of the dose that is higher than the approved dose. If drug plasma profile determination is not feasible, equivalence in systemic exposure may be demonstrated based on pharmacodynamic endpoints (e.g., adrenal suppression for inhaled corticosteroids). In accordance with the general considerations for a pharmacokinetic BE study, some design elements for orally inhaled and nasal drug product examples are provided below.

14.2.2.1 Dry Powder Inhalers

A pharmacokinetic BE study for dry powder inhalers is a single-dose, randomized, two-treatment, crossover study designed to compare test and reference drug products. This study design has been recommended in the recently published draft fluticasone propionate/salmeterol xinafoate dry powder inhaler, referencing Advair® Diskus® guidance (FDA 2013b). With advancement in bioanalytical methods, plasma profile measurements are generally feasible for drugs administered via dry powder inhalers (Daley-Yates et al. 2013).

In addition, since at present time the relationship among pharmacokinetic dose proportionality across multiple strengths of orally inhaled drug products, in vitro performance parameters (e.g., single actuation content and aerodynamic particle size distribution for dry powder inhalers) and product characteristics (e.g., formulation) are not well understood, a pharmacokinetic BE study is generally necessary for all strengths.

14.2.2.2 Metered Dose Inhalers

The study design for a pharmacokinetic BE study for metered dose inhalers is similar to that of dry powder inhalers, as described above. As per the draft albuterol sulfate metered inhalation aerosol, a single-dose, randomized, two-treatment, crossover pharmacokinetic BE study is recommended to demonstrate BE of test albuterol sulfate metered dose inhalers to the reference product guidance (FDA 2013b). In addition, like dry powder inhalers, it is essential to demonstrate equivalence in pharmacokinetics for all strengths of the metered dose inhalers as appropriate.

14.2.2.3 Nasal Sprays

As discussed earlier, pharmacokinetic BE study is essential for nasal sprays formulated as suspensions, and not considered as needed for nasal solution drug products. The study design for a pharmacokinetic BE study for nasal suspension products is also similar to that of dry powder inhalers and metered dose inhalers, as described above. The plasma profile measurements can be determined accurately following administration of drug via the nasal route with availability of sensitive assays (Ratner et al. 2011). Single-dose, two-way crossover pharmacokinetic BE studies have been used to approve several generic locally acting nasal suspension drug products including the generic fluticasone propionate nasal suspension sprays listed in Table 14.1.

Assessment of pharmacokinetic equivalence of test and reference orally inhaled and nasal drug products is based on the natural log-transformed area under the curve (AUC) and peak concentration (C_{max}) data using the average BE approach (FDA 2001). Two orally inhaled or nasal drug products are typically considered equivalent in pharmacokinetic if the 90 % confidence interval of the geometric mean ratio of AUC and C_{max} fall within 80.00–125.00 % (FDA 2003b). Keeping all study factors constant, the sample size required to demonstrate pharmacokinetic BE increases with increase in within-subject variability also referred to as percent coefficient of variation (%CV). Due to concerns regarding the relatively large sample size required for pharmacokinetic BE studies for highly variable drugs, defined as those with %CV $\geq$ 30 %, the FDA has implemented a reference-scaled average BE approach, which widens the BE limits in pharmacokinetic studies by employing a predetermined procedure (Haidar et al. 2008). For the reference-scaling approach, a replicate arm of the reference product is included in the pharmacokinetic BE study, which allows for determination of the within-subject %CV of the reference product. If the within-subject %CV of the reference product is $\geq$30 %, the BE limits are allowed to expand in proportion to the reference product variability. This approach is mainly used to reduce the sample size required for a BE study, while maintaining an adequate determination of BE. Considering the variability that is generally associated with orally inhaled and nasal drug products, the reference-scaling approach may also be applied to pharmacokinetic BE studies for these products.

14.2.3 Pharmacodynamic/Clinical Endpoint Bioequivalence Study

Although equivalence in pharmacokinetic BE studies,[6] coupled with in vitro studies, provides substantial weight-of-evidence to support equivalent product performance, there are still residual uncertainties regarding their relevance to local delivery at the site(s) of action. Moreover, direct sampling of drug concentrations in the lung/nose is currently not possible. For these reasons, an additional pharmacodynamic or clinical endpoint study is generally considered to support equivalence of local drug delivery for orally inhaled and nasal drug products formulated as suspension. The key aspects of the pharmacodynamic and clinical endpoint study are described below for dry powder inhalers and metered dose inhalers containing short-acting β-agonists, long-acting β-agonists and/or corticosteroids, and suspension nasal sprays containing corticosteroids.

14.2.3.1 Dry Powder Inhalers and Metered Dose Inhalers

14.2.3.1.1   Pharmacodynamic Bioequivalence Study

Pharmacokinetic metrics used for BE evaluation (e.g., AUC) usually display a linear relationship with doses within the clinical range, meaning that the observed difference in product performance between test and reference products on the response-scale (y axis) reflects a similar difference on the dose-scale (x axis), as shown in Fig. 14.3a.

In contrast, the dose–response relationship of pharmacodynamic endpoints or metrics for orally inhaled drugs formulated in dry powder inhalers and metered dose inhalers are generally nonlinear (Fig. 14.3b, c), and can be described reasonably well by the E_{max} model that describes a maximum level of the drug effect due to occupancy of all available receptor sites.

$$E_R = \phi_R(D_R) = E_{0R} + \frac{E_{maxR} * D_R}{ED_{50R} + D_R}, \qquad (14.1)$$

where E_R = response, D_R = administered dose, E_{0R} = placebo response in the absence of the drug, E_{maxR} = fitted maximum drug effect, and ED_{50R} = dose required to produce 50 % of the fitted maximum effect.

Due to such a nonlinear behavior, the observed difference on the response-scale due to test and reference products shows an altered magnitude of difference on the dose-scale. More importantly, this difference between two products on

---

[6]Although pharmacokinetic studies are conducted primarily for the safety reasons, a difference in pharmacokinetics may be related to differences in product characteristics and performance with respect to local drug delivery.

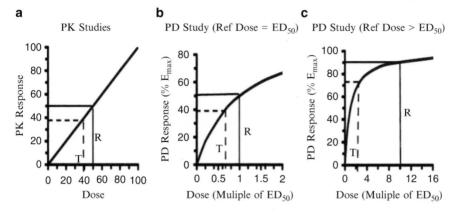

Fig. 14.3 Comparison of difference in the dose scale (*x*-axis) of hypothetical test and reference products demonstrating a 20 % difference in pharmacokinetic (**a**) and pharmacodynamic (**b, c**) responses. Plot (**a**) represents a linear dose–response relationship, while plots (**b** and **c**) represent nonlinear dose–response relationships (from the figure originally published in Singh and Adams (2005))

the response- and dose-scale progressively increases as the pharmacodynamic responses of test and reference products are near the shallow portion of the dose–response curve. In general, for adequate BE evaluation based on the dose-scale analysis as described below, it is essential to have the lowest dose of a reference product sufficiently close to the ED_{50} of the fitted dose–response curve (e.g., based on the above E_{max} model, Fig. 14.3b). However, if the lowest dose of a reference product is considerably large as compared to the ED_{50} (Fig. 14.3c), this type of dose–response relationship becomes unsuitable for BE evaluation, because a large change or fluctuation in the dose–response would result in a small change in the dose or vice versa.

Therefore, it is important to design a pharmacodynamic BE study showing an adequate dose–response effect (Fig. 14.3b), to assure that the planned study has proper capability to distinguish two inhalers that deliver drug differently to the lung. However, design of such a pharmacodynamic BE study can be challenging because the dose–response outcome can be a function of many factors such as the pharmacodynamic model, patient population and within- and between-subject variability, as described below.

Choice of the in vivo pharmacodynamic model for demonstration of equivalence in pharmacodynamics depends on the drug class in question. Dry powder inhalers or metered dose inhalers containing short-acting β agonists, like albuterol, are indicated for prevention and treatment of bronchospasm. Pharmacodynamic effects for this drug class can be determined by using either a bronchodilatation model or a bronchoprovocation model. In fact, these two models were used to approve abbreviated new drug applications for chlorofluorocarbon-based albuterol metered dose inhalers in the mid-1990s (Table 14.1).

Bronchodilatation Model: Determination of bronchodilatation effect in the airways involves measurement of increase in forced expiratory volume in 1 s (FEV_1) versus time profile, following administration of the bronchodilator drug. Although several metrics can be computed using the FEV_1 versus time profile, area under the effect versus time curve (AUEC, based on the trapezoidal rule calculations) and peak effect (FEV_{1max}) are used for BE assessment of test and reference orally inhaled and nasal drug products.[7] Since the duration of bronchodilatation, as measured by increase in FEV_1, may vary among different bronchodilators; AUEC is computed for a duration that is relevant to the clinical dosing regimen for the drug in question. For instance, since the product label for the albuterol sulfate metered dose inhaler recommends two inhalations to be repeated every 4–6 h, $AUEC_{0-4}$ and $AUEC_{0-6}$ are used as the metrics for determination of pharmacodynamic equivalence (FDA 2013a).

Selection of patient population plays a critical role to ensure demonstration of an adequate dose–response effect in bronchodilatation model. Patients with asthma are classified as mild, moderate, or severe asthmatics based on the severity of disease as outlined in the National Asthma Education and Prevention Program guidelines (NHLBI 2007). Because the bronchodilatation study is based on improvement in FEV_1, the dose–response effect for such a study depends on the severity of asthma in the study subjects. Hence, it is necessary to conduct this pharmacodynamic study in severe-to-moderate asthmatics, in order to provide a larger window of improvement in showing dose–response. The design of dose–response pharmacodynamic study also needs to take into consideration specifications for each study procedure to enhance the reproducibility of observations.

Bronchoprovocation Model: Determination of bronchoprovocation effect in the airways involves administration of differing doses of the bronchodilator (e.g., albuterol sulfate metered dose inhaler) on separate days in a cross-over study design coupled with subsequent challenge with a bronchoprovocation agent, such as methacholine, to provide a bronchoprotection dose–response curve. The provocative dose or concentration of methacholine challenge agent required to reduce the FEV_1 by 20 % following administration of differing doses of the bronchodilator (or placebo) (PD_{20} or PC_{20}, respectively) is used to support pharmacodynamic BE. Twenty percent reduction in FEV_1 is determined relative to the saline FEV_1, measured before the placebo or bronchodilator administration. The bronchial smooth muscle stimulation induced by inhalation of the bronchoprovocation agent, methacholine, results in airway narrowing and airway closure, which may pose an increasing risk for patients with moderate-to-severe asthma. Therefore, the bronchoprovocation studies are generally conducted in mild asthmatic patients.

The recently published draft albuterol sulfate metered inhalation aerosol guidance recommends either a bronchodilatation or a bronchoprovocation study to demonstrate pharmacodynamic equivalence for test albuterol sulfate metered

---

[7]Pharmacodynamic BE is evaluated using baseline-adjusted (pre-dose FEV_1) values for AUEC and FEV_{1max}.

inhalation aerosols to the reference product (FDA 2013a). The study design used for the above two in vivo models is a single-dose/separate day, randomized, crossover study consisting, at minimum, of four study arms, including two arms for two different doses of the reference product [$R1$ and $R2$, where 1 and 2 refer to a given dose], one dose of the test product [$T1$] and a placebo arm. At least two doses of the reference product are required to construct a dose–response curve. Additional doses of the reference product may be used to enhance precision of the dose–response. An adequate washout time between treatments is necessary to avoid any carry-over effects.

While bronchodilatation model measures responses using resting airway tone, bronchoprovocation model is able to translate latent airway disease into short-term bronchial reactivity, which is easily measurable. For this reason, the bronchoprovocation model has been reported to demonstrate considerable changes in bronchodilator activity towards airway responsiveness under conditions in which bronchodilatation has plateaued to its maximum achievable response (Ahrens 1984). In addition, the methacholine-induced bronchoprovocation model has been reported to be very reproducible (Juniper et al. 1978). Moreover, bronchodilator activity, as determined using bronchoprovocation model, is considered as clinically relevant as that obtained from bronchodilatation model, since methacholine-induced airway-responsiveness demonstrates good correlation to severity of asthmatic symptoms (Juniper et al. 1981) and exercise-induced bronchospasm (Anderton et al. 1979). Dose–response relationship based on data generated using bronchoprovocation model is generally steeper as compared to those from bronchodilatation model, which is expected to impart greater sensitivity to the bioassay to detect differences in drug delivery and local efficacy (Finney 1978).

To address the nonlinearity of a pharmacodynamic dose–response effect, a dose-scale approach based on the E_{max} model is utilized (Ahrens 2007; FDA 2010). This approach assesses equivalence in pharmacodynamic based on the ratio of the "delivered doses" of test and reference products, i.e., relative bioavailability (F). Details regarding the dose-scale approach can be found in the FDA's Guidance (2010): BE Recommendations for Specific Products: Orlistat Capsule (FDA 2010). In mid-1990s, the FDA used a BE limit of 67.00–150.00 % for the aforementioned pharmacodynamic BE studies to approve four Abbreviated New Drug Applications for chlorofluorocarbon-based albuterol metered dose inhalers using dose-scale analysis. The FDA continues to recommend this BE limit of 67.00–150.00 % for establishing pharmacodynamics equivalence for the hydrofluoroalkane-based albuterol sulfate metered dose inhalers using the dose-scale analysis (FDA 2013a).

14.2.3.1.2 Clinical Endpoint Bioequivalence Study

Unlike short-acting β-agonists, there are currently no established models that can demonstrate an adequate dose–response effect for inhaled corticosteroids such as fluticasone propionate and long-acting β-agonists such as salmeterol xinafoate. In case of inhaled corticosteroids, several pharmacodynamic BE models including induced allergen challenge, asthma stability model, sputum eosinophilia, and exhaled nitric oxide have been considered. For example, exhaled nitric oxide has been suggested as a possible biomarker to detect differences in local drug delivery of different doses of inhaled corticosteroids. It is a biologically relevant marker that is reported to increase in persistent asthma (Kharitonov et al. 1994) and decrease in a dose-dependent manner following administration of some inhaled corticosteroids such as beclomethasone dipropionate (Silkoff et al. 2001; Currie et al. 2003). Moreover, the methodology for measurement of exhaled nitric oxide is standardized and harmonized (ATS/ERS 2005). Thus, exhaled nitric oxide was initially considered to present itself as a possible option to measure dose–response of the inhaled corticosteroids, such as fluticasone propionate.

For the above reasons, the FDA sponsored an investigational study at National Jewish Health to explore the feasibility of using exhaled nitric oxide to assess the dose–response for inhaled corticosteroids in asthma patients. This study evaluated the effect of varying doses (44, 88, and 352 μg twice daily) of fluticasone propionate (Flovent® HFA) on fractional exhaled nitric oxide levels in a crossover study design. Mild-to-moderate asthmatics with fractional exhaled nitric oxide $\geq$45 ppb and $FEV_1 \geq 60$ % of predicted at screening and prior to start of the treatment were recruited in the study. Figure 14.4 shows the plot of the mean exhaled nitric oxide response as a function of daily dose, fitted with an E_{max} model.

Based on the results from this study, it was apparent that the dose–response relationship for fluticasone propionate was shallow, since the lowest daily dose, 88 μg, is much higher than the ED_{50} of 27 μg. In addition, not all patients enrolled in the National Jewish Health study exhibited a clear dose-dependent decrease in exhaled nitric oxide levels (data not shown). Besides, the patient recruitment and continuation in the National Jewish Health study was difficult due to stringent inclusion and exclusion criteria that are necessary for subject enrichment to improve the dose–response effect (i.e., only nine subjects completed the study). Furthermore, there were concerns over incomplete washout leading to carry-over effects in the crossover study. Therefore, contrary to the initial thinking, the above results and observations suggest that the exhaled nitric oxide model is not adequate for establishing BE of approved doses of FDA-approved orally inhaled drug products containing fluticasone propionate.

Similar results were obtained for approved doses of salmeterol xinafoate, using either bronchoprovocation or bronchodilatation pharmacodynamic BE model (Kemp et al. 1993; Palmquist et al. 1999). Figure 14.5 shows an example of a dose–response relationship of salmeterol using the bronchoprovocation model (Kemp et al. 1993). There was shallow or no noticeable dose–response effect observed for salmeterol using either of the two models. Specifically, the lowest approved daily dose of salmeterol (100 μg for dry powder inhalers and 84 μg for metered dose inhalers) appears to be near the upper flat portion of the dose–response curve.

Therefore, for these drug classes, a clinical endpoint BE study is used to support BE as part of the weight-of-evidence approach. In September, 2013, the FDA published the draft guidance on BE recommendations for fluticasone

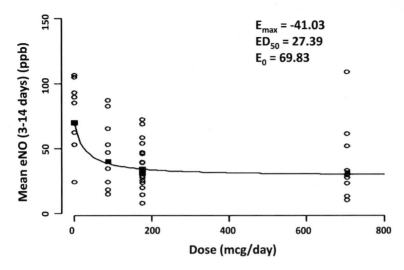

Fig. 14.4 Mean exhaled nitric oxide response as a function of daily dose of fluticasone propionate. Data collected through an FDA sponsored study with National Jewish Health Center

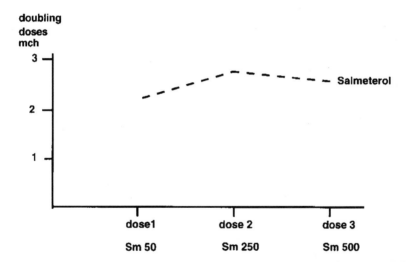

Fig. 14.5 Cumulative dose–response relationship for salmeterol (Sm) using a bronchoprovocation model. Sm 50, Sm 250, and Sm 500 represent the cumulative salmeterol dose via a Diskhaler[®], respectively (50 + 200 + 250 μg, total dose 500 μg) (from the figure originally published in Kemp et al. 1993)

propionate/salmeterol xinafoate dry powder inhaler referencing Advair[®] Diskus[®] [FP/SX BE guidance (FDA 2013b)]. This draft guidance recommends, among other things, a clinical endpoint BE study for supporting local drug delivery equivalence of a test product to Advair[®] Diskus[®], provided in vitro and pharmacokinetic equivalence is established for all approved dose strengths.

A clinical endpoint BE study typically consists of three treatment arms (reference product, test product, and placebo control). The use of placebo arm ensures the sensitivity of the study by demonstrating a significant difference ($p < 0.05$) between the placebo control arm with each of the two treatment arms containing test and reference products. In an attempt to further optimize the sensitivity of the study to detect potential differences between the test and reference products, the clinical endpoint study is generally conducted at the lowest labeled recommended dose. For example, the clinical endpoint BE study in the draft FP/SX BE guidance is based on the lowest recommended dose of 100 µg fluticasone propionate and 50 µg salmeterol powder for inhalation twice daily for the test and reference products.

In practice, test products rely on the scientific finding of the reference product being safe and effective at the approved doses for use in the intended indication and recipient population. Therefore, in the context of BE, it is only necessary to conduct a clinical endpoint BE study for one of the approved indications of the reference product, provided that equivalence in the in vitro and pharmacokinetic studies is established. For instance, the draft FP/SX BE guidance recommends a randomized, placebo-controlled, parallel group study of 4-week duration, preceded by a 2-week run-in period to establish pre-dose FEV_1 baseline, in patients with asthma (FDA 2013b), while Advair® Diskus® is indicated for patients with asthma and chronic obstructive pulmonary disease.

There are two clinical endpoints recommended in the draft FP/SX BE guidance to support BE of the fluticasone propionate and salmeterol xinafoate components. Demonstration of equivalence for the salmeterol xinafoate component is based on the area under the serial FEV_1-time curve calculated from time 0 to 12 h (AUC_{0-12h}) on the first day of the treatment. Since the fluticasone propionate component does not affect the FEV_1 on the first day of treatment, the (AUC_{0-12h}) relative to a pre-treatment baseline on the first day of treatment is considered to be contributed mainly by the salmeterol xinafoate component alone. An additional clinical BE endpoint is based on FEV_1 measured in the morning prior to the dosing of inhaled medications on the last day of a 4-week treatment period. The treatment duration of 4 weeks was chosen because the mean change from baseline in pre-dose FEV_1 reached steady state approximately at Week 4 following treatment with Advair® Diskus® (Advair). The change from baseline in pre-dose FEV_1 at Week 4 can be considered to be contributed by both fluticasone propionate and salmeterol xinafoate components of the drug product. Therefore, demonstration of equivalence for these two clinical BE endpoints, in conjunction with the in vitro and pharmacokinetic BE studies, will demonstrate BE for both fluticasone propionate and salmeterol xinafoate components.

Assuming both test and reference arms are superior to placebo arm, two orally inhaled and nasal drug products are considered bioequivalent in local delivery if the 90 % confidence intervals for the test/reference ratios for the two endpoints described above fall within 80.00–125.00 %. Details of the clinical endpoint study can be found in the FDA's draft FP/SX BE guidance (FDA 2013b).

14.2.3.2   Nasal Sprays

14.2.3.2.1   Clinical Endpoint Bioequivalence Study

Similar to inhaled corticosteroids, there are no established models for demonstrating dose–response effect for locally acting nasal corticosteroids for allergic rhinitis (Chowdhury 2001). Therefore, for suspension nasal sprays, a clinical (rhinitis) endpoint BE study (FDA 2003a) is also used to support BE of test and reference. The clinical rhinitis study is conducted in a randomized, placebo-controlled, parallel study in patients with seasonal allergic rhinitis, which is considered to extend to all indications in reference product labeling for locally acting nasal corticosteroids. The study is conducted using the lowest recommended dose of the drug and consists of a 1-week run-in phase, to establish a baseline and identify placebo responders, followed by a 2-week treatment period. Placebo responders, identified during the run-in phase, are recommended to be excluded from the study to increase the probability of showing a significant difference between the drug (test and reference products) and placebo treatment, and improve the study sensitivity to detect possible differences between test and reference products.

The clinical endpoints for BE evaluation are based on total nasal symptom scores (TNSS), which is a categorical variable classified into a number of discrete categories, and uses a four-point scale with signs and symptoms ordered by severity of symptom (runny nose, sneezing, nasal itching, and congestion[8]) from 0 (no symptoms) to 3 (severe symptoms). The primary BE endpoints analysis is based on the mean change from baseline in the reflective scores[9] for a 12-h pooled TNSS over the 2-week treatment period. Instantaneous scores[10] serve as a secondary endpoint. Clinical endpoint BE studies based on the TNSS score have been used to approve several generic locally acting nasal suspension drug products including the generic fluticasone propionate nasal suspension sprays listed in Table 14.1.

Given both test and reference groups are superior to placebo group, two locally acting nasal suspension products are considered bioequivalent in local delivery if the 90 % confidence intervals for the test/reference ratios for the primary endpoints fall within 80.00–125.00 %. Details regarding study design can be found in the FDA's draft Guidance: Bioavailability and Bioequivalence Studies for Nasal Aerosols and Nasal Sprays for Local Action (2003) (FDA 2003a).

---

[8]Addition of other nonnasal symptoms may be pertinent for certain drugs products.

[9]Reflective scores are made immediately prior to each dose to reflect the previous 12 h.

[10]Instantaneous scores are made immediately after dosing to know how the patient feels at the time of evaluation.

14.3   Device and Formulation Considerations

As described above, orally inhaled and nasal drug products comprise of formulation and device components. The performance of these drug products, in part, depends on the patient–device interactions. Therefore, with regard to the development of orally inhaled and nasal drug product, it is generally important that the design of device accounts for patient usability and acceptability, in addition to device performance with respect to the drug delivery. For the development of test orally inhaled and nasal drug products that are expected to be interchangeable with their reference counterparts, the differences in the device and its effect on the characteristics of emitted dose and consequent impact on the product safety and efficacy can be evaluated through BE studies, as described above. However, it is important to consider the switchability between the test and reference devices from a patient-use perspective. To address this device switchability issue, it is necessary to understand the current landscape of device features of FDA-approved orally inhaled and nasal drug products, as summarized below.

Dry Powder Inhalers: Dry powder inhaler devices currently marketed in the United States may differ considerably with respect to their interior design, appearance, and external operating principle. For instance, the basic operating principle for Diskus® consists of the following: (1) open inhaler, (2) slide lever until it clicks, (3) breathe quickly and deeply through the inhaler, (4) close the inhaler after use. However, the basic operating principle for HandiHaler® is very different from Diskus® and consists of the following: (1) open dust cap to expose mouthpiece, (2) open mouthpiece to expose the center chamber, (3) place Spiriva® capsule in the center chamber of the HandiHaler® device, (4) close mouthpiece until you hear a click, (5) press the piercing button, (6) breathe deeply though the device until you hear or feel the capsule vibrate, (7) open the mouthpiece and discard the used capsule.

Metered Dose Inhalers: Unlike dry powder inhalers, metered dose inhaler devices do not exhibit great diversity in their basic design and external operating principles. As shown in Fig. 14.1, metered dose inhaler device generally consists of a canister, actuator with a mouthpiece and a dust cap. The basic operating principle of metered dose inhalers generally comprises: remove dust cap, press down the canister while inhaling deeply and slowly, remove inhaler from mouth, and replace dust cap.

Nasal Sprays: Nasal spray devices also do not exhibit great diversity in their basic design and external operating principles. They usually consist of a bottle, a spray pump unit, an applicator tip and a dust cap, as shown in Fig. 14.1. The basic operating principle of nasal sprays comprises: remove dust cap, insert applicator tip to nostril and breathe in through the nose as you spray, and replace dust cap.

Therefore, the device switchability issue is primarily associated with dry powder inhalers. The development of a test dry powder inhaler device warrants consideration of the effect of design factors, such as energy source (e.g., active or passive (breath-actuated) device), metering principle (e.g., pre-metered multi-dose, device-metered multi-dose or pre-metered single-dose units), number of doses,

Table 14.5 Common excipients used in orally inhaled and nasal drug products

Drug product	Excipients	Function
Dry powder inhalers	Lactose, mannitol	Carrier
	Magnesium stearate	Dispersing agent
Metered dose inhalers	HFA-134a, HFA-227	Propellant
	Oleic acid	Surfactant
	Ethanol	Cosolvent
Nasal sprays	Microcrystalline cellulose	Suspending agent
	Glycerin	Tonicity agent
	Sodium citrate/citric acid	Buffer
	Polysorbate 80	Wetting agent
	Benzalkonium chloride	Preservatives

external operating principle, shape and size, on the patient handling relative to the reference dry powder inhaler device.

Inactive ingredients used in orally inhaled and nasal drug formulations have a considerable effect not only on the product performance, but also on local safety of the drug product (Pilcer and Amighi 2010). Table 14.5 provides examples of common types of excipients used in orally inhaled and nasal drug products. A test product, which contains different excipients or greater concentrations of the same excipients, as compared to the reference product, may raise local safety concerns (e.g., irritation).

In addition, due to the sensitivity of the performance of orally inhaled and nasal drug products to the nature and level of excipients (Bosquillon et al. 2001; FDA 2003a; Stein and Myrdal 2004 it is generally difficult to achieve equivalent performance with respect to local drug delivery, when these two formulation variables are considerably different between the test orally inhaled and nasal drug products and their reference counterparts. Therefore, to enhance the probability of establishing BE and eliminate the local safety concerns, the test orally inhaled and nasal drug products are generally Q_1 and Q_2 the same as their reference products.

14.4 Conclusion and Future

Despite the complexity associated with orally inhaled and nasal drug products, the FDA developed a weight-of-evidence approach to demonstrate BE for these locally acting drug products. This approach utilizes in vitro, pharmacokinetic, and pharmacodynamic or clinical endpoint studies to provide sufficient information to conclude equivalence in systemic exposure and local drug delivery between two products. As a result, the FDA published its first individual BE guidances for metered dose inhaler and dry powder inhaler in April and September, 2013, respectively (FDA 2013a, b).

As part of the critical path initiative, the FDA is also exploring more efficient methodologies to establish BE for orally inhaled and nasal drug products. In fact, the FDA recently sponsored a research study to assess if pharmacokinetic studies are able to provide information about the fate of a drug in the lung, specifically the possible relationship between the regional lung deposition of the orally inhaled drug and its time-dependent drug concentration in plasma. As explained in Sect. 14.2.2, equivalence in pharmacokinetic is currently required mainly to ensure safety of the test product. However, there is an emerging view that this downstream process may be related to the lung deposition of poorly soluble orally inhaled drugs that have very low bioavailability, like fluticasone propionate. It has been proposed that for such poorly soluble orally inhaled drugs, PK parameters, such as AUC and C_{max}, may be related to the central to peripheral (C/P) drug deposition ratio in the lung. If successful, the clinical endpoint BE studies may not be needed in the weight-of-evidence approach, without compromising the efficacy and safety of test orally inhaled and nasal drug products.

References

Adams W, Ahrens RC, Chen ML, Christopher D, Chowdhury BA, Conner DP, Dalby R, Fitzgerald K, Hendeles L, Hickey AJ, Hochhaus G, Laube BL, Lucas P, Lee SL, Lyapustina S, Li B, O'connor D, Parikh N, Parkins DA, Peri P, Pitcairn GR, Riebe M, Roy P, Shah T, Singh GJ, Sharp SS, Suman JD, Weda M, Woodcock J, Yu L (2010) Demonstrating bioequivalence of locally acting orally inhaled drug products (OIPS): workshop summary report. J Aerosol Med Pulm Drug Deliv 23:1–29

Advair Diskus Product Label. http://www.accessdata.fda.gov/drugsatfda_docs/label/2010/021077s042lbl.pdf. Assessed 2013

Ahrens RC, Bonham AC, Maxwell GA, Weinberger MM (1984) A method for comparing the peak intensity and duration of action of aerosolized bronchodilators using bronchoprovocation with methacholine. Am Rev Respir Dis 129(6):903–906

Ahrens RC (2007) Pharmacodynamic testing of test inhaler be: unresolved issues and potential solutions. In: Dalby RN, Byron PR, Peart J, Farr SJ, Suman JD (eds) RDD Europe 2007, vol 1. Davis Healthcare, River Grove, IL, pp 1–10, ISBN: 1-933722-07-X

Anderton R, Cuff MT, Frith PA, Cockcroft DW, Morse JLC, Jones NL, Hargreave FE (1979) Bronchial responsiveness to inhaled histamine and exercise. J Allergy Clin Immunol 63:315–320

ATS/ERS (2005) ATS/ERS recommendations for standardized procedures for the online and offline measurement of exhaled lower respiratory nitric oxide and nasal nitric oxide, 2005. Am J Respir Crit Care Med 171:912–930

Borgström L, Derom E, Ståhl E, Wåhlin-Boll E, Pauwels R (1996) The inhalation device influences lung deposition and bronchodilating effect of terbutaline. Am J Respir Crit Care Med 153:1636–1640

Bosquillon C, Lombry C, Préat V, Vanbever R (2001) Influence of formulation excipients and physical characteristics of inhalation dry powders on their aerosolization performance. J Control Release 23:329–339

Chowdhury B (2001) FDA Advisory Committee for Pharmaceutical Science: difficulties in showing a dose–response with locally-acting nasal sprays and aerosols for allergic rhinitis. http://www.fda.gov/ohrms/dockets/ac/01/slides/3763s1.htm. Assessed 2013

Currie G, Bates CE, Lee DK, Jackson CM, Lipworth BJ (2003) Effects of fluticasone plus salmeterol versus twice the dose of fluticasone in asthmatic patients. Eur J Clin Pharmacol 59:11–15

Daley-Yates PT, Mehta R, Chan RH, Despa SX, Louey MD (2013) Pharmacokinetics and pharmacodynamics of fluticasone propionate and salmeterol delivered as a combination dry powder from a capsule-based inhaler and a multidose inhaler in asthma and COPD patients. J Aerosol Med Plum Drug Deliv [Epub Ahead of Print]

Fardon T, Lee DKC, Haggart K, Mcfarlane LC, Lipworth BJ (2004) Adrenal suppression with dry powder formulations of fluticasone propionate and mometasone furoate. Am J Respir Crit Care Med 170:760–766

FDA (2001) Guidance for industry: statistical approaches to establishing bioequivalence. http://www.fda.gov/downloads/drugs/guidancecomplianceregulatoryinformation/guidances/ucm070244.pdf. Assessed 2013

FDA (2003a) Draft guidance for industry: bioavailability and bioequivalence studies for nasal aerosols and nasal sprays for local action. http://www.fda.gov/downloads/drugs/guidancecomplianceregulatoryinformation/guidances/ucm070111.pdf. Assessed 2013

FDA (2003b) Guidance for industry: bioavailability and bioequivalence studies for orally administered drug products-general considerations. http://www.fda.gov/downloads/drugs/guidancecomplianceregulatoryinformation/guidances/ucm070124.pdf. Assessed 2013

FDA (2010) Draft guidance for industry: bioequivalence recommendations for specific products: Orlistat capsule. http://www.fda.gov/drugs/guidancecomplianceregulatoryinformation/guidances/ucm075207.htm. Assessed 2013

FDA (2012) Draft guidance for industry: bioequivalence recommendations for specific products: budesonide suspension for inhalation. http://www.fda.gov/drugs/guidancecomplianceregulatory information/guidances/ucm075207.htm. Assessed 2013

FDA (2013a) Draft guidance for industry: bioequivalence recommendations for specific products: albuterol sulfate aerosol, metered/inhalation. http://www.fda.gov/drugs/guidancecompliance regulatoryinformation/guidances/ucm075207.htm. Assessed 2013

FDA (2013b) Draft guidance for industry: bioequivalence recommendations for specific products: fluticasone propionate and salmeterol xinafaote powder for inhalation. http://www.fda.gov/drugs/guidancecomplianceregulatoryinformation/guidances/ucm075207.htm. Assessed 2013

Finney D (1978) Statistical methods in biological assay. Charles Griffin & Co., London

Haidar S, Makhlouf F, Schuirmann DJ, Hyslop T, Davit B, Conner D, Yu LX (2008) Evaluation of a scaling approach for the bioequivalence of highly variable drugs. AAPS J 10:450–454

HandiHaler. http://www.inteda.net/handihaler.htm. Accessed 2013

Harris A, Svensson E, Wagner ZG, Lethagen S, Nilsson IM (1988) Effect of viscosity on particle size, deposition, and clearance of nasal drug delivery systems containing desmopressin. J Pharm Sci 77:405–408

Juniper EF, Frith PA, Dunnett C, Cockcroft DW, Hargreave FE (1978) Reproducibility and comparison of responses to inhaled histamine and methacholine. Thorax 33:705–710

Juniper E, Frith PA, Hargreave FE (1981) Airway responsiveness to histamine and methacholine: relationship of minimum treatment to control symptoms of asthma. Thorax 36:575–579

Kalaly K, Alhaleweh A, Velaga SP, Nokhodchi A (2011) Effect of carrier particle shape on dry powder inhaler performance. Int J Pharm 421:12–23

Kemp J, Bierman CW, Cocchetto DM (1993) Dose–response study of inhaled salmeterol in asthmatic patients with 24-hour spirometry and holter monitoring. Ann Allergy 70:316–322

Kharitonov S, Yates D, Robbins RA, Logan-Sinclair R, Shinebourne EA, Barnes PJ (1994) Increased nitric oxide in exhaled air of asthmatic patients. Lancet 343:133–135

Kublik H, Vidgren MT (1998) Nasal delivery systems and their effects on deposition and absorption. Adv Drug Deliv Rev 29:157–177

Kwok P, Glover W, Chan HK (2005) Electrostatic charge characteristics of aerosols produced from metered dose inhalers. J Pharm Sci 94:2789–2799

Lee S (2012) US regulatory considerations for test dry powder inhalers. In: Dalby RN, Byron PR, Peart J, Farr SJ, Suman JD (eds) Respiratory drug delivery 2012, vol 1. Davis Healthcare, River Grove, IL, pp 317–324, ISBN: 1-933722-57-6

Lee S, Adams WP, Li BV, Conner DP, Chowdhury BA, Yu LX (2009) In vitro considerations to support bioequivalence of locally acting drugs in dry powder inhalers for lung diseases. AAPS J 11:414–423

Lewis D, Johnson S, Meakin B (1998) Effects of orifice diameter on beclomethasonedipropionate delivery from a PMDI HFA solution formulation. In: Dalby RN, Byron PR, Peart J, Suman JD (eds) Respiratory drug delivery conference. Respiratory drug delivery program and proceedings, pp 363–364

Li B, Jin F, Lee SL, Bai T, Chowdhury B, Caramenico HT, Conner DP (2013) Bioequivalence for locally acting nasal spray and nasal aerosol products: standard development and test approval. AAPS J 15:875–883

Löfdahl C, Andersson L, Bondesson E, Carlsson LG, Friberg K, Hedner J, Hörnblad Y, Jemsby P, Källén A, Ullman A, Werner S, Svedmyr N (1997) Differences in bronchodilating potency of salbutamol in turbuhaler as compared with a pressurized metered-dose inhaler formulation in patients with reversible airway obstruction. Eur Respir J 10:2474–2478

Murthy T, Basal Vishnu Priya M, Satyanarayana V (2010) Performance of CFC free propellent-driven MDI of fluticasone propionate. J Sci Ind Res 69:866–871

Newman S, Busse WW (2002) Evolution of dry powder inhaler design, formulation, and performance. Respir Med 96:293–304

NHLBI (2007) Guidelines for the diagnosis and management of asthma (EPR-3). http://www.nhlbi.nih.gov/guidelines/asthma/. Assessed 2013

Palmquist M, Ibsen T, Mellén A, Lötvall J (1999) Comparison of the relative efficacy of formoterol and salmeterol in asthmatic patients. Am J Respir Crit Care Med 160:244–249

Pilcer G, Amighi K (2010) Formulation strategy and use of excipients in pulmonary drug delivery. Int J Pharm 392:1–19

Ratner P, Wingertzahn MA, Herzog R, Huang H, Desai SY, Maier G, Nave R (2011) An investigation of the pharmacokinetics, pharmacodynamics, safety, and tolerability of ciclesonide hydrofluoroalkane nasal aerosol in healthy subjects and subjects with perennial allergic rhinitis. Pulm Pharmacol Ther 24:426–433

Shur J, Lee S, Adams W, Lionberger R, Tibbatts J, Price R (2012) Effect of device design on the in vitro performance and comparability for capsule-based dry powder inhalers. AAPS J 14:667–676

Silkoff P, Mcclean P, Spino M, Erlich L, Slutsky AS, Zamel N (2001) Dose–response relationship and reproducibility of the fall in exhaled nitric oxide after inhaled beclomethasonedipropionate therapy in asthma patients. Chest 119:1322–1328

Singh G, Adams WP (2005) US regulatory and scientific considerations for approval of generic locally acting orally inhaled, and nasal drug products. In: Dalby RN, Byron PR, Peart J, Farr SJ, Suman JD (eds) RDD Europe 2005, vol 1. Davis Healthcare, River Grove, IL, pp 115–126, ISBN: 1-930114-80-X

Smyth H (2003) The influence of formulation variables on the performance of alternative propellant-driven metered dose inhalers. Adv Drug Deliv Rev 55:807–828

Stein S, Myrdal PB (2004) A theoretical and experimental analysis of formulations and device parameters affecting solution MDI size distributions. J Pharm Sci 93:2158–2175

Chapter 15
Bioequivalence: Modeling and Simulation

Xinyuan Zhang

15.1 General Considerations

Modeling is the process of establishing a mathematical model; a model is a representation of the construction and working of some system of interest (Maria 1997). A simulation of a system is the operation of a model of the system (Maria 1997). Models can be classified as empirical models, mechanism-based models, and hybrid models. Empirical models are developed based on experience or observations, and mechanism-based models are developed based on the underlying chemical, physical, biological, and pharmacological theories and principles of the target system. Empirical models are difficult to be extrapolated. Mechanism-based models require fully understanding the system and process, which is challenging and may not always be achieved. Hybrid models are combination of both models to maintain the advantages and overcome the limitations of both.

Modeling and simulation (M&S) is a powerful tool and plays an important role in the bioequivalence world, from product development to regulatory standards development. In this chapter, we will focus on some well-developed models over the past decades that have been gradually and widely adopted by industry and regulatory scientists. Specifically, these models include mechanistic models for oral drug absorption, in vitro and in vivo correlations (IVIVCs), and bioequivalence simulations. It should also be noted that the scope of this chapter is limited to small molecules.

Disclaimer: The views presented in this article by the authors do not necessarily reflect those of the US FDA.

X. Zhang (✉)
Office of Generic Drugs, Center for Drug Evaluation and Research,
Food and Drug Administration, Silver Spring, MD, USA
e-mail: Xinyuan.Zhang@fda.hhs.gov

L.X. Yu and B.V. Li (eds.), *FDA Bioequivalence Standards*, AAPS Advances
in the Pharmaceutical Sciences Series 13, DOI 10.1007/978-1-4939-1252-0_15,
© The United States Government 2014

15.2   Mechanistic Models for Oral Drug Absorption and Bioavailability

The majority of drug products on the market are formulated as oral dosage forms for patient convenience. Therefore, predicting bioavailability after oral administration is a necessary task.

It should be noted that oral drug absorption (Fa) and bioavailability (Fb) are different concepts. By definition, bioavailability means the rate and extent to which the active ingredient or active moiety is absorbed from a drug product and becomes available at the site of action (21CFR320.1, see http://www.accessdata.fda.gov/scripts/cdrh/cfdocs/cfCFR/CFRSearch.cfm?fr=320.1). Oral bioavailability is the fraction of an oral administered drug that reaches systemic circulation and can be calculated by Eq. (15.1),

$$Fb = Fa \cdot Fg \cdot Fh, \qquad (15.1)$$

where Fb is the oral bioavailability, Fa is the fraction of drug absorbed from the gastrointestinal tract (GIT) or oral drug absorption, which is the fraction of the total amount entering the cellular space of the enterocytes from the gut lumen, Fg is the fraction that escapes metabolism in the GI epithelial cells, and Fh is the fraction that escapes the liver extraction. The absorption process and the factors affecting the absorption of orally administered drugs are illustrated in Fig. 15.1. As depicted, drug absorption process (Fa and Fg) has an impact on the overall Fb.

Prediction of oral drug absorption is scientifically challenging because there are many factors that affect drug absorption including drug substance properties, formulation properties, and physiological properties. Mechanistic models for oral drug absorption integrate those components and have become more and more sophisticated over time with the significant progress made in scientific knowledge and computational techniques. Mechanistic oral absorption models have been classified into three categories: quasi-equilibrium models, steady-state models, and dynamic models, based on their dependence on spatial and temporal variables (Yu et al. 1996b). In this section, we will first discuss the factors that affect oral drug absorption and bioavailability, which ideally should all be included for mechanistic oral absorption model development. Then different types of mechanistic oral absorption models and their applications will be introduced in detail with the case examples.

15.2.1   Factors Affecting Oral Drug Absorption and Bioavailability

The releasing and dissolution of drug formulated in solid dosage forms are prerequisites for efficient absorption. The dissolved drug molecules transport in two

Portal Vein Epithelial Cells   Gastrointestinal Lumen

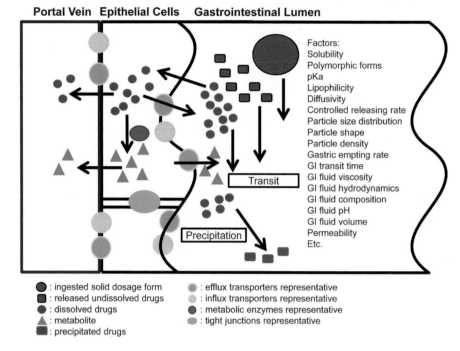

Factors:
Solubility
Polymorphic forms
pKa
Lipophilicity
Diffusivity
Controlled releasing rate
Particle size distribution
Particle shape
Particle density
Gastric empting rate
GI transit time
GI fluid viscosity
GI fluid hydrodynamics
GI fluid composition
GI fluid pH
GI fluid volume
Permeability
Etc.

Transit

Precipitation

● : ingested solid dosage form   ● : efflux transporters representative
■ : released undissolved drugs   ● : influx transporters representative
● : dissolved drugs   ● : metabolic enzymes representative
▲ : metabolite   ● : tight junctions representative
■ : precipitated drugs

Fig. 15.1 Schematic description of factors affecting drug absorption in the GI tract

directions: along the GIT, and perpendicular to the GIT, i.e., transport across the epithelia, enter the portal vein, go through the liver, and then enter the blood stream. Meanwhile, the unreleased and undissolved drug continuously transports along the GIT but is not absorbed. Mathematical equations that describe the rate of drug being released, dissolving, and transported will help us understand the factors that affect drug absorption. Those theoretical equations also serve as the fundamental basis for mechanistic oral absorption models.

Various mathematical models have been developed for drug dissolution (Siepmann and Siepmann 2013). One of the commonly used models is the Nernst–Brunner modified Noyes–Whitney equation (Eq. (15.2)) (Siepmann and Siepmann 2013),

$$\text{Dissolution rate} = \frac{dM}{dt} = \frac{DS}{\delta}(C_s - C_t), \tag{15.2}$$

where M is mass, t is time, D (area/time) is drug diffusion coefficient or diffusivity, S is the drug surface area available for dissolution, δ is the thickness of the unstirred liquid boundary layer, C_s is the solubility at the solid surface, and C_t is the bulk drug concentration at time t. From Eq. (15.2), it is evident that drug substance properties (e.g., diffusivity and solubility), and formulation properties (e.g., particle size distribution, shape, and density, which affect the drug surface area available for dissolution), are important factors affecting dissolution rate.

The absorption rate can be mathematically described by Eq. (15.3) (Amidon et al. 1995),

$$\text{Absorption rate} = \frac{dM}{dt} = \iint_A J_w dA = \iint_A P_w C_w dA, \quad (15.3)$$

where $J_w(x, y, z, t)$ is the drug flux (mass/area/time) through the intestinal wall at any position and time, $P_w(x, y, x, t)$ is the effective permeability of intestinal membrane, $C_w(x, y, z, t)$ is the drug concentration at the intestinal membrane surface, and A is the entire gastrointestinal surface (Amidon et al. 1995). The effective permeability (P_w) and local drug concentration (C_w) are time and location dependent. The underlying assumptions of Eq. (15.3) are (1) sink conditions exist for the drug inside the intestinal membrane; and (2) there is no luminal reaction. At that time intestinal membrane was treated as a film and intracellular reactions have not been introduced. Nevertheless, Eq. (15.3) indicates two important properties of drug substance that affect oral absorption including solubility (i.e., the limit of C_w), and permeability (i.e., the ability of drug substance transport across the intestinal membrane). Lipophilicity has been identified to be correlated with passive permeability (Yu et al. 1996b). Drug substance may exist as different species, e.g. ionized or neutral molecules. The fraction of each species can be calculated by the pH-partition theory and the pKa (acid dissociation constant) value(s). Each species may have different solubility and lipophilicity/permeability.

In addition to the above mentioned factors from drug product (i.e., drug substance and formulation factors), the physiological parameters related to GIT also affect drug absorption in various ways (Mudie et al. 2010). For instance, the pH values in the GI lumen affect drug ionization states and consequently solubility and diffusivity. The fluid viscosity, fluid hydrodynamics, fluid composition, and volume in the GIT affect drug dissolution rate. The properties of GI membrane affect the transport rate (e.g., influx/efflux transporters, channels, paracellular junctions, etc.), and the extent of absorption (e.g., metabolic enzymes). The gastric emptying time determines the residence time of a drug in the stomach and the rate at which the drug will be available at the small intestine. This parameter affects the onset of absorption since most drugs are absorbed in the small intestine but not in the stomach. The GI transit time or resident time affects the amount of time the drug substance has to dissolve and be absorbed.

To summarize this part, there are three categories of factors that affect drug absorption: drug substance properties (such as solubility, polymorphic forms, pKa, lipophilicity, and diffusivity), formulation properties (such as controlled releasing rate, particle size distribution, shape, and density), and physiology properties (such as gastric emptying rate, GI transit time, GI fluid viscosity, hydrodynamics, composition, pH, and volume, and permeability along the GIT). Figure 15.1 describes major events occurring in oral absorption. Ideally, a fully mechanistic model should include all known components and their interplays that affect oral absorption and bioavailability. However, due to the complexity of the whole process, some factors are simplified in some of the models for oral absorption. In the following sections, representative oral absorption models will be introduced.

15.2.2   Early Absorption Models

Yu et al. published a review article in 1996 that comprehensively discussed the utilities and limitations of early quantitative absorption models (Yu et al. 1996b). Details about the early absorption models can be found in the reference. Early absorption models were classified into three categories: quasi-equilibrium models, steady-state models, and dynamic models, based on their dependence on spatial and temporal variables (Yu et al. 1996b).

The quasi-equilibrium models are independent of the spatial and temporal variables. The steady-state models are independent of the temporal variable, but dependent on the spatial variable. The dynamic models are dependent on both temporal and spatial variables.

The quasi-equilibrium models employed pH-partition theory and provided a rough estimation for the fraction of dose absorbed (Eq. (15.4)) (Dressman et al. 1985)

$$AP = \log\left(P \cdot F_{non} \cdot \frac{S_0 \cdot V_L}{X_0}\right), \qquad (15.4)$$

where AP is the absorption potential as a predictor of the fraction absorbed, P is the 1-octanol–water partition coefficient that correlates with the permeability ratio (the permeability of gut wall to drug to the aqueous permeability of drug), F_{non} is the fraction in non-ionized form at pH 6.5, S_0 is the intrinsic solubility (aqueous solubility of the non-ionized species at 37 °C), V_L is the volume of the luminal contents, and X_0 is the dose administered (Dressman et al. 1985). This model was validated using seven drugs against their observed oral bioavailability. An "S" shape relationship was observed between the absorption potential and oral bioavailability. Although the absorption potential does not account for all process influencing oral drug absorption, the two important properties of drug substance, solubility and permeability, were included in the model. In addition, the model considered that solubility and permeability may change in different pH media due to ionization, and corrected the parameter values for pH 6.5, which is within the pH range of small intestine (Russell et al. 1993).

The steady-state models estimated the fraction of dose absorbed by mass balance approach. The small intestine was modeled as a cylinder with surface area of $2\pi RL$, where R is the radius and L is the length of the tube (Sinko et al. 1991). Assuming that the mass lost from the small intestine is due to absorption, the rate of mass absorbed can be described by Eq. (15.5) (Sinko et al. 1991)

$$-\frac{dM}{dt} = Q(C_0 - C_m) = \iint_A J_w dA, \qquad (15.5)$$

where C_0 is the inlet concentration, C_m is the outlet concentration, and Q is the volumetric flow rate. The right hand side of the equation is essentially Eq. (15.3). The fraction absorbed can be estimated by the ratio of concentration difference to initial concentration, i.e. $F_a = (C_0 - C_m)/C_0$, where F_a is the fraction absorbed. Combining Eq. (15.5) with the above F_a equation, and assuming the cylindrical geometry and constant permeability, the fraction absorbed can be expressed by Eq. (15.6) (Sinko et al. 1991),

$$F_a = 1 - \frac{C_m}{C_0} = \frac{2\pi RL}{Q} P_e \int_0^1 C_b^* dz^*, \qquad (15.6)$$

where P_e is the effective permeability, $C_b^* = C_b/C_0$ is the dimensionless concentration, and $z^* = z/L$ is the fractional length. C_b is the bulk drug concentration in the lumen, and z is the length from the inlet to the absorption site (Sinko et al. 1991).

Although those early absorption models were simplified absorption models and are not widely used currently, they have served as basis for more complicated, later developed absorption models, such as the compartmental and transit models. Complicated models dependent on both temporal and spatial variables were developed, such as different types of dynamic models (dispersion models (Ni et al. 1980; Willmann et al. 2007; Willmann et al. 2003b; Willmann et al. 2004), mixing tank models (Goodacre and Murray 1981; Dressman et al. 1984; Dressman and Fleisher 1986; Luner and Amidon 1993; Oberle and Amidon 1987), and compartmental models (Yu et al. 1996a; Yu and Amidon 1999; Yu 1999; Grass 1997; Parrott and Lave 2002; Agoram et al. 2001; Jamei et al. 2009b; Darwich et al. 2010)). Therefore, they can be used to predict the fraction of dose absorbed as well as plasma pharmacokinetic profiles (Yu et al. 1996b).

15.2.3 Compartmental Models

15.2.3.1 The Compartmental Absorption and Transit (CAT) Model

The CAT model was developed in the 1990s by Yu et al. (Yu et al. 1996a; Yu and Amidon 1999). The model treated the stomach as one compartment, the small intestine as seven compartments, and colon as one compartment. The drug transfers from one compartment to the next one in a first-order fashion (Yu and Amidon

1999). The model can be mathematically expressed by Eqs. (15.7)–(15.11) (Yu and Amidon 1999).

$$\frac{dM_s}{dt} = -K_s M_s, \tag{15.7}$$

$$\frac{dM_n}{dt} = K_t M_{n-1} - K_t M_n, \quad n = 1, 2, \ldots, 7, \tag{15.8}$$

$$\frac{dM_c}{dt} = K_t M_n, \quad n = 7, \tag{15.9}$$

$$\frac{dM_a}{dt} = K_a \sum_{n=1}^{7} M_n, \tag{15.10}$$

$$M_s + \sum_{n=1}^{7} M_n + M_c + M_a = M_0, \tag{15.11}$$

Equations (15.7)–(15.9) represent the drug mass change in the stomach, intestine, and colon lumen, respectively. In Eq. (15.8), when $n = 1$, the term $K_t M_{n-1}$ is replaced by $K_s M_s$. M_s, M_n, and M_c are the amount of drug in the lumen of stomach, the nth compartment of small intestine, and the colon, respectively. Equation (15.10) describes the rate of drug being absorbed from the small intestine into the plasma. M_a is the amount of drug absorbed. K_s, K_t, and K_a are the rate constants of gastric emptying, small intestine transit, and intrinsic absorption, respectively (Yu and Amidon 1999). Equation (15.11) is the overall mass balance.

The CAT model assumes that absorption from the stomach and colon is minimal compared with that from the small intestine, transport across the small intestinal membrane is passive, dissolution is instantaneous, and drug transit from one compartment to the next one follows first-order kinetics. In the model, K_a is proportional to the effective permeability. The model was able to describe the relationship between fraction absorbed and effective permeability for ten drugs covering a wide range of fraction absorbed (Yu and Amidon 1999). The number of compartments, seven, was selected based on residual sum of squares by comparing the simulated percentage of dose in colon with cumulative percentage of small intestinal transit time (Yu et al. 1996a). The CAT model coupled with a three-compartment PK model was able to predict the pharmacokinetic (PK) profiles of atenolol.

The original CAT model did not include components such as in vivo dissolution, transporter mediated transport, and intestinal metabolism. However, it served as the basic structure model for more complicated absorption models.

15.2.3.2   The Advanced Compartmental Absorption and Transit (ACAT) Model

The ACAT model adopted the CAT model structure, incorporated more components, and has been continuously developed and commercialized with a friendly user interface over the past two decades under the trade name GastroPlus™ (Agoram et al. 2001; Parrott and Lave 2002). The ACAT model has its advantages in two aspects: the GI physiology is more detailed, and more formulations can be simulated.

Besides gastric emptying rate and intestinal transit time, the ACAT model also includes pH, fluid volume, bile salt concentration, transporters, and metabolic enzymes, and pore radius in each GI compartment.

Unlike the original CAT model where instant dissolution was assumed, the drug product is treated as unreleased, undissolved, and dissolved forms in the ACAT model. The three forms can transit to the next lumen compartment. Dissolution models have been integrated in the model (e.g., Eq. (15.2)). Therefore, for immediate release formulations, it can simulate in vivo dissolution using formulation properties (such as particle size distribution, shape, and density), drug substance properties (such as solubility vs. pH profiles, and diffusivity). The model can also take the in vitro dissolution profile as a model input to predict absorption for modified release drug products (Lukacova et al. 2009).

15.2.3.3   The Advanced Dissolution, Absorption, and Metabolism (ADAM) Model

The ADAM model was also developed based on the CAT model and has been implemented in the commercial software Simcyp® (Darwich et al. 2010; Jamei et al. 2009a, b). Similarly, the ADAM model also consists of nine compartments for the GI tract (stomach, seven small intestinal compartments, and one colon compartment). Physiological parameters have been integrated in the model, such as pH in the lumen, fluid volume in different segments of GI tract, GI transit time, bile salts concentration, regional permeability, transporters expression, and metabolic enzymes. The model also integrated dissolution models and can be used to simulation absorption for modified release products.

15.2.3.4   The PK-Sim® Absorption Model

PK-Sim® is a comprehensive software tool for PBPK modeling and simulation. The original absorption model in PK-Sim® was a so-called "plug-flow-with dispersion" model, which incorporated the small intestine as single, continuous compartment with spatially varying properties (Willmann et al. 2003a, b, 2004, 2007). Recently, the absorption model in PK-Sim® was revised to include the large intestine, detailed mucosa for drug–drug interactions (DDIs), active transport, and gut wall metabolism simulation, and dissolution functions (Willmann et al. 2012; Thelen et al. 2011, 2012). Briefly, the absorption model includes 12 compartments representing the lumen of different GI tracts: stomach, duodenum, upper and lower jejunum, upper and lower ileum, cecum, colon ascendens, colon transversum, colon descendens, sigmoid, and rectum, and 11 compartments representing the intestinal mucosa (Thelen et al. 2011). Each mucosa compartment contains four subcompartments representing the intracellular, the interstitial, the red blood cells, and the plasma (Thelen et al. 2011). The model was further revised to account for dosage form dependent GI transit, disintegration, and dissolution processes of various immediate release and modified release dosage forms (Thelen et al. 2012).

15.2.3.5   Other Compartmental and Transit Absorption Model

Besides the commercial models discussed above, scientists also developed compartmental and transit absorption models internally used in drug development. Peters developed a generic PBPK mode using MATLAB® software which incorporated absorption, metabolism, distribution, and biliary and renal elimination (Peters 2008). The absorption model included a stomach compartment, seven small intestine compartments, and a colon compartment (Peters 2008). The drug product in the lumen exists as undissolved or dissolved forms (Peters 2008).

Sjogren et al. presented a compartmental and transit model developed with AstraZeneca, named "GI-Sim" (Sjogren et al. 2013). The GI-Sim model consists of one stomach compartment, six small intestine compartments, and two colon compartments (Sjogren et al. 2013). The GI-Sim model also includes algorithms describing permeability, dissolution rate, salt effects, partitioning into micelles, particle and micelle drifting in the aqueous boundary layer, particle growth and amorphous or crystalline precipitation (Sjogren et al. 2013).

15.2.4   Applications of Mechanistic Oral Absorption Models

As discussed above, mechanistic oral absorption model becomes increasingly complicated with the integration of more and more parameters. Commercially available software has facilitated the utility of mechanism-based oral absorption models due to the user friendly interface such as GastroPlus™, Simcyp®, and PK-Sim®. In this section, we will review some recently published applications of mechanistic oral absorption models illustrating what types of questions can be addressed by mechanistic oral absorption models after theoretically understanding the model.

Mechanistic oral absorption models can guide the research and development at different stages of drug development ranging from lead optimization in the drug discovery phase through clinical candidate selection and extrapolation to human to phase 2 formulation development (Parrott and Lave 2008; Heimbach et al. 2009; Peters et al. 2009). Sensitivity analysis can be used to explore the impact of uncertainties in critical formulation attributes, such as particle size distribution, density, and solubility (Parrott and Lave 2008; Zhang et al. 2011; Poulin et al. 2011; Jones et al. 2011; Parrot 2008). Extensive case examples have been published by pharmaceutical scientists.

15.2.4.1 Bioavailability Assessment

Mechanistic oral models have been used to study the bioavailability change relating to drug compounds' apparent solubility and particle size (Dannenfelser et al. 2004; Kuentz et al. 2006). The results showed that changes of those parameters (i.e., particle size and solubility) within two orders of magnitude hardly affected the oral bioavailability for a poorly soluble drug compound (Kuentz et al. 2006). In order to select the formation for Phase I studies, Kuentz et al. conducted mechanistic oral absorption simulations together with a statistically designed dog study. The simulation results showed that more sophisticated formulations would offer no significant advantages and they were subsequently abandoned to reduce the drug development cost (Kuentz et al. 2006).

Similarly, Kesisoglou et al. demonstrated cases where a mechanistic oral model was used to assess the effect of active pharmaceutical ingredient (API) properties on bioavailability during both formulation strategy setup as well as the development process to help with setting of specifications around the API (Kesisoglou and Wu 2008). Sjogren et al. applied the internally developed GI-Sim model to predict the fraction absorbed for 12 APIs with reported or expected absorption limitations in humans due to permeability, dissolution, and/or solubility (Sjogren et al. 2013). Overall, more than 95 % of the predicted pharmacokinetic parameters (Cmax and AUC) were within a twofold deviation from the clinical observations and the predicted plasma AUC was within one standard deviation of the observed mean plasma AUC in 74 % of the simulations, suggesting the predictive performance of oral absorption was high (Sjogren et al. 2013). GI-Sim was also able to capture the effects of dose and particle size, including nano-formulations on drug absorption (Sjogren et al. 2013). The authors concluded that the mechanistic oral absorption model predicted oral absorption event for challenging APIs and therefore, "could provide useful guidance in the development of oral formulation for challenging molecules leading to increased development efficiency by reducing trial and error approaches."(Sjogren et al. 2013). As indicated in the review by Kuentz (2008), mechanistic oral absorption models have "the potential to become an indispensable tool to guide the formulation development of challenging drugs, which will help minimize both risks and costs of formulation development." (Kuentz 2008).

15.2.4.2 Drug–Drug Interactions/Drug–Food Interactions

In recent publications, mechanistic oral absorption models were used to evaluate the potential of drug–drug interactions/drug–food interactions caused by the change of gastric pH, gastric volume, and gastric emptying time (Fotaki and Klein 2013; Wagner et al. 2013). DDIs studies have historically focused on the impact of transporters or metabolic enzymes leading to the change of pharmacokinetics (FDA 2012c). With mechanistic oral absorption models, the effect of the change of GI pH, fluid volume, GI transit time, and other GI physiology parameters caused by the co-administered drug substance(s) can be studied in mechanistic framework, such as for hypotheses testing, and for prediction of drug–drug interactions.

To summarize, mechanistic oral absorption modeling and simulation have been used in various stages of pharmaceutical development, to evaluate DDIs potential, to predict food effects (Jones et al. 2006; Parrott et al. 2009; Shono et al. 2009, 2010; Xia et al. 2013; Heimbach et al. 2013; Sugano et al. 2010), to study the gut metabolism and transport (Darwich et al. 2010; Wagner et al. 2013; Peters 2008), for early identification of drug-induced impairment of gastric emptying (Peters and Hultin 2008), for development of IVIVC (Kovacevic et al. 2009; Okumu et al. 2008, 2009; Wei and Lobenberg 2006), to support bioequivalence recommendation development (Lionberger et al. 2012), and potentially in abbreviated new drug application (ANDA) review (Jiang et al. 2011).

Despite the success that mechanistic absorption modeling and simulation have achieved, there are several aspects for which continuing development efforts and investigations are needed: (1) improve understanding of the GI physiology, including ethnicity, age, disease difference, and regional distribution of metabolic enzymes and transporters; (2) improve prediction for complex oral dosage formulations, such as nanoparticles, controlled-release, self-emulsifying drug delivery system (SEDDS), solid dispersions, etc.; and (3) improve prediction for drugs absorbed through colon (Sugano 2009).

15.3 In Vitro and In Vivo Correlations

15.3.1 IVIVC Definitions

There are extensive publications discussing the methodology, development, and application of IVIVC (Chilukuri et al. 2007). In this chapter, IVIVC for oral solid dosage forms will be summarized briefly. The audience is encouraged to follow up on the latest emerging research on IVIVC, such as IVIVC based on mechanistic oral absorption models.

IVIVC, as defined in the FDA Guidance to Industry: Extended Release Oral Dosage Forms: Development, Evaluation, and Application of In Vitro/In Vivo Correlations (FDA 1997), is a predictive mathematical model describing the relationship between an in vitro property of an extended release dosage form (usually the rate or extent of drug dissolution or release) and a relevant in vivo response, e.g., plasma drug concentration or amount of drug absorbed. Therefore, the main objective of developing and evaluating an IVIVC is to establish the dissolution test as a surrogate for human bioequivalence studies to reduce unnecessary human trials during drug development as well as approval processes.

There are four types of IVIVC defined in the guidance (FDA 1997). A level A correlation represents a point-to-point relationship between in vitro dissolution and the in vivo input rate (e.g., the in vivo dissolution of the drug from the dosage form). For a level B correlation, the mean in vitro dissolution time is compared either to the mean residence time or to the mean in vivo dissolution time. A level C correlation establishes a single point relationship between a dissolution parameter and a pharmacokinetic parameter. A multiple level C correlation relates one or several pharmacokinetic parameters of interest to the amount of drug dissolved at several time points of the dissolution profile. From a regulatory standpoint, a level A IVIVC is considered to be the most informative and is recommended, if possible (FDA 1997). The scenarios in which in vivo studies can be waived based on an established IVIVC are also described in the guidance, including Level 3 manufacturing site changes, non-release controlling excipient changes, Level 3 changes in the release controlling excipients, approval of lower strengths, approval of new strengths, changes in release controlling excipients, etc.(FDA 1997). An established IVIVC can also be used to set dissolution specifications (FDA 1997).

15.3.2 Mathematical Approaches

Various mathematical approaches have been developed to establish a level A IVIVC, including one-step approach and two-step approach. Both approaches require developing formulations with different release rates, such as slow, medium, and fast; and conducting in vitro dissolution studies and in vivo pharmacokinetic (PK) studies. The two-step approach involves deconvolution of in vivo PK profile to obtain in vivo absorption/dissolution, and convolution to establish a link model between in vivo absorption/dissolution and in vitro dissolution, while the one-step approach does not involve deconvolution and directly establishes a link model to connect in vitro profile with PK profile (Dunne et al. 1997, 1999; O'Hara et al. 2001; Jacobs et al. 2008; Gould et al. 2009). Despite the mathematical instability of deconvolution due to the involvement of derivatives (O'Hara et al. 2001), the two-step approach is still wildly used.

The in vitro dissolution data are usually collected as the fraction or percentage dissolved from each dosage unit at a series of time points. To parameterize tabulated dissolution profiles, various empirical models have been used to describe in vitro dissolution profiles such as zero-order, first-order, Higuchi, Peppas,

Table 15.1 Some of the empirical models for in vitro dissolution

Description	Model
Zero-order	$M_t/M_\infty = kt$
First-order	$M_t/M_\infty = \exp(kt)$
Higuchi	$M_t/M_\infty = kt^{1/2}$
Peppas/Power law	$M_t/M_\infty = kt^n$
Makoid–Banakar	$M_t/M_\infty = kt^n \exp(-ct)$
Weibull	$M_t/M_\infty = M_{max}\left(1 - \exp\left(\dfrac{-(t-T_{lag})^b}{a}\right)\right)$
Double Weibull	$M_t/M_\infty = M_{max}\left(1 - f1 \times \exp\left(\dfrac{-(t-T_{lag})^{b1}}{a1}\right) - f2 \times \exp\left(\dfrac{-(t-T_{lag})^{b2}}{a2}\right)\right)$

In all the models, M_t is the amount of drug released at time t; M_∞ is the mass dissolved at infinite time; t is time, and k, n, *and* c are constants. In the Weibull models, M_{max} is the maximum amount to be released, T_{lag} is a lag time, a, $a1$, and $a2$ are scale constants, b, $b1$, and $b2$ are shape constants, and $f1$ and $f2$ are fractions for phase 1 and phase 2, respectively. If M_∞ is assumed to be equal to the total dose, the term M_t/M_∞ represents the fraction of drug dissolved at time t. It should be noted that various versions may be reported for some of these models. Rate constants cannot be any values since M_t/M_∞ is less than or equal to 1

Makoid–Banakar, Weibull, and double Weibull models (Table 15.1) (Dokoumetzidis and Macheras 2006; Costa and Sousa Lobo 2001, 2003). Some of these models may be reported as alternative versions in the literature. Among all the models, Weibull models include the most number of parameters. In the work of Kosmidis et al. (Dokoumetzidis and Macheras 2006; Kosmidis et al. 2003a, b), it is demonstrated that the Weibull model is the most powerful tool for the description of release kinetics in either Euclidean or fractal spaces.

There are several ways to obtain in vivo dissolution/absorption profiles from in vivo plasma vs. time profiles, such as numeric deconvolution (Cutler 1978a, b; Pedersen 1980a, b; Iga et al. 1986; Lanao et al. 1992), Wagner–Nelson method (Wagner and Nelson 1963), and Loo–Riegelman method (Loo and Riegelman 1968). A linear pharmacokinetic system can be described by Eq. (15.12) (Chilukuri et al. 2007).

$$C(t) = \int_0^t i(\tau)r(t - \tau)d\tau, \tag{15.12}$$

where $C(t)$ is the observed plasma drug concentration as a function of time, $i(t)$ is the input function, and $r(t)$ is the unit impulse response. The impulse, which is known as the "reference," may be an intravenous bolus, an oral solution, or an immediate release dosage form. When the reference is an intravenous bolus, the deconvoluted input rate represents the in vivo dissolution and absorption. On the other hand, when the reference is an oral dosage form, the deconvoluted input rate represents the in vivo dissolution. The objective of deconvolution is to obtain the input function given plasma drug concentration and the unit impulse response. If the unit impulse response is presented by a one-compartment model with first-order elimination rate, the model becomes Wagner–Nelson method (Wagner and Nelson 1963), and if the unit impulse response is presented by a two-compartment model with elimination from the central compartment, the model becomes Loo–Riegelman method (Loo and Riegelman 1968; Chilukuri et al. 2007).

Although the numeric methods, Wagner–Nelson method, and Loo–Riegelman method are wildly used, the in vivo dissolution/absorption obtained through those methods is a composite function of in vivo dissolution, GI transit, permeation, and first-pass metabolism. Recently, with the development of mechanistic oral absorption models, scientists have started to explore deconvolution of the plasma concentration vs. time profiles against mechanistic absorption models to obtain in vivo dissolution (Saibi et al. 2012; Turner 2012; Zhang et al. 2011; Grbic et al. 2011; Lionberger et al. 2012). Deconvolution against physiologically based absorption models considers factors such as pH in the GI tract, GI transit, permeability, first pass, etc. that affect drug absorption (Fig. 15.1) and obtain a physiologically relevant in vivo dissolution profile.

The link model relates the in vivo dissolution/absorption to in vitro dissolution. Many models have been proposed including linear and nonlinear, time-independent, and time-dependent models (Chilukuri et al. 2007; Lu et al. 2011). The time-independent link models do not include time as a variant, which implies that the relationship is the same for slowly dissolving dosage forms as it is for rapidly dissolving dosage forms. For such models, the in vivo–in vitro relationship may vary as the dosage form passes through the changing environment in the GI tract, i.e., the in vivo–in vitro relationship changes with time (Chilukuri et al. 2007). To introduce the time variant to the link models, various approaches have been applied, such as time shifting and scaling, exponential attenuation, step function attenuation, sigmoid attenuation, and Michaelis–Menten type attenuation (Chilukuri et al. 2007).

Finally the established model should be validated, internally and externally. Percent prediction error (%PE), as described by Eq. (15.13), is usually used to evaluate the model performance.

$$\%PE = \frac{(\text{Observed value} - \text{Predicted value})}{\text{Observed value}} \times 100, \qquad (15.13)$$

The acceptance criteria defined in the guidance (FDA 1997) is no more than 10 % for average absolute %PE across all formulations and the %PE for each formulation should not exceed 15 %.

15.4 BE Simulations

Simulation is a next step of modeling which simulates and predicts the outcome of various scenarios based on established models to address the questions that are difficult to be answered by experiment due to the limitation of time, cost, feasibility, and other reasons, and to optimize BE trials. Simulations have been extensively used at various stages of pharmaceutical research and development, and in regulatory standard development and review. Discussion in this section will be focusing on the simulations that help solve BE related issues.

15.4.1 Development and Testing of the Properties of BE Approaches

The average bioequivalence (ABE) approach has been adopted by the Food and Drug Administration (FDA), Health Canada, and European Medicines Agency (EMA) for demonstration of BE for most drug products. Novel approaches have been developed for complicated drug products (Zhang et al. 2013). In the development of the approaches for demonstration of BE for highly variable drugs (HVDs), highly variable drug products (HVDPs), and narrow therapeutic index (NTI) drugs, BE simulations have played a significant role. HVDs are generally defined as those with great within-subject variability, often expressed in % coefficient of variation (%CV). The cutoff value (%CV) for HVDs from regulatory point of view is 30 % (Haidar et al. 2008b). Extensive simulation research suggested that "the 0.25 value for $\sigma w0$ (a constant set by the regulatory agency to define the scaled average BE limit) appears to provide the best results" (Haidar et al. 2008a, b). Modified Chi-square Ratio Statistic (mCSRS) has been proposed for demonstration of equivalence in aerodynamic particle size distribution (APSD), which is one of the key components for establishing bioequivalence of orally inhaled drug products. Simulations have been performed to test the robustness and sensitivity of the proposed metric, the median of the distribution of 900 mCSRSs (MmCSRS). It was demonstrated that MmCSRS was a robust metric and could potentially be useful as a test statistic for APSD equivalence testing (Weber et al. 2013a, b).

15.4.2 Assessment of the Appropriateness of BE Metrics

In addition to the traditional metrics, Cmax (the maximum concentration), AUCt (area under the concentration–time curve from time zero to the last quantifiable time point), and AUCinf (area under the concentration–time curve from time zero to time infinity), other metrics have been proposed to demonstrate BE, such as different measures of absorption rate and extent, mean residence time (MRT), truncated areas for drugs with long half-life, and partial AUCs (pAUCs). Simulations have been performed to assess the power, sensitivity, and robustness of those parameters in differentiating different formulations.

Based on one- and two-compartment, zero-order and first-order absorption PK models, Bois et al. tested the performance of various measures of absorption rate (Cmax, Tmax (time to Cmax), pAUCs computed up to Tmax, Cmax/AUCinf, Cmax/Tmax, Cmax/AUCtmax, featured slope, featured AUC, and plasma concentration at ¼, ½, and 1 times the average Tmax), and extent (AUCinf estimated by various methods) for demonstration of BE (Bois et al. 1994a, b). The authors concluded that different rate measures have demonstrated advantages and limitations depending on the PK properties of the drug, therefore, they recommended conducting BE simulations to assess the applicability of the rate measures within the context of a specific situation (Bois et al. 1994b).

The power of MRT as a BE metric was evaluated and compared to Tmax (Kaniwa et al. 1989). It showed comparable to or even higher power than those of Cmax and AUCt (Kaniwa et al. 1989). MRT also has enough sensitivity to variations in the absorption rate, and therefore, was considered a better metric than Tmax (Kaniwa et al. 1989).

Numerous BE simulations were performed to investigate the appropriateness of truncated AUC as a BE metric for long half-life drugs, and demonstrated that in general, truncated AUCs were a good measure of relative extent of bioavailability, particularly for drugs with long half-lives (Gaudreault et al. 1998; Jackson and Ouderkirk 1999; Sathe et al. 1999; Kharidia et al. 1999; El-Tahtawy et al. 2012). BE simulations also contributed to the development and evaluation of partial AUCs that have been recommended for zolpidem tartrate extended release tablets (FDA 2011; Lionberger et al. 2012), and methylphenidate hydrochloride modified release products (FDA 2012a, b; Fourie Zirkelbach et al. 2013).

15.4.3   Assessment of the Appropriateness of Analyte and Design for BE Trials

Whether the metabolite(s) should be measured in BE studies has been debated in the scientific community. Simulations were conducted to evaluate the role of metabolites in BE assessment by various groups. Although different models and assumptions were used, a general conclusion was drawn from different groups (Braddy and Jackson 2010; Chen and Jackson 1991; Jackson 2000; Fernandez-Teruel et al. 2009a, b; Karalis and Macheras 2010; Navarro-Fontestad et al. 2010) that the parent drug is more sensitive to detect the difference in the rate of absorption which reflects the differences in formulation.

BE simulations were also involved in the selection of single-dose (SD) vs. multiple-dose (MD) design. El-Tahtawy conducted Monte Carlo simulations for drugs such as indomethacin, procainamide, erythromycin, quinidine, and nifedipine (El-Tahtawy et al. 1994). They found that low accumulation indices (AI) drugs showed similar 90 % confidence interval (CI) of AUC and Cmax between SD and MD, while drugs with higher AI appeared to have smaller CI at steady state which decreased the probability of failing BE criteria (El-Tahtawy et al. 1994). The same group of authors also showed that MD design for HVDs does not always reduce intrasubject variability in Cmax or AUC, and AUC showed similar probabilities of failure for SD and MD BE studies (El-Tahtawy et al. 1998). Fernandez-Teruel et al. conducted simulation in a semi-physiological model for eight types of drugs (BCS class I to IV) with high or low intrinsic clearance to define the most sensitive analyte and study design (Fernandez-Teruel et al. 2009a, b). Their simulations showed that the SD design is usually more sensitive than the MD design for BE trials. There are several exceptional scenarios where MD study showed higher sensitivity, such as BCS class III drugs with low intrinsic clearance (Fernandez-Teruel et al. 2009a, b).

15.4.4 *BE Trial Simulations on Mechanistic Oral Absorption Models*

The previous examples of BE simulations were mostly based on empirical models. With the emerging of mechanistic oral absorption models, virtual BE simulation can be conducted based on mechanistic oral absorption models to address some scientifically challenging questions (Mathias and Crison 2012) although the number of publications on this topic is relatively small to date. For example, BE simulations have been conducted to evaluate the potential strategy to extend waiver of in vivo BE studies for BCS class III drugs (Crison et al. 2012; Tsume and Amidon 2010). Bioequivalence between different formulations can be estimated based on validated models (Zhang et al. 2011). Or if in vitro in vivo correlations have been established, virtual BE simulations can be conducted to select the most promising formulation for a pivotal BE study based on in vitro data.

15.5 Conclusions

Modeling and simulation have demonstrated their value and been recognized as a powerful tool in pharmaceutical development as well as regulatory review over the past years. In this chapter, we have mainly focused on modeling and simulation for oral drug products and introduced mechanistic oral absorption models and their applications, in vitro–in vivo correlations for oral solid dosage forms, and applications of BE trial simulations. Our vision for the future is that mechanistic modeling and simulation will be implemented routinely in formulation development, study design, risk analysis, and BE standard development. Although highly challenging, similar types of modeling and simulation activities are desired for non-oral dosage forms and complex drug products where conventional BE approaches are not appropriate (Zhang et al. 2013).

References

Agoram B, Woltosz WS, Bolger MB (2001) Predicting the impact of physiological and biochemical processes on oral drug bioavailability. Adv Drug Deliv Rev 50(suppl 1):S41–S67

Amidon GL, Lennernas H, Shah VP, Crison JR (1995) A theoretical basis for a biopharmaceutic drug classification: the correlation of in vitro drug product dissolution and in vivo bioavailability. Pharm Res 12:413–420

Bois FY, Tozer TN, Hauck WW, Chen ML, Patnaik R, Williams RL (1994a) Bioequivalence: performance of several measures of extent of absorption. Pharm Res 11:715–722

Bois FY, Tozer TN, Hauck WW, Chen ML, Patnaik R, Williams RL (1994b) Bioequivalence: performance of several measures of rate of absorption. Pharm Res 11:966–974

Braddy AC, Jackson AJ (2010) Role of metabolites for drugs that undergo nonlinear first-pass effect: impact on bioequivalency assessment using single-dose simulations. J Pharm Sci 99:515–523

Chen ML, Jackson AJ (1991) The role of metabolites in bioequivalency assessment. I. Linear pharmacokinetics without first-pass effect. Pharm Res 8:25–32

Chilukuri DM, Sunkara G, Young D (2007) Pharmaceutical product development: in vitro-in vivo correlation. Drugs and the pharmaceutical sciences. Informa Healthcare USA, New York, NY

Costa P, Sousa Lobo JM (2001) Modeling and comparison of dissolution profiles. Eur J Pharm Sci 13:123–133

Costa P, Sousa Lobo JM (2003) Evaluation of mathematical models describing drug release from estradiol transdermal systems. Drug Dev Ind Pharm 29:89–97

Crison JR, Timmins P, Keung A, Upreti VV, Boulton DW, Scheer BJ (2012) Biowaiver approach for biopharmaceutics classification system class 3 compound metformin hydrochloride using in silico modeling. J Pharm Sci 101:1773–1782

Cutler DJ (1978a) Numerical deconvolution by least squares: use of polynomials to represent the input function. J Pharmacokinet Biopharm 6:243–263

Cutler DJ (1978b) Numerical deconvolution by least squares: use of prescribed input functions. J Pharmacokinet Biopharm 6:227–241

Dannenfelser R-M, He H, Joshi Y, Bateman S, Serajuddin ATM (2004) Development of clinical dosage forms for a poorly water soluble drug I: application of polyethylene glycol-polysorbate 80 solid dispersion carrier system. J Pharm Sci 93:1165–1175

Darwich AS, Neuhoff S, Jamei M, Rostami-Hodjegan A (2010) Interplay of metabolism and transport in determining oral drug absorption and gut wall metabolism: a simulation assessment using the "Advanced Dissolution, Absorption, Metabolism (ADAM)" model. Curr Drug Metab 11:716–729

Dokoumetzidis A, Macheras P (2006) A century of dissolution research: from Noyes and Whitney to the biopharmaceutics classification system. Int J Pharm 321:1–11

Dressman JB, Fleisher D (1986) Mixing-tank model for predicting dissolution rate control or oral absorption. J Pharm Sci 75:109–116

Dressman JB, Fleisher D, Amidon GL (1984) Physicochemical model for dose-dependent drug absorption. J Pharm Sci 73:1274–1279

Dressman JB, Amidon GL, Fleisher D (1985) Absorption potential: estimating the fraction absorbed for orally administered compounds. J Pharm Sci 74:588–589

Dunne A, O'Hara T, Devane J (1997) Level A in vivo-in vitro correlation: nonlinear models and statistical methodology. J Pharm Sci 86:1245–1249

Dunne A, O'Hara T, Devane J (1999) A new approach to modelling the relationship between in vitro and in vivo drug dissolution/absorption. Stat Med 18:1865–1876

El-Tahtawy AA, Jackson AJ, Ludden TM (1994) Comparison of single and multiple dose pharmacokinetics using clinical bioequivalence data and Monte Carlo simulations. Pharm Res 11:1330–1336

El-Tahtawy AA, Tozer TN, Harrison F, Lesko L, Williams R (1998) Evaluation of bioequivalence of highly variable drugs using clinical trial simulations. II: comparison of single and multiple-dose trials using AUC and Cmax. Pharm Res 15:98–104

El-Tahtawy A, Harrison F, Zirkelbach JF, Jackson AJ (2012) Bioequivalence of long half-life drugs—informative sampling determination–using truncated area in parallel-designed studies for slow sustained-release formulations. J Pharm Sci 101:4337–4346

FDA (1997) Guidance for industry: extended release oral dosage forms: development, evaluation, and application of in vitro/in vivo correlations. http://www.fda.gov/downloads/Drugs/GuidanceComplianceRegulatoryInformation/Guidances/ucm070239.pdf. Accessed 9 Jan 2013

FDA (2011) Guidance on Zolpidem extended release tablets. http://www.fda.gov/downloads/Drugs/GuidanceComplianceRegulatoryInformation/Guidances/UCM175029.pdf. Accessed 9 Jan 2013

FDA (2012a) Draft guidance on methylphenidate hydrochloride extended release capsules: http://www.fda.gov/downloads/Drugs/GuidanceComplianceRegulatoryInformation/Guidances/UCM320005.pdf; http://www.fda.gov/downloads/Drugs/GuidanceComplianceRegulatoryInformation/Guidances/UCM281454.pdf. Accessed 9 Jan 2013

FDA (2012b) Draft guidance on methylphenidate hydrochloride extended release tablets. http://www.fda.gov/downloads/Drugs/GuidanceComplianceRegulatoryInformation/Guidances/UCM320007.pdf. Accessed 9 Jan 2013

FDA (2012c) Draft guidance for industry drug interaction studies—study design, data analysis, implications for dosing, and labeling recommendations. http://www.fda.gov/downloads/Drugs/GuidanceComplianceRegulatoryInformation/Guidances/ucm292362.pdf. Accessed 9 Jan 2013

Fernandez-Teruel C, Gonzalez-Alvarez I, Navarro-Fontestad C, Garcia-Arieta A, Bermejo M, Casabo VG (2009a) Computer simulations of bioequivalence trials: selection of design and analyte in BCS drugs with first-pass hepatic metabolism: Part II. Non-linear kinetics. Eur J Pharm Sci 36:147–156

Fernandez-Teruel C, Nalda Molina R, Gonzalez-Alvarez I, Navarro-Fontestad C, Garcia-Arieta A, Casabo VG, Bermejo M (2009b) Computer simulations of bioequivalence trials: selection of design and analyte in BCS drugs with first-pass hepatic metabolism: linear kinetics (I). Eur J Pharm Sci 36:137–146

Fotaki N, Klein S (2013) Mechanistic understanding of the effect of PPIs and acidic carbonated beverages on the oral absorption of itraconazole based on absorption modeling with appropriate in vitro data. Mol Pharm 10(11):4016–4023

Fourie Zirkelbach J, Jackson AJ, Wang Y, Schuirmann DJ (2013) Use of partial AUC (PAUC) to evaluate bioequivalence—a case study with complex absorption: methylphenidate. Pharm Res 30:191–202

Gaudreault J, Potvin D, Lavigne J, Lalonde RL (1998) Truncated area under the curve as a measure of relative extent of bioavailability: evaluation using experimental data and Monte Carlo simulations. Pharm Res 15:1621–1629

Goodacre BC, Murray PJ (1981) A mathematical model of drug absorption. J Clin Hosp Pharm 6:117–133

Gould AL, Agrawal NG, Goel TV, Fitzpatrick S (2009) A 1-step Bayesian predictive approach for evaluating in vitro in vivo correlation (IVIVC). Biopharm Drug Dispos 30:366–388

Grass GM (1997) Simulation models to predict oral drug absorption from in vitro data. Adv Drug Deliv Rev 23:199–219

Grbic S, Parojcic J, Ibric S, Djuric Z (2011) In vitro-in vivo correlation for gliclazide immediate-release tablets based on mechanistic absorption simulation. AAPS PharmSciTech 12:165–171

Haidar SH, Davit B, Chen ML, Conner D, Lee L, Li QH, Lionberger R, Makhlouf F, Patel D, Schuirmann DJ, Yu LX (2008a) Bioequivalence approaches for highly variable drugs and drug products. Pharm Res 25:237–241

Haidar SH, Makhlouf F, Schuirmann DJ, Hyslop T, Davit B, Conner D, Yu LX (2008b) Evaluation of a scaling approach for the bioequivalence of highly variable drugs. AAPS J 10:450–454

Heimbach T, Lakshminarayana SB, Hu W, He H (2009) Practical anticipation of human efficacious doses and pharmacokinetics using in vitro and preclinical in vivo data. AAPS J 11:602–614

Heimbach T, Xia B, Lin TH, He H (2013) Case studies for practical food effect assessments across BCS/BDDCS class compounds using in silico, in vitro, and preclinical in vivo data. AAPS J 15:143–158

Iga K, Ogawa Y, Yashiki T, Shimamoto T (1986) Estimation of drug absorption rates using a deconvolution method with nonequal sampling times. J Pharmacokinet Biopharm 14:213–225

Jackson AJ (2000) The role of metabolites in bioequivalency assessment. III. Highly variable drugs with linear kinetics and first-pass effect. Pharm Res 17:1432–1436

Jackson AJ, Ouderkirk LA (1999) Truncated area under the curve as a measure of relative extent of bioavailability: evaluation using experimental data and Monte Carlo simulations. Pharm Res 16:1144–1146

Jacobs T, Rossenu S, Dunne A, Molenberghs G, Straetemans R, Bijnens L (2008) Combined models for data from in vitro-in vivo correlation experiments. J Biopharm Stat 18:1197–1211

Jamei M, Marciniak S, Feng KR, Barnett A, Tucker G, Rostami-Hodjegan A (2009a) The Simcyp (R) population-based ADME simulator. Expert Opin Drug Metab Toxicol 5:211–223

Jamei M, Turner D, Yang J, Neuhoff S, Polak S, Rostami-Hodjegan A, Tucker G (2009b) Population-based mechanistic prediction of oral drug absorption. AAPS J 11:225–237

Jiang W, Kim S, Zhang X, Lionberger RA, Davit BM, Conner DP, Yu LX (2011) The role of predictive biopharmaceutical modeling and simulation in drug development and regulatory evaluation. Int J Pharm 418:151–160

Jones HM, Parrott N, Ohlenbusch G, Lave T (2006) Predicting pharmacokinetic food effects using biorelevant solubility media and physiologically based modelling. Clin Pharmacokinet 45:1213–1226

Jones HM, Gardner IB, Collard WT, Stanley PJ, Oxley P, Hosea NA, Plowchalk D, Gernhardt S, Lin J, Dickins M, Rahavendran SR, Jones BC, Watson KJ, Pertinez H, Kumar V, Cole S (2011) Simulation of human intravenous and oral pharmacokinetics of 21 diverse compounds using physiologically based pharmacokinetic modelling. Clin Pharmacokinet 50:331–347

Kaniwa N, Ogata H, Aoyagi N, Takeda Y, Uchiyama M (1989) Power analyses of moment analysis parameter in bioequivalence tests. J Pharm Sci 78:1020–1024

Karalis V, Macheras P (2010) Examining the role of metabolites in bioequivalence assessment. J Pharm Pharm Sci 13:198–217

Kesisoglou F, Wu Y (2008) Understanding the effect of API properties on bioavailability through absorption modeling. AAPS J 10:516–525

Kharidia J, Jackson AJ, Ouderkirk LA (1999) Use of truncated areas to measure extent of drug absorption in bioequivalence studies: effects of drug absorption rate and elimination rate variability on this metric. Pharm Res 16:130–134

Kosmidis K, Argyrakis P, Macheras P (2003a) Fractal kinetics in drug release from finite fractal matrices. J Chem Phys 119:6373–6377

Kosmidis K, Argyrakis P, Macheras P (2003b) A reappraisal of drug release laws using Monte Carlo simulations: the prevalence of the Weibull function. Pharm Res 20:988–995

Kovacevic I, Parojcic J, Homsek I, Tubic-Grozdanis M, Langguth P (2009) Justification of biowaiver for carbamazepine, a low soluble high permeable compound, in solid dosage forms based on IVIVC and gastrointestinal simulation. Mol Pharm 6:40–47

Kuentz M (2008) Drug absorption modeling as a tool to define the strategy in clinical formulation development. AAPS J 10:473–479

Kuentz M, Nick S, Parrott N, Rothlisberger D (2006) A strategy for preclinical formulation development using GastroPlus as pharmacokinetic simulation tool and a statistical screening design applied to a dog study. Eur J Pharm Sci 27:91–99

Lanao JM, Vicente MT, Sayalero ML, Dominguez-Gil A (1992) A computer program (DCN) for numerical convolution and deconvolution of pharmacokinetic functions. J Pharmacobiodyn 15:203–214

Lionberger RA, Raw AS, Kim SH, Zhang X, Yu LX (2012) Use of partial AUC to demonstrate bioequivalence of Zolpidem Tartrate Extended Release formulations. Pharm Res 29:1110–1120

Loo JC, Riegelman S (1968) New method for calculating the intrinsic absorption rate of drugs. J Pharm Sci 57:918–928

Lu Y, Kim S, Park K (2011) In vitro-in vivo correlation: perspectives on model development. Int J Pharm 418:142–148

Lukacova V, Woltosz WS, Bolger MB (2009) Prediction of modified release pharmacokinetics and pharmacodynamics from in vitro, immediate release, and intravenous data. AAPS J 11:323–334

Luner PE, Amidon GL (1993) Description and simulation of a multiple mixing tank model to predict the effect of bile sequestrants on bile salt excretion. J Pharm Sci 82:311–318

Maria A (1997) Introduction to modeling and simulation. In: Andradóttir S, Healy KJ, Withers DH, Nelson BL, eds. Proceedings of the 1997 winter simulation conference, Atlanta, Georgia

Mathias NR, Crison J (2012) The use of modeling tools to drive efficient oral product design. AAPS J 14:591–600

Mudie DM, Amidon GL, Amidon GE (2010) Physiological parameters for oral delivery and in vitro testing. Mol Pharm 7:1388–1405

Navarro-Fontestad C, Gonzalez-Alvarez I, Fernandez-Teruel C, Garcia-Arieta A, Bermejo M, Casabo VG (2010) Computer simulations for bioequivalence trials: selection of analyte in BCS drugs with first-pass metabolism and two metabolic pathways. Eur J Pharm Sci 41:716–728

Ni PF, Ho NFH, Fox JL, Leuenberger H, Higuchi WI (1980) Theoretical model studies of intestinal drug absorption V. Non-steady-state fluid flow and absorption. Int J Pharm 5:33–47

Oberle RL, Amidon GL (1987) The influence of variable gastric emptying and intestinal transit rates on the plasma level curve of cimetidine; an explanation for the double peak phenomenon. J Pharmacokinet Biopharm 15:529–544

O'Hara T, Hayes S, Davis J, Devane J, Smart T, Dunne A (2001) In vivo-in vitro correlation (IVIVC) modeling incorporating a convolution step. J Pharmacokinet Pharmacodyn 28:277–298

Okumu A, DiMaso M, Lobenberg R (2008) Dynamic dissolution testing to establish in vitro/in vivo correlations for montelukast sodium, a poorly soluble drug. Pharm Res 25:2778–2785

Okumu A, DiMaso M, Lobenberg R (2009) Computer simulations using GastroPlus to justify a biowaiver for etoricoxib solid oral drug products. Eur J Pharm Biopharm 72:91–98

Parrot N (2008) Application of physiologically based modeling in pre-clinical to clinical PK/PD prediction

Parrott N, Lave T (2002) Prediction of intestinal absorption: comparative assessment of GASTROPLUS and IDEA. Eur J Pharm Sci 17:51–61

Parrott N, Lave T (2008) Applications of physiologically based absorption models in drug discovery and development. Mol Pharm 5:760–775

Parrott N, Lukacova V, Fraczkiewicz G, Bolger MB (2009) Predicting pharmacokinetics of drugs using physiologically based modeling–application to food effects. AAPS J 11:45–53

Pedersen PV (1980a) Model-independent method of analyzing input in linear pharmacokinetic systems having polyexponential impulse response I: theoretical analysis. J Pharm Sci 69:298–305

Pedersen PV (1980b) Model-independent method of analyzing input in linear pharmacokinetic systems having polyexponential impulse response II: numerical evaluation. J Pharm Sci 69:305–312

Peters SA (2008) Identification of intestinal loss of a drug through physiologically based pharmacokinetic simulation of plasma concentration-time profiles. Clin Pharmacokinet 47:245–259

Peters SA, Hultin L (2008) Early identification of drug-induced impairment of gastric emptying through physiologically based pharmacokinetic (PBPK) simulation of plasma concentration-time profiles in rat. J Pharmacokinet Pharmacodyn 35:1–30

Peters SA, Ungell AL, Dolgos H (2009) Physiologically based pharmacokinetic (PBPK) modeling and simulation: applications in lead optimization. Curr Opin Drug Discov Devel 12:509–518

Poulin P, Jones RD, Jones HM, Gibson CR, Rowland M, Chien JY, Ring BJ, Adkison KK, Ku MS, He H, Vuppugalla R, Marathe P, Fischer V, Dutta S, Sinha VK, Bjornsson T, Lave T, Yates JW (2011) PHRMA CPCDC initiative on predictive models of human pharmacokinetics, part 5: Prediction of plasma concentration-time profiles in human by using the physiologically-based pharmacokinetic modeling approach. J Pharm Sci 100(10):4127–4157

Russell TL, Berardi RR, Barnett JL, Dermentzoglou LC, Jarvenpaa KM, Schmaltz SP, Dressman JB (1993) Upper gastrointestinal pH in seventy-nine healthy, elderly, North American men and women. Pharm Res 10:187–196

Saibi Y, Sato H, Tachiki H (2012) Developing in vitro-in vivo correlation of risperidone immediate release tablet. AAPS PharmSciTech 13:890–895

Sathe P, Venitz J, Lesko L (1999) Evaluation of truncated areas in the assessment of bioequivalence of immediate release formulations of drugs with long half-lives and of Cmax with different dissolution rates. Pharm Res 16:939–943

Shono Y, Jantratid E, Janssen N, Kesisoglou F, Mao Y, Vertzoni M, Reppas C, Dressman JB (2009) Prediction of food effects on the absorption of celecoxib based on biorelevant dissolution testing coupled with physiologically based pharmacokinetic modeling. Eur J Pharm Biopharm 73:107–114

Shono Y, Jantratid E, Kesisoglou F, Reppas C, Dressman JB (2010) Forecasting in vivo oral absorption and food effect of micronized and nanosized aprepitant formulations in humans. Eur J Pharm Biopharm 76:95–104

Siepmann J, Siepmann F (2013) Mathematical modeling of drug dissolution. Int J Pharm 453:12–24

Sinko PJ, Leesman GD, Amidon GL (1991) Predicting fraction dose absorbed in humans using a macroscopic mass balance approach. Pharm Res 8:979–988

Sjogren E, Westergren J, Grant I, Hanisch G, Lindfors L, Lennernas H, Abrahamsson B, Tannergren C (2013) In silico predictions of gastrointestinal drug absorption in pharmaceutical product development: application of the mechanistic absorption model GI-Sim. Eur J Pharm Sci 49:679–698

Sugano K (2009) Introduction to computational oral absorption simulation. Expert Opin Drug Metab Toxicol 5:259–293

Sugano K, Kataoka M, Mathews Cda C, Yamashita S (2010) Prediction of food effect by bile micelles on oral drug absorption considering free fraction in intestinal fluid. Eur J Pharm Sci 40:118–124

Thelen K, Coboeken K, Willmann S, Burghaus R, Dressman JB, Lippert J (2011) Evolution of a detailed physiological model to simulate the gastrointestinal transit and absorption process in humans, part 1: oral solutions. J Pharm Sci 100:5324–5345

Thelen K, Coboeken K, Willmann S, Dressman JB, Lippert J (2012) Evolution of a detailed physiological model to simulate the gastrointestinal transit and absorption process in humans, part II: extension to describe performance of solid dosage forms. J Pharm Sci 101:1267–1280

Tsume Y, Amidon GL (2010) The biowaiver extension for BCS class III drugs: the effect of dissolution rate on the bioequivalence of BCS class III immediate-release drugs predicted by computer simulation. Mol Pharm 7:1235–1243

Turner D (2012) Mechanistic IVIVC using the Simcyp ADAM model. http://www.pqri.org/workshops/ivivc/turner.pdf. Accessed 9 Oct 2013

Wagner JG, Nelson E (1963) Per cent absorbed time plots derived from blood level and/or urinary excretion data. J Pharm Sci 52:610–611

Wagner C, Thelen K, Willmann S, Selen A, Dressman JB (2013) Utilizing in vitro and PBPK tools to link ADME characteristics to plasma profiles: case example nifedipine immediate release formulation. J Pharm Sci 102:3205–3219

Weber B, Hochhaus G, Adams W, Lionberger R, Li B, Tsong Y, Lee SL (2013a) A stability analysis of a modified version of the chi-square ratio statistic: implications for equivalence testing of aerodynamic particle size distribution. AAPS J 15:1–9

Weber B, Lee SL, Lionberger R, Li BV, Tsong Y, Hochhaus G (2013b) A sensitivity analysis of the modified chi-square ratio statistic for equivalence testing of aerodynamic particle size distribution. AAPS J 15:465–476

Wei H, Lobenberg R (2006) Biorelevant dissolution media as a predictive tool for glyburide a class II drug. Eur J Pharm Sci 29:45–52

Willmann S, Lippert J, Sevestre M, Solodenko J, Fois F, Schmitt W (2003a) PK-Sim®: a physiologically based pharmacokinetic 'whole-body' model. Biosilico 1:121–124

Willmann S, Schmitt W, Keldenich J, Dressman JB (2003b) A physiologic model for simulating gastrointestinal flow and drug absorption in rats. Pharm Res 20:1766–1771

Willmann S, Schmitt W, Keldenich J, Lippert J, Dressman JB (2004) A physiological model for the estimation of the fraction dose absorbed in humans. J Med Chem 47:4022–4031

Willmann S, Edginton AN, Dressman JB (2007) Development and validation of a physiology-based model for the prediction of oral absorption in monkeys. Pharm Res 24:1275–1282

Willmann S, Thelen K, Lippert J (2012) Integration of dissolution into physiologically-based pharmacokinetic models III: PK-Sim®. J Pharm Pharmacol 64:997–1007

Xia B, Heimbach T, Lin TH, Li S, Zhang H, Sheng J, He H (2013) Utility of physiologically based modeling and preclinical in vitro/in vivo data to mitigate positive food effect in a BCS class 2 compound. AAPS PharmSciTech 14:1255–1266

Yu LX (1999) An integrated model for determining causes of poor oral drug absorption. Pharm Res 16:1883–1887

Yu LX, Amidon GL (1999) A compartmental absorption and transit model for estimating oral drug absorption. Int J Pharm 186:119–125

Yu LX, Crison JR, Amidon GL (1996a) Compartmental transit and dispersion model analysis of small intestinal transit flow in humans. Int J Pharm 140:111–118

Yu LX, Lipka E, Crison JR, Amidon GL (1996b) Transport approaches to the biopharmaceutical design of oral drug delivery systems: prediction of intestinal absorption. Adv Drug Deliv Rev 19:359–376

Zhang X, Lionberger RA, Davit BM, Yu LX (2011) Utility of physiologically based absorption modeling in implementing Quality by Design in drug development. AAPS J 13:59–71

Zhang X, Zheng N, Lionberger RA, Yu LX (2013) Innovative approaches for demonstration of bioequivalence: the US FDA perspective. Ther Deliv 4:725–740

Chapter 16
Bioanalysis

Sriram Subramaniam

16.1 Introduction

Bioequivalence (BE), pharmacokinetic (PK), and toxicokinetic (TK) studies involve assessment of drug exposure data that are vital to understand drug safety and efficacy. Generation of drug exposure data involves quantitation of the drugs and/or its metabolite(s) in biological matrix samples collected after drug administration. Therefore, quantitation of drugs and/or metabolites in biological matrices plays a vital role in the assessment and interpretation of BE, PK, and TK studies. Bioanalysis, a term which will be often used in this chapter, refers to the process of quantitation of drug and/or metabolites in biological matrices (i.e., blood, serum, urine, and tissues). Bioanalysis involves use of reliable bioanalytical methods to quantitate drugs and/or metabolites in samples from in vivo BE, PK, and TK studies. Hence, the quality of such studies is directly related to the quality of underlying bioanalytical methods and conduct. It is therefore imperative that the bioanalytical assays used in clinical and preclinical studies are validated for their intended use, and bioanalytical conduct is consistent and objective. Bioanalytical method validation (BMV) encompasses all of the procedures that demonstrate that a particular method used for quantitative measurement of analytes in a given biological matrix is reliable and reproducible for the intended use. This is especially important for bioanalytical methods used in clinical and nonclinical studies intended for submission to regulatory agencies, such as the United States' Food and Drug Administration ("FDA"), commonly referred to as regulatory bioanalysis. In fact, the United States' Code of Federal Regulations, Title 21 (21 CFR 320.29) require that bioanalytical methods used in BE studies are accurate, precise, and

S. Subramaniam (✉)
Center for Drug Evaluation and Research, U.S. Food and Drug Administration,
10903 New Hampshire Avenue, Silver Spring, MD 20993, USA
e-mail: Sriram.Subramaniam@fda.hhs.gov

L.X. Yu and B.V. Li (eds.), *FDA Bioequivalence Standards*, AAPS Advances
in the Pharmaceutical Sciences Series 13, DOI 10.1007/978-1-4939-1252-0_16,
© The United States Government 2014

sufficiently sensitive so that the actual concentration of the drug or its metabolite (s) achieved in the body can be measured (FDA CFR 2013). To address the expectations on bioanalysis to the pharmaceutical industry, the FDA published a guidance on BMV ("FDA BMV guidance") in 2001 (FDA 2001). In addition, the FDA recently issued a draft guidance (FDA 2013)[1] to reflect revisions to the existing FDA BMV guidance (FDA 2001). At this point, the revised FDA guidance is issued in draft form for public comments before it is finalized.

With the advancement of bioanalytical tools and techniques, and significant gains in scientific and regulatory experience over the years, there has been a critical examination of the current bioanalytical guidelines and practices. The third American Association of Pharmaceutical Scientists (AAPS)/FDA Bioanalytical Workshop in 2006 ("2006 AAPS/FDA Workshop") evaluated the current practices and clarified the FDA BMV guidelines (Viswanathan et al. 2007). This was followed by the 2008 AAPS Workshop ("2008 ISR Workshop") which further discussed issues raised during the 2006 AAPS/FDA Workshop (Fast et al. 2009). Since then, the recommendations of the 2006 AAPS/FDA Workshop (Viswanathan et al. 2007) and the 2008 ISR Workshop (Fast et al. 2009) have been discussed in several workshops and meetings (Timmerman et al. 2009; Savoie et al. 2009; Savoie et al. 2010; Garofolo et al. 2011; DeSilva et al. 2012), and have been the basis for the recent regulatory guidelines (EMA 2011; Health Canada 2012). Also, as mentioned earlier, the FDA has recently (2013) proposed revisions to the existing FDA BMV guidance (FDA 2001) in response to advancement in technology and changes in practices relating to BMV.

The focus of this chapter is to address the current best practices for BMV as it relates to BE studies. In addition to discussing the expectations of the FDA BMV guidance, the chapter will identify and evaluate recent bioanalytical practices, and highlight the potential challenges in bioanalysis based on review of scientific and regulatory articles, and white papers published since issuance of the FDA BMV guidance (2001). The chapter is not intended to describe in detail specific assay methods and resolution of bioanalytical issues, as these issues have been discussed in detail in current literature.

16.2   Bioanalytical Methods

Bioanalytical methods can be broadly classified as chromatographic and ligand binding methods. While a detailed description of the principles and procedures for the methods are beyond the scope of this chapter, a brief outline of the methods is provided below.

---

[1] This draft guidance is not for implementation. Since the draft guidance is issued for public review and comment, the recommendations in the guidance may be modified when finalized.

16.2.1 Chromatographic Methods

In chromatographic methods, the analyte of interest is isolated and separated using appropriate sample clean-up procedures and chromatographic conditions, respectively, and detected using a suitable detection system. Sample extraction, chromatography, and detection techniques are briefly discussed below.

16.2.1.1 Sample Extraction

Generally, prior to chromatography, sample clean-up is performed for method sensitivity. Proteins in biological matrices may bind to analyte of interest and can clog the chromatography columns. Blood contains intra- and extra-cellular proteins, plasma contains significant proteins, and urine and cerebrospinal fluids contain relatively less proteins but still require extraction to improve reliability (Mulvana 2010). In addition to proteins, endogenous compounds such as phospholipids and fatty acids, and exogenous components in biological matrices can potentially affect separation and detection of the analyte of interest (e.g., foul high performance liquid chromatography (HPLC) columns and contaminate MS source) (Singleton 2012). The purpose of sample clean-up is to extract out the analyte(s) of interest from biological matrices to minimize interference and maximize recovery. Consequently, sample clean-up reduces variability and inconsistencies during analysis. Different sample clean-up procedures are used depending on the choice of matrix, drug, chromatography, and detection systems. Broadly, sample clean-up procedures include, protein precipitation (PP), solid phase extraction (SPE), and liquid–liquid extraction (LLE).

In PP, miscible organic solvents (e.g., methanol or acetonitrile), often modified with buffer or acid and bases, are added to biological samples to denature proteins and consequently precipitate the samples. For example, if the analyte is highly protein bound, a volatile acid (e.g., formic acid) or base (ammonium hydroxide) is used to disrupt binding and increase analyte recovery. The precipitate is removed by centrifugation or filtration, and extract injected. Although PP is simple and fast, it does not necessarily yield clean extracts, as it may not remove endogenous components such as phospholipids, fatty acids, lipids (Van Eeckhaut et al. 2009; Mulvana 2010).

More efficient sample clean-up may be obtained from LLE and SPE. In LLE, immiscible organic solvents (e.g., diethyl ether, ethyl acetate, methyl-*tert*-butyl ether (MTBE), hexane) are used to extract the analyte of interest by partitioning it into an organic layer (Singleton 2012; Nováková 2013). Therefore, LLE can mitigate or avoid matrix effects as ionized compounds, including salts or phospholipids, do not partition into the organic layer (Nováková 2013). The advantage of LLE is mainly its ease of use, and requires no special instrumentation. A major limitation of LLE is its applicability to polar compounds (Nováková 2013). To transfer an ionizable analyte to organic solvent it first needs to be converted to a nonionic form in an aqueous medium at an appropriate pH, followed by selection of a suitable solvent to efficiently and selectively extract the analyte. Usually multiple extractions are necessary and final re-suspension in an aqueous medium at the original pH is needed, resulting in reduction in recovery of the analyte (Trufelli et al. 2011). Also, in LLE, there is a tendency to form emulsions at the interface between liquid layers (Trufelli et al. 2011; Singleton 2012; Nováková 2013). Further, LLE may require large solvent volumes. These problems have been reported to be minimized with new versions of LLE, such as supported LLE. In supported LLE, the entire sample is adsorbed on a solid support (i.e., diatomaceous earth), and an organic solvent is passed through the solid support resulting in partition of the analyte of interest into the organic solvent (Singleton 2012). Recently, LLE has been scaled down, requiring relatively low volumes of sample (50–100 μL) and organic solvent (0.6–2 mL) (Nováková 2013). Also, high throughput LLE versions using on-line extraction or 96-well plate arrangements are available. For other recent LLE techniques the reader is encouraged to refer to Singleton (2012) and Nováková (2013).

To further increase selectivity and clean-up, SPE is often employed. SPE can reduce sample volume, be easily automated, and used on-line with liquid chromatography separation. In SPE, the separation process is based on the affinity of the analyte to the stationary phase or sorbent. The sorbents are ion-exchange, normal phase, reverse phase or a combination to selectively retain the analyte of interest. The interfering matrix components either pass through unretained or are retained relatively longer than the analyte of interest. The choice of sorbent controls selectivity, affinity, and capacity (Nováková 2013), depending on the physiochemical properties of the analyte, biological matrix, and interaction between sorbent and analyte. The SPE usually involves a wash step to remove undesired components, and an elution step to extract the analyte of interest. Therefore, selection of the proper washing and elution solvents are important (Trufelli et al. 2011). It is reported that immunosorbents and molecularly imprinted polymer (MIP) sorbents can significantly increase selectivity of SPE (Nováková 2013). The drawbacks of SPE include, the time required for processing (manual SPE), expense, and lot-to-lot cartridge variability. Also, matrix effects have been reported to result from the sample pre-concentration step and the SPE procedure itself (i.e., from salts in buffers used) (Van Eeckhaut et al. 2009). However, the advantages of SPE overshadow the drawbacks. SPE remains one of the most widely used extraction techniques for routine bioanalysis. For recent SPE techniques, the reader is encouraged to refer to Mulavana (2010), Singleton (2012), and Nováková (2013).

16.2.1.2   Chromatography

The aim of chromatography is to assure that the analyte(s) of interest is adequately resolved from interfering components. Chromatographic separation is primarily based on the differences in physicochemical properties between the analyte and matrix components related to both mobile and stationary phases (Li et al. 2011).

The main factors (techniques) for chromatographic separation are hydrophobicity (reversed-phase), molecular charge (ion-exchange), and size (size exclusion) of the stationary phase (Bozovic and Kulasingam 2013). The choice of the separation technique depends on the characteristics of the analyte to be separated, and often a combination of techniques may be required.

In addition to adequately resolving the analyte of interest from those of other closely eluting compounds, an ideal chromatography technique should be able to measure the analyte at low levels, have short retention times, and be time and cost efficient. Reversed-phase chromatography is based on the reversible adsorption of molecules based on their polarity under conditions where the stationary phase is more hydrophobic than the mobile phase (Bozovic and Kulasingam 2013). This is the most popular and widely used liquid chromatography (LC) technique due to its robustness, efficiency, column stability and availability of several different phase chemistries that can be customized for a particular use.

In addition to LC column selection and mobile phase composition, factors such as gradient time, mobile phase pH, and column temperature need to be considered when dealing with unstable analytes (Li et al. 2011). Also, the purity of the solvent used to dissolve the analyte, and the compatibility of the solvent with mobile phase and ion source (i.e., if coupled to mass spectrometers) are important considerations. It is critical that buffers containing inorganic salts are avoided at all times, as well as inorganic acids, ion-pairing reagents, and nonvolatile buffers. Formate, acetate, and ammonia at low concentrations are frequently used additives, as they are compatible with mass spectrometric detection (Bozovic and Kulasingam 2013).

Increasing resolution efficiency, flow rate, and column temperature are some of the ways to improve run time. Gradient elution is the preferred mode of separation for small molecules, as it has a broader range of retentivity, higher peak capacity, and faster analysis compared to isocratic elution.

Over the years, development of stationary phases have evolved, including silica, phenyl, C8, or C18 columns that improve retention times, enhance column lifetime, and increase throughput (Mulvana 2010). Also, porous silica rod or MIP columns increase throughput and resolution. In addition, with the advent of columns with sub-2 μm particle size and liquid-handling systems that can operate such columns at high pressures, ultra-high performance liquid chromatography (UHPLC) has become increasingly popular in quantitative bioanalysis. UHPLC increases speed, resolution, sensitivity, and lower solvent consumption (Van Eeckhaut et al. 2009; Trufelli et al. 2011; Nováková 2013; Jemal et al. 2010). To prevent increase in back pressure and dirtying of columns, a pre-column is recommended for bioanalysis with UHPLC. Hydrophilic interaction liquid chromatography (HILIC) is another powerful, new technique for separation of small polar molecules that are weakly eluted or retained in conventional LC techniques. HILIC combines the use of bare silica or polar bonded stationary phases and mobile phase with high content of organic solvents (Van Eeckhaut et al. 2009). The higher content of organic solvents in HILIC increases selectivity, sensitivity, and efficiency of drug quantitation by effective retention of polar compounds, enhancing electrospray ionization (ESI), speeding separation under high flow rates or in columns with small particle size (due to low back pressure), and being compatible with elution solvent used in reversed phase-SPE (Van Eeckhaut et al. 2009; Trufelli et al. 2011; Nováková 2013). Therefore, HILIC has become very popular in bioanalysis, often in UHPLC arrangements (Nováková 2013).

16.2.1.3 Mass Spectrometry

Following sample clean-up and chromatography, the analyte(s) of interest is detected and quantitated using an appropriate detection system. Currently the most commonly used detection system for analysis of small molecules is mass spectrometry (MS). Therefore, this detection system is discussed briefly. Although MS detection is generally regarded as highly selective, chromatographic separation is still recommended to avoid problems with interferences in MS that can affect quantitation (Nováková 2013).

For detection by MS, the uncharged analytes eluting from the HPLC system have to be first transformed to ions. This occurs at the ionization source. Therefore, the ionization source serves as an interface between HPLC and mass spectrometer. There are various types of ionization sources. Currently, the most commonly used ionization sources are ESI and atmospheric pressure chemical ionization (APCI). Since ionization in ESI and APCI occurs at atmospheric pressure, the ESI and APCI sources are commonly referred to as c (API) sources. The effluent containing the analyte from the HPLC is nebulized. Nebulization occurs in ESI by the high voltage field resulting in charged droplets that are focused toward the mass analyzer and get smaller and smaller as they approach the entrance to the mass analyzer. As the droplets get smaller, individual ions emerge in a process referred to as "ion evaporation" (Niessen 2003). In APCI, nebulization occurs by spraying the mobile phase (containing the analyte) with a nebulizer gas in a heated vaporizer tube (350–500 °C) and the resultant aerosol cloud is ionized by a corona discharge needle (Niessen 2003). A newer ionization source, atmospheric pressure photoionization (APPI), vaporizes HPLC eluant like APCI, but uses photons from an ultraviolet (UV) lamp to initiate the ionization process (Korfmacher 2005).

Following ionization, the mass spectrometer analyzes the ion of the analyte of interest (i.e., precursor ion) based on its mass to charge ratio (m/z). However, for bioanalytical purposes, the MS response obtained for the precursor ion alone may not be suitable for quantitative analysis. This is because there may be many molecules in the matrix that produce ions of the same m/z as the target analyte, thus making the result nonspecific and often invalid. This limitation can be surmounted by tandem mass spectrometry (MS/MS). The most commonly used MS/MS in bioanalytical assay is the triple quadrupole mass spectrometer operated in selected reaction monitoring (SRM) or multiple reaction monitoring (MRM) mode (Niessen 2003; Korfmacher 2005).

The triple quadrupole mass spectrometer consists of three quadrupoles: the first (Q1) and third (Q3) quadrupoles are mass analyzers, and the second quadrupole (Q2) is the collision cell (Bozovic and Kulasingam 2013). When triple quadrupole mass spectrometer is operated in SRM or MRM mode, high selectivity is achieved due to two-stage mass filtering. Briefly, in the first stage, the selected precursor ion is resolved from coeluting components in Q1 based on its m/z, and accelerated into the collision cell, Q2, where it fragments by collision with a neutral inert gas (e.g., nitrogen or argon) in a process referred to as collision induced dissociation (CID). In the second stage, the analyte is further differentiated from interfering components in the third (Q3) quadrupole by monitoring unique fragment ion(s) (a.k.a., product or transition) of the precursor derived in Q2. This two-stage mass filtering of SRM or MRM increases the level of detection specificity, sensitivity, and throughput.

Selection of fragment ion(s) can be realized by careful tuning of the critical MS/MS parameters, such as collision energy, collision gas pressure, and cone voltage. Generally, to identify the precursor ion, a diluted solution of a pure compound can be directly introduced into the instrument (by flow injection analysis or split infusion) while the first quadrupole (Q1) is set to scan over a defined m/z range. The most abundant peak visible in the mass spectrum produced in this operating mode should represent the precursor ion (Bozovic and Kulasingam 2013). Precursor ions should be identified and the source parameters tuned to achieve the maximum peak intensity, without compromising signal-to-noise. Usually, once the precursor ion of the target analyte is identified, the mass spectrometer's ion optics and quadrupoles are tuned for the product ions. For selection of SRM transitions, Jemal et al. (2010) propose that at least two SRM transitions are utilized during method development as a coeluting metabolite or an endogenous compound may interfere with one or more of the selected SRM transitions.

LC coupled by an API source to MS/MS detection is currently considered the method of choice for quantitative analysis of small molecules in biological matrices. For more information on the factors to consider in development of LC-MS/MS bioanalytical methods, the reader can refer to excellent articles by Jemal and Xia (2006), Jemal et al. (2010), Mulvana (2010), and Li et al. (2011).

16.2.2 Ligand Binding Assays

Ligand binding assays (LBA) are immunoassays where an antigen–antibody reaction is used to capture the analyte of interest. Due to the advantages of LC-MS/MS methods to quantify small molecules, currently LBAs are not frequently used for low molecular weight compounds. However, LBAs are still the method of choice for quantitation of macromolecules and antibodies in complex biological matrices due to their high sensitivity and specificity. LBAs also play an important role in the detection and quantitation of biomarkers in clinical and nonclinical studies.

Immunoassays are broadly classified as homogeneous or heterogeneous assays (Findlay and Das 2006). In a homogenous assay all reagents are in solution, whereas in a heterogeneous assay at least one key reagent is immobilized and involves at least one washing step to remove excess analyte. Enzyme-linked immunosorbent assay (ELISA) is an example of heterogeneous assay. ELISA can be in a competitive or noncompetitive format. In a noncompetitive ELISA, the primary antibody to the analyte of interest is immobilized on microtiter or multi-well plate, and biological sample is introduced and incubated to facilitate binding of analyte to the immobilized antibody, and excess analyte is removed by washing. The immobilized antigen–antibody complex is then detected by directing an enzyme-labeled antibody specific to the analyte followed by addition of an enzyme-specific substrate probe. The resulting reaction is quantitated using an appropriate detection system depending on the type of the substrate probe. In the competitive ELISA, antigen is immobilized and competition is established between immobilized antigen and antigen in solution (i.e., analyte of interest) for fixed binding sites on the primary antibody in solution. After incubation and washing, an enzyme-labeled secondary antibody, directed against immunoglobins for the same species from which the primary antibody was created, is added. Following incubation and washing, an enzyme-specific substrate is added to generate a signal which is then quantitated.

The differences in regulatory requirements for chromatographic assays versus LBAs, and challenges involved in the conduct of the assays will be highlighted in subsequent sections.

16.3   Expectations for Validation of Bioanalytical Methods

According to the FDA BMV guidance (2001), validation involves documenting, through the use of specific laboratory investigations, that the performance character-istics of the method are suitable and reliable for the intended use. Method validation provides assurance that the bioanalytical method will perform reliably when used to analyze study samples. Therefore, during method validation, it is imperative that all the stress conditions and potential problems expected during analysis of the study samples are addressed to assure that the assay will perform as intended. This section describes the best practices for validation of bioanalytical methods.

16.3.1   Reference Standards

Reference standards are used to prepare stock solutions that are in turn used for the preparation of spiked samples (i.e., calibration standards and quality controls). Routinely, blank biological matrices are spiked with known concentrations of stock solutions to prepare calibrators and quality controls (QC). The calibrators and QCs are used to validate the performance of the method (see Sect. 16.3.4). Therefore, knowledge of the identity, purity, and stability of the reference standards is essential for reliable estimation of the analyte.

The FDA BMV guidance (2001) recommends that when possible reference standards are identical to the analyte of interest. Otherwise, an established chemical form (i.e., free acid/base, salt, or ester) of known purity can be used.

Reference standards can be broadly classified as (1) certified (e.g., U.S. Pharmacopeia (US)), (2) commercially available from a reputable source, and (3) in-house or custom-synthesized. Information for reference standards should include lot numbers, source, purity, storage, stability, handling, and expiration or recertification dates (Viswanathan et al. 2007). Usually, certificates of analysis (CoA) with the above information are available for reference standards. When CoAs are unavailable (e.g., rare metabolites) or reference standards are used beyond their expiration, FDA's recent draft guidance (FDA 2013) recommends that the purity and stability of the reference standards are demonstrated. CoAs or purity information is preferable for reference standards for internal standards (Sect. 16.3.2), however, lack of interference with the analyte of interest (Viswanathan et al. 2007), consistency between lots (e.g., when multiple lots are used) (DeSilva et al. 2012), or other suitability information may be demonstrated for internal standards. Also, sometimes the assays used by the vendors of reference standards may not be sensitive to assess purity (e.g., thin layer chromatography) and impurities (LC with ultraviolet detection). In such cases, purity determination using rigorous analytical methods may be necessary. Additional factors, including light sensitivity and moisture content may also need to be established for reference standards depending on the analyte.

Contrary to small molecules, macromolecules are usually not well characterized due to the nature of production. Macromolecular reference standards are often heterogeneous (Viswanathan et al. 2007), and therefore, lot-lot variability in purity and potency between preparations can be expected. It is therefore critical to use appropriate reference standards to validate an assay for macromolecules compared to the macromolecule used to dose the subjects.

In addition to reference standards, selection of reagents including ligand agents (e.g., antibody, antibody pairs), binding proteins, conjugated antibodies, and radioligands are critical in the development and validation of LBAs (Kelley and DeSilva 2007). Also, it is important that the reagents in LBAs have suitable specificity and selectivity for the intended use, and stable binding characteristics. Some reagents, including, conjugated antibodies and radioligands, have lot-to-lot variations. Therefore, for long-term studies, availability of a sufficient quantity of the reagents is necessary. Similar to the reference standards, reagents in LBAs are also macromolecules, hence assay sensitivity and robustness can be adversely affected due to instability. Therefore, appropriate storage and handling are paramount in maintaining the integrity of the reagents.

16.3.2 Internal Standards

To correct for analyte loss or variation during sample processing (e.g., extraction, evaporation, reconstitution), chromatographic separation, and instrumental performance (e.g., injection volume, ion suppression/enhancement), an internal standard (IS), which has the same or similar physical and chemical properties as the analyte, is added prior to sample processing to both spiked and study samples in equal concentrations. By using ratios of the response of analyte and IS in samples, variations in recovery and instrumental response can be corrected to improve the precision and accuracy of the methods. ISs are commonly used in chromatographic assays. ISs are less common in LBAs as sample clean-up is not as common as chromatographic methods.

Selection of IS is generally based on the following factors: (1) the physical and chemical properties (e.g., hydrophobicity, ionization properties) of the IS closely mimics the analyte during the analytical procedure, (2) purity of the IS is adequate, and (3) IS is stable during bioanalytical conduct.

There are two main types of IS: structural analogues and stable isotope labeled (SIL). The SIL ISs are compounds where atoms in the analyte are replaced with stable isotopes such as deuterium (^{2}H), ^{13}C, ^{15}N, or ^{17}O. For this reason, SIL ISs closely resemble the analyte to be measured and therefore are most effective to track variations in analyte response. SIL ISs are commonly used depending on availability and cost. Due to nearly the same physicochemical properties as the analyte of interest, SIL ISs, in theory, minimize the influence of matrix effects (Sect. 16.3.3) as the degree of ion suppression/enhancement caused by the coeluting matrix components must theoretically be the same for SIL ISs and its normal analyte counterpart (Viswanathan et al. 2007).

The selection of ISs depends on the extraction procedure, chromatographic separation, and analyte detection systems used. Also, the selection of ISs depends on which stages of analysis are critical for tracking the analyte. For example, if sample extracts are not clean, then tracking the analyte during MS detection is crucial to correct for matrix effects (Tan et al. 2012). Excellent articles by Tan et al. (2009, 2012) discuss the intricacies of IS selection.

Since ISs are used to correct for variations in analyte response, variations in IS response are expected. While excessive variations in IS response may affect quantitation, a high variation does not necessarily equate to unreliable data. Therefore, assessment of the impact of IS variations on quantitation is vital. There is no consensus on what constitutes an "excessive" IS response that affects quantitation. However, it is commonly accepted that monitoring IS response variations during sample analysis is a good practice. While the current FDA BMV guidance (2001) does not discuss IS variations, the recent FDA draft guidance (2013) recommends monitoring IS variations and establishing an objective, a priori criteria for abnormal IS variations. One of the common acceptance criteria for monitoring IS variations is setting a fixed percentage (e.g., ± 50 %) of mean IS response of spiked samples (i.e., calibrators and quality controls) within an analytical batch as an acceptable IS response range for the batch. Any sample with IS response outside the acceptable range in the batch will be flagged for reanalysis.

16.3.3 Matrix Effects

Although LC-MS/MS systems are generally considered to be very selective and sensitive, such methods do not automatically guarantee highly selectivity.

The transformation of uncharged molecules of the analyte to its charged components (i.e., ions) plays a key role in the detection of the analyte in LC-MS/MS systems. However, the efficiency of the formation of the desired ions is often perturbed by undetectable components in the incurred sample[2] matrix that coelute with the analyte(s) of interest. Hence, the efficiency of the formation of the analyte ions is matrix dependent. This phenomenon is referred to as "matrix effect" and results in reduction or enhancement of the ion intensity(ies) of the analyte(s) of interest, commonly referred to as "ion suppression" or "ion enhancement." Ion suppression or ion enhancement frequently is accompanied by a significant loss of precision and accuracy. Matuszewski et al. (2003) demonstrated that imprecision increased when the same method was validated with five different sources of plasma compared to a single source of plasma. Therefore, matrix effects may significantly affect assay performance. Appropriately, the FDA BMV guidance (2001) recommends that matrix effects are investigated and eliminated in LC-MS/MS methods. Excellent articles on matrix effects and its evaluation are available (Matuszewski et al. 2003; Van Eeckhaut et al. 2009; Trufelli et al. 2011). Estimation of matrix effect is discussed in Sect. 16.6.1.

Matrix effects can also arise in LBAs from interferences from unrelated compounds (from binding proteins, endogenous analogues, concomitant drugs, immunoglobulins) originating in the matrix (DeSilva et al. 2003; Kelley and DeSilva 2007). Therefore, validation of matrix effects in LBAs is extremely important when switching biological matrices.

16.3.4 Calibration Curve and Assay Performance

The minimum and the maximum known analyte concentrations used in an assay represent the lower limit of quantitation (LLOQ), and upper limit of quantitation (ULOQ), respectively, of the bioanalytical method. The LLOQ and ULOQ also describe the quantitation or calibration range of the bioanalytical method. In addition, the LLOQ describes the sensitivity of a bioanalytical assay (Sect. 16.3.4.2). Assessment of assay performance includes validation of the following components:

16.3.4.1 Calibration Curve

A calibration (or standard) curve describes the relationship between instrument response and known concentrations of the analyte. This relationship is essential to estimate the concentrations of the unknown samples. The FDA BMV guidance

---

[2] Samples collected from an animal or human dosed with drugs during drug development.

(2001) recommends that a calibration curve is prepared in the same biological matrix as the samples in the intended study (exception see Sect. 16.6.5) by spiking the matrix with known concentrations of the analyte. Also, it is recommended that the calibration range is based on the anticipated analyte concentration range in the BE study or studies (FDA 2001).

A calibration curve usually consists of a blank sample (i.e., matrix sample processed without analyte and internal standard), a zero standard (i.e., matrix sample processed without analyte but with internal standard), and at least six non-zero standards (i.e., matrix samples processed with analyte and internal standard) covering the expected range, including LLOQ and ULOQ (FDA 2001). The number of non-zero standards (or calibration standards) used is a function of the dynamic range and nature of the concentration-response relationship. A sufficient number of non-zero standards are often used to adequately define the relationship between concentration and response. The calibration standards can contain more than one analyte. Generally, it is good practice to use freshly prepared calibrators during validation to support that the method is sufficiently rugged.

Unlike chromatographic assays for small molecules, the standard curves for LBAs used to measure macromolecules are inherently nonlinear and therefore more non-zero standards may be recommended for LBAs. While the FDA BMV guidance (2001) recommends a minimum of six non-zero standards in duplicates, using additional calibrators is a good practice for LBAs. Kelley and DeSilva (2007) suggest including eight non-zero standards in duplicate. Also, due to the nonlinear response function, selection of non-zero standards to completely describe the calibration response becomes important for LBAs. In addition to non-zero standards, the FDA BMV guidance (2001) recommends anchoring points (above and below the established LLOQ and ULOQ: DeSilva et al. 2003) for LBAs to improve overall curve fit. While there is no consensus for acceptance criteria for anchor points, rejection of anchor points to force batch acceptance is discouraged (Savoie et al. 2010).

The FDA BMV guidance (2001) recommends that, except for the LLOQ, the back-calculated concentrations of the non-zero standards should be within 15 % of their nominal (theoretical) concentrations (20 % at LLOQ). For LBAs, the recent FDA draft guidance (2013) recommends that the back-calculated concentrations of the non-zero standards are within 20 % (25 % at LLOQ). Also, the recent FDA draft guidance (2013) recommends that at least 75 % of the non-zero standards are accurate, including the LLOQ, and the standards are excluded only for failure to meet the above acceptance criteria, or assignable causes (e.g., poor chromatogram, documented processing errors).

Usually, the standard curve fitting is determined by applying the simplest model that adequately describes the concentration-response relationship. The FDA BMV guidance (2001) recommends that selection of weighting and use of a complex regression model be justified. Also, it is important to assure that exclusion of an individual standard does not change the model used (FDA 2013). Since calibration response for LBAs shows nonlinear behavior, and their response-error relationship is not constant (i.e., highest precision does not always coincide with highest sensitivity), a weighted, nonlinear, least squares method with sufficient non-zero standards is recommended for LBAs (i.e., 4- or 5-parameter logistic model) (FDA 2001).

16.3.4.2 Assay Sensitivity

Assay sensitivity is often described by the LLOQ of the assay. It refers to the lowest concentration of the analyte that can be reliably quantitated by an analytical method, with acceptable accuracy and precision.[3] The FDA BMV guidance (2001) recommends that LLOQ is established using at least five QC samples at the LLOQ concentration in validation batches (see Sect. 16.3.4.4). The recent FDA draft guidance (2013) recommends that the accuracy does not deviate by more than ±20 % (±25 % for LBAs) of the theoretical concentration and the precision around the mean value does not exceed 20 % of the CV (25 % for LBA). The signal-to-noise ratio (S/N) at the LLOQ is recommended to be at least 5 (in other words the analyte response at the LLOQ is at least five times the response compared to blank response). Therefore, peak response in blanks or zero standards will be less than 20 % of LLOQ response. Peak response in blanks or zero standards greater than 20 % of LLOQ response is often referred to as interference and may affect accuracy and precision at the LLOQ. In addition, to control method error in LBAs, the consensus of the 2006 AAPS/FDA workshop was that total error[4] be less than ±40 % at the LLOQ (Viswanathan et al. 2007).

16.3.4.3 Selectivity

The terms "selectivity" and "specificity" are often mentioned in bioanalytical validation, sometimes interchangeably. Selectivity is a measure of extent while specificity is an absolute measure. In other words, specificity is the upper limit of selectivity, i.e., a method is specific when it is perfectly selective for an analyte or group of analytes (Rozet et al. 2011). For this reason, selectivity is used in this chapter.

According to the FDA BMV guidance (2001), selectivity is the ability of an analytical method to differentiate and quantify the analyte(s) of interest in the presence of interfering components in the matrix. Potential interfering

---

[3] The *accuracy* of an analytical method describes the closeness of mean test results obtained by the method to the actual concentration of the analyte. The *precision* (or imprecision) of an analytical method describes the random error of measurement, i.e., dispersion of the results around average value, often expressed as relative standard deviation (RS) or coefficient of variation (CV).

[4] Sum of absolute values of % accuracy and % precision.

substances in a biological matrix include endogenous matrix components, metabolites, decomposition products, concomitant medication, and other xenobiotics.

For selectivity, the FDA BMV guidance (2001) recommends analyses of blank samples of the appropriate biological matrix (plasma, urine, or other matrix) from at least six sources. It is recommended that each blank sample is tested for interference, and selectivity is assured at the lower limit of quantification (LLOQ). Routinely, interference is defined as peak response in blanks or zero standards equal to or greater than 20 % of LLOQ response.

The FDA BMV guidance (2001) recommends evaluating cross-reactivity of metabolites, concomitant medications, or endogenous compounds individually and in combination with the analyte of interest. This includes evaluation of expected concurrent medications that may potentially interfere with the analyte of interest. In certain situations, the European Medical Agency (EMA 2011) has proposed that the potential for interconversion of metabolite and parent drug during sample analysis be investigated, and its impact on quantitation determined (see Sect. 16.6.3).

Nonspecific binding should be determined for LBAs. Nonspecific binding can result from cross-reactivity with related compounds (e.g., metabolites, concomitant medications, or endogenous compounds), and interferences from matrix components. The guidance also recommends evaluation of parallelism for LBAs to detect matrix effect (FDA 2013). Parallelism shows that sample dilution response is parallel to standard concentration-response curve. It is important to note that parallelism is not the same as QC dilution linearity, as parallelism requires the use of incurred samples (DeSilva et al. 2012).

16.3.4.4 Precision and Accuracy

QCs at known concentrations are used to validate the precision and accuracy of a bioanalytical method. QCs are prepared by spiking known concentrations in the same blank biological matrix as intended for the study. It is a good practice to prepare QCs from an independent stock solution compared to the calibrations standards. When calibrators and QCs, are prepared from the same stock solution, it is a good practice to establish the accuracy of the stock solution against an independent stock solution.

In addition to LLOQ QC (see Sect. 16.3.4.2), QCs at a minimum of three concentrations, representing the entire range of the standard curve are recommended: one within $3 \times$ LLOQ (low QC sample), one near the center (middle QC), and one near the upper boundary of the standard curve (high QC) (FDA 2001). A minimum of five replicates per QC concentration is recommended (FDA 2001). It is recommended that the QC concentrations reflect the expected concentrations in the study (FDA 2013). A minimum of three to six validation batches are routinely used in method validation to assess assay precision and accuracy. Each validation batch usually consists of at least one set of calibration curve (i.e., blank, zero and non-zero standards) and a minimum of five QC replicates at each QC concentration.

Intra- and inter-batch precision and accuracy are determined based on the QC results. For acceptable performance, it is recommended that the assay accuracy be within 15 % of the nominal (theoretical) QC concentrations and the assay precision not exceed 15 % of the coefficient of variation (CV) at each QC concentration, with the exception of the LLOQ (for LLOQ criteria see Sect. 16.3.4.2) (FDA 2001). Due to greater variability for LBAs, the recent FDA draft guidance (2013) recommends acceptable accuracy and imprecision of ± 20 % (± 25 % at the LLOQ) for LBAs. In addition, to control method error in LBAs, recent FDA draft guidance (2013) recommends that the total error be less than ± 30 % for LBAs (± 40 % at the LLOQ; Viswanathan et al. 2007). It should be noted that precision and accuracy estimation requires inclusion of all QC data, including outlier data. Only data from QC samples with documented assignable causes (e.g., poor chromatogram, broken tube) can be excluded for precision and accuracy estimation.

In general, QC data from all precision and accuracy validation batches are necessary to provide a reliable estimation of precision and accuracy. Exclusion of batches not meeting QC acceptance may not be appropriate as it may bias precision and accuracy estimation (FDA 2013). Only validation batches with an assignable cause for failure are suitable for exclusion from precision and accuracy estimation (FDA 2013).

When multiple batches fail without an assignable cause, it is a good practice to investigate and resolve the reason for failure. In such situations, the nature of the batch failures (i.e., minor or major) should determine whether it is prudent to continue with method validation or return to method development.

In addition to precision and accuracy, recovery[5] of analyte(s) in a bioanalytical method needs to be validated. Recovery pertains to the extraction efficiency of an analytical method within the limits of variability. It is recommended that recovery experiments are performed by comparing the analytical results for extracted samples at three concentrations (low, medium, and high) with unextracted standards that represent 100 % recovery (FDA 2001). Recovery of the analyte need not be 100 %, but the extent of recovery of an analyte and of the internal standard must be consistent and reproducible (FDA 2013). Alternatively, to avoid matrix effect, recovery is also measured by comparing analyte extracted from matrix against analyte spiked to extracted blank matrix (Matuszewski et al. 2003).

16.3.4.5 Stability

The stability of the analyte must cover the expected storage and handling conditions of the samples during the study, including storage and handling conditions at the clinical site and during shipment. The storage and handling conditions include long-term (e.g., frozen) and short-term (e.g., bench-top, refrigerated) storage, and

---

[5] The extraction efficiency of an analytical process, reported as a percentage of the known amount of an analyte carried through the sample extraction and processing steps of the method.

freeze–thaw stability in the intended biological matrix, and stability of sample following extraction (e.g., extract stability). When the storage and handling conditions established in method validation are exceeded during the study, stability must be established under the actual study conditions. The consensus at the 2006 AAPS/FDA Workshop was that stability assessments are conducted in unaltered matrix intended for the study with same type of anticoagulant (Viswanathan et al. 2007). If a stabilizer is employed in the study, it is a good practice to evaluate stability with and without stabilizer in stability samples. It is recommended that all stability determinations be made using freshly prepared calibrators and/or QCs (FDA 2013). The FDA BMV guidance (2001) recommends stability evaluation at low and high QC concentrations, with at least three replicates at each QC concentrations. Stability acceptance criteria need to be established a priori, and are recommended to be within 15 % of nominal concentrations (FDA 2013).

Since it is difficult to predict the number times study samples will be frozen and thawed, it is recommended that freeze–thaw stability should be determined for at least three freeze–thaw cycles (FDA 2013). Also, freeze–thaw samples are recommended to be frozen at the intended frozen storage conditions of the study samples (e.g., −20, −70 °C), and completely thawed prior to freezing during freeze–thaw cycles (FDA 2013).

Long-term stability evaluations typically cover the expected time between the date of first sample collection and the date of last sample analysis (FDA 2013). If samples are stored at different temperatures during the course of the study, it is a good practice to assure stability at the different temperatures (Viswanathan et al. 2007). It is recommended that conditions used in long-term stability experiments reflect the same storage conditions intended for the study samples. For example, the long-term stability at higher temperature (e.g., −20 °C) may not be necessarily extrapolated to a lower temperature (e.g., −60 or −70 °C) at which study samples are stored (Andersson and Ehrsson 1995; Viswanathan et al. 2007). Although most compounds may show no difference in stability at different frozen storage temperatures, some may be more stable at a particular temperature.

While validation of stability using QC samples provides useful stability information, analysts need to be aware that this information sometimes may be limited as the complexities of incurred samples may not always be reflected in QCs (see Sect. 16.6.3).

Stability of analyte in stock solutions needs to be evaluated (FDA 2001). Typically, stock solutions of the analyte for stability evaluation are prepared in an appropriate solvent at known concentrations. When stock solution exists in a different buffer composition, the recent FDA draft guidance (2013) recommends that the stability of this stock for the duration of storage is demonstrated. The stock solutions for comparison against an older stock solution need to be prepared fresh from the reference standard.

For LBAs, assessments of analyte stability are recommended to be conducted in the matrix intended for the study (e.g., should not use a matrix stripped to remove endogenous interferences) (FDA 2001). Reagents including ligand agents (e.g., antibody, antibody pairs), binding proteins, conjugated antibodies and radioligands are critical in developments and validation of LBAs (Kelley and DeSilva 2007). Therefore, reagent stability is important for LBAs (Viswanathan et al. 2007). In addition to the reagents exhibiting specificity and selectivity, the stability binding characteristics are important (Kelley and DeSilva 2007). It is a good practice to store reagents under the designated conditions of the manufacturer, or at conditions for which stability data has been generated (Viswanathan et al. 2007).

16.3.4.6 Dilution

The FDA BMV guidance (2001) recommends that dilutions, if expected during the study, are validated by diluting QC samples with the same biological matrix as the study samples (FDA 2001). The dilution factor(s) intended for study sample analysis should be tested during validation. If dilution factors used during sample analysis are greater than those tested during validation, then validation of additional dilution factors may be necessary during sample analysis (Viswanathan et al. 2007). No within-study dilution QC samples are necessary if dilution is tested during validation and if the dilution of study samples is conducted with like matrix (human plasma for human plasma) (FDA 2001). The dilution integrity is demonstrated by accuracy and precision parameters during validation. While no specific criteria for dilution are recommended in the FDA BMV guidance (2001) or the recent FDA draft guidance (2013), the general consensus is that the dilution acceptance criteria do not exceed the assay accuracy criteria (see Sect. 16.3.4.4). The EMA (2011) has proposed that the accuracy and precision of the dilution QC samples be within ±15 %. However, one needs to be cautious that dilutions of QC samples may not always reflect dilution of incurred samples (DeSilva et al. 2012). Also, it is a good practice to dilute samples treated with enzyme inhibitors or stabilizers with enzyme inhibitor- or stabilizer-treated blank matrix.

16.3.4.7 Cross-Validation

Inter-bioanalytical method or inter-laboratory reliability needs to be established when two or more bioanalytical methods are used within the same study or across different studies, or when two or more laboratories are used for bioanalysis within a study. This is commonly referred to as cross-validation comparison. The FDA BMV guidance (2001) recommends conducting cross-validation with spiked matrix standards and subject samples at each site or laboratory when data within the same study are generated by two or more bioanalytical methods, or two or more laboratories. Cross-validation is also important when data are generated using different analytical techniques (e.g., LC-MS-MS versus ELISA) in different studies. While no specific criteria for cross-validation have been proposed in the FDA BMV guidance (2001) or the recent FDA draft guidance (2013), the EMA (2011) has proposed that the accuracy of QCs in two different methods is within 15 % (wider, if justified), and the difference in sample concentrations obtained from both methods is within 20 % of the mean value for at least 67 % of the repeats.

16.3.4.8 Partial Validation

Partial validation is recommended when changes are made to an already validated bioanalytical method (FDA 2001). Partial validation can range from one intra-assay accuracy and precision determination to a nearly full validation. The extent of partial validation depends on the type of modification to a bioanalytical method. The FDA BMV guidance (2001) provides examples of bioanalytical method changes that may require partial validation, including method transfers between laboratories or analysts, and changes in analytical methodology, anticoagulant in biological fluid, matrix within species or species within matrix, sample processing procedures, concentration range, instruments and/or software platforms, and sample volume. Also, partial validation may be necessary for demonstration of selectivity of an analyte in the presence of concomitant medications or specific metabolites.

16.3.4.9 Carry-Over

Carry-over can be related to autosampler or LC column. Carry-over can affect the reliability of quantitation, hence needs to be addressed during method validation (Viswanathan et al. 2007). Carry-over is commonly analyzed by injecting one or more blanks or zero standards immediately after a single or multiple injection of ULOQ calibrator or high QC samples (Viswanathan et al. 2007; Savoie et al. 2010). If carry-over exists, it is recommended that the source of carry-over is identified and eliminated. If carry-over is inevitable (e.g., highly retained compounds) or cannot be eliminated, it is a good practice to assess the extent of carry-over and its impact on quantitation, ascertain specific procedures to handle carry-over, and analyze study samples in their PK profile sequence without randomization (Viswanathan et al. 2007; Savoie et al. 2010). While the EMA (2011) has proposed carry-over criteria, at present there are no acceptance criteria for carry-over in the FDA BMV guidance (2001) and the recent FDA draft guidance (2013), and at the 2006 AAPS/FDA Workshop (Viswanathan et al. 2007).

16.3.4.10 Others

16.3.4.10.1 Multi-analyte

The recent FDA draft guidance (2013) recommends that samples involving multiple analytes should not be rejected based on the data from one analyte failing the acceptance criteria. The data from rejected batches need not be reported, but the FDA draft guidance (2013) recommends to document rejected batches and the reason(s) for failure. When samples are reassayed only for one analyte, the consensus is to collect and retain raw data collected for the other analytes (Viswanathan et al. 2007). Matuszewski et al. (2003) reported that matrix effect issues in LC-MS/MS methods simultaneously analyzing multiple analytes can be complex, consequently the absence of matrix effect for all individual analytes may need to be demonstrated.

16.4 Application of Validation Methods to Study Sample Analysis

16.4.1 Analytical Batch

Study samples are analyzed in analytical batches. Each analytical batch includes: (a) a calibration curve, consisting of blank sample, a zero standard, and at least six non-zero standards spanning the validated assay range, (b) at least duplicate QCs at three concentrations, and (c) study samples. The same regression model used in assay validation is employed for the calibration curve in all analytical batches. Also, similar to method validation, the three QC concentrations are selected based on the calibration range: one within $3\times$ the LLOQ (low QC sample), one near the center (middle QC), and one near the upper boundary of the standard curve (high QC) (see exceptions in Sect. 16.4.3). It is important that the QCs in the analytical batches represent the concentrations expected in the study.

The minimum number of QCs per batch recommended to ensure proper control of the analytical batch is at least 5 % of the number of study samples analyzed or a total of six QCs (i.e., duplicates at low, medium, and high QCs), whichever is greater (FDA 2001). In each analytical batch, it is imperative that calibrators and QCs are processed (preferably interspersed during processing) along with the subject samples under the same processing conditions (see Sect. 16.4.3 for special cases). It is recommended that all study samples from a subject be analyzed in the same batch when feasible. Study samples from multiple subjects may be analyzed in an analytical batch depending on the number of samples collected per subject, acquisition time, and a host of other factors. The storage of sample extracts prior to analysis need to be within the storage period validated for extract stability. Extrapolation of concentrations in study samples either below the LLOQ or above the ULOQ of the standard curve is not recommended (FDA 2001).

16.4.2   Acceptance Criteria

The criteria for batch acceptance should be established a priori and be objective. This includes criteria for acceptance of calibration standards, QCs, and interference. The FDA guidance (2001) recommends that 75 % of calibration standards in analytical batches are within 15 % of nominal value (20 % for LBAs), except at the LLOQ where the mean value is within 20 % of nominal value (25 % for LBAs). Only calibration standards outside the above-mentioned acceptance criteria or with documented assignable causes can be excluded. Based on extrapolation of the FDA BMV guidance's QC acceptance criteria, the recent FDA draft guidance (2013) recommends QC acceptance when at least 67 % of the total QCs and 50 % of the QCs at each level are within 15 % of their nominal concentrations in each analytical batch. The above QC acceptance criteria are independent of the number of QC levels and number of replicates at each QC level. Also, peak response in blanks or zero standards are recommended to be less than or equal to 20 % of LLOQ response to minimize interference.

Although the same QC acceptance criteria were recommended for LBAs in the FDA BMV guidance (2001), the LBAs are reported to have higher imprecision due to nature of the reagents and antibody–antigen reaction. Therefore, the FDA's recent draft guidance (2013) recommends that at least two-thirds of the total QCs and 50 % of the QCs at each level are within 20 % for LBAs, and any exception to this criteria is justified (Viswanathan et al. 2007; Kelley and DeSilva 2007; FDA 2013). This criteria has also been adopted by EMA (2011) for LBAs.

Typically, accuracy and precision of QC concentrations at each level from all successful analytical batches are evaluated to determine inter-batch accuracy and precision during the study. In-study assay performance is considered acceptable if the inter-batch accuracy and precision of QCs from successful runs are within 15 % (20 % for LBAs) of their nominal concentrations and 15 % CV (20 % CV for LBAs), respectively (FDA 2013). To understand the true assay performance, it is necessary to include inaccurate QC concentrations without any assignable causes in precision and accuracy estimation.

16.4.3   Analytical Conduct: Special Cases

Typically, the calibration range validated pre-study should be used in the analytical batches. However, in some situations, at the start of analysis, the study sample concentration range may be narrower than the expected concentration range. Consequently, the validated calibration range is too broad and QC concentrations may not be reflective of the study sample concentrations. In such instances, the recent FDA draft guidance (2013) recommends: (1) to narrow the calibration curve and modify QC concentrations, or (2) retain the original standard curve but include additional QC or new QC concentrations to reflect the study sample concentrations.

In either case, partial validation of the modifications is necessary. It is not necessary to reanalyze samples analyzed prior to modifying standard curve and/or QC concentrations as long as the partial validation is acceptable.

There may be situations when the bioanalytical method necessitates separation of an analytical batch into distinct processing batches (FDA 2013). Distinct processing batches include, but not limited to, extraction of finite samples due to limited capacity of SPE manifold, and processing of subject samples by multiple analysts due to large sample size. In such cases, the recent FDA draft guidance (2013) recommends that each distinct processing batch includes at least duplicates QCs at all QC levels (e.g., low, middle, and high) that are processed along with the study samples. Also, QC acceptance criteria are recommended for the analytical batch as a whole as well as the distinct processing batches (FDA 2013; also refer to Sect. 16.6.4).

For LBAs, replicate measurements during study sample analysis may not be necessary when replicate samples are used in validation and the method is demonstrated to be robust (DeSilva et al. 2012). Also, when using duplicate or triplicate determinations for samples in LBAs, exclusion of one or more of replicate determinations, if exercised, is based on pre-established, objective criteria.

Selection of samples for reanalysis and reporting of final values are recommended to be based on a priori, objective criteria (FDA 2013). It is a good practice to restrict sample reanalysis to samples with assignable causes that will invalidate the data (e.g., poor chromatogram, instrument failure, documented processing errors, samples below LLOQ or above ULOQ). Reanalysis of possible outliers (including PK, suspected, and confirmatory repeats) is discouraged, and when necessary needs to be justified with appropriate pre-established criteria.

It is not a good practice to re-inject failing analytical batches to bring them to acceptance. A high frequency of analytical batch failures needs to be investigated and resolved prior to continuing sample analysis. Also, following batch interruptions, the decision to continue analysis of the remaining samples or re-inject all the samples depends on the cause, duration, and resolution of the interruption. Generally, it is a good practice to have objective, pre-established criteria for analysis following batch interruption. Also, before re-injecting batches, it is important to establish re-injection reproducibility to determine whether an analytical batch can be reanalyzed.

Integration of chromatograms must be objective and consistent. When re-integration of chromatograms is normally discouraged, however, when performed the FDA's recent draft guidance (2013) recommends that the rationale for the re-integration is clearly described and documented, and audit trails maintained. It is recommended that objective procedures are established that specify the situations when re-integration is necessary and how it needs to be performed (FDA 2001). While modification of integration parameters may be necessary in some situations, it is a generally good practice to use the same integration parameters for all analytical batches on the same instrument for a given study provided the integration is valid and consistent.

The 2008 ISR Workshop (Fast et al. 2009) and the FDA's recent draft guidance (2013) recommends conduct of ISR for all BE studies (refer to Sect. 16.7 for details). This recommendation has also been adopted by the EMA (2011).

16.5   Documentation

The goal of documentation for regulated bioanalysis is to retrospectively construct events that transpired during method validation and study sample analysis. Therefore, contemporaneous recording of events is vital to good documentation. In addition to meeting the requirements of regulatory agencies, contemporaneous documentation is helpful to the firm to identify isolated problems or systemic issues retrospectively. The FDA BMV guidance (2001) recommends documenting summary (e.g., summary of methods, protocol, validation reports), method validation (e.g., complete method description, validation report of assay performance and stability, established procedures, and chromatograms), and study sample analysis (analytical report of in-study assay performance, reanalysis, deviations, unexpected events, chromatograms, and established procedures) information. A frame work for expected documentation at the analytical site for method validation and sample analysis, and essential information for validation and analytical reports, is provided in a tabulated format for easy reference in the 2006 AAPS/FDA Workshop whitepaper (Viswanathan et al. 2007). The paragraphs below highlight some important considerations for bioanalytical documentation.

The FDA BMV guidance (2001) recommends that analytical laboratory have established standard operating procedures (SOPs) that cover all aspects of analysis, from the time the samples reach the laboratory until the results of the analysis are reported. This includes SOPs for record keeping, security and chain of sample custody, sample preparation, and analytical tools such as methods, reagents, equipment, instrumentation, and procedures for verification of results. All study related communication within the analytical facility and between analytical facility and the sponsor or the clinical sites are part of study records, therefore are recommended to be retained (FDA 2013).

Records of contemporaneous entry of events constitute source records, and such records are recommended to be retained (FDA 2013). Source records for bioanalysis include, but are not limited to, laboratory notebooks, analysts' notes, receipt and storage of reference standards and samples, freezer log books, sample processing entries, instrument usage log and maintenance records, batch summary sheets, chromatograms, and audit trails. The recent FDA draft guidance (2013) recommends that sufficient information must be included in source records to re-construct the events described in the records. Acceptable data entry procedures include identifying the analyst recording the events, and the dating the entries. Investigation and resolution of all unexpected events are recommended to be documented (FDA 2013).

The recent FDA draft guidance (2013) recommends that records of all validation and analytical batches analyzed, including unsuccessful batches, batch summary sheets (should include sample IDs, analyte and IS response, response ratio, back-calculated concentrations, and record modification), chromatograms, and audit trails, are retained. When samples are reassayed for one analyte in a multi-analyte method, it is a good practice to retain the raw data collected for the other analyte(s) (data need not be processed). Also, it is good practice to retain records in the format it was acquired (i.e., electronic, paper).

Re-integration of chromatograms must be explicitly identified. The FDA's recent draft guidance (2013) recommends that the reason for re-integration and mode of integration are clearly documented, and re-integration is based on pre-established criteria. Also, the consensus from the 2006 AAPS/FDA Workshop was that original and re-integrated chromatograms, and audit trail of events during data processing are retained in the format it was acquired, and audit trail feature is enabled in the laboratory information management systems (LIMS) (Viswanathan et al. 2007).

The FDA's recent draft guidance (2013) recommends that the reason for rejecting batches should be clearly documented with supporting evidence. Also, reanalysis of analytical batches or samples is expected to be clearly identified, and based on pre-established procedures.

16.5.1 Validation and Analytical Reports

An outline of the necessary information in validation and analytical reports are provided below. Validation and analytical reports routinely include a brief description of the protocol and analytical method (analyte, IS, sample pretreatment, method of extraction and analysis) used, and identify the method SOP (and version). The FDA BMV guidance (2001) recommends that the reports indicate the lot number, purity, source, and expiration or retest dates of reference standards for drug and/or metabolites, and internal standards. Also, the guidance recommends the reports describe the procedures for preparation and storage of stock solutions, QCs, and calibrators, including preparation dates, and source and lot of blank matrix and reference standards used.

Summary tables listing all validation or analytical batches (successful and unsuccessful), dates of analysis, and reason for rejection are recommended for validation and analytical reports (FDA 2013). While unsuccessful batches need to be identified, reporting summary data for the batches is not required. Tabulation of the back-calculated concentrations of calibrators and QCs with inter-batch precision and accuracy information is recommended. In addition, validation reports are expected to include intra-batch accuracy and precision, necessary stability, extraction recovery, selectivity, and matrix effect information (FDA 2013).

Analytical reports are expected to include dates of study sample receipt, shipment temperature, sample integrity at the time of receipt, sample accountability, and storage location and temperature at the analytical site (v). Also, analytical reports are supposed to clearly identify the samples reanalyzed, the reason for reanalysis, and reporting of final values. All deviations from the protocol or procedures, and its impact on the study need to be detailed. The FDA's recent draft guidance (2013) recommends that ISR results, including samples reanalyzed, original and reanalyzed sample concentrations and their % difference, and the acceptability of ISR data, are included in the report. The assay procedure, protocol, and SOPs for re-integration, reanalysis, and acceptance criteria should be attached to the report. For pivotal BE studies for marketing, chromatograms from 20 % of subjects are recommended to be included in the report. Addendum to validation (e.g., partial validation, long-term stability) and analytical reports (e.g., investigations) if any, needs to be attached.

16.6 Challenges

One of the main issues in assuring adequate performance of the assays during study sample analysis stems from the challenges imposed by the matrix complexities of incurred samples. Although, for the most part, the contents of the matrix used to prepare QCs are the same as incurred samples, it is important to note that the matrix for QCs may not behave the same as incurred samples for several reasons (see Table 16.1). For example, QCs may not contain the same drug metabolites as

Table 16.1 Matrix differences between spiked (CS and QC) and incurred samples. (Reproduced from Tan et al. 2009)

	CS/QC	Incurred sample
Screening criteria for matrix sources	Usually loose	Usually specific and strict dependent on the objectives of a study, such as age 40–50 and nonsmoker
No. of lots/sources	Usually more than one source (pooled)	One single source
pH	Averaged due to pooling	More variable
Extra components associated with medication	None	Metabolite(s), co-medication and non-active ingredients in formulation
Amount collected	Usually large, e.g., 200 mL per collection	Usually small, e.g., 7 mL per sampling time
No. of freeze/thaw cycles prior to being extracted	Usually 2 or more	Usually 1
Storage tube and pre-use storage	Usually stored at $-20\,^{\circ}$ C and without special protection until being selected for a specific study	Could be collected under sodium light and stored at $-80\,^{\circ}$ C immediately after collection
Amount of anticoagulant	May be different because of different amounts collected	

CS calibration standard, QC quality control

incurred samples, which could be important as plasma metabolite concentrations generally are an order of magnitude higher compared to their parent drug. Also, compared to incurred samples, matrix used to prepare QCs may not contain drug isomers, have the same enzyme activity, or contain the same co-administered drugs or coeluting components. In addition, other factors may also affect assay performance during study sample analysis. This section highlights some of the factors that may affect quantitation during bioanalysis.

16.6.1 Matrix Effects

While the current FDA BMV guidance (2001) recommends that matrix effects be investigated and eliminated in LC-MS/MS methods, it does not specify procedure (s) to detect matrix effects. The consensus at the 2006 AAPS/FDA Workshop was that matrix factor (MF) can be used as a quantitative measure to ascertain matrix effect (Viswanathan et al. 2007). MF can be defined as ratio of analyte response in the presence of matrix ions to analyte response in the absence of matrix ions. In the absence of matrix effect, MF should be 1, while values below or above 1 may indicate ion enhancement or suppression. While absolute MF value is useful, it does not provide information of the variability in response in different incurred sample matrices (Viswanathan et al. 2007). Therefore, it has been proposed that the variability of MF be determined in six different matrix lots with an acceptable variability (as measured by the coefficient of variation) of <15 % (Viswanathan et al. 2007). Variations of MF include IS-normalized MF (i.e., ratio of MF of analyte to MF of IS, or analyte to IS ratio in the matrix extracts divided by analyte to IS ratio in the absence of matrix extract) (Viswanathan et al. 2007). The EMA guidance (2011) recommends the variability of the IS-normalized MF from six lots of matrix should not be greater than 15 % at low and high QC concentrations. Although matrix effects may extend to LC methods coupled to other detection systems (UV, fluorescence, electrochemical), matrix effects are usually linked to LC-MS/MS methods with simplified extraction procedures and minimal chromatographic separation, as such methods are popular for their high throughput. Also, the contribution of matrix effect for LC-MS methods vary depending on the ionization source (e.g., APCI versus ESI) of MS systems (Matuszewski et al. 2003). For bioanalytical methods that simultaneously analyze multiple analytes, it may be necessary to demonstrate lack of matrix effect for all individual analytes (Matuszewski et al. 2003). Table 16.2 provides a summary of various measures to eliminate or minimize matrix at different stages of bioanalytical methods (Nováková 2013). Recently, plasma phospholipids have been associated with matrix effects, which can be avoided by removing phospholipids during extraction and resolving the analyte from phospholipids during chromatography (Jemal et al. 2010). Excellent discussion of matrix effects and the various measures to eliminate or reduce matrix effects can be found in current literature (Matuszewski

Table 16.2 Approaches to minimize matrix effects (ME) at different stages of bioanalytical methods. (Reproduced from Nováková 2013)

A step of bioanalytical method	ME reduction approach	Examples of realization
Sample preparation	More extensive clean-up	SPE-based approaches with extensive and well optimized washing steps, RAM
		LLE-based approaches—ionized species do not partition into the organic layer
	Higher selectivity	SPE, MIP, immunoaffinity SPE
	Protein precipitation prior to SPE/LLE	
	Dilution of sample	
Chromatography	Higher separation efficiency	Fast/high resolution LC approaches, 2D-LC
	Nano-LC	Nano flow-rates, smaller droplets formed
	Change in selectivity	HILIC or other orthogonal chromatographic mode, change in mobile or stationary phase
	Gradient elution	Change in selectivity, enhancement of efficiency and also elution of highly retained interfering compounds
Mass spectrometry	Higher selectivity	Negative ion mode
	Ionization technique less susceptible to ME	APPI, APCI, EI-MS
Calibration data processing and other strategies	Appropriate calibration approach	Internal standard method, standard addition method, matrix-matched calibration
	Use of SIL-IS	13 C SIL-IS should be preferred over deuterium labeled compounds
	Echo peak strategy [122]	Elution very close to t_R of analyzed compounds ~ the same ME

LC liquid chromatography, *SPE* solid phase extraction, *LLE* liquid–liquid extraction, *RAM* restricted access materials, *MIP* molecularly imprinted polymers, *HILIC* hydrophilic interaction liquid chromatography, *APPI* atmospheric pressure photoionization, *APCI* atmospheric pressure chemical ionization, *EI-MS* electron ionization mass spectrometry, t_R retention time

et al. 2003; Jemal and Xia 2006; Van Eeckhaut et al. 2009; Jemal et al. 2010; Mulvana 2010; Li et al. 2011; Trufelli et al. 2011; Nováková 2013).

Nonspecific binding determination is important for LBAs. The guidance also recommends evaluation of parallelism for LBAs to detect matrix effect. Parallelism shows that sample dilution response is parallel to standard concentration-response curve. It is important to note that parallelism is not the same as QC dilution linearity as parallelism requires the use of incurred samples (DeSilva et al. 2012). Kelley and Desilva (2007) proposed testing for matrix effects in LBAs by comparing the concentration-response relationship of both spiked and unspiked samples of at least ten lots of the biological matrix to a comparable buffer solution.

16.6.2 Internal Standard

In addition to physicochemical factors, the concentration of IS is important (Tan et al. 2012; Mulvana 2010). Selection of optimum IS concentrations assures that the signal-to-noise ratio is adequate to obtain good sensitivity and precision, and minimizes or eliminates potential interference from unlabeled impurities in the reference standards of the IS or the analyte of interest.

While the use of IS acceptance criteria based on IS response range of spiked samples is a good practice for reanalyzing samples with abnormal IS response, it has to be used with caution in certain situations. For example, when IS variations in unknown samples and spiked samples are similar, IS variations do not affect the accuracy of the calibrators and QCs. In such situations, the need for reanalysis for IS variations may be moot. Also, in cases where IS variations in study samples are abnormally different from those in spiked samples, the IS acceptance criteria based on spiked samples may not be meaningful. In such cases, investigation should be conducted to confirm whether IS compensates for matrix effects (Tan et al. 2009; Savoie et al. 2010).

Abnormal variations in IS may occur for a number of reasons, including human errors (spiking twice or not spiking IS), imprecision of pipettes used to spike samples with IS (repeater pipettes), partial or complete blockage of autosampler needle (Table 16.3). Trends or patterns in variations in IS response may need to be investigated. Trends or patterns in IS variation include, but not limited to, contamination of the orifice or rods of MS due to incomplete or inadequate sample clean-up, matrix effects due to coeluting components, improper IS selection, incomplete solubility of IS in stock solution or extraction solvent, or inadequate mixing of IS (Tan et al. 2009). Therefore, it is a good practice to evaluate IS variations across an analytical batch, and investigate any abnormal patterns IS response in terms of its impact on the quantitation of unknown samples.

Although SIL ISs are preferred to develop a robust and accurate assay, the use of SIL ISs does not automatically guarantee accurate quantitation (Mulvana 2010). For example, it was shown that a deuterated IS may have a slightly different retention time compared to its normal counterpart, and thus may result in different degrees of matrix effects between the two analogues (a.k.a., deuterium isotope effect) (Wang et al. 2007). This may significantly affect analyte to IS ratio consistency. Also, the presence of normal analyte in SIL IS and its impact need to be assessed (EMA 2011).

16.6.3 Stability Issues

While assessment of freeze–thaw, short-term and long-term stabilities using QCs is useful to understand the stability of analyte in biological matrix, one should be aware that this information may be limited as QCs may not always mimic incurred

Table 16.3 Examples of abnormal internal standard (IS) response, reason for the response, and their impact on quantitation. (Reproduced from Tan et al. 2009)

Case	Observations	Root cause identified	Effect on quantitation or comments
1	Zero or nearly doubled IS response	Missed or double addition of IS	Yes
2	Random and sharp drop in IS response	Autosampler needle blockage	Usually no, unless S/N is too low
3	Gradual decrease of IS responses	Charging of mass spectrometer	Not in this case, but it usually depends on how well an IS follows an analyte
4	Random, sharp drop, and over-all downward trend in IS response	Autosampler needle blockage plus charging of mass spectrometer	It depends, but batch should be reinjected
5	Low IS responses for most of the extracted samples	Mixed usage of right and wrong caps in LLE	It depends, but samples should be reassayed by using correct materials
6	High IS responses observed for incurred samples only (usu-ally a whole subject)	Relatively less ion suppres-sion in subject samples than in CS/QC	It depends on how well an IS follows an analyte
7	High IS responses observed for incurred samples only (usu-ally a whole subject)	Recovery variation plus rela-tively less ion suppression in subject samples than in CS/QC	It depends on how well an IS follows an analyte
8	Low IS responses for incurred samples only (usually a whole subject)	Transfer of salt-containing intermediate layer in LLE	It depends, but samples should be reassayed
9	Less IS response variation with analogue IS than with deu-terated IS	Analogue IS did not follow analyte well	Quantitation affected with analogue IS and it should be changed
10	Gradual increase of IS responses	Insufficient mixing	Not in this case, but should be evaluated case by case
11	Randomly scattered low IS responses for incurred sam-ples only and not repeated during reanalysis	Not conclusive, but specu-lated as due to ascorbic acid and different cycles of F/T	Not in this case, but should be evaluated case by case
12	Deuterated IS not following the analyte and re-injection results not matching those of 1st injection	Not conclusive, but specu-lated as due to differential matrix effect between analyte and its deuterated IS	Yes in this case, but should be evaluated case by case

IS internal standard, *CS* calibration standard; *QC* quality control; *LLE* liquid–liquid extraction, *F/T* freeze and thaw

samples (Table 16.1). Therefore, during development of bioanalytical assays, a good understanding of the differences between QCs and incurred samples, the bioanalytical methods under consideration, and the physio-chemical and

Table 16.4 Examples of sources of instability and approaches to overcome instability

Causes of instability	Strategies to avoid instability	Examples of affected analytes
Enzymatic hydrolysis	Addition of enzyme inhibitors and/or freezing samples immediately after collection, or harvesting plasma at reduced temperature followed by immediate frozen storage	Olmesartan medoxomil, Capecitabine
Hemolysis	Depending on the analyte, testing the impact of different degrees of hydrolysis during method development. Factoring sample hemolysis during stability evaluations	Nitroglycerin, Fluvoxamine
Temperature	Lowering temperature during sample collection, processing, storage, extraction, reconstitution and analysis	Aspirin, Cisplatin, Acyl glucuronides
pH	Controlling pH within the desired range during sample collection, processing, storage, extraction, reconstitution and analysis	Cisplatin, Acyl glucuronides
Light	For photo-sensitive compounds, protection from light during sample handling is necessary, e.g., wrapping tubes in foil, using amber glass vials, or sample processing under yellow light or UV-filtered light	Nifedipine, Nisoldipine
Autooxidation	Addition of antioxidants to samples, e.g., ascorbic acid, sodium metabisulfite and ethylenediaminetetraacetic acid (EDTA)	Rifampin, Levodopa
Lactone/hydroxy acid interconversion	Decreasing pH and sample processing temperature or time	Atorvastatin, Simvastatin, Pravastatin
Adsorption to container walls	Using appropriate containers for sample collection, extraction, storage and analysis, e.g., silanized glass tubes. Addition of surfactants	Sufentanil, Tetrahydrocannabinol
In-source fragmentation/ transformation	Selecting suitable analyte-specific MS tuning of ionization conditions, assuring adequate chromatographic separation	Clozapine, Carboxylic acid metabolite

For references of the cited examples, refer to the appropriate sub-sections in Sect. 16.6.3

pharmacokinetics properties of the analyte(s) interest is essential (Jemal and Xia 2006; Mulvana 2010; Jemal et al. 2010; Li et al. 2011). Stability of the analyte of interest may be affected at different stages, from sample collection to sample analysis. Instability of analyte during bioanalysis can arise due to chemical or biological process following sample collection. These factors include photosensitivity, temperature, chemical reactivity, enzymatic degradation, hydrolysis of conjugated metabolites, interconversion under certain conditions (pH, enzymes, temperature), autooxidation, and transformation at the MS source (Table 16.4) (Jemal and Xia 2006; Briscoe and Hage 2009; Jemal et al. 2010; Silvestro et al. 2010; Yadav and Shrivastav 2011; Li et al. 2011). Several articles are available that discuss various factors that may impact instability (Chen and

Hsieh 2005; Jemal and Xia 2006; Briscoe and Hage 2009; Jemal et al. 2010; Mulvana 2010; Silvestro et al. 2010; Yadav and Shrivastav 2011; Li et al. 2011). A brief discussion of some of the factors is provided below.

16.6.3.1 Hydrolysis, pH, Temperature, Interconversion

The instability of analytes in biological fluids can be caused by enzymes. Esterases are the most prominent among hydrolases in plasma. Esterases catalyze the hydrolysis of esters and amide to their corresponding carboxylic groups (Chen and Hsieh 2005; Izhizuka et al. 2010). In addition to playing a vital role in the conversion of pro-drug to active drug, esterases can hydrolyze pro-drug or drug during sample collection, handling, and storage (Li et al. 2011). For compounds that are unstable in a biological matrix, taking adequate precautions during sample collection and/or handling are necessary to avoid instability of the analyte. These may include immediate freezing following sample collection, reduction in temperature during sample processing followed by immediate frozen storage at a very low temperature, thawing samples on wet ice, stabilizing samples by addition of enzyme inhibitors, or special treatment of samples such as acidification or protein precipitation (Guan et al. 2003; Besnard et al. 2008; Briscoe and Hage 2009; Mulvana 2010; Li et al. 2011). When an analyte of interest is unstable during blood sample collection, it is a good practice to confirm whole blood stability under the sample collection conditions. Also, when samples treated with enzyme inhibitors need to be diluted, the use of enzyme inhibitor-treated blank matrix is recommended. Also, one needs to be aware that stabilizers and inhibitors can cause interference or affect sample integrity (Mulvana 2010). Some enzyme inhibitors can be more prone to hydrolysis than the analyte of interest. In such cases, adding a relatively large amount of an analogue that is more sensitive to enzymatic degradation will prevent degradation of the analyte of interest (Li et al. 2011).

In general, lowering of temperature can substantially reduce degradation in biological matrix or solution (Chen and Hsieh 2005). For example, cisplatin is unstable at −25 °C, but stable at −70 °C (Andersson and Ehrsson 1995). The hydrolysis of acetylsalicylic acid (aspirin) can be controlled by thawing samples on ice, followed by extraction and analysis within 2 h after thawing. Storage at −20 °C for 11 days resulted in 20 % degradation of aspirin (Briscoe and Hage 2009).

The control of pH is also important for the analysis of most unstable analytes during sample collection, processing, storage, extraction, reconstitution and analysis, as pH within a narrow window is essential for most acid/base-catalyzed enzymatic and nonenzymatic reactions. pH can increase to 8.8 for unprocessed and untreated plasma samples stored at room temperature or at 37 °C, and to 9.5 during sample preparation (Fura et al. 2003). Therefore, maintaining pH in biological matrices and during sample processing is essential to prevent degradation of pH-sensitive compounds. Analytical methods for pH-sensitive compounds may need to include procedures to stabilize the pH of biological matrices and sample extracts at the desired pH range. Cisplatin is highly susceptible to pH changes. Andersson and Ehrsson (1995) showed that cisplatin rapidly degrades at pH 7.4 in plasma, blood and ultrafiltrate, but is stable at pH 5.5 in plasma.

Ether- or ester-glucuronides are formed by glucuronidation via oxygen, while N-glucuronide or N^+-glucuronide arises due to glucuronidation via a primary, secondary, or tertiary amine. While there is no general rule to predict the instability of ether- or ester-glucuronides, ester-glucuronides (acyl glucuronides) tend to be less stable than ether-glucuronides (Li et al. 2011).

Acyl glucuronides tend to be unstable, especially under alkaline conditions (~pH 7.4) and elevated temperature, resulting in back-conversion to parent form (Jemal and Xia 2006; Jemal et al. 2010). Therefore, for acyl glucuronides forming compounds including, telmisartan, clopidogrelat (metabolite of clopidogrel), and ibuprofen, control of pH is important. Hydrolysis of acyl glucuronides can be minimized under mildly acidic conditions (pH 3–5). Although this is true for most acyl glucuronides, there are exceptions (Li et al. 2011). Therefore, necessary evaluation to understand pH-dependent stability of acyl glucuronides should be conducted during method development.

For compounds like clopidogrel and enalapril with methyl and ethyl ester groups, respectively, use of methanol and ethanol are not preferable during sample extraction as acyl glucuronides can react with methanol or ethanol to back convert to the parent form under basic conditions (Jemal and Xia 2006; Briscoe and Hage 2009; Jemal et al. 2010; Li et al. 2011). This may lead to overestimation of the parent drug. In some situations, underestimation of the parent drug is also possible. For example, drugs containing ethyl ester group, like enalapril, can react with methanol to produce methyl ester analogue of the drug (Jemal and Xia 2006; Jemal et al. 2010).

For N-glucuronides, back-conversion of the parent drug under acidic/basic and/or physiological pH condition or at an elevated sample processing temperature is largely compound dependent.

The ex vivo interconversion between the lactone metabolite and hydroxyacid drug has been observed for statins, like atorvastatin, simvastatin, and pravastatin, with the two forms exhibiting different pharmacological activities. Increasing pH and sample processing temperature or time promotes the ex vivo conversion of the lactone metabolite to the drug. It was demonstrated for statins that interconversion can be minimized by lowering sample preparation temperature (e.g., storage on ice-bath) and pH (e.g., pH ~4.5) (Kearney et al. 1993; Jemal and Xia 2000; Jemal and Xia 2006; Jemal et al. 2010; Zhang et al. 2010). Also, under the right conditions, including physiological or extreme pH and temperature, all chiral compounds can undergo interconversion. The S and R isomers of thalidomide have different pharmacological activities and PK properties.

16.6.3.2 Chemical Instability, Photolability, Autooxidation

Many *N*-oxides are thermally labile, photo-sensitive, and unstable in solutions and/or biological matrices during sample extraction, especially under strong acidic or basic conditions (Li et al. 2011).

The photochemical sensitive moieties include carbonyl, nitroaromatic structures (as electrophilic radicals), *N*-oxide function, carbon–carbon double bonds (liable to *E* to *Z* isomerization and autooxidation), and aryl chloride groups (liable to hemolytic and/or heterolytic dechlorination) (Jemal and Xia 2006; Briscoe and Hage 2009; Jemal et al. 2010). For example, nisoldipine and its metabolite are extremely photolabile both in organic solvent and in plasma, with degradation half lives of 6.3–6.7 min in dichoromethane-pentane and 10.7–11.3 min in plasma (Van Harten et al. 1987). This photodegradation was prevented by handling samples under sodium light. Also, in the presence of light, nifedipine degrades by 15 % in whole blood after 1 h, but only by 5 % in plasma after 2 h (Abou-Auda et al. 2000). Light protection is necessary when handling photo-sensitive compounds, e.g., tubes wrapped in foil, use of amber glass vials, or sample processing under yellow light or UV-filtered light.

Many small molecules, especially those containing phenol (e.g., catechol) or alcohol groups, can be readily oxidized in biological samples or reconstituted sample extracts (Saxer et al. 2004). A simple addition of antioxidants has been found to be very effective for stabilizing those analytes. Addition of ascorbic acid (vitamin C), a potent antioxidant, can be used to prevent degradation of rifampin-spiked plasma samples. It is recommended that the sample be supplemented with ascorbic acid at the time of blood sampling to stabilize rifampin (Le Guellec et al. 1997). Autooxidation of levodopa and 3-methyldopa in human plasma samples was prevented by the addition of sodium metabisulfite and ethylenediaminetetraacetic acid (EDTA) (Saxer et al. 2004). In the presence of such additives, levodopa and 3-methyldopa samples were stable for 16 weeks at $-70\,^{\circ}\mathrm{C}$. It is has been reported that a combination of an antioxidants, such as EDTA, fluoride oxalate, sodium citrate, heparin, may work better than a single agent for stabilizing labile compounds in a biological matrix or biological sample extracts (Li et al. 2011).

16.6.3.3 In-Source Fragmentation and/or Transformation

In-source fragmentation refers to fragmentation of molecules during the ionization process prior to entry into the Q1 chamber of the MS/MS. This is frequently observed for *N*-oxides, *S*-oxides, and glucuronide- or sulfate-conjugated metabolites of the analytes of interest that are present in sample extract. In-source fragmentation of such molecules spontaneously produces ions identical to precursor ions of the analytes of interest (Tan et al. 2012). Also, certain analytes can transform to others in MS source via in-source transformation (Jemal and

Xia 2006). For example, lactonization of a carboxylic acid metabolite can generate the same precursor ion as the original lactone. Without proper chromatographic separation, the lactone may be over-estimated.

To maximize the ionization of the analyte of interest and minimize the in-source fragmentation, suitable analyte-specific MS tuning of ionization conditions is needed (Kruger et al. 2010). For example, with the APCI interface, clozapine-N-oxide produces two major "fragment" ions at the same m/z value as clozapine and its N-demethylation metabolite. However, these ions were not found with the ESI interface (Niederlander et al. 2006).

16.6.3.4 Anticoagulant

Anticoagulant and/or anticoagulant counter ions can have impact on compound stability (Li et al. 2011). Heparin and EDTA are two commonly used anticoagulants: heparin inactivates thrombin while EDTA chelates calcium ions and interrupts the clotting cascade at multiple points. EDTA can prevent the activity of calcium dependent phospholipases and ester hydrolases, while heparin may not be inhibitory. In general, EDTA is preferred to heparin as an anticoagulant in plasma samples (Sadagopan et al. 2003). Matrix-related irreproducibility appeared to be more pronounced with heparin as the anticoagulant than with EDTA (Smeraglia et al. 2002; Yue et al. 2008).

16.6.3.5 Nonspecific binding

Nonspecific binding or container surface adsorption of drug molecules in biological samples can occur. For example, tetrahydrocannabinol blood concentration level in the glass containers was reported unchanged after 4 weeks of storage at $-20\,°C$ but not in the plastic polystyrene container (Christophersen 1986). Also, sufentanil concentrations in plasma decreased in nonsilanized glass tubes but were stable in silanized glass tubes (Dufresne et al. 2001). Addition of Tween-80 or CHAPS to the matrix may prevent nonspecific binding or container surface adsorption (Li et al. 2010).

16.6.3.6 Hemolysis

Hemolysis (i.e., lysis of red blood cells) results in release of their contents (e.g., enzymes, hemoglobin, inorganic ions). There are instances where hemolysis may affect the stability of the drug (DeSilva et al. 2012; Bérubé et al. 2011). Therefore, accuracy of PK data can be compromised when several samples in BE studies are hemolyzed. Therefore, the impact of hemolysis should be investigated depending on the analyte of interest and the method. Also, study samples should be carefully monitored for hemolysis. The FDA BMV guidance (2001) or the FDA's recent draft guidance (2013) does not discuss hemolysis, and there is no established standard procedure to treat hemolyzed clinical samples. Also, the standard evaluation of degree of hemolysis in samples (i.e., visual) is a subjective determination (Bérubé et al. 2011; Garofolo et al. 2011). Therefore, for analyte or methods susceptible to hemolysis, the significance of hemolysis becomes a function of the extent of hemolysis (i.e., percentage of hemolyzed samples) or which samples are hemolyzed (i.e., around Cmax range).

Therefore, as described above, stability of analytes is of concern. Consequently, depending on the analyte of interest, it is recommended that the clinical and analytical sites coordinate precautionary measures (e.g., need for stabilizers, control temperature, protect samples from light) during sample collection, and post collection, processing, storage, and shipment conditions.

16.6.4   LBA Issues

In addition to reference standards, selection of reagents including ligand agents, binding proteins, conjugated antibodies, and radioligands are critical in the development and validation of LBAs (Kelley and DeSilva 2007). The reagents should allow for suitable specificity and selectivity, and stable binding characteristics. Similar to the reference standards, reagents in LBAs are also macromolecules, hence assay sensitivity and robustness can be adversely affected due to instability. Therefore, appropriate storage and handling are paramount in maintaining the integrity of the reagents.

LBAs generally use well plates (e.g., 96-well plates) for analysis of study samples. Each analytical batch may include several well plates. In such cases, the FDA's recent draft guidance (2013) recommends that sufficient replicate QCs are used in each plate to monitor accuracy, and acceptance criteria for the batch as well for individual plates are established. Some reagents including, conjugated antibodies and radioligands, and multi-well plates have lot-lot variations. Therefore, for long-term studies or studies with large sample size, sufficient quantity for a given lot is necessary. Also, if multiple lots of reagents or well plates are used, assessment of lot-to-lot variability and comparability may be necessary.

16.6.5   Endogenous Assays

Assays for macromolecules, commonly LBAs, are often used for quantification of macromolecule therapeutics that are recombinant or modified variants of endogenous proteins or peptides. Therefore, use of routine blank matrices will not guarantee accuracy of measurement of the therapeutic macromolecule, due to the presence of its endogenous counterpart. Therefore, the FDA's recent draft guidance (2013) recommends special considerations are made for matrix selection and conduct for such assays. One option is to use stripped matrix (e.g., charcoal, immunoaffinity) or alternate matrices (e.g., protein buffers) for calibrators, and unaltered matrix intended for the study for QCs. When using altered or alternate matrices, confirmation of the absence of measurable levels of the endogenous analyte is essential, preferably using an independent but sensitive and validated method (i.e., LC-MS/MS). Also, an attempt should be made to determine absence of matrix effects in altered or alternate matrices (DeSilva et al. 2003). For QCs in unaltered matrix, one can use the recovery experiment to assure accuracy (FDA 2013), wherein the recovery of spiked material is estimated from the unaltered matrix with quantifiable endogenous material, provided the endogenous and spiked analytes behave in an additive manner (DeSilva et al. 2003).

16.6.6 Diagnostic Kits

Kits (usually LBAs) routinely used for clinical diagnostics, are sometimes used in bioanalysis. Therefore, the FDA's recent draft guidance (2013) recommends demonstrating the reliability of such kits for quantitative determination. Some of the issues with diagnostic kits are briefly discussed below.

Manufacturers' validation data should not be relied on for diagnostic kits. Instead, the kits should be validated in-house, and a complete validation may be necessary if the kit is modified (see Sect. 16.3). Some kits may include sparse calibration standards (e.g., single- or two-point calibration curves). Therefore, it is a good practice to establish a calibration response curve with the required set of calibration standards (as described in Sect. 16.3.4) during validation and study sample analysis (FDA 2013). Also, the nominal concentrations of kit QCs are sometimes not provided, instead expressed as ranges. In such cases, in-house QCs with known nominal concentrations are recommended for use, independent of the kit-supplied QCs (FDA 2013). Proper justification and appropriate cross-validation experiments are required when standards and QCs supplied with the kits are prepared in a matrix different from the subject samples (see Sects. 16.3.4.7 and 16.6.5).

16.6.7 Automation/High Throughput Assays

Automation or high throughput analysis may require modification of validation procedures and in some cases, validation of additional factors. This may require increasing sample size of intra-batch validation, and/or additional validation batches. High throughput analysis may also require proper maintenance of instruments between batch analysis to prevent residual contamination.

16.6.8 Human Errors

Human errors can also contribute to lack of accuracy of study data. It has been shown that lack of homogeneity of study samples can result in errors in quantitation (Tan et al. 2009; Yadav and Shrivastav 2011; Nováková 2013). Also, inconsistent addition of IS to study samples may affect analyte/IS ratios and therefore its reported concentration (Tan et al. 2009). Switching samples during bioanalysis has been reported (Yadav and Shrivastav 2011). Therefore, adequate training of analysts and establishing ruggedness of methods are required.

16.7 Incurred Sample Reanalysis

Reproducibility issues in study samples from dosed subjects (i.e., incurred samples) are observed although the samples are analyzed using validated bioanalytical assays. Many of the assay issues discussed in the earlier section can contribute to reproducibility issues. Briefly, reproducibility issues may arise due to matrix effects, insufficiently validated method (e.g., improper extraction conditions, inadequate enzyme inhibitor, inadequate stability validation) and/or poor execution (inconsistent mixing, processing error) (Matuszewski et al. 2003; Jemal and Xia 2006; Wang et al. 2007; Besnard et al. 2008; Tan et al. 2009; Jemal et al. 2010; Silvestro et al. 2010; Meng et al. 2011; Yadav and Shrivastav 2011; Tan et al. 2012). Reanalysis of a subset of study samples to check for reproducibility in incurred samples is often referred to as incurred sample reanalyses (ISR). ISR is used to confirm that an analytical method performs as intended in clinical and nonclinical studies. Therefore, ISR serves as a confirmatory tool to ensure that all factors contributing to assay performance are in control during study sample analysis, and thereby assures the reliability of study data. ISR over the years has become an integral part of bioanalysis.

The concept of reanalysis of a subset of study samples was adopted by Health Canada in 1992 although the strategy for interpretation of the data was not clear, and the requirement was abandoned in 2003. The reanalysis of a subset of study samples as part of ISR and the need for implementation of ISR was first discussed at the 2006 AAPS/FDA Workshop (Viswanathan et al. 2007). The consensus for procedures for ISR conduct was reached at the 2008 ISR Workshop (Fast et al. 2009). Many of the ISR recommendations proposed at the 2008 ISR Workshop (Fast et al. 2009) have been adopted by the industry over the years, and recently by regulatory agencies including, the EMA (2011) and Health Canada (2012). In addition, the FDA has proposed similar ISR recommendations in their recently issued draft BMV guidance (2013). The consensus at the 2008 ISR Workshop was that ISR conduct is important for nonclinical and clinical studies where PK assessment is the primary end point, especially for all BE studies (Fast et al. 2009). Also, the consensus at the workshop was to use a sample size of 5–10 %

of the total study samples depending on size of the study (e.g., minimum of 5 % for a large study) for ISR. However, the recent FDA draft guidance (2013) recommends ISR sample size of 7 % of the total study samples. Sample selection for ISR includes selection of samples from individual subjects, with fewer samples (at the Cmax and elimination range of the PK profile) from more subjects. ISR is acceptable when at least 67 % of the reanalyzed concentrations are within 20 % (30 % for LBAs) of their original concentrations when normalized to their means of the original and reanalyzed concentrations (Fast et al. 2009; FDA 2013). It is a good practice to select samples from across the duration of sample analysis. If samples identified for reanalysis were diluted during the original analysis, then the same dilution factor used for the original result needs to be employed for ISR. Also, it is important that the number of replicates and the acquisition method used for ISR are the same as those used during the original analysis.

Acknowledgments The author thanks Drs. Brian P. Booth, Ethan M. Stier, Robert Lionberger, and Yongsheng Yang for their critical review of the manuscript.

Disclaimer The chapter reflects the views of the author and should not be construed to represent FDA's views or policies. No official endorsement by the FDA is intended or should be inferred.

References

Abou-Auda HS, Najjar TA, Al-Khamis KI et al (2000) Liquid chromatographic assay of nifedipine in human plasma and its application to pharmacokinetic studies. J Pharm Biomed Anal 22 (2):241–249

Andersson A, Ehrsson H (1995) Stability of cisplatin and its monohydrated complex in blood, plasma and ultrafiltrate-implications for quantitative analysis. J Pharm Biomed Anal 13:639–644

Bérubé ER, Taillon MP, Furtado M et al (2011) Impact of sample hemolysis on drug stability in regulated bioanalysis. Bioanalysis 3(18):2097–2105. doi:10.4155/BIO.11.190

Besnard T, Renée N, Etienne-Grimaldi MC et al (2008) Optimized blood sampling with cytidine deaminase inhibitor for improved analysis of capecitabine metabolites. J Chromatogr B 870 (1):117–120. doi:10.1016/j.jchromb.2008.05.040

Bozovic A, Kulasingam V (2013) Quantitative mass spectrometry-based assay development and validation: from small molecules to proteins. Clin Biochem 46:444–455. doi:10.1016/j. clinbiochem.2012.09.024

Briscoe CJ, Hage DS (2009) Factors affecting the stability of drugs and drug metabolites in biological matrices. Bioanalysis 1(1):205–220. doi:10.4155/BIO.09.20

Chen J, Hsieh Y (2005) Stabilizing drug molecules in biological samples. Ther Drug Monit 27 (5):617–624

Christophersen AS (1986) Tetrahydrocannabinol stability in whole blood: plastic versus glass containers. J Anal Toxicol 10(4):129–131

Desilva B, Smith W, Weiner R et al (2003) Recommendations for the bioanalytical method validation of ligand-binding assays to support pharmacokinetic assessments of macromolecules. Pharm Res 20(11):1885–1900

DeSilva B, Garofolo F, Rocci M et al (2012) 2012 white paper on recent issues in bioanalysis and alignment of multiple guidelines. Bioanalysis 4(18):2213–2226. doi:10.4155/BIO.12.205

Dufresne C, Favetta P, Paradis C et al (2001) Stability of sufentanil in human plasma samples. Ther Drug Monit 23:500–552

EMA (2011) Guideline on bioanalytical method validation. http://www.ema.europa.eu/docs/en_GB/document_library/Scientific_guideline/2011/08/WC500109686.pdf. Accessed July 2013

Fast DM, Kelley M, Viswanathan CT et al (2009) Workshop report and follow-up—AAPS workshop on current topics in GLP bioanalysis: assay reproducibility for incurred samples—implications of Crystal City recommendations. AAPS J 11(2):238–241. doi:10.1206/s12248-009-9100-9

FDA (2001) Guidance for industry: bioanalytical method validation. http://www.fda.gov/downloads/Drugs/GuidanceComplianceRegulatoryInformation/Guidances/UCM070107.pdf. Accessed July 2013

FDA CFR (2013) Code of Federal Regulations Title 21 Part 320. http://www.accessdata.fda.gov/scripts/cdrh/cfdocs/cfCFR/CFRSearch.cfm?CFRPart=320&showFR=1&subpartNode=21:5.0.1.1.7.2. Accessed July 2013

FDA (2013) Draft guidance for industry: bioanalytical method validation (revised). http://www.fda.gov/downloads/Drugs/GuidanceComplianceRegulatoryInformation/Guidances/UCM368107.pdf. Accessed Sept 2013

Findlay JWA, Das I (2006) Enzyme immunoassay and related bioanalytical methods. In: Swarbrick J (ed) Encyclopedia pharmaceutical technology. Informa HealthCare, New York, pp 1566–1579

Fura A, Harper TW, Zhang H et al (2003) Shift in pH of biological fluids during storage and processing: effect on bioanalysis. J Pharm Biomed Anal 32(3):513–522

Garofolo F, Rocci ML, Dumont I et al (2011) 2011 white paper on recent issues in bioanalysis and regulatory findings from audits and inspections. Bioanalysis 3(18):2081–2096. doi:10.4155/BIO.11.192

Guan F, Uboh C, Soma L et al (2003) Sensitive liquid chromatographic/tandem mass spectrometric method for the determination of beclomethasone dipropionate and its metabolites in equine plasma and urine. J Mass Spectrom 38(8):823–838

Health Canada (1992) Guidance to industry. Conduct and analysis of bioavailability and bioequivalence studies-part A: oral dosage formulations used for systemic effects. http://www.hc-sc.gc.ca/dhp-mps/alt_formats/pdf/prodpharma/applic-demande/guide-ld/bio/gd_cbs_ebc_ld-eng.pdf; http://faculty.ksu.edu.sa/64448/Documents/Guideline%20BA%20-%20BE%20Part%20A.pdf. 1371 Accessed Oct 2013

Health Canada (2003) Notice to industry-removal of requirement for 15% random replicate sample notice affecting guideline A and guideline B. http://www.hc-sc.gc.ca/dhp-mps/alt_formats/pdf/prodpharma/applic-demande/guide-ld/bio/gd_cbs_ebc_ld-eng.pdf. Accessed Oct 2013

Health Canada (2012) Notice to guidance document: conduct and analysis of comparative bioavailability studies. http://www.hc-sc.gc.ca/dhp-mps/alt_formats/pdf/prodpharma/applic-demande/guide-ld/bio/gd_cbs_ebc_ld-eng.pdf. Accessed July 2013

Izhizuka T, Fujimori I, Kato M et al (2010) Human carboxymethylenebutenolidase as a bioactivating hydrolase of olmesartan medoxomil in liver and intestine. J Biol Chem 285(16):11892–11902. doi:10.1074/jbc.M109.072629

Jemal M, Xia YQ (2000) Bioanalytical method validation design for the simultaneous quantitation of analytes that may undergo interconversion during analysis. J Pharm Biomed Anal 22(5):813–827

Jemal M, Xia YQ (2006) LC-MS development strategies for qualitative bioanalysis. Curr Drug Metab 7(5):491–502

Jemal M, Ouyang Z, Xia YQ (2010) Systematic LC-MS/MS bioanalytical method development that incorporates plasma phospholipids risk avoidance, usage of incurred sample and well thought-out chromatography. Biomed Chromatogr 24(1):2–19. doi:10.1002/bmc.1373

Kearney AS, Crawford LF, Mehta SC et al (1993) The interconversion kinetics, equilibrium, and solubilities of the lactone and hydroxyacid forms of the HMG-CoA reductase inhibitor, CI-981. Pharm Res 10(10):1461–1465

Kelley M, DeSilva B (2007) Key elements of bioanalytical method validation for macromolecules. AAPS J 9(2):E156–E163

Korfmacher WA (2005) Foundation review: principles and applications of LC-MS in new drug discovery. Drug Discov Today 10(20):1357–1367

Kruger R, Vogeser M, Burghardt S et al (2010) Impact of glucuronide interferences on therapeutic drug monitoring of posaconazole by tandem mass spectrometry. Clin Chem Lab Med 48 (12):1723–1731. doi:10.1515/CCLM.2010.333

Le Guellec C, Gaudet ML, Lamanetre S et al (1997) Stability of rifampin in plasma: consequences for therapeutic monitoring and pharmacokinetic studies. Ther Drug Monit 19(6):669–674

Li W, Luo S, Smith HT et al (2010) Quantitative determination of BAF312, a S1P-R modulator, in human urine by LC-MS/MS: prevention and recovery of lost analyte due to container surface adsorption. J Chromatogr B 878(5–6):583–589. doi:10.1016/j.jchromb.2009.12.031

Li W, Zhang J, Tse FLS (2011) Strategies in qualitative LC-MS/MS analysis of unstable small molecules in biological matrices. Biomed Chromatogr 25(10):258–277

Matuszewski BK, Constanzer ML, Chavez-Eng CM (2003) Strategies for the assessment of matrix effect in quantitative bioanalytical methods based on HPLC-MS/MS. Anal Chem 75:3019–3030. doi:10.1021/ac020361s

Meng M, Reuschel S, Bennett P (2011) Identifying trends and developing solutions for incurred sample reanalysis failure investigations in a bioanalytical CRO. Bioanalysis 3(4):449–465. doi:10.4155/BIO.10.212

Mulvana DE (2010) Critical topics in ensuring data quality in bioanalytical LC-MS method development. Bioanalysis 2(6):1051–1072. doi:10.4155/BIO.10.60

Niederlander HA, Koster EH, Hilhorst MJ et al (2006) High throughput therapeutic drug monitoring of clozapine and metabolites in serum by on-line coupling of solid phase extraction with liquid chromatography–mass spectrometry. J Chromatogr B 834(1–2):98–107

Niessen WM (2003) Progress in liquid chromatography-mass spectrometry instrumentation and its impact on high-throughput screening. J Chromatogr A 1000(1–2):413–436

Nováková L (2013) Challenges in the development of bioanalytical liquid chromatography-mass spectrometry method with emphasis on fast analytes. J Chromatogr A 1292:25–37. doi:10. 1016/j.chroma.2012.08.087

Rozet E, Marini RD, Ziemons E et al (2011) Advances in validation, risk and uncertainty assessment of bioanalytical methods. J Pharm Biomed Anal 55(4):848–858. doi:10.1016/j. jpba.2010.12.018

Sadagopan NP, Li W, Cook JA et al (2003) Investigation of EDTA anticoagulant in plasma to improve the throughput of liquid chromatography/tandem mass spectrometric assays. Rapid Commun Mass Spectrom 17(10):1065–1070

Savoie N, Booth BP, Bradley T et al (2009) The 2nd calibration and validation group workshop on recent issues in good laboratory practice bioanalysis. Bioanalysis 1(1):19–30. doi:10.4155/ BIO.09.11

Savoie N, Garofolo F, van Amsterdam P et al (2010) 2010 white paper on recent issues in regulated bioanalysis & global harmonization of bioanalytical guidance. Bioanalysis 2(12):1945–1960. doi:10.4155/BIO.10.164

Saxer C, Niina M, Nakashima A et al (2004) Simultaneous determination of levodopa and 3-O-methyldopa in human plasma by liquid chromatography with electrochemical detection. J Chromatogr B 802(2):299–305. doi:10.1016/j.jchromb.2003.12.006

Silvestro L, Gheorghe MC, Tarcomnicu I et al (2010) Development and validation of an HPLC-MS/MS method to determine clopidogrel in human plasma. Use of incurred samples to test back-conversion. J Chromatogr B 878(30):3134–3142. doi:10.1016/j.jchromb.2010.09.022

Singleton C (2012) Recent advances in bioanalytical sample preparation for LC-MS analysis. Bioanalysis 4(9):1123–1140. doi:10.4155/BIO.12.73

Smeraglia J, Baldrey SF, Watson D (2002) Matrix effects and selectivity issues in LC-MS-MS. Chromatographia 55(suppl 1):S95–S99

Tan A, Hussain S, Musuku A et al (2009) Internal standard response variations during incurred sample analysis by LC-MS/MS: case by case trouble shooting. J Chromatogr B 877(27):3201–3209. doi:10.1016/j.jchromb.2009.08.019

Tan A, Boudreau N, Lévesque A (2012) Internal standards for quantitative LC-MS bioanalysis. In: Xu QA, Madden TL (eds) LC-MS in drug bioanalysis. Springer, New York, pp 1–32

Timmerman P, Luedtke S, van Amsterdam P et al (2009) Incurred sample reproducibility: views and recommendations by the European bioanalysis forum. Bioanalysis 1(6):1049–1056. doi:10.4155/BIO.09.108

Trufelli H, Palma P, Famiglini G et al (2011) An overview of matrix effects in liquid chromatography–mass spectrometry. Mass Spectrom Rev 30(3):491–509. doi:10.1002/mas.20298

Van Eeckhaut A, Lanckmans K, Sarre S et al (2009) Validation of bioanalytical LC-MS/MS assays: evaluation of matrix effects. J Chromatogr B 877(23):2198–2207. doi:10.1016/j.jchromb.2009.01.003

Van Harten J, Lodewijks MT, Guyt-Scholten JW et al (1987) Gas chromatographic determination of nisoldipine and one of its metabolites in plasma. J Chromatogr 423:327–333

Viswanathan CT, Bansal S, Booth B et al (2007) Quantitative bioanalytical methods validation and implementation: best practices for chromatographic and ligand binding assays. Workshop/conference report. AAPS J 9(1):E30–E42

Wang S, Cyronak M, Yang E (2007) Does a stable isotopically labeled internal standard always correct analyte response? A matrix effect study on a LC/MS/MS method for the determination of carvedilol enantiomers in human plasma. J Pharm Biomed Anal 43(2):701–707. doi:10.1016/j.jpha.2006.08.010

Yadav M, Shrivastav PS (2011) Incurred sample reanalysis (ISR): a decisive tool in bioanalytical research. Bioanalysis 3(9):1007–1024. doi:10.4155/BIO.11.76

Yue B, Pattison E, Roberts WL et al (2008) Choline in whole blood and plasma: sample preparation and stability. Clin Chem 54(3):590–593. doi:10.1373/clinchem.2007.094201

Zhang J, Rodila R, Gage E et al (2010) High-throughput salting-out assisted liquid/liquid extraction with acetonitrile for the simultaneous determination of simvastatin and simvastatin acid in human plasma with liquid chromatography. Anal Chim Acta 661(2):167–172. doi:10.1016/j.aca.2009.12.023

Abbreviations

AAPS	American Association of Pharmaceutical Scientists
ABE	Average bioequivalence
ACAT	Advanced compartmental absorption and transit
ADAM	Advanced dissolution, absorption, and metabolism
AEs	Adverse events
AMA	American Medical Association
AMP	Adenosine monophosphate
ANDAs	Abbreviated new drug applications
ANOVA	Analysis of variance
AOA	Administration on aging
APCI	Atmospheric pressure chemical ionization
API	Active pharmaceutical ingredient
APSD	Aerodynamic particle size distribution
ASEAN	Association of South East Asian Nations
AUC	Area under the curve
AUEC	Area under the effect–time curve
BA	Bioavailability
BCS	Biopharmaceutics classification system
BDDCS	Biopharmaceutics drug disposition classification system
BE	Bioequivalence
CAT	Compartmental absorption and transit
CFR	Code of Federal Regulations
CHMP	Committee for Human Medicinal Products
CI	Confidence interval
C_{max}	Peak plasma concentration
CoA	Certificates of analysis
COPD	Chronic obstructive pulmonary disease
CSF	Cerebrospinal fluid
CTP	Child-Turcotte-Pugh

L.X. Yu and B.V. Li (eds.), *FDA Bioequivalence Standards*, AAPS Advances
in the Pharmaceutical Sciences Series 13, DOI 10.1007/978-1-4939-1252-0,
© The United States Government 2014

CV	Coefficient of variation
DDIs	Drug–drug interactions
DEPC	Dierucoylphosphatidylcholine
DESI	Drug efficacy study implementation
DMD	Dermal microdialysis
DMPC	Dimyristoyl phosphatidylcholine
DMPG	Dimyristoyl phosphatidylglycerol
DOPC	Dioleoylphosphatidylcholine
DPI	Dry powder inhalers
DPK	Dermatopharmacokinetics
DPPG	Dipalmitoylphosphatidylglycerol
DR	Delayed release
EAEC	Enteroaggregative *E. coli*
ED50	Median effective dose
EDTA	Ethylenediaminetetraacetic acid
ELISA	Enzyme-linked immunosorbent assay
EMA	European Medicines Agency
ESI	Electrospray ionization
ETEC	Enterotoxigenic *Escherichia coli*
FDA	Food and Drug Administration
FEV1	Forced expiratory volume within 1 s
FFD&C Act	The Federal Food, Drug, and Cosmetic Act
FPM	Fine particle mass
GABA	Gamma-aminobutyric acid
GE	Glucose excursion
GI	Gastrointestinal
GIT	Gastrointestinal tract
GMR	Geometric mean ratio
GSD	Geometric standard deviation
HE	Hepatic encephalopathy
HHS	Health and Human Services
HILIC	Hydrophilic interaction liquid chromatography
HPA	Hypothalamic–pituitary–adrenal axis
HPLC	High performance liquid chromatography
HPMC	Hydroxypropyl methylcellulose
HSPC	Hydrogenated soy phosphatidylcholine
HV	Highly variable
HVDs	Highly variable drugs
IBE	Individual bioequivalence
ICS	Orally inhaled corticosteroids
INDs	Investigational new drug applications
INR	International normalized ratio
IR	Immediate release
IS	Internal standard
ISR	Incurred sample reanalyses

ITT	Intent-to-treat
IVIVC	In vitro–in vivo correlation
IVRT	In vitro drug release testing
LABAs	Long-acting β2 adrenoceptor agonists
LBA	Ligand binding assays
LC	Liquid chromatography
LD50	Median lethal dose
LIMS	Laboratory information management systems
LLE	Liquid–liquid extraction
LLOQ	Lower limit of quantitation
LMWH	Low molecular weight heparin
mCSRS	Modified chi-square ratio statistic
MD	Multiple dose
MDD	Maximum daily dose
MDP	Muramyl dipeptide
ME	Matrix effects
MEC	Minimum effective concentrations
MELD	Model for end stage liver disease
MHE	Minimal hepatic encephalopathy
MIP	Molecularly imprinted polymer
MITT	Modified intent-to-treat
MMAD	Mass median aerodynamic diameter
MmCSRS	The median of modified chi-square ratio
MMF	Mycophenolate mofetil
MPA	Mycophenolic acid
MR	Modified release
MRM	Multiple reaction monitoring
MRT	Mean residence time
MS	Mass spectrometry
MTC	Minimum toxic concentrations
MVL	Multivesicular liposomes
NCEs	New chemical entities
NDAs	New drug applications
NIR	Near infrared
NME	New molecular entity
NO	Nitric oxide
NSAID	Non-steroidal anti-inflammatory drug
NTI	Narrow therapeutic index
OGD	Office of Generic Drugs
OIP	Orally inhaled products
OTA	Office of Technology Assessment
pAUCs	Partial area under the curve
PBC	Population bioequivalence criteria
PBE	Population bioequivalence
PBPK	Body physiologically based pharmacokinetic

PC20	Provocative concentration that produces a 20 % decrease in FEV1
PD	Pharmacodynamic
PD20	Provocative dose that produces a 20 % decrease in FEV1
PDR	Population difference ratio
PEG	Polyethylene glycol
P-gp	P-glycoprotein
PK	Pharmacokinetic
PP	Per-protocol
PP	Protein precipitation
PSD	Particle size distribution
PSE	Portal systemic encephalopathy
PT	Prothrombin
QbD	Quality-by-design
QC	Quality controls
R	Reference
RES	Reticuloendothelial system
RLD	Reference listed drug
RMSE	Root mean square error
RSABE	Reference-scaled average bioequivalence
SABAs	Short-acting $\beta2$ adrenoceptor agonists
SAC	Single actuation content
SD	Single dose
SD	Standard deviation
SEDDS	Self-emulsifying drug delivery system
SIF	Simulated intestinal fluid
SIL	Stable isotope labeled
SONIC	Spectrum of neurocognitive impairments in cirrhosis
SPE	Solid phase extraction
SRM	Selected reaction monitoring
T	Test
TD	Travelers' diarrhea
TE	Therapeutically equivalent
TEWL	Transepidermal water loss
TK	Toxicokinetic
TLUS	Time to last unformed stool
T_{max}	Time to peak concentration
TNSS	Total nasal symptom scores
UCB	Upper-confidence bound
UHPLC	Ultra-high performance liquid chromatography
ULOQ	Upper limit of quantitation
US-FDA	US Food and Drug Administration
UV	Ultraviolet
VCA	Vasoconstrictor assay
WHO	World Health Organization
WSV	Within-subject variability

Index

Printed by Printforce, the Netherlands